Professionals in food chains

Professionals in food chains

EurSafe 2018
Vienna, Austria
13 – 16 June 2018

edited by:
Svenja Springer
Herwig Grimm

EAN: 9789086863211
e-EAN: 9789086868698
ISBN: 978-90-8686-321-1
e-ISBN: 978-90-8686-869-8
DOI: 10.3920/978-90-8686-869-8

First published, 2018

Acknowledgments

This book contains contributions presented at the 14th Congress of the European Society for Agricultural and Food Ethics. Under the topic 'Professionals in Food Chains: Ethics, Roles and Responsibilities' the congress and this volume aim to bring together a broad range of disciplines in order to discuss and reflect on fundamental issues and current challenges professionals are confronted with in food chains.

We would like to thank the EurSafe board for their confidence and support during the preparation of the congress. Furthermore, our sincere thanks go to the University of Veterinary Medicine, Vienna, the Messerli Research Institute and the University of Vienna. All these institutions provided resources to make the congress possible.

Special thanks go to the Messerli Foundation which generously financed scholarships for researchers and PhD students from South-Eastern European countries to participate in the congress. Further, we would like to thank the Austrian Federal Ministry of Health and Women's Affairs, the Austrian Federal Ministry of Education, Science and Research, the City of Vienna, the Austrian Chamber of Veterinary Surgeons and the Lower Austrian Animal Health Services. They did not only support the idea of the congress but also helped realize it with their financial support.

The editors are grateful to Wageningen Academic Publishers and particularly to the many colleagues who voluntarily provided their expertise and constructive criticism during the review process.

Moreover, we would also like to thank the 82 authors for sharing their thoughts and written contributions. Without the effort to write the short academic papers presented in this book, it would not be possible to distribute the ideas and topics discussed at the congress to a wider audience. We hope that this book will stimulate further debates and provide a source for academic exchange.

Reviewers

We would like to thank all reviewers for their efforts and invested time during the peer review process. Their constructive criticism provided all authors with valuable support for the drafting of their short academic papers, as well as during the preparation of their presentations for the congress.

S. Aerts
J. Benz-Schwarzburg
J. Beusmann
B. Bovenkerk
S. Camenzind
M. Di Paola
D.E. Dumitras
C. Dürnberger
M. Eggel
L. Escajedo San-Epifanio
M. Gjerris
H. Grimm
S. Gunnarsson
K. Hagen
M. Huth
A. Inza-Bartolomé
A. Jelenkovic
J.I.M. Jitea
M. Kaiser
P. Kaiser
A. Kallhoff
N.C. Krafyllis
J.L. Lassen
M. Magalhães-Sant'ana
F.L.B. Meijboom
S.M. Meisch
K. Millar
S. Monsó
I.C. Muresan
L. Neto
I.A.S. Olsson
I. Pali-Schöll
F. Pirscher
C.B. Pocol
T. Potthast
E. Rebato
H. Röcklinsberg
P. Sandin
M. Schörgenhumer
C.G. Schwarz-Plaschg
B. Skorupinski
S. Springer
P.B. Thompson
O. Varga
P.M.R. Vaz-Pires
L. Voget-Kleschin
A. Wallenbeck
K. Weich

Table of contents

Section 2.
Sustainable food production

Section 3.
Ethics of production and consumption

Section 4.
Food ethics

Section 5.
Food politics: policy and legislation

Section 6.
Veterinary ethics: methods, concepts and theory

Section 7.
Veterinary ethics: in practice

Section 8.
Veterinary ethics: in teaching

Section 9.
Media, transparency and trust

Section 10.
Animal ethics

Section 11. Animal research

Section 12. Biotechnology

Section 13. Aquaculture

Section 14.
Water ethics

Posters

Foreword: an ethical focus on the role of professionals in the food chain

It is a delight to introduce this valuable EurSafe conference book that explores themes that are presented by researchers working in the interdisciplinary field of agricultural, veterinary and food ethics.

Focusing on the ethical dimensions of agriculture and food production, over the last 19 years the European Society of Agricultural and Food Ethics (EurSafe) has hosted conferences that bring together academics, policy-makers and members of both public and private organisations working in agriculture and food. However, this is the first conference, and the first volume I am aware of, where the topic of professionals and professional practice are discussed in the context of agricultural and food ethics. On reflection this has been an omission within the field of food ethics. Although papers on professional ethics have become part of the suite of contributions with the field, this is the first instance where there has been a concert effort to bring together ethicists, sociologists, lawyers and other professionals who are specified focus on examining the role of professionals and professions in the food chain.

One of the drivers for this conference and this substantial volume is to explore whether in order to get a 'better understand ... *of* current and future problems in the food chain, it is essential that we pay greater attention to the role and position of professionals.'. As such the contributions within this volume examine some of the common themes of EurSafe members, such as mapping the key ethical challenges, analysing the value-based nature of complex food and agriculture problems and reflecting on the expectations and concerns of different moral actors. Yet the contributions in this book go beyond some of these themes and ask questions about the role of participating players and what are and should be there areas of responsibility. In addition some the contributions define and analysis professional values and how these values can shape professional responses. The importance of not only talking about the substantive and procedural ethical issues that arise in the food chain but also the role of 'the professional' is timely and it is excellent to see this Society initiate a great examination of these issues.

As well as this topic being very timely, it is very appropriate that the first EurSafe Conference on ethics and professions is hosted by the Messerli Research Institute based at the University of Veterinary Medicine, Vienna and led by Professor Herwig Grimm, Head of the Unit of Ethics and Human-Animal Studies, Messerli Research Institute and Ms Svenja Springer, Research Assistant and doctoral student also based in the Unit of Ethics and Human-Animal Studies. Members of the Messerli Unit of Ethics and Human-Animal Studies have been working on the ethical aspects of the role of professionals for a number of years. Professor Grimm and members of this unit have a particular interest in veterinary ethics and professionalism and as a result the team has been conducted leading work in the area of veterinary ethics. The Society is delighted that the interdisciplinary Messerli Research Institute is hosting the EurSafe Conference as the society has always been seen as a welcoming home for interdisciplinary research and the conferences are organised to be a meeting place for different professionals who work in the broad areas of food and agriculture.

This book presents a much needed contribution to the interdisciplinary field of agricultural, veterinary and food ethics and extends existing work by shining an important ethical torch on the role of the professional. Congratulations to the Vienna team for all of their hard work and for editing such a valuable volume which we, the conference participants, will receive on the first day of the EurSafe conference. I hope you all enjoy this conference proceedings and for those of you who are unable to join us in Vienna in June 2018, I hope that your reading of this book may encourage you join us at a future EurSafe event, we are always delighted to warmly welcome new professional colleagues.

Dr Kate Millar,

President of the European Society of Agricultural and Food Ethics and
Director, Centre for Applied Bioethics, Schools of Biosciences and Veterinary Medicine and Science, University of Nottingham, UK

Keynote contributions

1. Protecting society: the value of the professional regulatory model

S.A. May
Royal Veterinary College, University of London, Hawkshead Lane, North Mymms, Hatfield, Hertfordshire AL9 7TA, United Kingdom; smay@rvc.ac.uk

Abstract

Society requires market regulation to assure citizens of the quality and safety of goods and services. Western models include the so-called capitalist 'free' (efficient) market and professional peer regulation. The free market, justified by the 18th century observations of Adam Smith, has dominated government thinking in the last 40 years, and led to calls for deregulation in all market activities, including those controlled by professional regulation. However, a catastrophic failure of the banking system in 2007-08 has drawn attention to the importance of prioritising customer interests, including maximising their knowledge and understanding of goods and services being provided, and holding suppliers accountable for failures and outright fraud. All these important principles for efficient and fair markets are firmly embedded in the social contract that underpins the legitimacy of the professions. This is not, in itself, an argument for greater professionalization of market sectors, but rather an indication of the need for effective government regulation of the free market that emphasises the key features of the efficient market: accountability, transparency and recognition of 'externalities' that must be accounted for in any comprehensive and fair market system.

Keywords: free market, regulation, professionalism

Introduction

For the last 50 years, as in other jurisdictions, the professional, peer regulatory model has protected the UK public and assured them of high standards of veterinary care (May, 2013). In parallel, over the same period, a so-called 'free market' paradigm has been developed and championed by economists as a superior way of protecting the public good. Many remain seduced by this model, despite its catastrophic failure in the 2007-08 banking crisis (Bragues, 2010; Jaffer *et al.*, 2014a,b). The free (efficient) market philosophy (Karnani, 2011) is based on modern interpretations of Adam Smith's (1776/1970) famous conclusion that it is not in direct service of 'the public good' that individuals best promote that good but paradoxically in the pursuit of self-interest (James and Rassekh, 2000). This continued justification in the face of market failure is even harder to understand when it is considered how Smith's observations, based on 18th century business, in which individuals were known to their customers, could be held accountable, and were thus concerned about their long term reputations, have been assumed to hold for the complexity of 20th century society, in which many companies are multinational, working on a global scale. Therefore, a comparison of the two models provides a timely and fruitful opportunity to understand the assumptions on which each is based and define principles on which regulation and good governance can be established, whatever the prevailing political fashions.

The efficient market

Although 'self-interest' is often interpreted as 'selfishness' (Teulon, 2014), it is clear that for Smith it was a more complex concept involving a combination of personal ambition and recognition of the individual's dependence on their community's approbation for them to succeed (Fiori, 2005). Effective competition would give the public opportunities to access high quality goods at low prices, through their knowledge of the provenance and nature of the products, the long term interests that the vendor, and often their family, had in the business, and the ability to feed back on any problems with the products. Crucially, it

EurSafe 2018 – DOI 10.3920/978-90-8686-869-8_1, © Wageningen Academic Publishers 2018

can be seen that Smith's conclusions will be undermined if society allows businesses to achieve levels of market power that prevent effective competition, information is asymmetric, so purchasers cannot judge the quality of the product or know all about its origins, and vendors do not have long term interests in their enterprises and cannot be held accountable for deficiencies in their behaviours and products. Additionally, the market will fail society when there are costs ('externalities'), often environmental, to its business activities that are not borne by the businesses themselves (Karnani, 2011). Thus, Smith's model is seen to be optimised for the context that generated it, a local business serving its local community, rather than a global business, involving multiple players, often with long supply chains, crossing national borders and a range of different legislative regimes.

Within local markets, even quite complex longer term issues involving resource restriction, short term profiting at the expense of others, and externalities associated with individual selfishness, could be handled by open communication, understanding and a sense of fair play (Dietz *et al.*, 2003). So the 'tragedy of the commons' is avoided by transparency and the sort of self-interest, moderated by a consciousness of the approval of one's neighbours, recognised by Adam Smith.

To a certain extent, the limitations of Smith's model were managed up until the 1970's by a willingness of governments to create external regulatory frameworks that set standards for the quality of various products, and gave legal rights to consumers to seek redress if there were problems with their purchases. In the case of some services, such as banking, there were also legal requirements for businesses to look after the interests of their client, particularly where complexity of the service means the client may lack the knowledge to judge its quality. However, from the 1970's onwards, under the influence of economists such as Milton Friedman (1970), there was an increasing belief 'that markets are self-adjusting and the role of government should be minimal' (Stiglitz, 2009). This has been so pervasive that it has even extended to interpretation of the law on fiduciary responsibility (Getzler, 2014). Crucially, fiduciary law aligns strong powers with strong duties and, as far as possible, identifies conflicts of interest that prevent alignment of fiduciaries' and beneficiaries' interests. Central to this is a requirement for accountability that goes beyond the norms of tort and contract law, which are linked to harms caused, to potential surrender of gains related to poor decision-making on behalf of a client. However, from the 1980's onwards, terms of contract were seen to trump any fiduciary duties, which came to be increasingly seen as voluntary duties associated with contractual agreements. In the case of banks, the regulatory approach created as part of this culture is seen with the benefit of hindsight, as having 'failed to secure genuine accountability', 'failed to deal effectively with borders', failed in that 'information about asset securities and derivatives was insufficient for investors to assess value and risk' and supported the growth of companies that were 'too big to fail' (Jaffer *et al.*, 2014).

The professional self-regulatory model

The professions grew out of the recognition that expert groups could provide and assure the quality of services of value to society through setting standards around processes and holding members accountable for the achievement of minimum standards (May, 2013; Rollin, 2006). Despite the challenges over its legitimacy and continued relevance in the latter half of the 20th century (Evetts, 2003), peer-regulation remains as a defensible way in which the public can be protected and assured of the standards of the regulated professions (Rollin, 2006), including medicine, veterinary medicine and the law. A Hobbesian-type 'social contract' exists in which the profession is given a monopoly on certain activities in exchange for ensuring that client interests are well-served (May, 2013), assuring the type of alignment observed by Smith in his local communities/economies. This is achieved through agreement on the corpus of knowledge and skills necessary to the profession, and limitation of the area of activity to recognised members of the profession (Thistlethwaite and Spencer, 2008). Codes of practice ensure that, in contrast to the marketplace, members prioritise the interests of their clients, alongside self-interest, and recognise limitations to their knowledge and skills that could threaten their ability to deliver the quality of service expected. Departure from the code in the form of acting when incapacitated, recklessly or way beyond an individual's known area of competence constitutes grounds for action by the profession to protect the public against malpractice. The professional model deals with the asymmetry of information, through an emphasis on informed and shared decision-making (Kon, 2010), and the challenge to members of the public of judging the quality of complex decision-making, through peer review, when concerns are voiced about a practitioner (Rollin, 2006). The emphasis, until recently, on personal service, particularly in the clinical professions, has meant that the professions have not been challenged in the same way by global regulatory issues, although technology in support of telemedicine has started to become a concern for professional regulators (RCVS, 2017).

In terms of externalities, a challenge that does exist for a member of a profession is the tension between the group interests of a public that empowers them and individual clients that pay for their services. For instance, the treatment of individual animals might not be in the overall interests of a domestic animal population where rapid elimination of the affected individual minimises the risk to the group as a whole, so governments often intervene to proscribe selfish self-consideration and incentivise, through compensation schemes, behaviours that recognise the broader public interest by removing individual disadvantage (May, 2013).

Managing markets

For all the reasons outlined, in a complex, global world, at the level of individual organisations, 'self-regulation is preposterous' (Stiglitz, 2009) within a free market environment. Contract law is also a very weak way of balancing supplier and consumer interests. At the very least, genuine individual accountability and, in complex areas, enforceable fiduciary and/or professional duties are essential to consumer protection. Individual reputations and assets must be at stake, if duties to consumers are to be protected. In addition, wherever possible, to help consumers achieve the best possible judgements on goods and services, there needs to be maximum transparency about provenance and the nature of the product, with clear means of verification of claims being made (Brooks *et al.*, 2017). For very complex processes and decision-making, it is hard to see how anyone other than the experts themselves can make judgements about the processes, but there do need to be appropriate safeguards in place to ensure that professionals are not seen to defend their own (Rollin, 2006).

Food systems

Food systems, and their successes and failures, act as valuable case studies of the functioning of virtually every aspect of the market. The lack of verifiability of products associated with extended multinational food chains has led to consumers having no ability to confirm that a product is as it is labelled (Moore *et al.*, 2012) or even as the vendor expects it to be (Elliott, 2014)! The separation of most citizens in the developed world from food production, in particular animal production, creates challenges in terms of assuring animal welfare. These have been enhanced by the ability to overcome the natural limitations on intensification through medication (Rollin, 2006). Thus, a crucial externality, that must be balanced against livestock productivity, and regulated, is animal welfare (FAO, 2018), as without this the temptation will be to source food of animal origin at the lowest cost and sell it to consumers who have limited understanding of the distinctions between different farming systems, even if the food is labelled in some way. In addition, recycling of waste products and the search for low cost foodstuffs has precipitated human health crises, for instance bovine spongiform encephalopathy (Smith and Bradley, 2003). And, with the notable exception of those involved in a regulated profession, there is a lack of clear fiduciary duties to consumers and accountability of individuals involved in food systems when it comes to minimum standards of performance.

Conclusion

Globalisation brings with it very clear regulatory challenges for governments and citizens. In the light of various examples of market failure, it is clear that Adam Smith's observations related to efficient markets are bounded by the context within which they were made. Without clear enforcement of the individual elements that contributed to the efficient functioning of Smith's local communities, markets are doomed to continue to fail. The focus of the professional model on individuals working within their areas of competence and being accountable to peers when they fall short of required standards and the civil courts for redress provides a defensible check to the selfishness that otherwise might ensue. Therefore, far from being an inferior regulatory model to modern interpretations of the efficient market, it is clear that the principles which underpin the professional model are important to the long-delayed revisions to government regulation of the free market. However, this is not an argument for professionalization beyond current levels, but rather an argument for much greater knowledge and control of global supply chains, on the part of suppliers, their peers, governments and consumers, so that the customer is fully informed and their interests are placed first when it comes to standards and safety. This must be subjected to effective audit, particularly in the case of multinational companies, that goes beyond the rather inadequate financial safeguards that are the current market norm (Brooks, *et al.*, 2017; O'Brien, 2014). It is clear that currently in various market sectors, including that related to the food industry (Barnett *et al.*, 2015), the consumer has very low expectations that their interests will be taken into account. In the case of failure, it is also important that individuals can be identified as accountable and held to fiduciary standards in their business activities, rather than being allowed to evade the consequences through resort to the law of contract.

References

Barnett, J., Begen, F., Howes, S., Regan, A., McConnon, A., Marcu, A., Rowntree, S. and Verbeke, W. (2016). Consumers' confidence, reflections and response strategies following the horsemeat incident. Food Control 59: 721-730.

Bragues, G. (2010). Leverage and liberal democracy. In: Kolb, R.W. (ed.) Lessons from the financial crisis: causes, consequences, and our economic future. John Wiley & Sons, Hoboken, New Jersey, pp. 3-8.

Brooks, S., Elliott, C. T., Spence, M., Walsh, C. and Dean, M. (2017). Four years post-horsegate: an update of measures and actions put in place following the horsemeat incident of 2013. Npj Science of Food 1(1):5. Available at: https://doi.org/10.1038/s41538-017-0007-z. Accessed 29 March 2018.

Dietz, T., Ostrom, E. and Stern, P. C. (2003). Struggle to govern the commons. Science 302: 1907-1912.

Elliott, C., Copson, G., Scudamore, J., Troop, P., Steel, M. and Walker, M. (2014). Elliott review into the integrity and assurance of food supply networks – final report. Available at: https://www.gov.uk/government/uploads/system/uploads/attachment_data/file/350726/elliot-review-final-report-july2014.pdf. Accessed 29 March 2018.

Evetts, J. (2003). The sociological analysis of professionalism: occupational change in the modern world. International Sociology 18: 395-415.

FAO (2018). Shaping the future of livestock: sustainably, responsibly, efficiently. The 10th Global Forum for Food and Agriculture, Berlin, 18-20 January 2018. Available at: http://www.fao.org/publications/card/en/c/I8384EN. Accessed 3 April 2018.

Fiori, S. (2005). Individual and self-interest in Adam Smith's Wealth of Nations. Cahiers D'économie Politique / Papers in Political Economy 2(49): 19-31. Available at: https://doi.org/10.3917/cep.049.0019. Accessed 29 March 2018

Friedman, M. (1970). The social responsibility of business is to increase its profits. The New York Times Magazine, September 13, 1970.

Getzler, J. (2014). Financial crisis and the decline of fiduciary law. In: Morris, N. and Vines, D. (eds.) Capital failure: rebuilding trust in financial services. Oxford University Press, pp. 193-208.

Jaffer, S., Morris, N., Sawbridge, E. and Vines, D. (2014a). How changes to the financial services industry eroded trust. In: Morris, N. and Vines, D. (eds.) Capital failure: rebuilding Trust in Financial Services. Oxford University Press, pp. 32-64.

Jaffer, S., Knaudt, S. and Morris, N. (2014b). Failures of regulation and governance. In: Morris, N. and Vines, D. (eds.) Capital failure: rebuilding trust in financial services. Oxford University Press, pp. 100-126.

James, H.S. and Rassekh, F. (2000). Smith, Friedman, and self-interest in ethical society. Business Ethics Quarterly 10: 659-674.

Karnani, A. (2011). Doing well by doing good. California Management Review 53(3): 69-86.

Kon, A.A. (2010). The shared decision-making continuum. The Journal of the American Medical Association 304: 903-4.

May, S.A. (2013). Veterinary ethics, professionalism and society. In C. M. Wathes, S. A. Corr, S.A. May, S.P. McCulloch, and M.C. Whiting (eds.) Veterinary and animal ethics. Wiley-Blackwell, Oxford, pp. 44-58.

Moore, J.C., Spink, J. and Lipp, M. (2012). Development and application of a database of food ingredient fraud and economically motivated adulteration from 1980 to 2010. Journal of Food Science 77(4): R118-R126.

O'Brien, J. (2014). Professional obligation, ethical awareness, and capital market regulation. In: Morris, N. and Vines, D. (eds.) Capital failure: rebuilding trust in financial services. Oxford University Press, pp. 209-233.

RCVS (2017). Telemedicine: examining the digital future. RCVS News, March 2017. Available at: https://www.rcvs.org.uk/news-and-views/publications/rcvs-news-march-2017/rcvsnews-mar17.pdf. Accessed 29 March 2018.

Rollin, B.E. (2006). An introduction to veterinary medical ethics (2nd ed.). Blackwell Publishing, Oxford, 352 pp.

Smith, A. (1776/1970) The wealth of nations, books I-III. Penguin Books, London, 537 pp.

Smith, P.G. and Bradley, R. (2003). Bovine spongiform encephalopathy (BSE) and its epidemiology. British Medical Bulletin 66: 185-198.

Stiglitz, J.E. (2009). Capitalist fools. Vanity Fair, January 2009, pp. 48-51.

Teulon, F. (2014). Ethics, moral philosophy and economics. IPAG Business School, Working paper series. Available at: http://www.ipagcn.com/wp-content/uploads/recherche/WP/IPAG_WP_2014_288.pdf. Accessed 29 March 2018.

Thistlethwaite, J. and Spencer, J. (2008). Professionalism in medicine. Radcliffe Publishing, Abingdon, 200pp.

2. Beyond technocratic management in the food chain – towards a new responsible professionalism in the Anthropocene

V. Blok
Social Sciences Group, Wageningen University, Hollandseweg 1, 6707 KN Wageningen, the Netherlands; vincent.blok@wur.nl

Abstract

In this contribution, we argue that three related developments provide economic, environmental and social challenges and opportunities for a new responsible professionalism in the food chain: (1) the Anthropocene; (2) the bio-based economy; (3) precision livestock farming. These three interrelated developments indicate a transition in the way we understand the role and function of the food chain on the micro-, the meso- and the macro-level. This transition can be understood in two fundamental different ways, namely either as an extension of technocratic management beyond the micro level to the meso- and macro-level of the food chain, or as a transition to a new responsible professionalism. We argue that the technocratic approach is not able to address the socio-ethical issues that come along with these three developments, and explore various competencies and abilities that constitute a new responsible professionalism in the food chain.

Keywords: food chain, Anthropocene, biobased economy, precision livestock farming, responsible professionalism

In this contribution, we argue that three related developments provide economic, environmental and social challenges and opportunities for a new responsible professionalism in the food chain:

1. The changed conditions under which the contemporary food chain operates can be conceptualized in terms of the Anthropocene; the Anthropocene is a new geological epoch, in which the human has become the most influential 'terraforming' factor on earth (Blok 2017). Climate change is one of the main characteristics of the Anthropocene. On the one hand, it shows the dependency of the food chain on Earth's carrying capacity to feed the world. On the other hand, this calls for the transition to a more sustainable food chain. A shift in the food chain is required from a primary focus on market value to a focus on earth-oriented value, that provides huge challenges and opportunities for a new responsible professionalism in agri-food practices (cf. Bovenkerk and Keulartz, 2016).
2. One way to operationalize this shift is the transition to the bio-based economy (BBE). The BBE can be defined as a 'production paradigm that rely on biological processes and, as with natural ecosystems, use natural inputs, expend minimum amounts of energy and do not produce waste as all materials discarded by one process are inputs for another process and are reused in the ecosystem' (European Commission, 2012). Because the transition to the BBE provides a paradigm in which earth-oriented values become central in economic considerations, while it at the same time raises ethical issues of inter- and intra-generational equity (Murray *et al.*, 2017), the BBE provides huge challenges and opportunities for a new responsible professionalism in the food chain.
3. One way to operationalise the BBE in practice is the introduction of precision lifestock farming (PLF). PLF can be defined as 'the management of livestock production using the principles and technology of process engineering. ... PLF treats livestock production as a set of interlinked processes, which act together in a complex network' (Wathes *et al.*, 2018). By the integration of smart technology and the internet of things – in which computers, sensoring devices, GPS systems but also robots and even animals communicate with one another and function autonomously in an integrated farm management system – farmers can reduce farm inputs (fertilizers and pesticides) and increase yields, while reducing emissions to the environment (Bos and Munnichs, 2016). In PLF, the

Svenja Springer and Herwig Grimm (eds.) ***Professionals in food chains***
EurSafe 2018 – DOI 10.3920/978-90-8686-869-8_2, © Wageningen Academic Publishers 2018

internet of things is extended to farm animals (Blok and Gremmen, 2018). Because PLF provides concrete strategies to manage and control sustainable production in the food chain, while it at the same time raises ethical issues associated with the increased corporatization and industrialization of the agricultural sector, PLF provides both opportunities and challenges for a new responsible professionalism in the food chain.

These three interrelated developments indicate a transition in the way we understand the role and function of the food chain on the micro-, the meso- and the macro-level. This transition can be understood in two fundamental different ways, namely either as an extension of technocratic management beyond the micro level to the meso- and macro-level of the food chain, or as a transition to a new responsible professionalism.

The notion 'Anthropocene' does not only describe the current geological epoch in which humanity determines the face of the Earth, but makes us sensitive for the idea that we have to take responsibility for Earth's life-support systems on Earth (Kolbert, 2011). This responsibility is often understood in terms of planetary engineering and management of the Earth (Baskin, 2015; Blok, 2017). In the context of the BBE, we can think of mitigation strategies to improve resource efficiency of agri-food products via biotechnology, feed ingredient optimisation, waste reduction and better control of reproduction processes (European Commission 2012). In the context of PLF, we can think of the introduction of integrated farm management systems that enable further intensification of livestock farming and the emergence of megastalls in general, and the monitoring and control of animal growth, milk or egg production and greenhouse gas emissions in particular (Blok and Gremmen, 2018; Bos and Gremmen, 2013; Harfeld, 2010).

We can frame these strategies in the BBE and PLF as technocratic management. Technocracy is the idea of management and control of the food chain by technical experts. The farmer is no longer primarily the producer of food products, but a data processor and data manager that controls automated farm management systems. By exchanging information about oestrus detection, health, feed intake, waste material, etc., food chain actors are enabled to optimize coordination and efficiency throughout the supply chain (Blok and Gremmen, 2018). The advantage of the technocratic approach of the BBE and PLF is the eco-efficiency of agri-food production and consumption it can achieve, while contributing to economic returns as well.

At the same time, the technological fixes of the technocratic approach obscure the complexity of the ecological problems we face in the Anthropocene and the socio-technological systems in which they appear. A technological fix can be defined as follows: 'Recasting all complex social situations either as neatly defined problems with definite, computable solutions or as transparent and self-evident processes that can be easily optimized – if only the right algorithms are in place' (Morosov, 2013). It neglects the 'wickedness' of problems like climate change in the Anthropocene, the normative dimension of BBE and the conflicting values involved in PLF.

Climate change can be considered a wicked problem, i.e. a problem that is ill structured, difficult to pin down and unsolvable in a way. If we consider the famous definition of sustainable development from the Brundlandt report – 'Our common future' – (World Commission on Environment and Development, 1987) it seems to be quite simple: the use of resources today should not constrain the use of (non-renewable) resources in the future. But if we take the carrying capacity of the earth as point of departure and, with this, the fact that every resource will eventually be exhausted if we continue our current patterns of production and consumption, it becomes clear that climate change is a sustainability related eco-system failure that is difficult to pin down and highly complex, just like its solution (Blok, 2018; Peterson, 2009). In fact, it is not possible to satisfy the needs of the current generation without changing the conditions for future generations. Because climate change is a problems that has no closed form and concerns the eco-systems of planet Earth in which cause and effect are uncertain or unknown, no simple solutions exists for such a problem. 'On the one hand, we cannot propose definite solutions if we do not have a definite problem description. On the other hand, all proposed solutions remain finite and provisional compared to the complexity and depth of the sustainability problem itself. In this sense, we can never reduce sustainability to a finite set of particular 'problems', nor say that these problems can definitely be solved, i.e. that sustainability is fully achieved' (Blok *et al.*, 2016). Because of its focus on partial problems and efficient solutions, it is highly questionable whether a technocratic approach is sufficient to address wicked problems like climate change in the Anthropocene.

While current developments in the BBE are promising, most strategies are focussing on the technical issues to improve eco-efficiency in the food chain, whilst the normative dimension of the transition to the BBE are not taken into account. Sustainable development is a normative notion because it doesn't describe the world as it is, but as it should be. This means that the normative and socio-economic barriers to the transition towards the BBE should be taken into account. The transition to the BBE raises for instance ethical issues concerning inter-and intra-generational equity. The demand for biofuel for instance has resulted in the replacement of tropical forests by soy fields (Farigone *et al.*, 2008), putting pressure on food production in poor countries (cf. Murray *et al.*, 2017). Because costs and benefits or risks and rewards are difficult to calculate for future generations, if not impossible, it is questionable whether a technocratic approach is sufficient to address these type of questions in current research and policy making processes (Schlaile *et al.*, 2017), or requires a fundamental reflection on the relation between the biosphere and the economic sphere in the BBE (Zwier *et al.*, 2015).

While current developments in PLF are promising, it also raise ethical issues associated with the further intensification and industrialisation of livestock farming, like the emergence of megastalls with various socio-ethical consequences (Bos and Gremmen, 2013). More in general, it may result in the alienation between animals, farmers and citizens because of the robotisation and digitalization of farm management systems (Bos and Munnichs, 2016; Blok and Gremmen, 2018). Because the solution of these social-ethical issues cannot be calculated by experts but require public engagement, it is questionable whether a technocratic approach is sufficient to address the socio-ethical issues that make society reluctant to accept PLF.

Because the three interrelated developments – Anthropocene, biobased economy and precision livestock farming – indicate a transition in the way we understand the role and function of the food chain on the micro-, the meso- and the macro-level, but the technocratic approach is not able to address the socio-ethical issues that come along with it, we argue for a new responsible professionalism in the food chain.

We argue for a new responsible professionalism in the food chain, because the complexity and wickedness of global problems like climate change can no longer be solved by the unilateral application of ethical norms in the Anthropocene. The wickedness of global problems like climate change means that professionals in the food chain have to deal with imperfect foresight. Our knowledge of the unintended or even irreversible and harmful consequences for future generations of new technologies related to BBE and PLF is principally imperfect and therefore insufficient to distinguish between good and bad strategies in the food chain. The unilateral application of ethical norms is further complicated because multiple actors have differing and often conflicting norms and value frames.

We argue for a new responsible professionalism in the food chain, because the three developments show the centrality of the normative dimension in the Anthropocene, and the economic, social and environmental challenges and opportunities it provides for responsible professionalism. Only by integrating this normative dimension in current practices, we are able to take full advantage of the shift in the food chain from a primary focus on market value to a focus on earth-oriented value in the Anthropocene.

We argue for a new responsible professionalism in the food chain, because the three economic, social and environmental developments require a new type of professional in the Anthropocene, who is able to take responsible action within the context of imperfect foresight. This professionalism consists in the normative competence as ability to apply, negotiate and reconcile norms and principles in the food chain, which is unique in every situation because of the differences between the norms and interests of multiple stakeholders involved. The socio-ethical issues have to be weighed and revised over and over again because of changing circumstances or new insights, and the role of the professional involved in the food chain is to decide which norm to work with in a given situation. Next to normative competence, this requires action competence as the ability, based on critical thinking, reflection and incomplete knowledge, to actively involve oneself in responsible actions to improve the solution of the social-ethical issues involved in the food chain in general, and in BBE and PLF in particular. The combination of normative competence and action competence can be combined in a virtuous competence (Blok *et al.*, 2016; cf. Ploum *et al.*, 2017) that characterizes the new responsible professionalism in the food chain in the Anthropocene.

Contrary to the expert-engineer, who is the central figure in the technocratic approach, the new responsible professional doesn't consider partial technological fix solutions. He or she takes a food chain-perspective, is able to integrate the socio-ethical dimension in the further development of the BBE and PLF, doesn't only consult experts but all relevant stakeholders involved in the food chain, and takes the lead in the transition from a primary focus on market value to a focus on earth-oriented value in the food chain in the Anthropocene. Although much more research is needed to operationalise this new responsible professional, we propose the following competencies and abilities (cf. Blok, 2018):

- to apply, negotiate and reconcile norms and principles in the food chain (normative competence);
- to take responsibility as actor involved in the food chain, based on the capability to reflect and engage in critical thinking (action competence);
- to enhance collaborative action with multiple market and non-market oriented stakeholders;
- to take risk by exploring and exploiting radical uncertain opportunities in the food chain; the risks and uncertainties involved do not only concern the economic risk, but also the social-ethical risks associated with the food chain;
- to engage in satisficing solutions that are satisfactory and sufficient to maintain the life support systems of planet earth in the Anthropocene, and are always open for future subversions, revisions and improvements at the same time.

Acknowledgement

This working paper contains work in progress for my keynote speech during the EURSAFE conference 2018, Vienna (Austria). The building blocks of the current draft are heavily dependent on previous work that I was engaged in over the years together with many colleagues in Wageningen, and my team of students, PhD's, Post-docs in particular, that enabled me to develop this vision on a new responsible professionalism in the food chain in the Anthropocene. I have acknowledged these contributions as much as possible in the references provided in the text.

References

Baskin, J. (2015). Paradigm Dressed as Epoch: The Ideology of the Anthropocene. *Environmental Values* 24: 9-29.

Blok, V., Gremmen, B., Wesselink, R. (2016). Dealing with the Wicked Problem of Sustainable Development. The Role of Individual Virtuous Competence. *Business & Professional Ethics Journal* 34(3): 297-327 (DOI: 10.5840/bpej201621737).

Blok, V. (2017). Earthing Technology: Towards an Eco-centric Concept of Biomimetic Technologies in the Anthropocene. *Techne: Research in Philosophy and Technology.* 21(2-3): 127-149 (DOI: 10.5840/techne201752363).

Blok, V. (2018). Information Asymmetries and the Paradox of Sustainable Business Models: Toward an integrated theory of sustainable entrepreneurship. In: Idowu, S.O. *et al.*, *Sustainable Business Models: Principles, Promise, and Practice.* Dordrecht: Springer (in press).

Blok, V. and Gremmen, B. (2018). Agricultural technologies as living machines: toward a biomimetic conceptualization of technology. *Ethics, Policy and Environment* (in press).

Bos, J. and Munnichs, G., (2016). *Digitalisering van Dieren. Verkenning Precision Livestock Farming.* Den Haag: Rathenau.

Bos, J. and Gremmen, B., (2013), Does PLF turn animals into objects? *Precision Livestock Farming* 2013. Leuven: EC-PLF.

Bovenkerk, B. and Keulartz, J. (Eds.) (2016), *Animal Ethics in the Age of Humans.* Springer: Dordrecht.

European Commission (2012). *Innovating for sustainable growth – a bioeconomy for Europe* Available at: http://ec.europa.eu/research/bioeconomy/pdf/bioeconomycommunicationstrategy _b5_brochure_web.pdf).

Farigone, J., Hill, J., Tilman, D., Polasky, S., and Hawthorne, P. (2008). Land clearing and the biofuel carbon debt. *Science*, 319, 1235-1238.

Harfeld, J., (2010). Husbandry to industry: Animal Agriculture, Ethics and Public Policy. *Between the Species* 10: 132-162.

Kolbert, E. (2011). Enter the Anthropocene – age of man. *National Geographic* (March).

Morosov, E. (2013). *So save everything. Click here. The folloy of technological solutionism.* Penquin.

Murray, A., Skene, K., and Haynes, K. (2017). The Circular Economy: An Interdisciplinary Exploration of the Concept and Application in a Global Context. *Journal of Business Ethics* 140: 369-380.

Peterson, C. (2009). Transformational supply chains and the 'wicked problems' of sustainability: aligning knowledge, innovation, entrepreneurship, and leadership. *Journal of Chain and Network Science,* 9(2): 71-82.

Ploum, L., Blok, V., Lans, T. and Omta, O. (2017). Toward a validated competence framework for sustainable entrepreneurship. *Organization & Environment* DOI: 10.1177/1086026617697039 (*in press*).

Schlaile, M., Urmetzer, S., Blok, V., Andersen, A.D., Timmermans, J., Mueller, M., Fagerberg, J. and Pyka, A. (2017). Innovation Systems for Transformations towards Sustainability? Taking the Normative Dimension Seriously. *Sustainability* (DOI: 10.3390/su9122253

Wathes, C.M. (2009). Precision livestock farming for animal health, welfare and production. *Sustainable Animal Production: The Challenges and potential Developments for Professional Farming* 411-419.

World Commission on Environment and Development (1987). *Our common Future.* Oxford: Oxford University Press

Zwier, J., Blok, V., Lemmens, P. and Geerts, R.J. (2015). The Ideal of a Zero-Waste Humanity: Philosophical Reflections on the demand for a Bio-Based Economy. *Journal of Agricultural & Environmental Ethics,* 28(2): 353-374 (DOI: 10.1007/s10806-015-9538-y).

3. Should we help wild animals suffering negative impacts from climate change?

C. Palmer
Department of Philosophy, Texas A&M University, College Station, TX 77843 USA; c.palmer@tamu.edu

Abstract

Should we help wild animals suffering negative impacts from anthropogenic climate change? It follows from diverse ethical positions that we should, although this idea troubles defenders of wildness value. One already existing climate threat to wild animals, especially in the Arctic, is the disruption of food chains. I take polar bears as my example here: Should we help starving polar bears? If so, how? A recent scientific paper suggests that as bears' food access worsens due to a changing climate, we should consider supplementary feeding. Feeding starving bears could meet ethical obligations to help wild animals suffering from climate change. But supplementary feeding may also cause harms, and lead to park-like management of some bear populations – a concern for those who care about the value of wildness. While this situation is in many ways intractable, I'll make a tentative suggestion of a possible way forward for wildlife managers.

Keywords: animal welfare, animal ethics, wildlife management, supplementary feeding

Introduction

Impacts of anthropogenic climate change are now being felt globally, both by humans, and by non-humans. However, while questions of climate ethics and justice between human beings have been widely discussed, very little has been written about our obligations to sentient animals in dealing with climate change. Yet many wild animals are already suffering from heat stress, flooding, hunger and increased parasitic infestation, at least in part on account of anthropogenic climate change. Do we owe assistance to wild animals in this situation?

Some assumptions of this paper

First, I will assume both that climate change is anthropogenic and that humans are now not just causally, but also morally responsible (Nolt, 2011) although the details of the distribution of moral responsibility remain contested. Second, the harms/suffering I'm discussing here are caused to individual sentient animals. I'll assume the (widely held) view that sentient animal welfare matters morally, and I'll focus on welfare in terms of animals' subjective experience – that is, that what happens to them matters in terms of how it feels; and they have an interest in not suffering. Third, I'll assume that relevant degrees of climate change cannot be prevented. One likely objection to this paper is that what would help wild animals most would be reducing greenhouse gas emissions. Of course, this *is* the ideal strategy. However, significant overall global drops in emissions are not likely soon; and, even if they occurred, there would still be a timelag before any benefit was felt; we are locked into a warming climate with impacts on wild animals, for the foreseeable future.

Should we help wild animals suffering from climate change impacts?

Although the general claim that we should prevent and relieve the suffering of wild animals was once regarded as a *reductio* of an animal liberation position, recently it has been seriously defended. Delon and Purves (2018, forthcoming), for instance, note that if we think suffering is intrinsically bad – as

EurSafe 2018 – DOI 10.3920/978-90-8686-869-8_3,

we should – from a number of independently plausible moral principles, we have a *pro tanto* obligation to intervene in wild animal suffering. Horta (2013), the most prominent defender of this view, argues that many wild animals have extremely bad lives in which suffering outweighs pleasure, and that we should intervene either to make bad lives better, or to ensure that there are fewer bad lives, if we can do so without creating more overall suffering. This position can obviously be justified from a number of consequentialist ethical approaches, but it is not confined to consequentialists (for instance, Nussbaum, 2007 endorses a similar view). On these views we should prevent and relieve wild animal suffering when we can, unless doing so would make matters worse overall, whether the cause is anthropogenic or not. So, we should help wild animals negatively impacted by climate change where we can, just because they are suffering.

However, many animal ethicists from non-consequentialist traditions reject *general* intervention into the wild to relieve suffering, based on a distinction between harming and assisting. Regan (1984) for instance, defends wild animals' negative rights not to be seriously harmed or killed by moral agents, but does not claim that animals have general positive rights to be assisted against disease or predators. Palmer (2010) argues that while we have *prima facie* general duties not to harm sentient animals, we do not have similar positive duties to assist them. But such arguments against general duties of assistance to wild animals doesn't rule out special obligations. And on many rights and other deontological views, prior human harms or injustices that cause animal suffering can create just such obligations. For instance, Palmer (2010) argues that where sentient animals are made vulnerable or suffer human-caused harms, as in the case of climate change, if it's possible for wild animals to be assisted, we should do so. Many non-consequentialist views that take animals' interests or rights seriously maintain that harms, injustices or rights violations give us a moral responsibility to respond, unless doing so would create new harms, injustices or rights violations.

Some philosophers are more reluctant to commit to assisting wild animals even if they accept that prior human injustices have occurred. For instance, Donaldson and Kymlicka's (2011, 2014) political theory of animal rights proposes that sentient wild animals should be recognized as members of self-organizing, sovereign wild animal communities. 'Sovereignty' is assigned to protect these communities from human incursions of all kinds, including climate change, which violates individuals' rights and undermines 'the ecological fabric they [wild animals] depend on' (Kymlicka, 2014). Because these communities are sovereign, though, we should not automatically move to assist, if there's a likelihood this would undermine sovereignty. One-off or small-scale interventions are unlikely to undermine sovereignty, but sustained and continuous actions may well do so. In terms of climate interventions, then, for Donaldson and Kymlicka, key questions concern whether the interventions are sustained and continuous, and whether they further violate animals' rights.

Finally, there are in-principle objections to assisting wild animals, based on the preservation of wildness value. Hettinger (2014), for instance, endorses a principle of 'Respect for an Independent Nature'. On this view, human interventions into wild nature should be minimized, allowing Nature its own 'self-willed' evolutionary and ecological developments as free as possible from human constraint and guidance. While animal suffering may matter, for Hettinger, it doesn't matter as much as respecting wildness value. And while Hettinger accepts that climate change compromises wildness, he also argues that wildness comes in degrees. Systematically assisting wild animals threatened by climate change is likely to reduce the wildness of nature even more than climate change itself. So, we should not normally assist, even where wild animals are suffering from anthropogenic causes.

Wildness value aside, however, ethicists and political philosophers from a number of ethical traditions agree that there's good reason in principle to help wild animals suffering from climate change, either because we should, generally, relieve suffering when we can or because the suffering is anthropogenic and therefore can be viewed as a harm, an injustice, a rights violation, or a human encroachment on sovereign communities; and as such it creates special obligations or requires compensation or reparation. But restrictions on such assistance are also clear. Assistance should not cause more suffering than it relieves; it should not violate more rights or cause major harms; and for Donaldson and Kymlicka, it should not compromise the autonomy of wild animal sovereign communities. These restrictions, though – especially in the case of climate change – can make assisting wild animals, in practice, very difficult to do. I'll illustrate this by looking more closely at a specific case where wild animals are suffering negative climate impacts: the case of polar bears in Arctic food chains.

Polar bears, climate change, and Arctic food chains

About 26,000 polar bears live in the Arctic; they depend on sea ice as the base from which they hunt ringed seals, their main food source. In summer, when sea ice melts, many bears live on land, surviving on stored fat and scavenging; they begin to hunt seals again when sea ice forms in early autumn. However, climate change is already disrupting Arctic food chains, as sea ice is melting earlier and re-forming later, requiring bears to live off their own fat, on land, for longer. This has already resulted in a worsening of bears' body condition in some populations (Stirling and DeRocher, 2012). But as the Arctic continues to warm, in an approaching year, sea ice is likely to form so late that some bears starve. DeRocher *et al.* (2013: 370), a group of leading polar bear scientists, conclude: 'Malnutrition at previously unobserved scales may result in catastrophic population declines and numerous management challenges.' What – if anything – can and should be done to meet these challenges likely to confront wildlife managers? DeRocher *et al.* (2013) propose several possible 'proactive conservation and management options', including translocations (which they do not recommend; so I won't pursue this here); feeding bears, both to divert them from human communities and as a supplement to 'provide sufficient short term energy to help individuals survive periods of food deprivation'; and intentionally euthanizing starving bears, as the most humane option for 'individual bears that are in very poor condition and unlikely to survive'. Let's take these as the likely options, along with doing nothing.

We've already seen that many animal ethicists agree we should help wild animals suffering from the impacts of climate change, unless doing so creates other significant moral concerns. For many of these the *prima facie* best solution would most likely be to offer supplementary food; and therefore I will focus on this. Is it a good solution? What are the potential ethical concerns about doing so? I'll consider two major worries here: first that feeding bears would harm bears or other animals; and second, concerns about wildness and the sovereignty of wild animal communities.

Would feeding bears cause harms?

First: Would feeding bears harm the bears themselves? An initial starvation event would probably be brief, and only some populations would need short-term feeding. In this eventuality, food could be air-dropped to bears on a one-off basis; this would be unlikely to harm bears. However, given sea-ice decline, after a few years, supplementary feeding is likely to involve more populations for longer every year, eventually perhaps for several months. It's here that more significant questions arise. Studies of human intentional supplementary feeding across a range of wild species show mixed benefits, and some hazards, to wild animals' welfare (Dubois and Fraser, 2013). Feeding stations can be places of conflict or disease transmission; food can be nutritionally inadequate, and animals' natural behaviour can be disrupted, for instance by a loss of fearfulness or by habituation to being fed. In the bear case, some risks look easier for wildlife managers to avoid than others; conflicts and diseases can, to some degree at least, be avoided. Disruption of bears' natural behaviour however, is inevitable; at the very least, consuming supplementary food requires different behaviours from hunting seals on sea-ice. Should we be concerned about this? Earlier, I chose to focus on welfare in terms of subjective experiences, such as suffering. Since the goal of feeding hungry bears, after all, is to relieve and prevent suffering, negative welfare effects don't seem an issue in the short term. But is it likely that bears' dependence on and habituation to human provision would cause them more suffering over time? At the moment, we lack clear answers to those questions. Habituation to supplementary feeding could change bears' behaviours, such that they lose the ability to hunt seals independently on the sea ice, though this seems unlikely. Feeding may also make bears more vulnerable to human policy changes – for instance, if sponsors decide that feeding is too expensive to continue. Certainly, feeding now does risk future suffering. But without it, these particular bears will starve. So, even if the bears' future is uncertain with feeding, for these particular bears, it's their only hope of continuing to live at all, let alone living with reasonable welfare.

Second: Would feeding bears harm other sentient animals? There's one way in which feeding bears seems inevitably to harm other animals: in terms of what they are fed. Polar bears require extremely high levels of fat; they are 'the most carnivorous of all bears' (IUCN, 2015). The most obvious foodstuffs would either be ringed seals, killed and air dropped; or zoo polar bear chow. Feeding bears with ringed seals, however, would raise serious problems for all the ethical views we're considering here. First, it involves humans killing seals, thus harming them/violating their rights. Second, ringed seals are also threatened by climate change, as lair collapses due to early snowmelt are making pups much more vulnerable. Justice-based views, in particular, could not accept compensating victims of injustice by killing other victims of injustice. So, would commercial bear chow be better? It includes fishmeal from menhaden (highly fatty fish), pork fat and bone meal, probably derived from parts of pigs outside the human food chain, rendered into animal by-products. In sum, fish would be directly killed to feed polar bears; and while pigs may not be directly killed, the production of bonemeal and meat byproduct is obviously part of the meat industry. From the ethical views considered here, is this a reason not to feed bears, even if bears will otherwise suffer and starve due to anthropogenic climate change? Different ethical perspectives may diverge here, but there's at least a basis for arguing that feeding polar bear chow can be justified, despite harms to other animals, in such a non-ideal situation, where there are only 'imperfect solutions to horrific problems' (Emmerman, 2014). Intensive animal agriculture and fishing is not going away any time soon; most people will not be vegetarian – nor will their pets. So, to pick out polar bears – obligate carnivores, unlike people and dogs – already unjustly deprived of the ability to feed themselves, and insist that they should starve or be euthanized because feeding them harms other animals, potentially adds a further layer of injustice to the ones from which bears are already suffering. Polar bear chow, even though implicated in harming other animals, looks like a permissible food for starving polar bears, even on the ethical views with which I'm working here.

The problem of wildness value and 'semi-managed bear parks'

However, even if it's permissible to feed bears chow, ethical issues remain. Feeding bears raises (related) concerns about wildness and the creation of 'semi-managed bear parks'. Let's start with Hettinger's (2014) idea of 'Respect for Independent Nature'. For Hettinger, naturalness value is measured by the degree of independence and autonomy of a being or a place from humanity; and, he maintains, this value has only become more important as human influence has increased. As noted above, climate change may have reduced such wildness, but it has not entirely eliminated it. In particular, Hettinger maintains, we can still treat animals in ways that are problematically 'unnatural'. Feeding bears would be 'unnatural' in exactly this way: it makes bears less autonomous, dependent on people, potentially in the long term, for significant parts of the year. The only circumstance in which Hettinger might accept feeding polar bears would be if they were on the edge of extinction, since he maintains that species are valuable, and that land from which a native species is entirely lost due to human activity is less 'natural' than land containing that native species, even if humans have had to launch a rescue operation to keep the species there. But this exception doesn't apply to the polar bear case, since polar bears are not imminently facing extinction; we're talking about feeding bears for the *bears'* sake, not because they are the last bears.

While Hettinger's worry is not identical to Donaldson and Kymlicka's (2011, 2014) both focus on particular ideas of autonomy. Hettinger is concerned with the autonomy of what he calls 'Nature'. Donaldson and Kymlicka are concerned with the autonomy both of individual bears, and of wild animal communities, where animals 'should be recognized as having the right to live autonomously on their own territories, and hence as exercising their own sovereignty'. But even the scientists discussing supplementary feeding are concerned that doing so will create a 'semi-managed bear park' (DeRocher *et al.*, 2013). This sounds like exactly the kind of intervention Kymlicka rejects: 'a kind of permanent paternalistic management in which we take over responsibility for feeding and sheltering them' where 'we are basically turning wilderness into a zoo' (Kymlicka, 2014).

So, does the proposal to feed starving bears, then, create a rift between views that primarily emphasize welfare and suffering (such as Horta and Palmer) and so are likely to support feeding, and those that emphasize autonomy, either of Nature (Hettinger) or individual bears and wild animal sovereign communities (Donaldson and Kymlicka) and so oppose it? Not necessarily. It does seem clear that Hettinger would oppose supplementary feeding, unless it was necessary to save the species 'polar bear', at least in the wild. But Donaldson and Kymlicka have reason to support such feeding. First, it's not clear that feeding would violate *individual* bears' autonomy. Starving bears would autonomously choose to eat what humans provide; and, after all, dead bears have no autonomy at all. Second, Donaldson and Kymlicka's objection to sustained interventions focuses on essentially 'competent' communities, where members of the community can normally meet their own needs. However, the food chain disruption caused by climate change is rendering wild Arctic communities as a whole (not just polar bears, but many species of wild animals) essentially 'incompetent', or 'failing' and inflicting persistent injustices on individual wild community members. Ideally, in such a case, on Donaldson and Kymlicka's theory, climate impacts on wild animal sovereign communities should force humans sharply to reduce emissions, thus restraining human incursions into wild communities. But in the non-ideal political circumstances that currently prevail, this won't happen; and in any case, there's enough time-lagged warming already built in that some bears will likely starve in the next few years. Where climate change impacts are so severe, continuing to defend wild sovereignty would undermine exactly what it was established to protect: wild animals' interests. Here, animals' interests are most likely to be protected by sustained intervention that reduces bear suffering, keeps bears alive in the face of severe anthropogenic threats, and ultimately retains the possibility that at some point in the future Arctic wild animal communities may become sovereign again. And while the idea of semi-managed bear parks is troubling to *us*, it's not something that bears themselves can resent.

Conclusion and a tentative proposal

In a nearby autumn, if we do nothing, polar bears in some wild populations will starve to death because of anthropogenic climate change. When this situation arises, we can either do nothing, relieve their suffering by euthanizing them, or we can feed them, though this entails killing other animals, a loss of wildness value, and risking the creation of 'semi-managed bear parks'. This is clearly a non-ideal situation; no good options are available. There's little chance to reframe this problem without a significant reduction in greenhouse gas emissions globally, (or perhaps geoengineering). This isn't going to happen in time to save at least some bears from starving. The best options seem involve either supplementary feeding or euthanizing bears when they reach extreme levels of hunger. But there is much we don't know about this situation – in particular, how quickly bears get habituated to being fed, how far it will affect their behaviour, and how far it will affect their offspring's behaviour. Perhaps feeding won't create semi-permanent bear parks. One possible approach, I tentatively suggest, is for wildlife managers to try supplementary feeding (that also diverts bears from human communities) in the first population that becomes seriously at risk, to see how bears respond, how quickly they return to the sea ice, and whether they come back to feeding stations once ice has formed (and also whether other hazards manifest themselves, such as unmanageable disease outbreaks at feeding stations). How this turns out could give further guidance and a better basis for wildlife managers to make decisions as sea-ice continues to shrink across the Arctic. It's not a strategy that's very palatable to anyone (and may be absolutely rejected by those who prioritize wildness value over animal suffering). But it may be the best strategy available given the short- and medium-term disruption of food chains in the Arctic.

References

Delon, N. and Purves, D. (2018). Wild animal suffering is intractable. Journal of Agricultural and Environmental Ethics 34/1 (In press).

Derocher, A.E., Aars, J., Amstrup, S.C., Cutting, A., Lunn, N.J., Molnár, P.K., Obbard, M.E., Stirling, I., Thiemann, G.W., Vongraven, D., Wiig, Ø. and York, G. (2013). Rapid ecosystem change and polar bear conservation. Conservation Letters, 6: 368-375.

Donaldson, S. and Kymlicka, W. (2011). Zoopolis: A Political Theory of Animal Rights. Oxford University Press, New York, USA.

Dubois, S. and Fraser, D. (2013). A Framework to Evaluate Wildlife Feeding in Research, Wildlife Management, Tourism and Recreation. Animals 3: 978-994.

Emmerman, K. (2014). Sanctuary, Not Remedy: The Problem of Captivity and the Need for Moral Repair. In Gruen, L. (ed.) The Ethics of Captivity. Oxford University Press, New York, USA pp. 213-230.

Hettinger, N. (2014). Valuing Naturalness in the 'Anthropocene': Now More than Ever. In Weurthner, G., Crist, E. and Butler, T. (eds) Keeping the Wild: Against the Domestication of Earth. Island Press, Washington DC pp. 174-179.

Horta, O. (2013). Zoopolis, Interventions and the State of Nature. Law, Ethics and Philosophy 1: 113-125.

IUCN. (2015). Ursus Maritimus. IUCN Red List of Threatened Species. Available at: doi.org/10.2305/IUCN.UK.2015-4.RLTS.T22823A14871490.en. Accessed 5 January 2018.

Kymlicka, W. (2014). Will Kymlicka on Animal Denizens and In the Wilderness. Interview with Arianno Mannino. Available at: http://gbs-schweiz.org/blog/will-kymlicka-on-animal-denizens-and-foreigners-in-the-wilderness-interview-part-2. Accessed 20 January 2018.

Nolt, J. (2011). Non-anthropocentric climate ethics. WIRES Climate Change 2/5 701-711.

Nussbaum, M. (2007). Frontiers of Justice. Harvard University Press, Cambridge, USA.

Palmer, C. (2016). Saving Species but Losing Wildness? Midwest Studies in Philosophy 40: 234-251.

Regan, T. (1984). The Case for Animal Rights. University of California Press, Berkeley.

Stirling, I. and DeRocher, A. (2012). Effects of climate warming on polar bears: a review of the evidence. Global Change Biology 18/9: 2694-2706.

Section 1.
Professional responsibility in the food chain

4. Connecting parties for improving animal welfare in the food industry

M.R.E. Janssens[1] and F. van Wesel[2]*
[1]Rotterdam School of Management, Erasmus University, Burgemeester Oudlaan 50, 3062 PA Rotterdam, the Netherlands; [2]Department of Methodology & Statistics, Faculty of Social and Behavioural Sciences, Utrecht University, Padualaan 14, 3584 CH Utrecht, the Netherlands; janssens@rsm.nl

Abstract

Communication by companies plays an important role in corporate (social) responsibility (CSR). One of the optional CSR topics is animal welfare. Our exploratory qualitative study reveals which factors of communication stimulate an attitude of responsibility towards animals in the animal-based food industry. It shows that a manager who is made responsible for animal welfare can strengthen the company's ethical position in two ways. The first one is to connect with stakeholders within and outside the company. The second way is to facilitate, as a moderator, connections between these stakeholders in which the manager is not involved per se. In both cases, if these connections take the form of personal meetings, this is extra helpful for enhancing a responsible attitude towards animals, because that is how insight, trust and collaboration are gained and sustained.

Keywords: communications, CSR, animal ethics, food, manager

Introduction

Communication by companies plays an important role in corporate (social) responsibility (CSR) (Golob *et al.*, 2013). Without communication, customers or corporate buyers would have no extra stimulus to buy responsibly produced products or services. On the other hand, CSR communication can lead to accusations of greenwashing and window dressing, even if it is honest and accurate (Schlegelmilch and Pollach, 2005). Fear of this type of accusation may be one of the reasons for window blinding: keeping silent about a decent CSR performance (Mauser, 2001; Nielsen and Thomsen, 2009).

Animal welfare as a CSR issue seems to be a blind spot for many, in business ethics as well as in business practice (Janssens and Kaptein, 2016). We will focus on communication in relation to corporate responsibility for animals in the food industry. As a starting point for our analyses, we will use the Laswell Formula for ethical communications: 'Who says what to whom in which channel and with what effect?' (Schlegelmilch and Pollach, 2005). In this case we start from the effect: we will look specifically for those factors which strengthen an ethical stance of the corporation towards animals. In addition, we will focus on who and in which channel, so that we can explore the communication routing. Analysing the content (what) is beyond the scope of this paper.

In this paper we define animal welfare as it is used in common language, referring to any positive moral status of animals, including issues concerning the way they live, their happiness and pain, their intrinsic value, their worthiness of being protected, the duration of their lives, the way they die, and the issue of actively taking their lives. During the study we were open to references to this broad concept of animal welfare from any philosophical, religious, or other approach.

EurSafe 2018 – DOI 10.3920/978-90-8686-869-8_4, © Wageningen Academic Publishers 2018

Method

In this exploratory study, we use part of the data from a former qualitative study, collected in 2015 and 2016 during nine interviews with managers responsible for animal welfare, whom we called responsible managers (RMs), of 9 large (1000+ employees), internationally operating, Western-Europe based companies in the animal-based food production chain (authors, under review). The companies are producers, processors, wholesalers and retailers. The methodology used was based on grounded theory (Boeije, 2010; Corbin and Strauss, 2015). We searched online and in real life for fitting companies, and were directed to new ones by the participants (snowball sampling). In total, we invited thirteen companies to participate, of whom nine accepted: seven based in the Netherlands, one based in the United Kingdom, and one based in Switzerland. The interviews were semi-structured, which provides the richest source of data for theory building (Boeije, 2015; Corbin and Strauss, 2015). For reasons of data triangulation, we combined the interview data with data from the companies' CSR reports and their websites. As the different sources mentioned several communication channels, but hardly ever social media, a medium we did not want to overlook, we collected additional data from Twitter, YouTube, LinkedIn, Facebook and Instagram. By comparing and analysing the extended data set through a communicative lens, we identified a communicative model, which we presented and explained to the RMs by email, for reasons of validation. Their comments were incorporated in our discussion and conclusions.

Findings

Based on our data, we found several communicative drivers that enhance the level of responsibility for animals taken by the company. By grouping these drivers, and drawing communicative connections between the stakeholders involved, we designed a model explaining how these elements interact (Figure 1). Furthermore, we made an inventory of the communication channels involved in each of the communicative connections in the model: Table 1.

We stress that our model and table do not reflect the average situation. We combined all positive influencers that were shown to enhance corporate responsibility for animals. Therefore, we view our model and table as an overview of what works well in animal-based food companies, and how these elements could interact if they were all put into practice in a variety of combinations. Consequently, the model reflects a combination of drivers, all of which can be found in practice. All companies in the study use a selection of the connections and channels. We will now describe the RM as a connector and as a moderator. Our data show that in both roles, it is extra helpful if an RM as a person is open, sensible and accessible.

The RM connecting to stakeholders

We found 5 connections (dark arrows in Figure 1) by which the RM communicates directly with other stakeholders, inside and outside the company.

For connecting with employees, RMs use channels for dialogue by which they can connect directly. Channels we found in our data are intranet, meetings, and live events such as celebrations of milestones.

The complexity of animal welfare as a CSR issue (e.g.: what is best for animals?) makes it important for the RM to communicate with colleagues from adjacent departments, especially those responsible for reputation, strategy, marketing, issue management, and product quality, including the management department. If the RM is open to these departments' views, and incorporates them in the change process, more can be achieved.

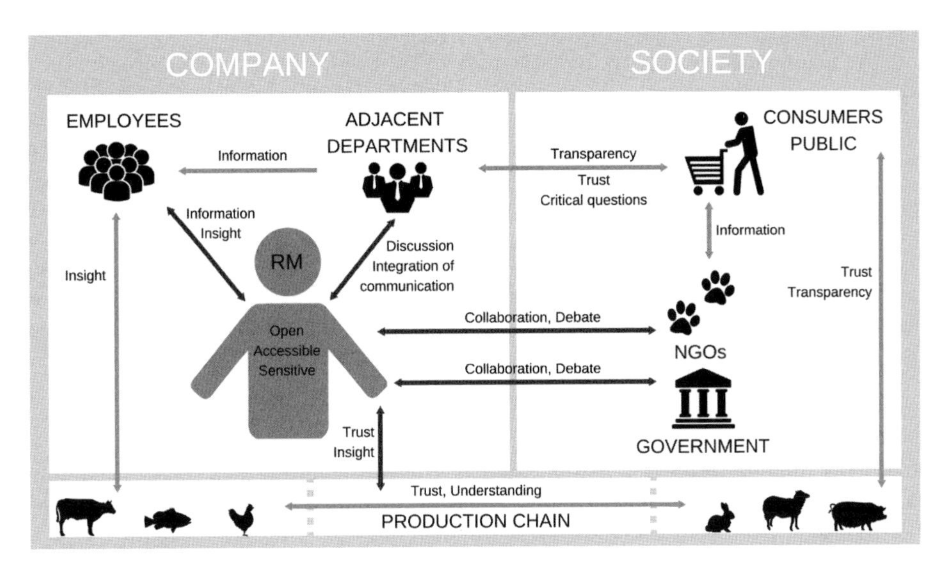

Figure 1. Communicative drivers for corporate responsibility for animals in the food industry. RM = Responsibility Manager (CSR manager or any other manager responsible for animal welfare); UPPERCASE TEXT + icon = stakeholder involved in communication; Dark arrow = direct communicative connection with the RM; Light arrow = indirect communicative connection in which the RM plays moderating role; Arrow point = direction of the communication; Lowercase text = what is exchanged or gained through the connection.

Table 1. Communication channels used by managers responsible for animal welfare in food companies.

Communicative connections	Communication channels used
Connections of the responsibility manager (RM) with stakeholders	
RM ↔ employees	Intranet, meeting, event, game
RM ↔ adjacent departments	Meeting
RM ↔ NGOs	Event, award, campaign, stakeholder debate
RM ↔ government	Meeting
RM ↔ production chain	Visit, meeting, extranet, newsletter, stakeholder debate
Connections between stakeholders, in which the RM plays a moderating role	
Employees ↔ production chain	Visit, events
Adjacent departments → employees	Newsletter, intranet, company website, press release, social media
Adjacent departments ↔ consumers/public	CSR report, packaging, website, press release, social media, traceability system
NGOs ↔ consumers/public	NGO website, company website, social media, joint press release, label
Production chain ↔ consumers/public	Visits, website
Production chain ↔ production chain	Visit, meeting

Outside the company, RMs connect with NGOs. Programs for working together on, for example, meeting the requirements for an animal welfare label are helpful for getting companies to take a responsible stance towards animals.

Communication plays a modest role in the complex relationship between government and company. A dialogue, which is effectuated through meetings, is necessary to start innovative programs. An RM who is willing to join the debate, and is supported to do so by corporate management, is influential.

There are strong connections between RMs and companies in the production chain. In addition to a rather impersonal channel such as a newsletter, and a quasi-personal channel such as an extranet, live meetings and farm visits are important, because they can establish mutual trust and understanding. For example, RMs visit farmers to see how things work from their perspective.

The RM moderating connections between stakeholders

Apart from communicating with the stakeholders, the RM connects stakeholders. We found 6 connections (light arrows in Figure 1) inside and outside the company, in which the RM is involved as a kind of moderator, stimulating stakeholders to communicate about animal welfare issues.

RMs connect employees to parties in the production chain by organising farm visits, which increases employees' understanding of animal welfare issues.

The RM can initiate and facilitate connections between adjacent departments and other stakeholders, for example employees. The communications department can write about animal welfare in newsletters, the intranet, social media and the company's website.

Social media and websites connect as well to the public, including consumers. Other relevant channels between departments and the consumer are packaging (where product information is disclosed) and product traceability systems. Communication through these channels is extra effective if it is integrated in overall CSR or sustainability communication, and helps to establish transparency and trust. This is sometimes done by storytelling, which makes complex issue of animal welfare come alive. Social media are used to receive and answer critical consumer questions. This helps to manage the company's reputation.

NGOs use several channels to express themselves on animal welfare in relation to the food industry. RMs use these existing channels by being in close contact with these NGOs. Through NGOs they communicate to the public, for example about products, labels and NGO awards. According to our data, the channels used are NGO websites, company websites, social media, joint press releases, and animal welfare labels on packaging.

One of the special roles of RMs is to connect the production chain with consumers and the public. What holds for employees, holds for the latter groups as well: the topic comes alive as soon as one meets farmers (and their animals) in real life. Therefore, RMs initiate participation of farms in activities such as open-door days. Activities like these extend transparency, establish trust, and take away negative prejudices.

Finally, we found data about connections between parties within the production chain, facilitated by RMs. Views on animal welfare and corporate responsibility can vary considerably between chain partners. According to the data, trust and mutual engagement are crucial for achieving progress. Both are established by live meetings between parties within the chain. Therefore, the use of live communication channels, such as visits and other meetings (e.g. focus groups, program meetings and strategic sessions), is a positive influencer. Sometimes communication is embedded in partnerships, to make sure that the chain partners meet, connect and collaborate.

Discussion, conclusions, and recommendations

Our model (Figure 1) and the accompanying table (Table 1) show a network of interactions from which elements are used by several RMs and stakeholders. Central concepts that play a role are openness, trust, collaboration, and meetings. Each stakeholder is approached through appropriate communication channels. Personal contact works out well with partners in the production chain and with NGOs. It would probably work with consumers and the public too, but is not feasible, except during events like Open Farm Days. Authentic storytelling, as well as tuning in to trends and NGO campaigns with genuine commitment, can work as a substitute for personal contact. Sometimes, differences in channels can be explained from the different positions companies have in the production chain. It is more important for a retailer to inform the public thoroughly than for a processor.

The RM acts as both a partner in connections and a moderator of connections. Both types of connections are intertwined. An RM will join an employee visit to a farm and affirm the personal relationship with the farmer, or will provide the marketing department of content for the public website.

Although quantity is not an issue in our study, we note that references to animal welfare on social media channels were sparse. For CSR issues in general, Etter (2013) concludes that, fearing the potential negative publicity, companies are hesitant to proactively communicate them on Twitter, although some solve the paradox of engagement and risk management by starting a special CSR account.

The success of working with adjacent departments points towards embedding animal welfare communication in broader, existing channels of (CSR) communication. This is in line with what Vanhonacker and Verbeke (2014) conclude. In addition, Cornelissen (2004) states that CSR communication needs to be integrated in business activity.

We have shown that employee communication and consumer communication are supportive for a responsible stance towards animals. The latter is confirmed for agribusiness by Luhmann and Theuvsen (2016). They state that it is important for a company to share its view to increase transparency, knowledge, trust and reputation. The importance of communication and gaining trust is also confirmed by Schlegelmilch and Pollach (2005), and especially for the food sector by Gössling (2011) and Vanhonacker and Verbeke (2014).

Our study is of an exploratory nature, and the small number of participants is a limitation. We stopped interviewing when saturation in the diversity of topics occurred, which means that some of the influencers we found were only experienced by one or two of the interviewees. Thus, we cannot make statements about the strength or frequency of the influencers found. Therefore, we recommend that future research further explores our findings.

There is much that companies and their RMs can do to strengthen a responsible stance of the company by communicating. Our list of channels offers plenty of tools. We recommend exploring them by taking the following steps.

- If no manager is responsible for animal welfare yet, make it the explicit responsibility of the CSR manager, the quality manager, or any other fitting manager.
- Explore the 5 opportunities for connecting with stakeholders, and the 6 opportunities for facilitating connections between stakeholders. Strengthen existing connections and initiate new ones.

- Use existing (CSR or other) communication channels and incorporate animal welfare as one of the themes the company is concerned about and wants to account for honestly (in terms of aims, issues, achievements, failures, etc.).
- Add personal contacts to these channels where possible and explore substitutes like storytelling (e.g. on video).
- Add new channels, like apps and gaming.

References

Boeije, H. (2010). Analysis in qualitative research. London: SAGE Publications Ltd.

Corbin, J.M., and Strauss, A.L. (2015). Basics of Qualitative Research, Techniques and Procedures for Developing Grounded Theory. California, Sage Publishing.

Etter, M. (2013). Reasons for low levels of interactivity; (Non-) interactive CSR communication in twitter. Public Relations Review 39, 606-608.

Golob, U., Elving, W.J., Nielsen, A.E., Thomsen, C. and Schultz, F. (2013). CSR communication: quo vadis? Corporate Communications 18(2), 176-192.

Gössling, T. (2011). Corporate social responsibility and business performance. Theories and evidence about organizational responsibility. Cheltenham, Edward Elgar Publishing.

Janssens, M.R.E. and Kaptein, M. (2016). The Ethical Responsibility of Companies Towards Animals: A Study of the Expressed Commitment of the Fortune Global 200. The Journal of Corporate Citizenship 63(3), 42-72.

Luhmann, H. and Theuvsen, L. (2016). Corporate Social Responsibility in Agribusiness: Literature Review and Future Research Directions. Journal of Agricultural and Environmental Ethics 29, 673-696.

Mauser, A.M. (2001). The greening of business. Environmental management and performance evaluation: an empirical study of the Dutch dairy industry. Delft, Eburon Publishing.

Nielsen, A.E. and Thomsen, C. (2009). Investigating CSR communication in SMEs: a case study among Danish middle managers. Business Ethics, A European Review 18(1), 83-93.

Schlegelmilch, B.B. and Pollach, I. (2005). The Perils and Opportunities of Communicating Corporate Ethics. Journal of Marketing Management 21(3-4), 267-290.

Sumner, L.W. (1996). Welfare, Happiness & Ethics. Oxford, Clarendon Press.

Vanhonacker, F. and Verbeke, W. (2014). Public and consumer policies for higher welfare food products: Challenges and opportunities. Journal of Agricultural and Environmental Ethics 27(1), 153-171.

5. Roles and responsibilities in transition? Farmers' ethics in the bio-economy

Z. Robaey, L. Asveld and P. Osseweijer*
Biotechnology & Society, Department of Biotechnology, Delft University of Technology, Van der Maasweg 9, 2629 HZ, Delft, the Netherlands; z.h.robaey@tudelft.nl

Abstract

Concepts of bio-economy and the circular economy come at a time where technological solutions are increasingly needed to move away from a fossil-fuel based economy in the context of climate change and other rising environmental concerns such as waste disposal. These technological solutions rely not only on the use of biotechnology but also on the use of biomass produced by farmers world-wide. The origins of this biomass might become increasingly specialised, for instance through the use of energy crops, or existing edible crops might find to have multiple uses for varied industries. Farmers, then, become an important provider of a resource that might be needed by many, and that is not food or feed. What arable land is used for and how it is used is a question of moral significance (Kline *et al.*, 2017). The role of farmers in the bio-economy is of moral significance. What are the roles and responsibilities of farmers in the context of this transition? What is a farmer's ethics in the bioeconomy? Most studies on farmers seem to focus on behavior and policy incentives impact on behavior. Recently, Meijboom and Stafleu (2016) suggested that entrusting farmers with professional moral autonomy increases the chance of them formulating innovative answers to ethical issues. This ties in with three recent arguments in the ethics of technology. First, Asveld (2016) argues that experimenting and thereby learning (as in moral learning, learning about impacts and institutional learning) is necessary for the bioeconomy. Second, Robaey (2016a) argues that in order to be morally responsible for new technologies they use, farmers must have access to knowledge and develop their epistemic virtues (2016b). Last but not least, the suggestion that the bio-economy and its technologies might improve the life of farmers invites looking into Oosterlaken's (2015) arguments on the capabilities approach and its relation to new technologies. In this paper, we want to articulate the relation between farmer's values, virtues, and capabilities in order to first flesh out the implications of the professional moral autonomy argument and thereby to provide a framework for conceptualizing their changing roles and responsibilities in the context of the bio-economy.

Keywords: virtues, values, capabilities, professional moral autonomy, bioeconomy

Introduction

Concepts of bioeconomy and the circular economy come at a time where technological solutions are increasingly needed to move away from a fossil-fuel based economy. In order to redress our environmental bill and fight climate change new production systems are needed. Currently, our economy produces externalities, also known as pollution and wastes that themselves create new environmental problems, such as carbon emissions responsible for greenhouse gases, and micro-plastics in the food chain whose effects remain unknown.

In the context of the bio-economy, and the circular economy, new sources of materials are needed whose impact on the environment should be less than those of fossil fuels. The first major example of the bio-economy was that of bio-fuels. After a great deal of enthusiasm, significant criticism rose on the food vs fuel debate. The Nuffield Council formulated an ethical framework addressing these issues (2011) and in the meantime, an array of technological developments have emerged for producing biofuels from different sources of biomass, including left-overs of agriculture, and energy crops.

EurSafe 2018 – DOI 10.3920/978-90-8686-869-8_5, © Wageningen Academic Publishers 2018

There are multiple innovations in this field, and multiple stakeholders aiming to set up sustainable biobased value chains, for instance in the marine fuel sector, in the jet fuel sector, etc. These chains have also appeared as new potential output for the forestry sector. Technologies that allow the transformation of biomass therefore provide new avenues for their use. As they are gaining in momentum, questions on what other products could be made from biomass arise such as materials, or nutraceuticals. This is referred to as diversification, and it results in the creation of new supply chains. We speak here of value chains because how these chains are set up and how value is shared is of ethical relevance. In other words, low cost materials are transformed into high-value products through the use of new technologies. Who benefits from them, how and why are object to ethical inquiry. Indeed, the idea of sustainability therefore also applies to the arrangements made along the chains.

These innovations rely most and foremost on use of biomass produced by farmers worldwide. Depending on the technologies used, and the value chains, the origins of the biomass needed might become increasingly specialised. Energy crops could be used solely for the production of biofuels, oil rich plants for the production of nutraceuticals, etc. Edible plants could also be processed to these ends. Farmers, then, become an important provider of a resource that might be needed by many, and that is neither food nor feed. The transformation of low cost material intro high-value products also suggests that farmers may have the opportunity to earn more money on their harvest, and have more choices as to what they want to plant to what ends, and to whom they sell.

Considering today's environmental challenges, what arable land is used for and how it is used is a question of moral significance (Kline *et al.*, 2017). Considering the role of farmers and the choices they will be brought to make regarding what crops to plant, these are of moral significance as well. In this paper, we ask: what are the roles and responsibilities of farmers in the context of this transition? Or in other words, what is a farmer's ethics in the bioeconomy?

What is a farmers' ethics?

The first challenge with this question is that it assumes there is a farmer's ethics. In the sense that farmers are moral agents and practice a certain profession, we can speak of a farmer's ethics. However, a family farmer in the Netherlands, a large-scale farmer in the United States and a subsistence farmer in Malawi have very different realities and face very different choices. Yet, this does not inform us as to what they value as part of their ethics. If anything, their context informs us as to what their capacities are and what these might afford in terms of making choices. This rejoins the crux of the literature on farmer's preferences, or attitudes. While this will have an impact on an individual farmer's ethics, it is not in itself a farmer's ethics.

There are two conflicting trends observed in the literature: agriculture is becoming increasingly technologized, forcing farmers to keep up with developments in order to stay competitive, and at the same time, agriculture and its technologization raise more and more ethical issues (Meijboom and Stafleu, 2016; Buckhart, 1988; Hendrickson and James, 2004). This is actually described as a narrowing of options, contrary to what we suggest in the introduction that the bioeconomy might create more options and leave farmers with difficult choices.

As Meijboom and Stafleu (2016) underline, farmers have value, moral beliefs and a voice to contribute on debates of morally relevant debate around agriculture in society. At the same time, they underline how actors most active in those debates are removed from the reality of farming. So how can we give farmer's ethics a place in the debates, as well as the opportunity to deal with these moral issues? Meijboom and Stafleu (2016) suggest that entrusting farmers with professional moral autonomy (PMA) increases the chance of them formulating innovative answers to ethical issues.

So here's the crux to a farmers' ethics and PMA: being able to make choices and base them on certain elements of justification (could be principles, rights, values, etc.). This implies that a farmer is not in acting from a place of constraint. To do so, Meijboom and Stafleu (2016) suggest a number of institutional solutions after making the case that farming is a profession: a code of ethics, ethics education where agriculture is being taught, ethics as an integral part of farmers' organization. These elements would anchor farmer's moral responsibilities and support their PMA, or in other words, capacity to act in a given agricultural moral dilemma. However, the bioeconomy present several challenges for such institutional solutions like the existence of different institutions in different countries and the related difficulties of coordination. Also, the rapid technological developments create new value chains in which farmers, as producer of biomass, could have the opportunity to participate if they were autonomous.

What are farmer's roles and responsibilities?

The account presented in the previous section presents many encouraging elements. However, these seem still far removed from the reality of the farmer in her context. Also, these require working and reliable institutions, with a high degree of agreement amongst farmers. In addition, in the context of the bioeconomy, and increasing globalization, such a governance structure for farmers' ethics might prove insufficient to deal with the challenges of institutionalizing a farmers' ethics locally and then globally.

In the coming section, we provide a new way of thinking about farmers' ethics and their professional moral autonomy in the context of the bioeconomy. These ideas come as a complement to the institutional suggestions by Meijboom and Stafleu (2016).

Roles

One way of thinking of a farmers's ethics would be through the idea of role responsibility (Abbarno, 1993). In a transition to the bioeconomy, farmers are no longer solely producers of food, or feed, or fibre. They could be producers of biomass for fuel, nutraceuticals, or materials. The notion of role responsibility might be appropriate for medical practitioners, lawyers because their relationship to their patients or clients is well defined. With the advent of possibilities with biomass, what is then the role of farmers? If instead of trying to define roles, and code of ethics, which tend to be fixed in time, we defined a farmers' ethics as the ability to make choices and justify them.

Rights

The Nuffield Council for Bioethics posits six principles where the first one is that biofuels 'should not be produced at the expense of people's essential rights (including access to sufficient food and water, health rights, work rights and land entitlements)'. What about the farmers own access to these rights? What if choosing to produce something else than food, feed or fabric meant the possibility to access other essential rights? Rights based approaches are seen in movements such as 'Via Campesina' (https://viacampesina.org/en/what-are-we-fighting-for/human-rights) but have not been yet observed in the context of the bio-economy. Also, rights-based approaches might be, just as roles, too static to grasp the changing context of the bio-economy.

Values

Another way to think of PMA of farmers is to help identifying their values. Values are important goals shared by society, like freedom, or equality, or sustainability.

Asveld (2016) argues that experimenting and thereby learning (as in moral learning, learning about impacts and institutional learning) is necessary for the bioeconomy. For biofuels, the two competing values of sustainability and economic benefits for farmers first seemed to go hand in hand. Rapidly after, issues of indirect land use change (whereby a change in agricultural production can change the amount of CO_2 released in the atmosphere) made these two values confront each other: it was sustainability or economic benefits. According to Asveld, had these two values been made more explicit, the societal debate would have had the opportunity to address conflicting issues earlier on.

What is the place of farmers' values in this debate? What are these values? It is not always simple to explicitly state what matters to us. Also, values can change through time, or the meaning of a value can change. For instance, sustainability means different things for different people according to their worldviews (see for instance Asveld and Stemerding 2017). Looking at values would perhaps allow a more dynamic understanding of a farmers' ethics.

Epistemic virtues

Having a professional moral autonomy means being able to make choices that will lead to desirable ends in a profession. In a context of transition, there is a need to learn about change. What are the new technologies available? What are the risks and benefits for the professional, and for society at large? If a farmer chooses to use a technology after learning about these effects, who will support a continuous learning? In other work, Robaey (2016a) suggests that responsibility in using new technologies can only happen if learning happens. Also, when a technology is acquired, the responsibility to learn must be shared, supported and encouraged along the value chain (Robaey, 2016b).

Cultivating epistemic virtues and being supported to do would allow farmers to learn about the transition: about their values, about impacts, and about institutions. Epistemic virtues are for instance not being biased in assessing information, being curious, sharing knowledge with the community. They are the qualities of someone who learns. Now the challenge is that different types of farmers have different ways of learning about their land and their crops. Some might have gone through training, or some might have inherited the knowledge via their families.

In order to avoid a situation where rapid technologization leaves farmers with less options and on a race to optimization, learning should be supported within farming communities and also by other actors in the value chain. For this to be successful, learning has to happen in a way that makes sense in their existing practice. Singleton describes how regulations to improve farm safety have actually created more burdens on farmers rather than helped them (2010). This is the kind of situation we want to avoid so epistemic virtues need to be cultivated in a context specific manner.

Capabilities

Oosterlaken (2015) suggests that we can design technologies for capabilities. In the case of seeds, there is an increasing trend over the past 100 years that seeds are developed by seeds breeders and not by farmers. Perhaps with the bioeconomy, and increasing interests in frugal innovations, farmers could redefine their roles. I would like to suggest that they should choose for or develop technologies that best allow them to realize their capabilities. Explicating values, and cultivating epistemic virtues could be a way to realize capabilities. I expand on two of Nussbaum's central capabilities as reported by Oosterlakken (2015).

First, 'Practical reason. Being able to form a conception of the good and to engage in critical reflection about the planning of one's life (this entails protection for the liberty of conscience and religious observance)'. Here, explicating what values mean, and learning about a technology by exercising one's epistemic virtues will allow a farmer to formulate a conception of the good, and thereby make choices for her field, and plan for the next season and maybe more.

Second, 'Other species. Being able to live with concern for and in relation to animals, plants, and the world of nature.' Learning about the impact of a technology can help a farmer understand what is good for her land, and good in terms of keeping up good crop production. For this, agreements have to be made that do not put great pressures on the farming practices, or on the land. This in turn, might allow explicating what values are at stake when for instance scaling up, or changing crop or end product. For instance, if a farmer learns that a certain crop has not been beneficial for her soil and sees a decrease in yield, she might choose to switch to a crop that is better suited. That will mean that her value chain might change if other actors in the value chain do not allow for certain capabilities either via institutional or technical means.

Responsibility

What the responsibilities of farmers are, is a question that remains open. If farmers try to explicate their values, cultivate and exercise their epistemic virtues, and strive to realize their capabilities, then one could speak of their forward-looking responsibility, Robaey 2016a speaks of forward-looking moral responsibility for uncertain risks as 'epistemic-virtue responsibility. What that responsibility should be for could depend on the context: soil stewardship, agricultural biodiversity, consistent yields, or even a responsibility to make income for their families. This question cannot be addressed without considering the responsibilities of other actors in the value chain, and neither can the question of backward-looking moral responsibility.

Conclusion

The focus of this paper is on farmers' ethics and their professional moral autonomy. Instead of aiming for institutionalizing roles and codes, we need a more fluid understanding of professional moral autonomy. This understanding is focused on fundamental aspects of values, virtues, and capabilities, which offer a flexible and dynamic translation and interpretation of these fundamental aspects. This fluidity and dynamism is primordial for learning in the context of a transition and making choices that can be justified as good. In the case of the bioeconomy good can mean different things for citizens, businesses, and the environment.

Acknowledgements

This paper was written under the MVI-NWO funded project 'Inclusive bio-based innovation securing sustainability and supply through farmers'.

References

Abbarno, J.M. (1993). Role Responsibility and Values. The Journal of Value Inquiry 27: 305-316.

Asveld, L. (2016). The Need for Governance by Experimentation: The Case of Biofuels. Science and Engineering Ethics 22, no. 3: 815-830.

Asveld, L. and Stemerding, D. (2017). Algae Oil on Trial. Den Haag: Rathenau Instituut.

Burkhardt, J. (1988). Biotechnology, Ethics, and the Structure of Agriculture. Agriculture and Human Values 5, no. 3: 53-60.

Hendrickson, M. and James, H. S. Jr. (2004). The Ethics of Constrained Choice: How the Industrialization of Agriculture Impacts Farming and Farmer Behavior. Department of Agricultural Economics Working Paper No. AEWP 2004-03. Available at: https://ageconsearch.umn.edu/bitstream/26040/1/wp040003.pdf. Accessed 22 January 2018.

Kline, K. L., Msangi, S., Dale, V. H., Woods, J., Souza, Glaucia M., Osseweijer, P., Clancy, J. S., Hilbert, J. A., Johnson, F.X., McDonnell, P. C. and Mugera, H. K. (2017). Reconciling food security and bioenergy: priorities for action. GCB Bioenergy, 9: 557-576.

Meijboom, F. L. B., and Stafleu, F.R. (2016). Farming Ethics in Practice: From Freedom to Professional Moral Autonomy for Farmers. Agriculture and Human Values 33, no. 2: 403-414.

Nuffield Council on Bioethics (2011). Biofuels: Ethical Issues. Available at: http://nuffieldbioethics.org/wp-content/uploads/2014/07/Biofuels_ethical_issues_FULL-REPORT_0.pdf. Accessed 22 January 2018.

Oosterlaken, I. (2015). Human Capabilities in Design for Values. In: J. van den Hoven, P. E. Vermaas, and I. van de Poel (eds) Handbook of Ethics, Values, and Technological Design: Sources, Theory, Values and Application Domains, 221-50. Dordrecht: Springer Netherlands,

Robaey, Z. (2016a). Gone with the Wind: Conceiving of Moral Responsibility in the Case of GMO Contamination. Science and Engineering Ethics 22, 3: 889-906.

Robaey, Z. (2016b). Transferring Moral Responsibility for Technological Hazards: The Case of GMOs in Agriculture. Journal of Agricultural and Environmental Ethics, 29, 5: 767-786.

Singleton, V. (2010). Good Farming: control or care? In A. Mol, I. Moser, J. Pols (eds.) Care in Practice: On Tinkering in Clinics, Homes and Farms pp. 235-256.

6. Linking professionals in the food chain: a modified ethical matrix to debate zootechnical interventions

H. Grimm[1*], *S. Fromwald*[2] *and S. Springer*[1]
[1]*Unit of Ethics and Human-Animal-Studies, Messerli Research Institute, Vienna, Veterinaerplatz 1, 1210 Vienna, Austria;* [2]*Fromwald Consulting, Castellezgasse 27/27, 1020 Vienna, Austria; herwig.grimm@vetmeduni.ac.at*

Abstract

Zootechnical interventions, such as castrating and tail-docking in piglets, or dehorning calves and goat kids, have given rise to an important societal debate. This has led to an increasing pressure to adjust the legal framework regarding these interventions. In Austria, the Federal Ministry of Health (FMH) has the responsibility to draft the relevant legal regulations for farmed animals, which are then enforced by the parliament. In preparation for this, the FMH approached the Messerli Research Institute to organize a participatory process for nominated stakeholder representatives and researchers to describe pros and cons of various alternatives to the aforementioned zootechnical interventions. Based on Ben Mepham's (1996) idea of the ethical matrix, a modified matrix (MM) was used developed to analyse and compare possible alternatives to the currently performed interventions. The modified matrix comprised four domains: (1) impacts on animal welfare; (2) impacts on the farmer/economy (3) issues related to implementing a particular alternatives; (4) 'other aspects.' In the first phase of the project the matrix was developed and embedded into an online questionnaire that was sent out to nominated participants. In a period of two weeks, they had the possibility to fill out the modified matrix based on their expertise and specialised knowledge. The content of the filled out matrices provided the basis of discussion for the subsequent face-to-face meetings in the second phase. By means of the modified matrix, a systematic and consensual description of alternative measures could be reached, even though the background of stakeholder representatives varied from animal protection groups, chamber of agriculture, chamber of veterinarians, animal breeders, retailers, consumer protection groups, industry representatives to veterinary scientists. In this paper we will (1) describe the background of the project; (2) elaborate on the development of the adopted matrix and methodology of the workshops; (3) demonstrate its usefulness to structure debates; (4) outline the results of the workshops. In summary, this ethical tool provides a well-structured and promising way to consensually describe the normative dimensions of the zootechnical interventions in question.

Keywords: ethical tools, participatory processes, animal welfare, economy, implementation

Introduction

Practices such as castrating or tail-docking in piglets, dehorning calves and goat kids, have given rise to an important societal debate during the last years in Austria and other countries. Although these painful interventions are often carried out systematically and on a regular basis, citizens do often not know that these zootechnical interventions take place on farms. Shifting light to these practices triggered campaigns and debates, which has led to an increasing pressure to adjust the legal framework. E.g. in Austria a campaign has been launched under the title 'Who has the balls to stop this?', picturing a piglet to be castrated looking at the recipient and a bloody scalpel lying under it. Such campaigns not only raise awareness but also increase the pressure for improvement and change. Changes in the agricultural sector challenge value decisions of the past, which were often made with less consideration of animal interests. Aiming e.g. for more animal welfare in the agricultural sector threatens established structures

EurSafe 2018 – DOI 10.3920/978-90-8686-869-8_6, © Wageningen Academic Publishers 2018

and brings about a new distribution of winners and losers. Hence, conflicts among affected parties like representatives of farmers and animal protection groups is typically linked to these debates.

Against the background of the increased pressure to change the Austrian legislation and the risk of conflict amongst stakeholders, the Federal Ministry of Health (FMH) approached the Messerli Research Institute to organize a participatory process for nominated stakeholder representatives and researchers to describe pros and cons of various alternatives to the aforementioned zootechnical interventions. Since the FMH is the responsible ministry to draft the legal regulations for farmed animals, which are then potentially enforced by the parliament, the task was to document the advantages and challenges of different possible alternatives to the current zootechnical interventions. To analyse the various alternatives, the ethical matrix, originally developed by Ben Mepham in 1996, was used and turned it into a modified matrix (MM). This matrix was embedded into an online questionnaire that was sent to the nominees in the first phase of the project to gather the information provided by stakeholder groups in order to prepare for face-to-face meetings in the second phase.

Method and materials

Phase 1: The first phase of the project started at the end of 2014, when representatives of different stakeholder groups were invited to contribute to the process. A top down approach was applied. The FMH invited various stakeholder organisations in a first step for a meeting on zootechnical interventions. At this meeting, the project idea was presented from the representative of the FMH and the project leader (University Professor for Ethics and Human-Animal Studies). As a main point of the presentation, it was made clear that the project does not aim for a final decision on the best solution but to provide a solid basis for political decision-making and a consequent reform of the given legislation. Representatives from the chamber of agriculture, the chamber of veterinarians, the economic chamber, the chamber for labour, the umbrella organisation for Austrian animal protection groups, the representative of the Austrian animal welfare ombudspersons, farmer organisations, retailers, and scientists were invited. After agreeing on the basic terms of the project, they were asked to nominate three persons each. Those three nominees would be each a member of one working group. In total, three groups were convened based on the species to be dealt with in the process: pigs, cattle, goats. The group dealing with pigs worked on two topics: castrating and tail-docking piglets. The second group focussed on cattle, and on debudding calves in particular. The third group on goats scrutinizing the issue of debudding goat kids. A top down nomination was applied where the invited representatives nominated three persons with relevant experience in their organisation. This nomination strategy was also chosen to establish groups not bigger than 12 persons in order to be able to work efficiently.

Besides the participants, an organizational team was brought in place, consisting of the project leader (University Professor for Ethics and Human-Animal Studies) and his scientific assistant, the head of the Austrian Tierschutzrat (advising body of the FMH in animal protection issues), one assistant person from the FMH.

The nominees were informed that the project consists of two phases: a questionnaire based online-phase including the MM and face-to-face meetings where participants had to attend workshops in Vienna at the MRI. The questionnaire was sent two weeks before the first face-to-face meetings. The deadline to return the filled out questionnaire was three days before the workshop. The questionnaire was structured in the following way:

I Information on background and aim of the project
II Questions to be answered in preparation to the face-to-face meetings
II.i Case: Which zootechnical intervention will you be working on?

II.ii Problem formulation: which problems are solved with the current zootechnical intervention? Background: current interventions solve a particular problem. Alternatives have to solve the problem in one or the other way as well.
II.iii Search for alternatives: please list alternatives to the current practice without going into details. Background: only if thinkable alternatives are explicit, the room for possible solutions and negotiations becomes clear.
II.iv Matrix: analysis and comparison of alternatives: which advantages and disadvantages are related to the alternatives? Please use the matrix below to analyse the alternatives with regards to animal welfare, economic issues, and implementation.
Background: the pros and cons of the various alternatives should be made explicit and challenges of implementation identified.
II.v Example of a filled out matrix for illustrational reasons.
II.vi Blank matrix to be filled out (Figure 1).

Description of the MM

In order to tailor the original ethical matrix to the task, a workshop was carried out to develop the MM. It became clear that Mepham's matrix would not easily serve the task. Therefore, the matrix was designed to address the most important challenging dimensions that cover normative issues and the alternatives to be compared. To fit this into one matrix, the structure described in Figure 1 has been developed.

The provided matrix allowed comparing alternative interventions with regard to the four domains: (1) animal welfare/animal protection; (2) economic effects/effects on the farmer; (3) issues related to implementing alternatives. The last domain (4) 'other aspects' was also integrated to allow for a broad and extensive description and to provide the opportunity to mention aspects, which did not fall under the domain 1-3). For instance, the wider public's perception of a particular alternative or detailed veterinary aspects. The MM was designed in such a way that the nominees could fill in the open fields of the MM directly and save it as a PDF file and send it back to the project organizers. This step resulted in matrices from nominees that compared the pros and cons of possible alternatives identified by them. In preparation to the face-to face meetings, all information sent by the nominees in the matrices was

Matrix	Zootechnical intervention: Solution to the following problem:			
	Effects on animal welfare/ animal protection	Economic effects/ effects on the farmer	Aspects of implementation	Others
Current intervention				
Alternative 1:				
Alternative 2:				
Alternative 3:				
...				

Figure 1. Modified ethical matrix to compare alternatives on zootechnical interventions.

compiled in one matrix. This comprehensive matrix had the identical structure but listed all alternatives with all the information sent by the nominees in one document.

In general, participants had no difficulty in dealing with the questionnaire and matrix. Quite the contrary, most of the matrices were well developed and e-mailed to the project leaders in time so that a fully developed, comprehensive matrix could be prepared before the face-to-face meetings in phase 2.

Phase 2: The second phase of the project aimed at a matrix that only contained information on which all stakeholders could consent. To achieve this goal, a very pragmatic approach was used: All contents of the boxes in the comprehensive matrix were coloured red and discussed until the point at which consent on the validity, correctness and relevance of the presented information was obtained. If consensus was achieved, the passage under debate was turned into black. In general, no time limit was set for the discussions. It turned out that for each intervention three four-hours workshops were necessary. Although such a massive time investment was neither intended nor foreseeable, the participants of all groups agreed to continue and proceed in the second and even the third workshop until the goal was achieved.

The responsibility to organize and hold the workshops successfully was within the organizing team (consisting of four people) with the following roles and functions:

i *Chairing the workshop*. The project leader chaired the workshops and led a strict list of speakers.
ii *Documentation of results*. A scientific assistant of the project leader documented results in the matrix under the guidance of the chair. At all times the comprehensive matrix was presented via Power Point on the screen. Hence, all changes done in the matrix were transparent to the group.
iii *Documentation of process*. One person from the FMH took notes throughout the entire workshop.
iv *Feedback on the process*. In order not to lose track and keep the process focused and efficient, the chair of was given feedbacks by the head of the Austrian Tierschutzrat who also participated as an observer of the workshop. She and the chair had a short update before each workshop, feedback rounds during breaks and after the workshops on the process.

Since the participants had very different backgrounds – e.g. chamber of agriculture and animal rights groups – the process was at risk to turn into conflict constantly. However, under the strict regime of the chair and by means of the structured MM as a working basis, the discussions were intense but except once where a participant threatened to leave the room, highly effective and constructive.

Each group reached the intended result, which was a black and white matrix that contained consensual information on pros and cons of the various alternatives. Since the relevant stakeholders and scientists were all on the table, not only information on veterinary treatments, animal welfare issues, monetary costs, management challenges, legal challenges, etc. could be elaborated in depth, but also scientific literature on the alternatives was referenced and integrated into the matrices. The final version of the four matrices can be accessed via a web-link (Matrices, 2018).

Before the final version was sent to the FMH, all participants had the chance to add final comments. Finally, four matrices were sent to the ministry. The matrix on castrating male piglets comprised the detailed description of six alternatives, the matrix on debudding calves described eight alternatives, and the matrix on debudding goat kids elaborated two alternatives. Although the matrix on tail-docking in piglets compared the standard intervention with two alternatives as well, it was highlighted in the preamble of the MM that tail-docking of piglets carried out systematically is illegal.

Conclusions

The modified matrix helped to reach a systematic and consensual description of alternatives for current standard zootechnical interventions. Despite the participation of representatives of stakeholders as diverse as animal protection groups, chamber of agriculture, chamber of veterinarians, animal breeders, retailers, consumer protection groups, industry representatives, and scientists, the information sent to the FMH was consensual. The MM linked the participants with different backgrounds in the best sense and allowed for a constructive mined dialogue. In summary, the MM provided not only a well-structured and promising way to consensually describe normative dimensions of the zootechnical interventions but rather offered a safe space to voice all aspects and factors, which were considered of relevance by the participants. The information provided to the FMH in form of the final matrices allowed for well-informed political decision-making based on the workshop results.

Acknowledgements

We would like to thank all participants and contributors of this project for their support, their positive attitude, and their time investment.

References

Matrices (2018). Available at: https://www.vetmeduni.ac.at/de/messerli/forschung/forschung-ethik/projekte/eingriffe-an-landwirtschaftlich-genutzten-tieren-ein-diskussionsprozess. Accessed 20 January 2018.

Mepham, B. (1996). Ethical analysis of food biotechnologies: an evaluative framework. In: Mepham, B. Foodethics. Routledge.

7. Negotiating welfare in daily farm practice – how employees on Danish farms perceive animal welfare

I. Anneberg[1] and P. Sandøe[2]*
[1]Department of Animal Science, Aarhus University, Blichers Alle 20, 8830 Tjele, Denmark; [2]Department of Food and Resource Economics and Department of Veterinary and Animal Sciences, University of Copenhagen, Rolighedsvej 25, 1958 Frederiksberg C, Denmark; inger.anneberg@anis.au.dk

Abstract

Little is known about how employees on husbandry farms perceive animal welfare and about factors influencing the relationship between them and the animals in daily work. Today, Danish farms are mainly family-owned, and the employees are often of other nationalities, and one third are unskilled. The aim of this paper is to document how employees perceive animal welfare and to discuss how they deal with ethical assumptions in daily work. The paper reports the findings of qualitative interviews with 23 employees from five Danish farms (mink, dairy and pig production). Employees emphasise physical aspects of animal welfare relating to feed, water and health. However, some employees described naturalness, which is known to be of importance to the public, as an area that could be negotiated. Some issues, like pain, were also negotiated, especially pain imposed on the animals by the employees themselves. Getting used to impose pain in daily work was described as a working condition in the job which one had to accept. A negative relationship among employees and managers as well as lack of credit also related to animal welfare and were described as creating a worse situation for the animals.

Keywords: interviews, perception, farm workers, ethical assumption

Introduction

Research has shown that farmers tend to equate animal welfare with basic health and access to necessities such as food and water (Te Velde *et al.*, 2002; Vanhonacker *et al.*, 2008). However, little is known about how the employees on farms with livestock perceive animal welfare and about what factors influence the relationship between the employees and the animals in daily work. Therefore, the aim of this study is to document how employees perceive animal welfare and to analyse how they deal with ethical issues in daily work such as animal welfare versus profit efficiency.

Danish farms today are mainly family owned. The Danish farms, the estimated number is around 36,000 in total for all types of farms, have around 35,000 employees (Anonymous 1, 2, 2015). Today, the producers often hire employees of different nationalities of which one third comes from Eastern Europe and one third is unskilled (Anneberg and Sørensen, 2016). No specific courses on animal welfare are needed to be employed on farms with animals. All skilled Danish employees are educated at Danish agricultural colleges, and some have further education in economy and management. Very little is known about the educational background of employees from Eastern Europe. Previous research found that students at agricultural colleges perceive good animal welfare as a means of achieving a profitable production (Lassen *et al.*, 2016).

We formulated the following research question to guide the research process: How do employees on Danish husbandry farms perceive animal welfare, and how do social relations on the farms affect interest for animal welfare among employees?

Svenja Springer and Herwig Grimm (eds.) ***Professionals in food chains***
EurSafe 2018 – DOI 10.3920/978-90-8686-869-8_7, © Wageningen Academic Publishers 2018

Method

The paper draws on a qualitative study of employees (n=23) on five farms in Denmark, covering mink, dairy cattle and pigs. The five farms were chosen to ensure diversity regarding size, employees (nationality, gender) and animal species. The study comprised interviews and observations on the five farms. The aim of the observation was to obtain a contextual understanding of the employees' perception of animal welfare. All interviews followed an interview guide, and interviews with employees from Eastern Europe were mainly done in English if the employees wanted that. Use of interpreter was deselected for practical reasons. The guide included a section on the employee's background to work with animals and with a particular focus on the welfare of pigs, cattle and mink. The guide also included a ranking exercise concerning the so-called five freedoms.

All interviews were recorded and subsequently transcribed verbatim. The subsequent analysis included a thematic coding followed by a meaning-condensation where the essence was extracted from the coded sections. Finally, patterns in the condensations were identified and described.

Results

The observations showed that animal welfare was a well-known, often used concept among the employees. It was especially used in relation to production results and health, but also when discussions arose about how the world outside the farm understood farming and in relation to the demands from the market. Observations also showed that everyday dilemmas about animal welfare would occur in certain situations, e.g. whether to use stomach tubes to deliver raw milk to new-born calves or whether to use extra time to give straw to pigs.

When elaborating on these points of views and arguing for the need for animal welfare, the employees referred to three different sets of justifications: (1) Animal welfare was backed by concerns about production and health and could be negotiated – specially in relation to naturalness. (2) Employees' view on animal welfare (e.g. imposing pain on animals) was affected by working conditions on which they had no influence. (3) Animal welfare was connected to the working conditions on the farm, e.g. a negative relationship between workers and managers and a lack of appreciation could create a worse situation for the animals.

Production as central – and naturalness can be negotiation

One of the employees mentioned fodder and water as *cornerstones* – this is what everything builds on when it comes to animal welfare. To most of the employees, this was central. Fodder, water and a safe environment to live in, keeping the animal healthy, were seen as more important to production results than fulfilling the behavioural need of the animals. Fulfilling these needs, e.g. by having loose-housed sows, could affect the economy of the farm in a negative direction.

> Having the dry sows loose during oestrus increases the risk of having to return to oestrus. Therefore, it is about a financial risk. It has to be in balance – and after all, keeping them locked in the box in that unit is not for a long time. (...) Where I work now, I can see they do enjoy to be loose in deep bedding in that period, so I guess it matters, but where I worked before, they were not loose during oestrus, and I never thought about it (16).

Having animals in captivity, like minks, was seen as safer for the animals. Not being stressed, living in comfort and getting fodder was associated with a precondition for production of the fur but also with a good life for the animal.

> They (the minks) are not stressed about anything. They have everything they need – food, toys, a shelf (...) and they can still express themselves. If we do see stereotypies, it relates to stress, and then we do not keep them. They are destroyed. It is seldom, but we do see minks that cannot handle this life. Like sometimes, people cannot handle it. Of course, it is a big change for them compared to nature where they can hunt. Hunting is the only thing they cannot do in a cage. However, they can run. It is like humans doing fitness, when they jump around. This is how I see it; others may have a different opinion (19).

Negotiating advantages and disadvantages concerning animals in relation to natural living was also seen on a dairy farm, where an employee described how the dairy cows walked differently in the indoor environment than on pasture, and that his vet had told him that none of the cows in his stable walked normally. The employee mentioned in the interview that nothing around the dairy cows was 'normal' any longer. He did not think that a natural life for a cow was a relevant discussion.

> Because, well, it is not normal today that they give 70 kilos of milk, is it? It is not normal but happens because we have boosted them to do this. With our expertise. It has been a long time ago since a cow was like an aurochs; we have passed that long ago. I do not think that we should look at that at all (10).

Imposing pain or distress – getting used to it

Although viewing animals in the context of production was a dominant perspective, employees also brought forward other aspects, e.g. the importance of preventing suffering. Some described handling of animals that would cause pain or suffering or shouting at them as very negative. However, when it came to pain, employees expressed the dilemma that they themselves had to impose pain, e.g. when castrating or tail-docking piglets. Castration was a job that employees on pig farms disliked but also talked about as something that they had to get used to. Thus, one will have to overcome one's objections to stay in this job.

Furthermore, employees did not always agree about whether this action was painful or not. A young male employee was sure it was extremely painful and related this to his own gender, but he also felt that he simply had to cope with this part of his job, even though he was convinced that pigs could feel pain. A young female employee related this unpleasant part of her job to the fact that consumers in Denmark did not want the risk of getting smelly boar-meat, and it was a part of her job to prevent this. In addition, she had been told that piglets do not feel pain the first week. An experienced employee had heard a similar explanation in an agricultural college but still felt that modifying the animals was wrong.

> I feel that there is a reason why they are created the way they are. In addition, it must be essential to change them as little as possible. And I do feel that it is annoying to take part in this (13).

The same employee also said that discussing this dilemma about changing or not changing the animals was not an issue on the farm where he worked, as he felt he could not change the situation anyway. Being unable to manoeuvre within the framework conditions made employees accept imposing pain or discomfort or changing the animals as a daily term in their job. In dairy farming, a dilemma was seen when employees tube-fed the new-born calves with raw milk on a regular basis. Some of the employees disliked this procedure and would rather use a bottle or let the calf suck the cow, but working conditions, e.g. an expectation to be effective, prevented this.

> It has something to do with being effective. When we get many calves, it is really hard. They must have their raw milk within 1-2 hours, and that would take us 3-4 hours for each calf to make them drink what they need. Using the tube is much more effective (...) I would prefer to see the cow and calf together, but it is not up to me to decide. The owner is the producer. Sucking the cow is what I see as normal, but on a farm like this you can drop this idea (...) it is nice for the calf but ... (2)

Working conditions and animal welfare

Being acknowledged by your boss or by your colleagues and working in an environment with a positive atmosphere may affect animal welfare. This came out as a theme among the employees in several descriptions. Some employees mention the importance of being able to make a mistake without being criticised by the boss.

> It is an important lesson that I have learned where I work now. That it is not bad to make a mistake. It is bad if you do not learn from it. In the country where I came from, a mistake could really be the end of the world. That is worse, because it makes you afraid, and then you start making more mistakes (2).

To work without feeling acknowledged was mentioned several time as a problem that could affect the animals.

> (...) Perhaps you get frustrated. You feel that you work without being acknowledged. You get your payment but this little extra... You really want to be acknowledged when you contribute to improving things (at the farm). In addition, if you are in a conflict with your boss, then you might feel that the boss is not the best to target. So instead, you target the animals, you go and spoil a tractor, or (...) I think humans react like this. Frustrations will grow, and all you want is to go home, sleep, eat and do something completely different (20).

Furthermore, a bad social climate could affect animal welfare directly, and the animals would respond to the mood of the employees.

> In case there is lack of confidence between me and a colleague or between me and the management, it could be a real catastrophe. In case you cannot trust each other getting the job done in the way you agreed, then things might start not to be done at all. You might then think: well, she has to find out herself what is wrong with that calf! Therefore, instead of saving the calf, it might end up dying. That of course influences animal welfare (10).

> If you are in bad mood, it affects the animals, but if you feel well, like the place where I am now, you also share your good feelings and your behaviour with the animals. The animals will feel it immediately if you are upset or angry, even though you do not talk about it and are just quiet. They sense it, and they get worried (11).

Discussion

For the employees, a key argument in favor of good animal welfare is that it is a way of achieving a profitable production. A safe environment, fodder, water and health are central elements. As such, they are in line with the producers' understanding of animal welfare (Vanhonacker *et al.*, 2008). In addition, when it comes to natural living, they also hold an instrumental approach, negotiating behavioral needs of the animals, seeing them as important when it relates to production, but of less interest if it increases the costs. This is in contrast to the views of the public, where many justify animal welfare with reference to the animal itself and emphasises freedom, space and naturalness (Lassen *et al.*, 2006).

One of the major differences between farmers and employees when it comes to the perception of animal welfare relates to the fact that the employees do not themselves define the terms of their work. Some of them had ethical concerns about imposing pain on the animals or changing the body of the animal (tube feeding, tail docking, castration). They found it unpleasant to do but also (to a certain extent) painful and unpleasant for the animal, and they had difficulties getting used to it, though they argued that this was what they had to do. Furthermore, one of them talked about his unwillingness to change the body of the animal through mutilation such as taildocking.

In this situation, a focus on animal welfare defined in terms of feelings seems to be in conflict with a view that emphasises natural living. However, at the same time, the employees did not find it possible to discuss these dilemmas during work. Not having any influence when it comes to changing the production system was one argument for not bringing the ethical dilemmas into the open, for instance in a discussion with colleagues or the farm owner. Another was giving up on the discussion, because 'this is what the market is asking for', e.g. castration.

Finally, employees argued that the job on the farm was framed by a need to be effective, because working conditions demanded this. This was seen in relation to tube feeding of calves, an act that is not allowed to do routinely according to Danish legislation. Nevertheless, it has become the normal way of handling newborn calves in some large dairy productions. For some employees, the distress of such a dilemma would not be brought into the open, because they left it to the owner to decide.

A big issue in Danish farming today is difficulties in recruiting employees, whether from Denmark or from abroad. Especially the pig production has problems hiring employees but also mink and dairy productions are challenged. Under these circumstances, the need for an open dialogue among the present employees about daily dilemmas could be important. Recruiting young people to do farm work today has to reflect the ongoing societal discussions about animal welfare – a discussion that should not be left to the surrounding society.

Furthermore, the results of the study show that good leadership is of great importance for the employees. Not feeling acknowledged by the boss may lead directly to lack of welfare among the farm animals. Here, Danish agriculture also has a challenge as many farm owners have no education in management. The results also show a combination of employees who are not used to discuss the dilemmas they see in their daily work and a leadership that may not realise the importance of a dialogue and a positive motivation. This could lead to an even bigger gap between what goes on at farm level and the expectations of the surrounding society than we see today.

Acknowledgement

We would like to thank the anonymous employees at the five farms for their participation in this project and Lotte Hansen from Aarhus University for proofreading.

References

Anonymous 1 (2015). Statistic Denmark, Available at: https://www.dst.dk/da/Statistik/nyt/NytHtml?cid=21902. Accessed at 25 January 2018.

Anonymous 2 (2015). Statistic Denmark, Available at: https://www.statistikbanken.dk/statbank5a/SelectVarVal/Define.asp?MainTable=BDF7&PLanguage=0&PXSId=0&wsid=cftree. Accessed at 25 January 2018.

Anneberg, I., Sørensen J.T. (2016). Medarbejderne i dansk husdyrbrug. Hvem er de, og hvad er deres rolle i sikring af god dyrevelfærd? DCA Rapport nr. 080, Aarhus Universitet.

Lassen, J., Sandøe P., and Anneberg, I. (2016). For the sake of production. How agricultural colleges shape students' views on animal welfare. In: Food futures: etchics, science and culture (pp.126-136) Wageningen Academic Publishers.

Lassen, J., Sandøe, P. and Forkman, B. (2006). Happy pigs are dirty! – conflicting perspectives on animal welfare. Livestock Science, 103(3):221-230.

Te Velde, H.T., Aarts, N., Van Woerkum, C. (2002). Dealing with ambivalence: farmers' and consumers' perceptions of animal welfare in livestock breeding. Journal of Agricultural and Environmental Ethics. Agric. Environ. Ethics 15, 203-219.

Vanhonacker, F., Verbeke, W., Van Poucke, E., Tuyttens, F. (2008). Do citizens and farmers interpret the concept of farm animal welfare differently? Livestock Science. 122, 126-136.

8. How to use theory to elucidate values rather than pigeonhole professionals in agriculture?

O. Shortall
Social, Economic and Geographical Sciences, The James Hutton Institute, Craigiebuckler, Aberdeen AB15 8QH, Scotland, United Kingdom; orla.shortall@hutton.ac.uk

Abstract

Researchers within agricultural ethics can contribute to a better understanding of controversies in agriculture by exploring debates in terms of paradigms of agriculture, such as industrial and alternative agriculture. Within the industrial paradigm the purpose of agriculture is seen as producing commodities as efficiently as possible, whereas in the alternative paradigm agriculture is seen as having a wider significance for society beyond commodity production. The use of theory can help elucidate values underpinning debates. However, when does theorising different positions in terms of paradigms of agriculture and their theoretical lineage serve to pigeonhole professionals in agriculture rather than open up the complexity of values involved in debates? Professionals, including farmers and stakeholders within the agricultural sector, often hold a multiplicity of values about agriculture which may get lost if we theorise agriculture in terms of overarching paradigms. This paper explores this question in relation to indoor dairy farming in the UK. Indoor dairy farming is a growing though controversial trend in the UK. It can be seen to be within the industrial paradigm of agriculture as it is driven by economies of scale and greater use of scientific and technological innovation. However, it is often defended with appeal to the importance of the farmer-animal relationship which is framed as more important to animal welfare than the type of production system. This argument appeals the importance of care within livestock farming and traditional norms of good stock keeping which, in theory, are not an obvious fit within industrial agriculture. This paper argues that understanding debates in terms of paradigms of agriculture can provide theoretical insights which help elucidate values for professionals involved and wider stakeholders. But that such theorising must be responsive to the complexity of professionals' different values and identities.

Keywords: theorising agriculture, dairy farming, values, intensification

Introduction

Agricultural bioethics can contribute to debates about controversies in agriculture by helping to elucidate values underpinning different points of view. Debates based on empirical arguments about facts can be difficult to resolve if the facts are underpinned by incommensurate values and perspectives (Sarewitz, 2004). Exploring the theoretical origins and implications of different viewpoints can shed light on why people disagree and where potential commonalities lie (Thompson, 2012).

One way to explore debates is using different philosophies or paradigms of agriculture. A philosophy of agriculture can be understood as a vision for how the production, distribution and consumption of agricultural produce should be organised (Thompson, 2012). Agriculture has been theorised in numerous ways (Wilson, 2007) and this paper will focus on the distinction between industrial and alternative philosophies of agriculture. The industrial agriculture philosophy is the view that agriculture is another sector of the industrial economy whose aim is to produce commodities as efficiently as possible, through scientific and technological innovation (Thompson, 2010). This is tied up with the philosophy of productivism: that the fundamental aim of agriculture is to produce more. At the other end of the philosophical spectrum is the position of alternative agriculture which comes in different forms, but

Svenja Springer and Herwig Grimm (eds.) ***Professionals in food chains***
EurSafe 2018 – DOI 10.3920/978-90-8686-869-8_8, © Wageningen Academic Publishers 2018

holds that agriculture has a greater significance for society beyond its material contribution. This could be its role in shaping the countryside, environment, character of a nation or through its visceral role in connecting our bodies to the land through eating. This means that we should bring other values and considerations to bear on agriculture such as care for the environment and animals, tradition, aesthetics and forbearance (Thompson, 2010).

Debates about agriculture can take place without the participants being aware that they are drawing on different philosophies that involve a particular theoretical and historic lineage. However, the reality of agricultural production on the ground is more complicated than theoretical distinctions between different philosophies might lead us to belief (Shortall *et al.*, 2015). This paper will ask the question when theorising controversies in terms of philosophies is helpful and when it could serve to entrench positions and polarise views further. It will explore this questions in relation to indoor dairy farming in the UK. Indoor dairy farming is an interesting case study to explore because it is a relatively new and highly contested phenomenon in the UK. It raises question about the role of agricultural bioethics research in exploring both rhetoric and practice around agriculture and how theory can be used in relation to discrepancies between these.

This paper is part of a three years fellowship project exploring indoor and outdoor dairy farming in the UK and Ireland. The research presented in this paper is based on preliminary analysis of documents and web resources about the dairy sector from non-governmental organisations (NGOs), academics, industry and government.

This paper will first provide some background information on indoor dairy farming in the UK, followed by analysis of indoor dairy farming in terms of philosophies of agriculture, highlighting the contradictions and uncertainties this gives rise to. It will conclude by reflecting on next steps for research in this area.

The dairy sector in the UK and philosophies of agriculture

Animal production in industrialised countries has been undergoing a process of intensification and industrialisation for decades. For the dairy sectors in countries such as the USA and Saudi Arabia this process is far advanced, with the majority of production taking place in large, indoor housing units (Barkema *et al.*, 2015). The UK in contrast is characterised by a diversity of production systems and visions about the future of the sector. Around a third of dairy farmers in the UK use the traditional system of summer grazing and winter housing, whereas the rest are moving cows indoors for more of the year (March *et al.*, 2014). In indoor systems the animals are housed all year round, rather than grazing during the summer months as in traditional systems. The trend towards more indoor housing is in response to financial difficulties dairy farmers have been under due to sustained low milk prices (March *et al.* 2014).

Indoor dairy farming in the UK can undoubtedly be seen as to be within the industrial paradigm of agriculture. The primary motivation behind housing cows for more of the year is a productivist one: to produce more milk. Cows are kept indoors for longer periods so they can consume more high energy concentrate which results in increased yields. Productivism is pursued within industrial agriculture through economies of scale and involves the consolidation of production on fewer individual farms (Thompson, 1995). The trend towards indoor farming is also linked to a trend towards bigger dairy farms: when farms reach a certain size it becomes logistically very difficult to keep cows outdoors on grass and bring them in twice or three times a day for milking in a central parlour. Industrial agriculture is also seen to be bound up with scientific and technological innovation (Vanloqueren and Baret, 2009). Indoor dairy farming involves and necessitates more high tech production systems compared to smaller, mixed indoor and outdoor systems. The introduction of robotic milking which requires minimal human involvement has also allowed the increase in size of dairy farms and trend towards more indoor housing. Industrial agriculture involves the globalisation of agricultural supply chains, which can also be seen in systems that involve more indoor housing. In traditional summer grazing and winter housing dairy systems the cows eat grass during the summer months and silage made from grassland on the farm, supplemented with other feed sources in the winter. Indoor systems involve a greater proportion of feed sources which are not produced on the farm or may be imported from other countries. In this way indoor dairy systems move closer to the concentrated animal feeding operations (CAFO) which dominate the pig and poultry sectors in European agriculture and away from farming systems which use the land base surrounding the operation.

Indoor dairy farming has not been without controversy and opposition in the UK. Much of the opposition can be seen to be underpinned by alternative philosophies of agriculture. Campaigning NGOs have attacked indoor dairy farming in the UK using the language of 'intensive' and 'factory' farming (World Animal Protection, 2016). Criticisms of indoor dairy farming centres on the impacts on animal welfare and the environment (Viva, 2016). The intensification of agriculture is seen as pushing animals to produce at an unsustainable level which impacts their health and causes them suffering as well as restricting their natural behaviour (CIWF, n.d.). A report on the future of dairy farming in the Republic of Ireland also points out that intensive, indoor systems involve considerable investment and claims that experiences in other countries show that this can result in increased stress for the farmer and mental health problems as their lifestyle changes and their farm carries a large amount of debt (Hurley and Murphy, n.d.). Thus the root of problems with indoor dairy farming is seen to lie in the philosophy of industrial agriculture which puts productivity before other considerations such as animal welfare and farmer lifestyle, dispenses with tradition and heritage and weakens the link between agriculture and nature.

Some defend the future of indoor dairy farming through a modern productivist justification: the need to feed 9 billion people by 2050 (NFU, 2010). Many within the dairy industry also contest claims that indoor dairy farming is bad for animal welfare, interestingly, using arguments that involve features of the alternative philosophy of agriculture. Commentators argue that the type of system is not the main determinant of animal welfare, but rather the care and expertise farmers bring to animal husbandry can ensure good animal welfare in any system (Farming UK, 2016; NFU, 2010, 2016). Previous research has shown that care and skill in animal husbandry are part of traditional notions of what makes a 'good farmer' (Burton *et al.*, 2012; Shortall *et al.*, 2017). This includes having an in depth knowledge of the health and behaviour of one's own animals; being able to judge animal health and wellbeing by eye; and experience in working with and treating animals (Shortall *et al.*, 2017). These arguments can be seen to counter claims that indoor systems alienate farmers and animals from each other, from tradition and from nature by appealing to the importance of the human element in the system and the value of care in the relationship between farm workers and animals.

Theorising indoor dairy farming

We can now ask how researchers could use theory to elucidate debates such as this one rather than pigeon hole professionals involved. The aim is to present preliminary findings about debates and also questions for reflection in pursuing this area of research.

We could theorise the defence of indoor dairy farming described above as involving a crossover of industrial and alternative philosophies of agriculture. Previous research has also shown that distinctions between alternative and industrial systems are not easy to draw in practice (Sonnino and Marsden, 2006). Farmers' identities are complex and those involved in the agricultural sector are likely to hold many, potentially competing values around how food should be produced. Those who defend indoor dairy farming move the argument away from defending the more industrial features of indoor systems: the size of the farm, the disconnection with the surrounding land and the technology used on the farm, to focus on the farmer-animal relationship as the determining factor in the animals' health and wellbeing. Thus this argument suggests that those who defend indoor systems are resisting these systems being painted as a step change in the intensification and industrialisation of dairy farming in the UK. As long as dairy farming still involves farmers there will be continuity with more 'traditional' systems.

As researchers involved in this arena we risk losing credibility if we do not reflect the nuance and complexity in these debates. Those who defend indoor dairy farming state that criticisms of it in terms of industrial, factor farming overlook this complexity. Even if we can bring an understanding of the historical and contemporary context around debates that take place in terms of industrial and alternative philosophies, will this help if proponents within the debate are trying to move the discussion beyond these binary ideologies?

However, we could also contend that our role as researchers is not only to reflect and contextualise claims with theory but to interrogate and question these claims. On one hand arguments that indoor dairy farming still involves a relationship of care between the farmer and the animal can be seen to reflect the complexity of farming on the ground. It can also be seen as a rhetorical way to distance indoor dairy farming from negative associations with industrial, factory farming which make it unpopular with the general public. These claims can be countered with the point that large, indoor dairy farms involve a different farmer-animal relationship than smaller, traditional farms because of the scale of the farm and types of technology involved (Butler and Holloway, 2015). Dairy farmers and stock keepers interact with the animals at each milking and keep dairy cows on the farm for several pregnancy cycles. This means they can become familiar with animals as individuals and assess their health and wellbeing based on pre-existing knowledge of each animal (Burton *et al.*, 2012). On larger, indoor farms, in contrast, individual knowledge of each animal is not possible because of the number of cows, and health and wellbeing is monitored through key performance indicators collected through precision agriculture technologies (Butler and Holloway, 2015). A review on animal health and welfare in indoor systems claims that the type of system is a determining factor in animal health and welfare and indoor systems do result in worse outcomes for animals (Arnott *et al.*, 2017).

Thus, based on this example we can see that philosophies of agriculture can be used to explore rhetoric in debates about indoor dairy farming. This produces interesting results in itself: it is interesting that a defence of indoor dairy farming is being made using ideas from within alternative agriculture to complicate the picture painted by NGOs of indoor dairy farming as factory farming. As researchers we can also challenge the coherence of these arguments. We can also carry out further research to explore the reality on the ground: how exactly does the farmer-animal relationship change within indoor dairy systems? What implications do these changes have for traditional notions of good farming and farming skills in the management of health and welfare? Can philosophies of agriculture help to make sense of these changes and the farmers' role and identity?

It is also important however to be reflexive and interrogate our own position as researchers. We cannot claim that using theory to explore and question arguments and practices within indoor dairy farming gives us scientific objectivity. The choices of which arguments to deconstruct and question, and how to frame our analysis carry normative weight. Even though our stated aim is to explore indoor dairy farming in terms of philosophies of agriculture rather than decide whether it is a good idea or not, we could choose to question NGO arguments opposing it or industry arguments in favour of it. Our research could be underpinned by an implicit assumption that alternative agriculture or industrial agriculture is the better philosophy.

In addition to choices of what we analyse and question we can also consider how we balance our understanding of the issues with that of professionals involved. If a large scale indoor dairy farmer still claims what he or she is doing is quite closely aligned with traditional dairy farming, do we have a right to disagree based on our knowledge about the philosophy and history of agriculture? Do we understand the context better than the participants themselves? These are issues and questions to consider when we seek to use theory to open up rather than close down debates, and to elucidate rather than pigeonhole professionals' roles, values and identities within the agricultural sector.

This paper is based on preliminary findings from UK documents. The questions can be explored in future through more in depth interview analysis with stakeholders. Comparisons between countries, for instance between the UK and Ireland where dairy farming is much more uniformly extensive can yield results on the questions of how philosophies of agriculture can shed light on our understanding of visions of the future of dairy farming, and how we as researchers could use these theories to explore rhetoric within debates, practices and 'realities' on the ground and the self representations and identities of professionals involved.

Acknowledgements

This work was funded by a British Academy early career fellowship and the Scottish Government's Centre for Expertise on Animal Disease Outbreaks (EPIC), project 2.4.1.

References

Arnott, G., Ferris, C. P. and O'Connell, N. E. (2017). 'Review: welfare of dairy cows in continuously housed and pasture-based production systems', *Animal*, 11(2), pp. 261-273.

Barkema, H. W., von Keyserlingk, M. a. G., Kastelic, J. P., Lam, T. J. G. M., Luby, C., Roy, J.-P., LeBlanc, S. J., Keefe, G. P. and Kelton, D. F. (2015). 'Invited review: Changes in the dairy industry affecting dairy cattle health and welfare', *Journal of Dairy Science*, pp. 7426-7445.

Burton, R. J. F., Peoples, S. and Cooper, M. H. (2012). 'Building 'cowshed cultures': A cultural perspective on the promotion of stockmanship and animal welfare on dairy farms', *Journal of Rural Studies*, 28(2), pp. 174-187.

Butler, D. and Holloway, L. (2015). 'Technology and Restructuring the Social Field of Dairy Farming: Hybrid Capitals, 'Stockmanship' and Automatic Milking Systems', *Sociologia Ruralis*, 56(4).

CIWF (no date) *Report on the welfare of EU dairy cows*. Surrey.

Farming UK (2016). *'Cows never go outside': Intensive dairy farm report 'false' says NFU, Farming UK*. Available at: https://www.farminguk.com/News/-Cows-never-go-outside-Intensive-dairy-farm-report-false-says-NFU_41892.html. Accessed: 5 January 2018.

Hurley, C. and Murphy, M. (no date). *Building a Resilient, Flourishing, Internationally Competitive Dairy Industry in Ireland*. Dublin.

March, M. D., Haskell, M. J., Chagunda, M. G. G., Langford, F. M. and Roberts, D. J. (2014). 'Current trends in British dairy management regimens.', *Journal of dairy science*. 97(12), pp. 7985-94.

NFU (2010). *Dairy farming systems in Great Britain*. Available at: http://www.thedairysite.com/articles/2549/dairy-farming-systems-in-great-britain. Accessed: 14 September 2016.

NFU (2016). *British dairy farming – does size matter?* Available at: https://www.nfuonline.com/sectors/dairy/dairy-news/british-dairy-farming-does-size-matter. Accessed: 5 January 2018.

Sarewitz, D. (2004). 'How science makes environmental controversies worse', *Environmental Science and Policy*, 7(5), pp. 385-403.

Shortall, O. K., Raman, S. and Millar, K. (2015). 'Are plants the new oil? Responsible innovation, biorefining and multipurpose agriculture', *Energy Policy*, 86, pp. 360-368.

Shortall, O., Sutherland, L.-A., Ruston, A. and Kaler, J. (2017). 'True cowmen and commercial farmers: Exploring vets' and dairy farmers' contrasting views of 'good farming' in relation to biosecurity', *Sociologia Ruralis*, doi: 10.1111/soru.12205.

Sonnino, R. and Marsden, T. (2006) 'Beyond the divide: rethinking relationships between alternative and conventional food networks in Europe', *Journal of Economic Geography*, 6(2), pp. 181-199.

Thompson, P. (2012). 'Agriculture, Food and Society – Philosophy to Nanotechnology', *WCDS Advances in Dairy Technology*, 24, pp. 53-65.

Thompson, P. B. (1995). *The Spirit of the Soil*. Routledge, Abingdon, UK, 196 pp.

Thompson, P. B. (2010). *The Agrarian Vision: Sustainability and Environmental Ethics*. University of Kentucky Press, Lexington, USA, 323 pp.

Vanloqueren, G. and Baret, P. V (2009) 'How agricultural research systems shape a technological regime that develops genetic engineering but locks out agroecological innovations', *Research Policy*, 38, pp. 971-983.

Viva (2016). *Intensification and zero-grazing*. Available at: https://www.viva.org.uk/dark-side-dairy/intensification-and-zero-grazing. Accessed: 5 January 2018.

Wilson, G. A. (2007). *Multifunctional agriculture: a transition theory perspective*. CABI, Wallingford, UK, 374 pp.

World Animal Protection (2016). *Dairy farming in the UK*. Available at: https://www.worldanimalprotection.org.uk/campaigns/animals-farming/dairy-farming-in-uk. Accessed: 5 January 2018.

9. Richard Haynes and the views of professionals in the animal welfare science community

J. Deckers
School of Medical Education, Newcastle University, Newcastle-upon-Tyne, NE2 4HH, United Kingdom; jan.deckers@ncl.ac.uk

Abstract

Richard Haynes has argued that professionals in the animal welfare science community have appropriated the term 'animal welfare' to discredit the views of others. He claims that this appropriation would be illegitimate because alternative views have not received a fair hearing in policy discussions and animals have been exploited because of it. In this publication, Haynes's critique and his alternative will be examined with the aim to address the role of professionals in discussions of animal welfare. I argue that professionals in moral philosophy have a significant role to play in discussions of the meaning and moral relevance of animal welfare, and that my own account provides a superior alternative to address moral issues in the two key areas addressed in Haynes's work: the use of animals for biomedical science, and their use for human food.

Keywords: animals, ethics, expertise, philosophy

Introduction

Many will know Richard Haynes, who died in 2014, as the former editor-in-chief of the Journal of Agricultural and Environmental Ethics. Haynes worked as a professor in philosophy at the University of Florida until 2007. Shortly after his retirement, he published 'Animal Welfare. Competing Conceptions and Their Ethical Implications' (Haynes, 2008). Whilst this work was his magnum opus, it has been claimed that it is unlikely that it will receive the recognition that it deserves for freeing 'animal welfare from the narrow and self-serving definition widely disseminated by the animal welfare science community' and replacing it with a richer alternative (Hoch, 2009: 289). As this Congress engages with the roles of professionals in diverse communities of expertise, it provides an ideal opportunity to explore Haynes's critique and proposed alternative, which thus far have not been subjected to systematic scrutiny.

Haynes traces the birth of the science of animal welfare to particular events that took place in the United States of America and the United Kingdom. Whilst I shall not recount these, I explore Haynes's (2008: ix) claim that animal welfare scientists (AWS) appropriated the concept of animal welfare 'illegitimately' to foster particular normative goals that are at odds with what he claims animal welfare should be about. Haynes (2008: ix) associates AWS primarily with animal scientists interested in improving the conditions of animals used for science or for food, with interest in the latter being triggered to a large extent by the publication of Ruth Harrison's (1964) book 'Animal machines' and the Brambell (1965) Commission Report that followed shortly afterwards. The charge of illegitimacy rests in alternative understandings of animal welfare not being given a fair hearing and, consequently, in animals being wronged. Haynes' critique will be analysed in order to shed light on the role of professionals in discussions of animal welfare.

The views of professionals in the animal welfare science community and the role of philosophical expertise

With regard to the former aspect of his critique, Haynes (2008: 6-35) argues convincingly that the legal and advisory documents that have been drawn up to guide the relationships of humans with other

Svenja Springer and Herwig Grimm (eds.) ***Professionals in food chains***
EurSafe 2018 – DOI 10.3920/978-90-8686-869-8_9, © Wageningen Academic Publishers 2018

animals in relation to the welfare of the latter have been decided primarily by animal welfare scientists (AWS). This is problematic for a number of reasons. Firstly, many of these experts are users of animals or their institutional colleagues. They may therefore have a vested interest in the continued use of animals. The definitions of animal welfare that they advocate are therefore likely to be biased by their interests in continued use. Secondly, and relatedly, many of these prefer self-regulation, foreclosing external scrutiny. Thirdly, Haynes takes particular issue with the fact that AWS claim to possess a more objective account than the more sentimental account provided by those who seek to limit the human use of other animals more severely. In the front cover of the first handbook published by the Universities Federation of Animal Welfare, for example, it is written that 'the duty which man owes to his fellow creatures cannot be adequately discharged if it is approached in an emotional or sentimental spirit, and UFAW accordingly seeks to build up a realistically humane policy based on objective fact' (Worden and Dalling, 1947). Haynes (2008: 8) could have done more to explain why this is problematic, but he appeared to adopt the view that it is impossible to adopt policies on the basis of facts that would not be tainted by one's values or sentiments. The suggestion that AWS, merely by virtue of being scientists, are experts in the ethical relevance of animal welfare is problematic as science, even if it is ultimately reliant on ethical conceptions of what science ought to be, is logically distinct from ethics. Put succinctly, one might say that the role of science lies in the art of observation, whilst the role of moral philosophy lies in determining how to observe and how observations should inform conduct.

This takes us to the latter aspect of Haynes's critique. Haynes argues that animals have been exploited because AWS have adopted an overly narrow view of animal welfare. Many may, for example, be interested in good animal health because healthy animals may be better models to study human disease. Haynes (2008: xix) argues even that 'an 'experts discourse' about animal welfare' has emerged, 'whose primary function ... has been to construct and normatize the ideal animal research subject (or food animal?)'. Thus, those who are influenced by this discourse would be blind to more holistic understandings of animal health that might be undermined by their use. The expertise of AWS might perhaps be more trustworthy if they were preoccupied with the question how animals could be provided with 'more opportunities to engage in enjoyable activities', which is what Haynes (2008: 63) argues in chapter 6 that a critical theory of animal welfare should be concerned with.

Whilst acknowledging that it is not easy to determine what might be in the interests of a nonhuman animal, Haynes argues, inspired by Sumner and Nussbaum, that animals must have adequate opportunities to flourish authentically in order to fare well. Straddling the controversy between subjectivist and objectivist theories of welfare, he argues that an animal must be 'justifiedly satisfied with its life' (Haynes, 2008: 128). Contrary to most AWS, as well as Nussbaum, Haynes (2008: 54, 124) argues that this means that the human infliction of death should normally be classed as an avoidable moral harm. In this, he is influenced by the work of Sapontzis (1987: 166-170, 209-229), who argues that life has instrumental value for organisms with interests, where depriving them of life is problematic as it deprives them of a future wherein they could pursue their interests. Whereas Haynes does not think that health is the ultimate good, I have argued elsewhere that this concern with a flourishing life can equally be understood in terms of a moral duty to safeguard health or welfare interests (Deckers, 2016: 4).

Whereas my own view of welfare is closely aligned with that of Haynes, the question must be asked whether Haynes's account contains any greater expertise than the 'experts discourse' of AWS. Whilst Haynes's views about animal welfare may be less likely to be influenced by the desire to use animals for research or for food, this is insufficient to warrant greater expertise. Nonetheless, the sheer fact that moral philosophers are trained in the art of thinking might make it more likely that they possess greater expertise than those who have much less time to ponder moral issues. I think the role of philosophical expertise lies at least in part in providing clarity in relation to the different values that are at stake. Whereas both institutionalised and other philosophers will also be biased in their definitions, it is at least their job to try to expose and discuss (hidden) values. Whether philosophers also have expertise in deciding what counts as a moral value and how different values should be weighed against each other is another, and hotly contested issue (Tobia *et al.*, 2013). If we adopt the view that philosophers engage in sustained reflection, and that some reflection on potential candidates for moral values is better than the uncritical adoption of such values, the view that there is room for philosophical expertise seems plausible.

Contingent expertise

However, let us not get carried away. Perhaps the proof is in the pudding. Rather than accept what professional philosophers say because they have been trained in the art of thinking, the theories that they come up with should be subjected to critical scrutiny. When they survive such scrutiny, we might then say that a particular philosopher possesses expertise on a particular subject. What Haynes (2008: vi) came up with is a theory that allows the use of animals in biomedical research only when the research is likely to benefit the animals in question and that prohibits the use of animals for human food 'unless they are already dead, or unless killing them painlessly when their lives have reached a point where they are sufferings'. To be more precise, I think he may have meant to say that animals should never be killed in order to feed humans, but that they could be eaten where they die naturally or accidentally, as well as in situations where they are killed in order to spare them from intolerable suffering. In the remainder of this piece I shall endeavour to explore whether this theory stands up to scrutiny, turning my attention first to the use of animals in biomedical research, and then to the use of animals for human food, the two main areas that Haynes focused on.

With regard to the use of animals in biomedical research, Haynes (2008, xii) explains that he 'can imagine situations where their use [in] research is not harmful to their interests'. Whilst it is not clear what situations he had in mind, we may think of veterinary research here, for example where a cow suffers from cancer, no known drug is available that could save the cow, but an experimental drug could be tried. It would be hard to object to this, at least where no non-experimental drug is available, the drug's chemical profile suggests that it might have a therapeutic effect, and safety tests on cells and tissues had been favourable. If continued life and good health can be said to be in the interests of the cow, and if the cow's life and health are being undermined by her illness, it might indeed be concluded that the research would not be harmful to the cow's interests.

Haynes's (2008: 54-55) general concern with nonhuman animal research seems to be that most researchers are prepared to sacrifice the interests of other animals when they are convinced that 'the social gain outweighs the cost to the animals', whereas he speaks of the importance of 'constructing consent', whereby we step into the shoes of the animals, 'as advocates for incapacitated humans' may do. If we do so, I think that Haynes is right to suggest that we are unlikely to agree to sacrificing the life of an animal for some social gain. What is less clear, however, is how much Haynes would be prepared to sacrifice. Haynes (2008, 55) accepts that animals can be used for social gain in some situations, if they are 'adequately compensated for their work, including any discomfort imposed on them'. In this context, he seems to model the human-nonhuman relationship on that of an employer to an employee, where animals might get a 'good deal' by 'exchanging services to their employers for the care that is provided for them' and by humans 'limiting their use and retiring them after they have earned certain use credits' (Haynes, 2008: 54-55).

A first question that must be asked is whether it is appropriate to cause discomfort to animals when the research is unlikely to benefit the animals themselves. When we consider human children who lack capacity or human adults who have lost capacity, it might be argued that they should not be enrolled in research that is unlikely to benefit themselves and that would cause them discomfort. One might argue that it would be much better to carry out research on those who can consent. The problem that arises here, however, is that some research may not be able to be carried out on those who have capacity. Imagine, for example, a study on human infants that required blood pressure monitoring. Such monitoring might be associated with minor discomfort and may not be expected to provide any benefits for the infants themselves. In spite of this, I am not convinced that the study would necessarily be unacceptable. Imagine that the same study could also be done on nonhuman infants. Once again, it might be argued, pace Haynes, that minor discomforts should be allowed to be inflicted on them as they are outweighed by the possibility of obtaining social gains.

A second question is whether Haynes is right to believe that a necessary condition for the use of nonhuman animals in research is the provision of compensation. When it concerns the use of human children for research, one might argue that any children who are bruised because of blood pressure monitoring equipment should be compensated for any discomfort caused, for example by the provision of ointment. One might argue that the same should apply to the use of nonhuman animals. This seems uncontroversial. The compensation is merely provided for any damage that may have been done. However, one might argue that children should be provided with ointment that might soothe the pain associated with their bruises regardless of whether or not these bruises were inflicted on them through research or through anything else that they had been subjected to by other human beings. The provision of ointment, therefore, should not be conditioned on whether or not children had sustained bruises through human fault. I believe that the same could be said with regard to other animals who are held within our care: we owe them a duty of care, regardless of whether or not they are used in research.

This exposes a central inconsistency in Haynes's reasoning. Haynes was unwilling to adopt the view that harm should be allowed to be imposed on some to benefit others, and recognised that some research imposes harm whilst not providing any benefits. Rather than turn his back on such research, he was willing to endorse some of it, provided that adequate compensation is provided. I have argued that, whereas it makes sense to 'compensate' animals for any harms that they have suffered as a direct consequence of research, for example through soothing their bruises, it would be better to think of this compensation as an aspect of the normal care that we owe to animals rather than as something we owe them because they were used in research.

If we should take good care of animals regardless of whether they are used in research, the question remains whether we should go along with Haynes's reluctance to allow any research that might cause harm and that is unlikely to yield any benefits for the animal. This reluctance is, once again, clear where he writes: 'The criterion should be whether the benefits to the animal being used outweigh the costs to that animal' (Haynes, 2008: 68). It might be countered that we do allow human beings to be involved in this kind of research, and that we should allow other animals to be used as well. A significant obstacle that threatens the validity of this analogy is the argument that this kind of research should only be allowed if those who are being used also provide their consent, where nonhuman animals may lack the capacity to do so. Whilst some nonhuman animals can be trained to accept some discomfort in exchange for something else, we must be careful not to construct this as consent. The reason for this is that it can hardly be said that an animal who is willing to forgo food in exchange for not experiencing electric shocks is consenting to forgo food. Likewise, an animal who is willing to undergo the measurement of blood pressure because the equipment provides nice sensations to the animal's limb does not consent to participation in research. The possibility that an animal may experience such pleasure or that they might be thought to experience little discomfort with particular research methods, however, would be morally relevant. Whilst I do not have the space to develop my arguments here, I accept the view that some non-beneficial research that is likely to cause harm may be acceptable where the research could not justifiably be done on organisms who are able to consent.

This raises the question whether we should adopt the view that, if we do not accept that a particular harm is imposed on human beings who are unable to consent, we should not impose a similar harm on a nonhuman being either. When the harm in question is death, Haynes (2008: 54) argues that, 'if we can justify sacrificing the life of an [animal] for some larger social good, then we should also be willing to use severely retarded humans for the same purpose'. I have argued elsewhere that we have a moral duty to adopt speciesism, which demands that human moral agents attribute moral relevance to their interest in prioritising members of our own species (Deckers, 2016). Whereas this does not imply that the lives of nonhuman animals can always be sacrificed for the well-being of human animals, it does imply that we should be biased towards human interests. In this light, Haynes's (2008: 54) theory of distributive justice is problematic, as it claims 'that life is a good equally desired by all sentient living things and that all deserve an equal amount of it, limited only by natural life expectancy and the demand on resources to support life'. Haynes (2008: 146) also refers to Nussbaum in support of this egalitarian position: 'Nussbaum's version is that all animals have an equal right to lead a flourishing life. This claim seems intuitively sound to me. The principle implies that humans ought not to interfere in an animal's ability to flourish.'

I think he is not only mistaken about the implication, but also to think that Nussbaum (2004: 315) adopts the view that all animals have 'an equal right to lead a flourishing life', as she uses the fact that 'animals will die anyway in nature' to suggest that it may be appropriate to kill them for food, provided that they are not 'extremely young'. This interpretation is also consistent with the fact that Nussbaum consumes animal products regularly (personal communication). More importantly, this egalitarian position should not be maintained. Imagine that we invented a new chemical that might be a suitable candidate to kill the Plasmodium parasite, who is responsible for transmitting malaria to people. If we adopted the view that the parasites in question had an equal right to life as human beings, we should allow them an equal right to thrive. Consequently, testing or using the chemical to kill them would be objectionable. Pace Haynes, I am not convinced that it is.

Haynes's (2008: vi) reluctance to deal with conflicting situations is also clear from his reflections upon the use of animals for food, where he argues that their use is precluded 'unless they are already dead' or unless they could be killed 'painlessly when their lives have reached a point where they are sufferings' or when their lives are no longer 'worth living', adding that he adopts the position of Gary Comstock (2000: 116) in this regard. Whereas Haynes (2008: xii) claims that this position makes him an 'abolitionist in regard to using animals ... for food', I have argued elsewhere that this cannot be concluded, as it fails to point out any concerns with the human consumption of animals who are killed justifiably, as well as with those who die naturally or accidentally (Deckers, 2016). From the theory of qualified moral veganism that I have developed, it must also be concluded that Haynes's position ought not to be maintained (as it would condemn a large number of people who either cannot or should not rely on consuming alternatives to animal products to malnutrition).

Conclusion

I have argued that Haynes is right to question the expertise of AWS in defining animal welfare and in determining its moral significance and that moral philosophy is the proper field of such expertise, but that Haynes's own account fails to provide it. To conclude, I would like to highlight that this failure may also stem from Haynes's dualistic ontology, which separates sentient from insentient beings. Whereas the adoption of such an ontology does not imply the view that human beings should not use other sentient beings, but are welcome to use insentient beings, it may both reinforce and be reinforced by it. In the panexperientialist ontology that I have set out elsewhere, by contrast, there is no luxury of drawing a dividing line between sentient and insentient beings (Deckers, 2016: 70-71). Those who adopt this ontology may recognise more clearly that sentient beings must kill other sentient beings to sustain themselves. As I have argued in detail elsewhere, the question which sentient beings should be allowed to be killed should be settled not only by taking account of differences in sentient capacities, but also by the degree to which a being is related to a human being, an extension of speciesism that has been labelled as evolutionism (Deckers, 2017; Paez, 2017). Whilst this theory has been subjected to critical scrutiny by other scholars and has its own problems (Laestadius, 2017; Mancilla, 2016; Paez, 2017), it provides a superior account of animal ethics to that developed by Haynes. It is inspired, however, by the same 'search for love' that was a central feature of Haynes's (2012) life.

References

Brambell, F. (1965). Report of the technical committee to enquire into the welfare of animals kept under intensive husbandry systems. Cmnd 2836. HM Stationery Office, London.

Comstock, G. (2000). An alternative ethic for animals. In: Hodges, J. and Han, K. (eds.) Livestock, ethics, and quality of life. CABI, Wallingford, pp. 99-118.

Deckers, J. (2016). Animal (de)liberation: Should the consumption of animal products be banned?. Ubiquity Press, London.

Deckers, J. (2017). Why 'Animal (de)liberation' survives early criticism and is pivotal to public health. Journal of Evaluation in Clinical Practice 23: 1105-1112.

Harrison, R. (1964). Animal machines: the new factory farming industry. Vincent Stuart, London.

Haynes, R. (2008). Animal welfare. Competing conceptions and their ethical implications. Springer, Dordrecht.

Haynes, R. (2012). Journey to Agali land: An autobiography with my poems and short stories. Self-published.

Hoch, D. (2009). 2009, Review of 'Animal welfare. Competing conceptions and their ethical implications. Springer, Dordrecht'. Journal of Agricultural and Environmental Ethics 22: 285-290.

Laestadius L. (2017). Self-interest for the greater good. Review of Deckers, J., Animal (de)liberation: should the consumption of animals be banned? London: Ubiquity Press. Journal of Evaluation in Clinical Practice 23: 1101-1104.

Mancilla, A. Veganism. In: Thompson, P. and Kaplan, D. (eds.) Encyclopedia of food and agricultural ethics. Springer, Dordrecht.

Nussbaum, M. (2004). Beyond 'compassion and humanity.' Justice for nonhuman animals. In: Nussbaum, M. and Sunstein C. (eds.) Animal rights. Current debates and new directions. Oxford University Press, Oxford, pp. 299-320.

Paez, E. (2017). The pitfalls of qualified moral veganism. A critique of Jan Deckers's holistic health approach to animal ethics. Journal of Evaluation in Clinical Practice 23: 1113-1117.

Sapontzis, S. (1987). Morals, reason, and animals. Temple University Press, Philadelphia, PA.

Tobia, K., Buckwalter, W., and Stich, S. (2013). Moral intuitions: Are philosophers experts? Philosophical Psychology 26: 629-638.

Worden, A. and Dalling, T. (1947). The UFAW handbook on the care and management of laboratory animals. London, Bailliere, Tindall and Cox.

10. Modernising the Kenyan dairy sector?

C.J. Rademaker[1*], *S.J. Oosting*[2] *and H. Jochemsen*[1]
[1]*Wageningen University, Social Sciences Group, Hollandseweg 1, 6700 EW Wageningen, the Netherlands;* [2]*Wageningen University, Animal Production Systems Group, De Elst 1, 6708 WD Wageningen, the Netherlands; corne.rademaker@wur.nl*

Abstract

The track record of livestock development interventions in promoting sustained poverty reduction is believed to be meagre. Could it be that the track record of livestock development interventions is so meagre because of the influence of a reductionistic worldview? The Cartesian worldview has been extremely influential in Western culture and broader. Central in this worldview is a thinking in terms of traditional versus modern and subject(ive) versus object(ive). It has been argued that in development cooperation this has given rise to projectivistic thinking and acting where reality, including the realities of farmers and others, is only meaningless material, to be freely given shape based on a rational design. This is accompanied by an attitude in which our by and large individualistic Western society with its (presumed) institutional mechanicism – such as the market mechanism – is seen to be the model for other societies as well. Finally, we can note a marginalisation of religion and worldview in the public debate. The aim of this contribution is thereforeto analyse in a qualitative way whether such traces of the Cartesian worldview can be linked to development programmes' successes or failures in achieving impact. To this end, a set of eight impact evaluations of dairy development interventions in Kenya are considered, spanning the period 2007-2014. We (1) analyse what positive and negative effects – as emerging from the evaluation documents – have resulted from these interventions; (2) normatively reflect on what the evaluation documents posit as positive and negative effects; (3) analyse whether we can speak of a significant influence of the Cartesian worldview in such interventions; and (3) establish whether there is a relationship between the degree to which projects and programs embody a Cartesian worldview and their respective success or failure. Even though results are not available yet, and the relationship between the influence of a Cartesian worldview and a project or program's success is hypothetical, we at least expect to provide a typology of positive and negative effects of Kenyan dairy development interventions, and be able to show to which degree projects and programs are influenced by the Cartesian worldview.

Keywords: modernity, evaluation, worldview, dairy development, Kenya

> If already in the Netherlands the classical image of progress in animal husbandry lies scattered, what is the point of all our projects in the Third World where we display ourselves as the experts because we would know so much about animal husbandry?
>
> Jan Douwe van der Ploeg (1992, 99; our translation)

Introduction

In a 1992 public lecture on livestock husbandry in developing countries, the sociologist Jan Douwe van der Ploeg (1992) tells about a conversation he had a few days before with the director of the Dutch Association for Artificial Insemination *Oost*. In the conversation, the director had made clear his vision of the 'dairy cow of the future': the Dutch cows should become smaller, much smaller than the present-day Holstein Friesian cows, with a lower milk productivity and a better meat quality. Moreover, dairy cows should be able to produce on low-quality roughage. In addition, the dairy cows should become more tolerant to diseases.

EurSafe 2018 – DOI 10.3920/978-90-8686-869-8_10, © Wageningen Academic Publishers 2018

The director's vision was for Van der Ploeg the sign that the classical image of genetic progress among Dutch breeders had come into crisis.[1] That is, in the early nineties how to redefine genetic progress became a big question. Given this crisis in the classical image of genetic progress, Van der Ploeg posed the question where we can possibly find a justification for the idea of transfer (of knowledge and technologies) from The Netherlands to developing countries. If there is so much discussion possible on how to define and materialise genetic progress, how can we claim to know for others what genetic progress means in their case?

The illusion that we do know for sure what genetic progress – and with that, progress in livestock farming in general – means, can be linked to what Van der Ploeg (2003) in another place calls the Cartesian theatre. With this phrase he refers to the worldview in which the world is a machine that functions mechanistically. Everything in that world is taken to be determined by natural laws and changes go back unilinearly to definable (physical) causes. Human agency is irrelevant, except perhaps to understand and utilise (via technologies) natural laws.

Now, the Cartesian (and dominant) reading of the Dutch dairy farming history is that after the Second World War the art of cattle breeding finally changed into the science of breeding, based on scientific, quantitative breeding techniques. That is, 'the scientists' rational, objective approach', exemplified by quantitative measurement of milk yield, finally won over 'the intuitive, subjective methods of the breeders' that focused on the qualitative evaluation of an animal's type and outward characteristics (Theunissen, 2012, 279). Importantly, on this reading, milk productivity increase is taken to be an objective, economic necessity. Understandably, progress then becomes bound up with dairy cows and dairy farms that have even higher milk yields. Technology is to play a key role in this regard to sustain the further milk yield increases. All in all, on this view, there is therefore simply no alternative to the use of the newest technologies, at least if one wants to stay in business.

Yet, as Theunissen argues, the changeover in dairy cattle breeding in The Netherlands cannot be conceived of as an inevitable result of modernisation. Economic considerations played a role, but so did 'an amalgam of [other] considerations, as diverse as they were disputable, pertaining to practical experience, methods, scientific insights and technologies, ideas about responsible farming, aesthetic preferences, [...] and government policies.' (Theunissen, 2012, 309) The changeover can therefore be better understood by the winning over of one culture of breeding over another, i.e. one normative view about good dairy farming won over another (Theunissen, 2008, 2012).

Not only in relation to the historical development of Dutch dairy farming, but also in relation to the historical development of agriculture in developing countries, the idea of modernisation has been criticised (e.g. Long and Van der Ploeg, 1994). Yet, the question is whether in present-day development practice this idea of modernisation has really disappeared. Interestingly, earlier work by Hoksbergen (1986) showed that USAID interventions in rural road infrastructure and water supply were dominated by a neoclassical economic paradigm – characterised by, a.o. a view of humans as self-interested, rational maximisers, acting in a world inherently moving towards equilibrium – which we may see as an 'economistic reduction' of a broader Cartesian worldview (Goudzwaard, 1979, 10-32). Yet, this research is dated and more research into present-day development practice is needed.

The question into the worldviews operative in dairy development interventions is all the more pertinent given that 'the track record of livestock sector development interventions in promoting sustained poverty reduction is weak' (Otte *et al.*, 2012; LID 1999). Could it be that this failure is inherent to

[1] With genetic progress we refer to the process of improving the genetic make-up of farm animals through breeding and selection. This improvement is measured against particular goals, the breeding goals.

the way development cooperation is conceptualised in the Cartesian worldview which has had such a formative influence in Western culture? That is, could the root cause for the failure of livestock sector development cooperation, even if partly, be found in the dominant Cartesian worldview which has been so influential in Western culture? Obviously, we think that the Cartesian worldview cannot do justice to reality as we experience it. Yet, whether such a Cartesian worldview is really operative in present-day dairy development cooperation is open for question.

The aim of the present paper is therefore to analyse a set of eight evaluations of dairy development interventions in Kenya to see whether characteristics of the Cartesian worldview can be found in the way those development interventions have been conceived and implemented, and in how far such an influence relates to the respective success or failure of the development intervention.

Theory and methodology

To be able to qualitatively test the success or failure of dairy development interventions in relation to the influence of a modern, Cartesian worldview in the conception of dairy development interventions, we will have to mention some criteria that characterise this worldview. To start with, we can consider René Descartes (1596-1650) – hence the term Cartesian – as one of the representatives of this worldview, even though, as Kołakowski (1997, 8) notes, 'he codified philosophically a cultural trend that had already paved its way before him'.[2] Answers to all kinds of questions provided by 'the' tradition are radically put in question. Radically, because Descartes wanted to pursue 'contextless doubt' (MacIntyre, 2006, 8) and construct new, rational knowledge, unhindered by tradition. Thus, Descartes took himself to stand outside of all tradition. Since then, the opposition of tradition to modernity has been firmly embedded in the Cartesian tradition of thought, as well as in Western culture broadly. The first criterion is therefore the opposition between tradition and modernity where the latter is taken to be inherently progressive (criterium 1).

Related to the opposition of tradition to modernity, is the opposition of subject to object. If we look at Descartes again, Descartes thought he had found absolute certain knowledge in his own thinking existence: *Cogito, ergo sum*. This absolute starting-point was to provide a sure fundament for the reconstruction of the edifice of knowledge. Yet, as Geertsema (2008) points out, this also had practical consequences, because Descartes (also) took the mathematical method of natural science to be the surest way to build the edifice of knowledge, which application in practical life would lead to the improvement of life. Yet, with this move, all that is outside the thinking subject started to appear as object to be understood and controlled with the mathematical method. The practical result of this opposition of subject to object is that everything outside the subject is seen to be meaningless. It is material for rational reconstruction, where meaning is given by the subject(s) doing the reconstruction. Objective reality, in contrast, is without intrinsic value and without normative limits to how it can be dealt with (Geertsema, 2008).

Finally, in (agricultural) development cooperation, this has given rise to projectivistic thinking and acting (criterium 2); as if reality, including the realities of farmers and others, is only meaningless material, to be freely given shape based on a rational design and for an instrumental end (Goudzwaard and Verweij, 2001, 144; cf. Long and Van der Ploeg, 1989; Scott, 1998). To this is related a marginalisation of religion and worldview to the subject-pole in the public debate; hence, any truth claims originating

[2] And, we may add, several logical consequences of his worldview only showed up later with his followers and in Western culture in general (for instance the role of faith and God in relation to knowledge). Do note, therefore, that Descartes is used here as a pivotal figure in the development of a broader modern worldview. 'Cartesian worldview' therefore does not refer so much to the worldview Descartes held himself, but to a broader and at the same time less nuanced cultural current.

in religion and worldview are considered personal preference and as such without legitimacy in a public context (marginalisation; criterium 3). Second, it has given rise to a view in which the individual is the building block of (or for) a society (individualism; criterium 4) (Hoksbergen, 1986), where interactions among those individuals is mediated by mechanisms – such as the market mechanism (institutional mechanicism; criterium 5)(Goudzwaard and Verweij, 2001, 144).

While the first component in this research is the testing of dairy development programmes[3] on characteristics of the Cartesian worldview, the second component involves establishing success and failure of these programmes. Success and failure – of dairy development interventions – are established relative to standards. In a multiple-case study of the impact of citizen engagement on democratic and developmental outcomes, Gaventa and Barrett (2010) coded outcomes as positive or negative, dependent on whether they contributed to democratic and developmental outcomes. Yet, this foregoes any possible intrinsic value of citizen engagement and participation. That is, Gaventa and Barrett used an external standard because they assess the outcomes in light of some (their own?) understanding of development and democracy. Yet, in our view, it would be fairer to *start* coding effects as either positive or negative in light of the project or programme's overall goal(s). That is, to make use of an internal standard and judge the programme on its own merits. In a later phase, after being acquainted with the data and the context of the programme, one may shift to a critique of the programme's goals (Fischer, 2006) and evaluate the programme's effects in light of another (interpretation of the) standard. The latter we will pursue in the second part of the paper. Yet, in the initial phase of the study we code effects of dairy development interventions as being positive or negative based on whether the professional evaluator judged these effects as positive or negative. Here, we assume that the professional evaluator's judgment is a judgment informed by the programme's overall goal(s).

Because the aim of the present study was not to empirically meta-evaluate dairy development interventions – with the concomitant claim for methodological rigour – but rather to provide for normative reflection, a limited amount of case studies was collected based on expert knowledge of recent Kenyan dairy development programmes. The focus on Kenya was due to the fact that the first author was involved with another research project on Kenyan dairy development (Rademaker *et al.*, 2016). Evaluations that were included had to evaluate whole projects or programmes and not one specific variable.

Program evaluation documents were imported as PDF document into the software programme ATLAS.ti 7.[4] A limited code scheme consisting of the main labels MODERNISM, PROJECTIVISM, INDIVIDUALISM, MECHANICISM, and EFFECTS was created (categorical coding). The sublabels for EFFECTS were left to emerge from the text documents (open coding). As a whole, the study therefore employed mixed coding (Gough *et al.*, 2012).

[3] Where *policies* respond to broad and often interrelated problems in a particular field and provide a vision for development in that field, *programmes* are directed at specific situations within that field (Fischer 2006, 28). *Projects* can be seen as parts of programmes. Here we are specifically interested in programmes and projects. For brevity, in the following we will use the term 'programme' to refer to both of them, except when otherwise stated.

[4] ATLAS.ti is a software program designed for qualitative data analysis. Imported text documents can be coded and a range of analysis tools can be applied to those coded text fragments. ATLAS.ti is a registered trademark of ATLAS.ti Scientific Software Development GmbH, Berlin.

Findings

Table 1 provides an overview of the dairy projects and programmes that were included in this study, as well as some characteristics of them. Unfortunately, the results of the analysis are not yet available and presentation of the results will have to wait until the EurSafe 2018 conference, coming June 13-16. We do expect, however, to present a typology of positive and negative effects from Kenyan dairy development interventions. Whether or not a relationship can be established between successful programs – i.e. programs with predominantly positive effects – and a lesser dominance of a Cartesian worldview remains to be seen.

Table 1. List of evaluated Kenyan dairy programmes as included in this study.

Project/Program name	Abbreviation	Time period	Funding agencies	Main implementing agencies
Improving Livelihoods in the Smallholder Dairy Sector in Kenya	ILSD	2007-2010	UK Department for International Development (DfID)	Traidcraft Exchange, SITE Enterprise Promotion
National Agriculture and Livestock Extension Programme Phase II	NALEP II	2007-2011	Swedish International Development Cooperation Agency (Sida)	Kenyan Ministry of Agriculture, Kenyan Ministry of Livestock Development
Smallholder Dairy Commercialization Programme	SDCP	2007-2014	International Fund for Agricultural Development (IFAD), Government of Kenya	Kenyan Ministry of Agriculture, Livestock and Fisheries
Kenya Dairy Sector Competitiveness Program	KDSCP	2008-2013	United States Agency for International Development (USAID)	Land O'Lakes International Development
East African Dairy Development Project	EADD	2008-2013	Bill and Melinda Gates Foundation (BMGF)	Heifer International, TechnoServe, International Livestock Research Institute, World Agroforestry Centre (ICRAF), African Breeders Services-Total Cattle Management
Core Subsidy Funded Dairy Program Kenya	CSFDP	2009-2012	Dutch Directorate-General for International Cooperation (DGIS)	SNV Netherlands Development Organisation
East African Agricultural Productivity Programme	EAAPP	2010-2014	World Bank	Association for Strengthening Agricultural Research in Eastern and Central Africa, Kenyan Ministry of Agriculture, Kenyan Ministry of Livestock Development, Kenya Agriculture and Livestock Research Organisation
Farmers Fighting Poverty/Producenten Ondersteuning Programma	FFP	2012-2014	DGIS	Agriterra

References

Fischer, F. (2006). *Evaluating Public Policy*. Mason, OH: Thompson Wadsworth.

Gaventa, J. and Barrett, B. (2010). 'So What Difference Does It Make? Mapping the Outcomes of Citizen Engagement.' *IDS Working Papers* 2010 (347): 01-72. https://doi.org/10.1111/j.2040-0209.2010.00347_2.x.

Geertsema, H.G. (2008). 'Knowing within the Context of Creation.' *Faith and Philosophy* 25: 237-60.

Goudzwaard, B. (1979). *Capitalism and Progress. A Diagnosis of Western Society*. Translated by Josina Van Nuis Zylstra. Toronto, ON / Grand Rapids, MI: Wedge Publishing Foundation / William B. Eerdmans Publishing Company.

Goudzwaard, B. and Verweij, M. (2001). 'Kleine Autonomie van Het Ontwikkelingsbegrip.' In *Als de Olifanten Vechten... Denken over Ontwikkelingssamenwerking Vanuit Christelijk Perspectief*, edited by Govert J. Buijs, 138-54. Verantwoording 17. Amsterdam: Buijten & Schipperheijn.

Gough, David, Oliver, S. and Thomas, J. (2012). *An Introduction to Systematic Reviews*. London: SAGE Publications.

Hoksbergen, R. (1986). 'Approaches to Evaluation of Development Interventions: The Importance of World and Life Views.' *World Development* 14 (2): 283-300. https://doi.org/10.1016/0305-750X(86)90060-4.

Kołakowski, L. (1997). *Modernity on Endless Trial*. Chicago and London: University of Chicago Press.

LID. (1999). 'Livestock in Poverty-Focused Development.' Crewkerne, UK: LID (Livestock in Development).

Long, N. and Van der Ploeg, J.D. (1989). 'Demythologizing Planned Intervention: An Actor Perspective.' *Sociologia Ruralis* 29 (3-4): 226-49. Available at: https://doi.org/10.1111/j.1467-9523.1989.tb00368.x.

Long, N. and Van der Ploeg, J.D. (1994). 'Heterogeneity, Actor and Structure: Towards a Reconstitution of the Concept of Structure.' In *Rethinking Social Development: Theory, Research and Practice*, edited by David Booth. Harlow, England: Longman Scientific & Technical.

MacIntyre, A. (2006). *The Tasks of Philosophy: Selected Essays. Volume 1*. Cambridge: Cambridge University Press.

Otte, J., Costales, A., Dijkman, J., Pica-Ciamarra, U., Robinson, T., Ahuja, V., Ly, C. and D. Roland-Holst, D. (2012). 'Livestock Sector Development for Poverty Reduction: An Economic and Policy Perspective. Livestock's Many Virtues.' Pro-Poor Livestock Policy Initiative. Rome: FAO (Food and Agriculture Organization of the United Nations).

Rademaker, C.J., Omedo, B., van der Lee, J., Kilelu, C. and Tonui, C. (2016). 'Sustainable Growth of the Kenyan Dairy Sector: A Quick Scan of Robustness, Reliability and Resilience.' Wageningen University & Research, Wageningen Livestock Research. Available at: http://library.wur.nl/WebQuery/wurpubs/508760.

Scott, J.C. (1998). *Seeing Like a State: How Certain Schemes to Improve the Human Condition Have Failed*. Yale Agrarian Studies Series. New Haven and London: Yale University Press.

Theunissen, B. (2008). 'Breeding Without Mendelism: Theory and Practice of Dairy Cattle Breeding in the Netherlands 1900-1950.' *Journal of the History of Biology* 41 (4): 637-76. https://doi.org/10.1007/s10739-008-9153-0.

Theunissen, B. (2012). 'Breeding for Nobility or for Production? Cultures of Dairy Cattle Breeding in the Netherlands, 1945-1995.' *Isis* 103 (2): 278-309. https://doi.org/10.1086/666356.

Van der Ploeg, J.D. (1992). 'Perspectieven.' In *Veeteelt in Ontwikkelingslanden*, edited by Merel Langelaar, Baukje Schat, and Jan Weerdenburg, 97-104. Studium Generale 9110. Utrecht: Bureau Studium Generale, Universiteit Utrecht.

Van der Ploeg, J.D. (2003). *The Virtual Farmer: Past, Present and Future of the Dutch Peasantry*. Assen: Uitgeverij Van Gorcum.

Section 2.
Sustainable food production

11. On the ethics and sustainability of intensive veal production

S. Aerts[1*] *and J. Dewulf*[2]
[1]*Odisee University College, Ethology and Animal Welfare, Hospitaalstraat 23, 9100 Sint-Niklaas, Belgium;* [2]*Ghent University, Faculty of Veterinary Medicine, Department of Obstetrics, Reproduction and Herd Health, Salisburylaan 133 D4, 9820 Merelbeke, Belgium; stef.aerts@odisee.be*

Abstract

In this paper we analyse the intensive veal production system as currently found in the Dutch-Belgian border region using the common sustainability framework. In the social acceptability pillar we will include animal welfare and other ethical aspects that may relate to the production system as it is currently deployed. One of the major arguments in favour of veal production as it is currently known, is the assertion that it is a way to prevent male offspring of dairy cattle to be considered mere by-products that need to be disposed of. In such dairy production systems these male calves are quickly removed and culled. This has economic and ecological benefits: avoiding food waste by adding value to a product. Slaughtering young calves for veal production wouldn't be an issue to be discussed in a strong animal rights ethics as in this perspective any kind of killing is considered unethical. In an ethics that integrates proportionality, it is certainly not morally neutral, maybe even problematic. Veal production is one of those sectors that are said to be highly instrumentalising to animals. Not only are calves a 'by-product' of milk production, some also indicate the unnaturalness of the system. There is an interesting parallel between intensive veal production and the production of Belgian Blue cattle. In both cases, veterinarian measures are taken to ensure animal welfare, where problems are almost determined by the context. Veal production is extremely antibiotics dependent: virtually 100% of calves are treated with antibiotics. This could be considered a case of extreme instrumentalisation, and results in several negative effects, such as development of antibiotic resistance in animal and human medicine. The latter is problematic in both consequentialist and anthropocentric theories. The long term effects thereof seem disproportionate compared to the limited positive effects of veal production. We conclude that intensive veal production as it is currently known, is highly debatable from an ethical point of view.

Keywords: antibiotics, sustainability, instrumentalisation

Introduction

Different animal production systems have been discussed and analysed from a sustainability viewpoint, but the intensive veal production system as currently found in the Dutch-Belgian border region has not been extensively discussed in international literature. This may be in part due to the relatively regional nature of this particular type of production. There is, however, some peer-reviewed literature available. We will not repeat all the information provided by Pardon *et al.* (2012a,b; 2014), but it seems appropriate (and useful) to summarise here the most important characteristics of the intensive veal industry. The veal production that is referred to in this paper, is the traditional 'white veal' (according to Regulation EC 566/2008) that is produced using calves between 0 and 8 months, raised on milk or special feed. There are other types defined in EU legislation (rosé veal and meat from bovines aged 8-12 months), and bobby veal (meat from calves slaughtered within a week from birth; not often found in the EU). These we will not consider here, as the production systems are significantly different.

According to different studies cited in Pardon *et al.* (2014), global veal production is a predominantly European activity (over 4/5 of production), with France, The Netherlands, and Italy as the major producing countries. Belgium produces about the same amount (6% of global production) as the US, Germany and Canada. In 2016, Belgium produced 376,036 heads, amounting to 62,746 tons of veal

EurSafe 2018 – DOI 10.3920/978-90-8686-869-8_11, © Wageningen Academic Publishers 2018

(Statistics Belgium, 2017). The bulk of this sector is situated close to the Dutch border, and the sectors on both sides of the border have evolved similarly in the 20th century (Pardon *et al.*, 2014).

Economy

Truyen (2011) estimated the total annual turnover of the veal industry (this is including post-farm actors) at 600 million euro (at that time about less than 350,000 calves were produced). Comparing this to the 286 veal herds in 2011, it is clear this is a highly specialised, and industrialised production sector. The sector has evolved at the beginning of the 20th century as a way to valorise the excess of male dairy calves that came available when the Western European agricultural sector shifted towards animal production (Lips, 2004; Pardon *et al.*, 2014). Before that, male calves were usually slaughtered shortly after birth (Pardon *et al.*, 2014). One of the major arguments in favour of veal production is the assertion that it is a way to prevent male offspring of dairy cattle to be considered mere by-products that need to be disposed of. The white veal system is, of course, not alone in this; the alternative veal systems, as in many European countries, resulting in the production of rosé meat where calves are maintained longer and receive a mixed ration with milk and roughage, can play a similar role. The fact that it adds value to by-products of the dairy industry, and in fact plays an important role in regulating milk markets, remains a strong economic argument. For example, veal calves account for about one third of dairy herd slaughter (Sans and de Fontguyon, 2009). The enormous turnover in Belgium seems to contradict such a role, but looking at the composition of the herd, there is some truth in this. About 60% of the veal calves are purebred dairy calves, about 25% are crossbreds (Belgian Blue × Holstein Friesian), and only 15% is purebred Belgian Blue (Pardon *et al.*, 2014).

Ecology

A similar line of thought could be used within the ecological component of sustainability. In countries without a significant veal production sector the male calves from dairy production systems are quickly removed and culled as bobby calves (e.g. Australia and New Zealand). A practice that could be considered a typical example of what De Tavernier and Aerts (2016) called 'killing as collateral damage'. Below we will see there are ethical aspects to be discussed with regard to this assertion, but there is also the ecological aspect: avoiding wasting resources. The possibilities and merits of the use of sexed sperm in dairy production as a means to avoid unwanted male (dairy) calves have been discussed for more than a decade (see e.g. De Tavernier *et al.*, 2003). However, it is impossible to sustain a dairy sector without an annual calving in each cow. We can therefore modulate the number of (purebred male) dairy calves, but not the total number of excess calves. Realistically, these animals cannot be raised within the dairy sector. Thus, they will need to be marketed as bobby veal, or transferred to the veal (or beef) sector. In this aspect, the economic line of reasoning is paralleled by the ecologic argument. Within the current animal production sector organisation, the highly integrated veal sector succeeds in bringing together the excess animals and the available milk and protein sources (Pardon *et al.*, 2014; Sans and De Fontguyon, 2009).

Social acceptance

The assertion by De Tavernier and Aerts (2016) that the public discomfort after the disclosure of some standard practices (such as culling of day old chicks) reveals the great tension between the low societal acceptance of these practices and the fact that most people are (in)directly involved in the activities that lead to this killing, could probably be applied to veal production as well.

However, the issues raised during the different agricultural crises of the past three decades are not limited to the issue of killing animals. Some were indeed animal (welfare) concerns – for example about caesarean section, tail docking, unstunned slaughter – but others were mostly human (health) concerns – such as

BSE, avian influenza, hormones – and some were a mixture. In the following paragraphs, we will discuss some of these issues. We will discuss animal welfare in veal production, the other animal ethics issues, and some of the societal and ethical concerns that go beyond animal ethics, such as the development of antimicrobial resistance (AMR). We will argue that there is an interesting parallel between intensive veal production and caesarean section in the Belgian Blue cattle breed.

Animal welfare

Pardon *et al.* (2014) cited a number of studies that show decreased animal welfare in these animals, based on higher levels of stereotypies and a high incidence of abomasal ulcerations. It also appears that raising (purebred) Belgian Blue veal calves is more difficult than dairy or crossbred calves, given the significantly higher mortality rate in this breed than in HF and crossbreds (Pardon *et al.*, 2012b).

In order to be able to produce white veal are housed and fed in such a way that low iron intake is ensured (preventing colouring of the meat). This is often cited as the source of (most of) the problems in veal production, and it does have different side effects. Iron deficiency results in reduced appetite, and impaired immune function (Pardon *et al.*, 2014). European Directives 91/629/EC and 97/2/EC (prescribing a minimum iron level in liquid feed) have been put in place to safeguard against the worst of these effects.

Housing is another source of potential welfare problems in veal production. The traditional crate housing has been phased out in the EU since 2007 (Directive 91/629/EEC, now replaced by 2008/119/EC) and the sector is moving in the same direction in the US as well (American Veal Association, 2018).

Group housing is generally better for animal welfare (Babu *et al.*, 2004, and many other studies such as those cited in Pardon *et al.*, 2014), based on social interaction and other behaviour. There are however, behavioural abnormalities that are more prominent in groups housing (Babu *et al.*, 2004; Le Neindre, 1993). Although better, it seems group housing is not yet the ideal housing system. Individual housing is still allowed during the first (eight) weeks. This has drawbacks with regard to behavioural freedom), but the results by Brscic *et al.* (2012) show it also reduces the disease risks substantially (we will refer to disease occurrence again later). The net welfare effect of individual housing during the first months is therefore not straightforwardly negative.

Animal ethics beyond animal welfare

There are ethical issues within animal production that go beyond animal welfare. Not in all ethical frameworks these are the same issues, but it is not within the scope of this paper to discuss this in depth. Suffice to say that in a utilitarian or consequentialist ethic, actions towards animals are acceptable if there is a proportional reason. In a deontological approach of ethics, proportionality is not used. So, slaughtering young calves for veal production wouldn't be an issue to be discussed in an animal rights ethics as in this perspective any kind of killing would be considered unethical. In an ethics that integrates proportionality, it is certainly not morally neutral, but not necessarily problematic.

Aerts and De Tavernier (2016) have given the culling of male dairy calves as an example of collateral killing. They specifically indicated that they do not include veal production within their analysis, but both the two criteria they use could also be relevant here. Therefore, we will also look at age at killing and instrumentalisation. As they do, in instrumentalisation we will include the idea that the animal is becoming a by-product of milk production, and the unnaturalness of the system. Note that the notion of 'by-product' appears again in this analysis, but in a very different context. Both criteria are intuitively important, but notoriously difficult to use within an ethical analysis that allows for the use and killing of animals for human ends. It is difficult, for example, to identify the essential difference between killing an animal directly after birth, after a few months, or a few years when its natural lifespan would be over a decade. De Tavernier and Aerts (2016) put it this way: 'In fact, in all animal production systems animals are killed at relatively young ages, and very few – if any – will reach their 'natural' maximum age.' And a similar argument is constructed for the other criteria: '[these criteria] are all prone to the accusation of being arbitrary statements. Indeed, it is difficult to identify which practice is right and which one is wrong when these practices differ only gradually over a very wide and continuous scale.' In this case: if it is wrong to cull a male dairy calf after birth, is it then 'less wrong' to slaughter it a few months later, or is it 'better'? And if slaughtering a bull at 24 months is acceptable, is veal production then 'less acceptable', or 'wrong'?

Societal ethics

Another ethical issue, that is at least as relevant for humans as it is for animals, is that of the potential development of AMR. The (white) veal production sector is probably one of (if not) the most extremely antibiotics dependent animal production sectors. Pardon *et al.* (2012a) have quantified antibiotics use in white veal production in 2007-2009. They found that the average Treatment Incidence based on the Animal Defined Daily Dose – i.e. TI_{ADD} – of antimicrobial treatments was 416.8 ADD per 1000 animals at risk. This means that an average white veal calf would be treated with a daily dose of antibiotics during 41.6% of their life expectancy. Using another measure (Defined Daily Doses or DDD_{veal}) we see that Dutch veal has a DDD_{veal} of 28.6 and Belgian veal has a DDD_{veal} of 61.0 of antimicrobials per year. This is significantly higher than dairy, for example, that has a DDD of 5.8. Based on these numbers, and other data in Pardon *et al.* (2012a), we can fairly state that in white veal production – at least as it is found in Belgium – virtually 100% of calves are treated with antibiotics at some stage in production. Also in the Netherlands the veal calve production is animal sector with one of the highest antimicrobial usage levels with a limited reduction in use over the last years (in contrast to other production sectors) (SDA, 2017). This is quite a problem, though, as it has been irrefutably demonstrated that antimicrobial treatment is the strongest driver for AMR selection. This is a normal evolutionary reaction, nicely illustrated in Figure 3 in Ohlsen (2009). There is also sufficient evidence that higher use results in higher resistance incidence, at farm and country level (Callens *et al.*, 2011; Chantziaras *et al.*, 2014). As a consequence, the resistance levels retrieved in Belgian veal calves are among the highest of all food producing animals (Callens *et al.*, 2017). This is an important issue for the animal production sector, but unfortunately, the types of resistance that are found (in veal production) are relevant for human health as well (Pardon *et al.*, 2012a; and for an overview of other literature, see Pardon *et al.*, 2014). The conclusion by Pardon *et al.* (2014) is certainly correct, although probably an understatement: 'Given the current organization of the industry, which implies commingling of neonatal, recently transported calves at high stocking densities, reduction of antimicrobial use forms a huge challenge.'

Discussion

If we take a broader look at the issues presented here, we see that with respect to the economic and ecological aspects, the veal sector is revealed as a useful addition to the animal production sector as a whole. Regarding the animal welfare and related ethical aspects the sector – since its fundamental reorganisation due to societal pressure and following legislation – does not present exceptional problems beyond what is found in many other animal production sectors. The question whether this general situation is acceptable *per se*, is certainly relevant, but does not fall within the scope of this paper. What is significantly different from most sectors, is the use of antimicrobials. This situation results in several negative effects, such as AMR in animal and human medicine. We would like to draw the parallel with another exceptional animal production sector from the same region: the purebred Belgian Blue. A full description thereof is beyond the scope of this paper, but it is analogous in that both sectors have developed into a sector in which systematic medical assistance is needed to be able to sustain the *status quo*. Almost 100% of Belgian Blue calves are delivered by caesarean section, and the entire veal production sector is dependent on antibiotics. In both cases one is acting to protect (the welfare of) an animal, but in a preventive way – no longer curatively – in a context that causes these problems quite predictively. Both sectors are confronted with societal pressure, but one can also question whether the positive effects of the sector (valorisation of by-products, meat production, etc.) are not outweighed by the now-present negative effects (development of AMR, ...). There are, of course, differences. In the Belgian Blue, the problem is a surgical procedure aimed at a single individual in an acute situation, whereas in the white veal sector, it is a medical treatment given through feed (milk), aimed at a (whole) group of animals to remedy a chronical situation. Whether the current course of action is right depends on the time frame used. In the short term, which is the situation such as it presents itself now, there is no doubt that medical care should be administered to the animals. This is true within anthropocentric ethics (as it is a way to safeguard the culinary and economic benefits of veal production) as well as zoocentric ethics (in which the protection of animal health and welfare is important).

The more fundamental, long term question is whether the production model itself is not flawed. Both sectors are cases of extreme instrumentalisation (in the sense that not only the lives of individual animals, but also the mere continuation of the sector depend on our assistance), resulting in several negative effects. Certainly in the veal sector these consequences are severe, for animals as well as humans, as AMR is impacting both animal and human medicine. When analysed from a consequentialistic account, this is problematic. The impact on human medicine means that this is true, even in strictly anthropocentric theories where animal welfare is not considered relevant. The long term effects thereof seem disproportionate compared to the positive effects of veal production. In more deontological and zoocentric interpretations the level of instrumentalisation (in the sense indicated above) is indicative of the problematic nature of this sector.

Conclusion

Within the broad scope of a preliminary sustainability analysis of (white) veal production, the sector seems to score quite well on most accounts (or at least not worse than most animal production sectors). The non-animal ethical considerations are exceptions. Veal production is extremely antibiotics dependent: virtually 100% of calves are treated with antibiotics. This could be considered a case of extreme instrumentalisation (at population level), and results in AMR and other negative effects in animal and human medicine. The latter is problematic in consequentialist, deontological, anthropocentric, and zoocentric theories alike. The long term effects seem disproportionate compared to the limited positive effects of veal production. The question whether this general situation is acceptable *per se*, is certainly relevant, but does not fall within the scope of this paper. We conclude that intensive veal production as it is currently known, is highly debatable from an ethical point of view.

References

Aerts, S. and De Tavernier, J. (2016). Killing animals as a matter of collateral damage. In: Meijboom, F.L.B. and Stassen, E.N. (eds.). The end of animal life: a start for ethical debate. Wageningen Academic Publishers, The Netherlands, pp. 167-186.

American Veal Association (2018). AVA confirms 'mission accomplished'. Available at: http://www.americanveal.com/ava-policies/2018/1/7/ava-confirms-mission-accomplished-member-companies-and-farms-complete-transition-to-group-housing. Accessed 25 January 2018.

Babu, L.K., Pandey, H.N. and Sahoo, A. (2004). Effect of individual versus group rearing on ethological and physiological responses of crossbred calves. Applied Animal Behaviour Science, 87: 177-191.

Brscic M., Leruste H., Heutinck L.F., Bokkers E.A., Wolthuis-Fillerup M., Stockhofe N., Gottardo F., Lensink B.J., Cozzi G. and Van Reenen C.G. (2012). Prevalence of respiratory disorders in veal calves and potential risk factors. Journal of Dairy Science, 95: 2753-2764.

Callens, B., Boyen, F., Persoons, D., Maes, D., Haesebrouck, F., Butaye, P. and Dewulf, J. (2011). Relation between antimicrobial use and resistance in Belgian pig herds. 4th Conference on Antimicrobial Resistance in Animals and Environment (ARAE), 27-29 June 2011, Tours, France.

Callens, B., Sarrazin, S., Cargnel, M., Welby, S., Dewulf, J., Hoet, B., Vermeersch, K. and Wattiau, P. (2017). Associations between a decreased veterinary antimicrobial use and resistance in commensal *Escherichia coli* from Belgian livestock species (2011-2015). Preventive Veterinary Medicine. Accepted. https://doi.org/10.1016/j.prevetmed.2017.10.013.

Chantziaras, I., Boyen, F., Callens, B. and Dewulf, J. (2014). Correlation between veterinary antimicrobial use and antimicrobial resistance in food-producing animals: a report on seven countries. Journal of Antimicrobial Chemotherapy, 69(3): 827-834.

De Tavernier J., Lips D., Delezie E., Aerts S., Van Outryve J., Spencer S. and Decuypere E. (2003). Ethical considerations on the usage of sexed sperm in stock breeding. In: Jolivet, E. (ed.). Ethics as a dimension of agrifood policy (Proceedings of the 4th Congress of the European Society for Agricultural and Food Ethics). INRA, France, pp. 174-175.

Le Neindre P. (1993). Evaluating housing systems for veal calves. Journal of Animal Science, 71: 1345-1354.

Lips, D. (2004). Op zoek naar een meer diervriendelijke veehouderij in de 21ste eeuw. Aanzet tot het ontwikkelen van win-winsituaties voor dier en veehouder. PhD. Katholieke Universiteit Leuven, Belgium, 184 pp.

Ohlsen, K. (2009). Novel Antibiotics for the Treatment of *Staphylococcus aureus*. Expert Review of Clinical Pharmacology, 2(6): 661-672.

Pardon, B., Catry, B., Boone, R., Theys, H., De Bleecker, K., Dewulf, J. and Deprez, P. (2014). Characteristics and challenges of the modern Belgian veal industry. Vlaams Diergeneeskundig Tijdschrift, 83: 155-163.

Pardon, B., Catry, B., Dewulf, J., Persoons, D., Hostens, M., De Bleecker, K. and Deprez P. (2012a). Prospective study on quantitative and qualitative antimicrobial and anti-inflammatory drug use in white veal calves. Journal of Antimicrobial Chemotherapy, 67(4): 1027-38.

Pardon, B., De Bleecker, K., Hostens, M., Callens, J., Dewulf, J. and Deprez, P. (2012b). Longitudinal study on morbidity and mortality in white veal calves in Belgium. BMC Veterinary Research, 8: 26.

Sans, P. and De Fontguyon, G. (2009). Veal calf industry economics. Revue de Médecine Vétérinaire, 160(8-9): 420-424.

SDA (2016). Het gebruik van antibiotica bij landbouwhuisdieren in 2016. Autoriteit Diergeneesmiddelen, Utrecht, 93 pp.

Statistics Belgium (2017). Kerncijfers landbouw. Belgische landbouw in cijfers. Algemene Directie Statistiek – Statistics Belgium, Belgium, 52 pp.

12. Organic animal production – a tool for reducing antibiotic resistance?

S. Gunnarsson[1] and A. Mie[2]*
[1]Swedish University of Agricultural Sciences (SLU), Dept. of Animal Environment and Health, P.O. Box 234, 532 23 Skara, Sweden; [2]Karolinska Institutet, Dept. of Clinical Science and Education, Södersjukhuset, Sjukhusbacken 10, 118 83 Stockholm, Sweden; stefan.gunnarsson@slu.se

Abstract

Antibiotics are today an integral part of intensive animal production, and farm animals may act as important reservoirs of resistant genes in bacteria. The therapeutic and preventive use of antibiotics in intensive livestock production is closely linked to the animal housing and rearing conditions. The transmission dynamic of antibiotic resistance is complex and not fully understood, but still antibiotics should be used with cautiousness. The WHO has launched an action plan aiming for 'reduction in nontherapeutic use of antimicrobial medicines in animal health'. Thus, the risk factors associated with intense production needs to be tackled in order enable animal production with minimal antibiotics usage. In contrast to most other concurrent commercial animal husbandry in Europe and in the world, organic livestock production today essentially complies with this position. It has been postulated that lower need and use of antibiotics in organic livestock production will diminish the risk of development of antibiotic resistance, and this has also been demonstrated in organic pigs and poultry. However, the differences between species and countries are large. Organic production is always certified according to rigorous standards, which means that transparency for consumers and authorities throughout the food chain should be guaranteed. This transparency may be fruitful in order to acquire knowledge and methods for combating the elevating problems of transmissions of antimicrobial resistance within the food production. In summary, organic animal husbandry essentially fulfils today the demands on restrictive use of antibiotics made by the WHO and by the European Parliament, for counteracting the development and spreading of antibiotic resistance. Knowledge dissemination between conventional and organic production may be important steps in the right direction. However, transition into organic production for the whole livestock sector would, on its own, only be part of a solution of the antibiotic resistance issue, because factors outside animal production, such as the use in humans, will be unaffected.

Keywords: antibiotic resistance, morbidity, animal husbandry, pig, poultry, dairy

Antibiotic use in animal production

Antibiotics constitute an integral part of intensive animal production today, and farm animals may act as important reservoirs of resistant genes in bacteria. It is reported that a substantial proportion (50 to 80%) of antibiotics are used for livestock production worldwide, for therapy, prophylaxis or growth promotion (Cully, 2014). On a 'per kg biomass' basis, in 2014 the amount of antimicrobial drugs consumed by farm animals was slightly higher than the antimicrobial drugs used for humans in the 28 EU/EEA countries surveyed, with substantial differences between countries regarding volumes and types of substances (European Food Safety Authority (EFSA), 2017).

In recent decades, there have been increasing concerns that the use of antibiotics in livestock would contribute to impairing the efficiency of antibiotic treatment in human medical care (WHO, 2016). Despite the lack of detailed information on transmission routes for the vast flora of antibiotic-resistant bacteria and resistance genes, there is a global need for action to reduce the emerging challenges

EurSafe 2018 – DOI 10.3920/978-90-8686-869-8_12, © Wageningen Academic Publishers 2018

associated with the reduced efficiency of antibiotics and its consequences for public health, as well as for the environment more generally (Laxminarayan *et al.*, 2013).

The use of antibiotics may increase the economic outcome of animal production (Hao *et al.*, 2014), but the spreading of multi-resistant genes is not just a problem for the animal production sector alone: negative effects are affecting parts of society not directly associated with livestock production. This means that the costs of side effects are borne by society in general and not primarily by the agricultural sector. However, the generalization cannot be made that all antibiotic treatment in farm animals represents a hazard to public health (Mather *et al.*, 2013).

The use of antibiotics in contemporary intensive livestock production is closely linked to the housing and rearing conditions of farm animals. Specific conditions for conventional livestock farming in different countries, as well as farmers' attitudes, may differ between countries, e.g. conventional pig production at above EU animal welfare standards and farmers' attitudes in Sweden (Bruckmeier and Prutzer, 2007). Conventional production is typically aiming for high production levels with restricted input resources such as space, feed, etc., and these conditions may cause stress in the individual animal as it is unable to cope with the situation, e.g. in pig production. This means that higher stocking density, restricted space and barren environment are factors increasing the risk of the development of diseases, and therefore it is more likely that animals under these conditions need antibiotic treatments. Overly prevalent prophylactic use of antibiotics in animal production is an important factor contributing to increasing human health problems due to resistant bacteria.

Organic production and antibiotic use

In studies investigating conventional and organic productions, benefits like increased biodiversity (Bengtsson *et al.*, 2005) have been found, whereas it has been argued that the environmental impact per unit produced may be larger in organic production, due to lower efficiency and increased total land use (Foley *et al.*, 2005). Regarding organic animal production, the decreased efficiency and increased sensitivity to disturbances have been put forward previously. However, it is important to preform scientific studies in order to investigate conditions that have been taken for granted, e.g. epidemiological studies from Sweden could not find that the exposure to Salmonella in outdoor poultry production was higher than in the indoor production (Wierup *et al.*, 2017). Furthermore, food safety issues related to organic animal food have been found, such illness due to unpasteurized organic milk and other items that may contain disease agents (Harvey *et al.*, 2016). However, the risk factors for these problems in organic animal production may not inevitably be associated with the organic standards (e.g. lack of pasteurisation).

In a recent review of human health implications of organic food, Mie and co-worker (Mie *et al.*, 2017) found that organic food consumption may reduce the risk of allergic disease and of overweight and obesity. However, the evidence is not conclusive due to likely residual confounding factors. Furthermore, they found that epidemiological studies have reported adverse effects of certain pesticides on children's cognitive development at current levels of exposure, but these data have so far not been applied in formal risk assessments of individual pesticides.

Contrary to conventional animal production, antibiotics use is strongly restricted in organic husbandry, which instead aims to provide good animal welfare and enough space in order to promote good animal health. This is the main reason why organic production aims for less intensive animal production. Generally, this means that the organic animals have access to a more spacious and enriched environment, access to an outdoor range and restricted group sizes, and other preconditions. This would ultimately decrease the need for preventive medication of the animals as they can perform more natural behaviours and have more opportunity to maintain a good health. However, in practice, the health status of organic livestock is complex and disease prevention needs to be adapted to the individual farm (Kijlstra and Eijck, 2006). A report on the consequences of organic production in Denmark demonstrates that meeting the requirements of organic production has several positive consequences in relation to animal welfare and health (Jespersen *et al.*, 2017).).

According to EU regulations, routine prophylactic medication of animals in organic production is not allowed. However, diseases should be treated immediately to avoid suffering, and the therapeutic use of antibiotics is allowed, but with longer withdrawal periods than in conventional production (Council of the European Union, 2007). Furthermore, products from animals treated more than three times during 12 months, or, if their productive lifecycle is less than one year, more than once, cannot be sold as organic (European Commission, 2008). This means that therapeutically the same antibiotics used in conventional farming may be used in organic farming, but under different conditions. For example, antibiotics mainly used for sub-therapeutic treatment as prophylaxis are never considered in organic production.

While the organic regulations aim for a low use of antibiotics in livestock production, the actual use of antibiotic drugs in European organic compared to conventional animal husbandry is not comprehensively documented. Scattered studies indicate that the antibiotic use generally is substantially higher in conventional compared to organic systems, especially for pigs (approximately 5 to 15-fold higher). In studies from Denmark (Bennedsgaard *et al.*, 2010) and the Netherlands (Kuipers *et al.*, 2016), the antibiotic use in dairy cows was 50 and 300% higher in conventional compared to organic systems, although a Swedish study found no differences in disease treatment strategies between organic and conventional dairy farms, e.g. for mastitis (Fall and Emanuelson, 2009). Although only sparingly documented (Snary *et al.*, 2006), there is only little use of antibiotics in EU organic broiler production. This is a consequence of regulations prohibiting prophylactic use and prescribing long withdrawal periods before slaughter, in conjunction with the fact that it is not feasible to treat single animals in broiler flocks. However, in conventional broiler production, antibiotic use is common.

Antibiotic resistance and organic animal production

Recently, gene sequencing has revealed that the routes of transmission of resistance genes between human and farm animal reservoirs seem to be complex. Nevertheless, a recent EFSA report found that 'in both humans and animals, positive associations between consumption of antimicrobials and the corresponding resistance in bacteria were observed for most of the combinations investigated' (European Food Safety Authority, 2015), which has subsequently been strengthened (European Food Safety Authority (EFSA), 2017). In addition to direct transmission between animals and humans *via* contact or *via* food, resistant strains and resistance genes may also spread into the environment (Kemper, 2008).

Previously, it has been postulated that a reduced need and use of antibiotics in organic livestock production will diminish the risk of development of antibiotic resistance (Aarestrup, 2005), and this has also been demonstrated with regard to resistant *Escherichia coli* in organic pigs compared to conventional pigs (Osterberg *et al.*, 2016). It has also been found that the withdrawal of prophylactic use of antibiotics when poultry farms are converted from conventional to organic production standards leads to a decrease in the prevalence of antibiotic-resistant *Salmonella* (Sapkota *et al.*, 2014).

Resistant bacteria may be transferred within the production chain from farm to fork. It has been found that organic livestock products are less likely to harbour resistant bacteria in pork and chicken meat (Smith-Spangler *et al.*, 2012). In pig production, particular attention has been paid to methicillin-resistant *Staphylococcus aureus* (MRSA), and in Dutch and German studies, for example, MRSA has been isolated in 30 and 55% respectively of all pigs tested (De Neeling *et al.*, 2007; Fromm *et al.*, 2014). However, the prevalence of MRSA in pig production may differ between conventional and organic farms, and in a meta-study in 400 German fattening pig herds, the odds ratio (OR) for MRSA prevalence was 0.15 (95% CI 0.04, 0.55) in organic (n=23) compared to conventional (n=373) pig farms (Fromm *et al.* 2014). Multivariate adjustment for potential risk factors rendered this association non-significant, suggesting that it was carried by other factors, including factors that are regulated in or associated with organic production, such as non-slatted floors, no use of antibiotics, and farrow-to-finish herd types. Factors that may be prevalent in low-input farming that is not organic, but rarely seen in intense conventional production. Furthermore, even if there are considerable differences in antibiotic use between countries, it has been found that antibiotic resistance is less common in organic pigs compared to conventional pigs in France, Italy, Denmark and Sweden (Gerzova *et al.*, 2015).

Although it is rare for conventional farms to adopt knowledge about management and housing from organic production except when converting farms in line with organic standards, there may be options to improve animal health and welfare by knowledge transfer to conventional farms in order to reduce the use of antibiotics (Gleeson and Collins, 2015). However, in several aspects conventional dairy farming would implement some of the housing features used in organic dairy farming, such as loose housing and rough feeding commonly used, and outdoor ranging and grazing in Sweden. Furthermore, space allowances all types of systems could be adjusted in order to improve animal welfare. Improved disease surveillance, considerate use for medication and use of non-drug prophylaxis may easily be adapted from organic into conventional farming, although organic crop and feed production may not be possible to use.

Within organic production, labelling requires full traceability in all steps in order to guarantee the origin of the organic products being marketed (Council of the European Union, 2007). Application of the general principle of organic regulations about transparency throughout the food chain can be used to mitigate emerging problems of transmission of antimicrobial resistance. Furthermore, the traceability would enable development of HACCP based system that would monitor antibiotic use, as well as, the incidence of antibiotic resistance. Even if some surveillance programs in conventional animal farming may exist in some countries, the organic certification is developed and performed consistently in EU. Therefore, the organic labelling has an advantage in transparency and traceability compared to conventional production, where monitoring of antibiotic use and surveillance of antibiotic resistance is rudimental. However, transition to organic production for the whole livestock sector would, on its own, be only part of a solution to the antibiotics resistance issue, because factors outside animal production, such as their use in humans, will be unaffected.

The labelling gives consumers options to choose, and thereby contributes to a decrease in antibiotic use. However, in some countries, e.g. the US, some organic animal food may be market as anti-biotic free, which may seem attractive to consumers. Thus, a non-antibiotic strategy towards infection diseases in farm animals may unintentionally impose bad animal welfare, i.e. sick animal are not getting the appropriate treatment. In some countries, conventional animal farmers have strived for improved health prophylaxis without antibiotics, e.g. by increasing space allowance and enriched barren environments with litter material. These improvements in housing and management may have contributed to that no differences in health problems in dairy cows were found (Fall and Emanuelson, 2009). Furthermore, the improvements may have contributed to the low prevalence of antibiotic resistance also found in conventional farming (Osterberg *et al.*, 2016).

Conclusions

With respect to the development of antibiotic resistance in bacteria, organic animal production may offer a way of restricting the risks posed by intensive production, and even decreasing the prevalence of antibiotic resistance. Organic farm animals are less likely to develop certain diseases related to intensive production compared to animals on conventional farms. As a consequence, less antibiotics for treating clinical diseases are required under organic management, where their prophylactic use also is strongly restricted. This decreases the risk for development of antibiotic resistance in bacteria. Furthermore, the transparency in organic production may be useful for acquiring knowledge and methods to combat the rising issues around transmission of antimicrobial resistance within food production.

References

Aarestrup, F.M. (2005). Veterinary drug usage and antimicrobial resistance in bacteria of animal origin. Basic Clinical Pharmacology and Toxicology 96, 271-281.

Bengtsson, J., Ahnstrom, J. and Weibull, A. C. (2005). The effects of organic agriculture on biodiversity and abundance: a meta-analysis. Journal of Applied Ecology 42, 261-269.

Bennedsgaard, T.W., Klaas, I.C. and Vaarst, M. (2010). Reducing use of antimicrobials – Experiences from an intervention study in organic dairy herds in Denmark. Livestock Science 131, 183-192.

Bruckmeier, K. and Prutzer, M. (2007). Swedish pig producers and their perspectives on animal welfare: a case study. British Food Journal 109, 906-918.

Council of the European Union (2007). Council Regulation No 834/2007 of 28 June 2007 on organic production and labelling of organic products and repealing Regulation (EEC) No 2092/91. Official Journal of the European Union, L 189/1 (20.7.2007).

Cully, M. (2014). Public health: The politics of antibiotics. Nature 509, S16-17.

De Neeling, A.J., Van Den Broek, M.J. M., Spalburg, E.C., Van Santen-Verheuvel, M. G., Dam-Deisz, W. D. C., Boshuizen, H. C., De Giessen, A. W. V., Van Duijkeren, E. and Huijsdens, X. W. (2007). High prevalence of methicillin resistant *Staphylococcus aureus* in pigs. Veterinary Microbiology 122, 366-372.

European Commission (2008). COMMISSION REGULATION (EC) No 889/2008 of 5 September 2008 laying down detailed rules for the implementation of Council Regulation (EC) No 834/2007 on organic production and labelling of organic products with regard to organic production, labelling and control. Official Journal of the European Union, L 250/1 (18.9.2008).

European Food Safety Authority (2015). Scientific report of ECDC, EFSA AND EMA ECDC/EFSA/EMA first joint report on the integrated analysis of the consumption of antimicrobial agents and occurrence of antimicrobial resistance in bacteria from humans and food-producing animals. EFSA Journal 13(1): 4006, 114 pp.

European Food Safety Authority (2017). ECDC/EFSA/EMA second joint report on the integrated analysis of the consumption of antimicrobial agents and occurrence of antimicrobial resistance in bacteria from humans and food-producing animals. EFSA Journal 15(7): 4872, 135 pp.

Fall, N. and Emanuelson, U. (2009). Milk yield, udder health and reproductive performance in Swedish organic and conventional dairy herds. Journal of Dairy Research 76, 402-410.

Foley, J.A., Defries, R., Asner, G. P., Barford, C., Bonan, G., Carpenter, S.R., Chapin, F.S., Coe, M. T., Daily, G.C., Gibbs, H.K., Helkowski, J.H., Holloway, T., Howard, E.A., Kucharik, C.J., Monfreda, C., Patz, J.A., Prentice, I.C., Ramankutty, N. and Snyder, P.K. (2005). Global Consequences of Land Use. Science 309, 570-574.

Fromm, S., Beißwanger, E., Käsbohrer, A. and Tenhagen, B.-A. (2014). Risk factors for MRSA in fattening pig herds – A meta-analysis using pooled data. Preventive Veterinary Medicine 117, 180-188.

Gerzova, L., Babak, V., Sedlar, K., Faldynova, M., Videnska, P., Cejkova, D., Jensen, A.N., Denis, M., Kerouanton, A., Ricci, A., Cibin, V., Österberg, J. and Rychlik, I. (2015). Characterization of Antibiotic Resistance Gene Abundance and Microbiota Composition in Feces of Organic and Conventional Pigs from Four EU Countries. PLOS ONE 10(7): e0132892.

Gleeson, B.L. and Collins, A.M. (2015). Under what conditions is it possible to produce pigs without using antimicrobials? Animal Production Science 55, 1424-1431.
Hao, H., Cheng, G., Iqbal, Z., Ai, X., Hussain, H. I., Huang, L., Dai, M., Wang, Y., Liu, Z. and Yuan, Z. (2014). Benefits and risks of antimicrobial use in food-producing animals. Frontiers in Microbiology 5, 288.
Harvey, R. R., Zakhour, C. M. and Gould, L. H. (2016). Foodborne Disease Outbreaks Associated with Organic Foods in the United States. J Food Prot 79, 1953-1958.
Jespersen, L.M., Baggesen, D.L., Fog, E., Halsnæs, K., Hermansen, J.E., Andreasen, L., Strandberg, B., Sørensen, J.T. and Halberg, N., (2017). Contribution of organic farming to public goods in Denmark. Organic Agriculture 7: 243-266.
Kemper, N. (2008). Veterinary antibiotics in the aquatic and terrestrial environment. Ecological Indicators 8, 1-13.
Kijlstra, A. and Eijck, I. a. J.M. (2006). Animal health in organic livestock production systems: a review. NJAS – Wageningen Journal of Life Sciences 54, 77-94.
Kuipers, A., Koops, W. and Wemmenhove, H. (2016). Antibiotic use in dairy herds in the Netherlands from 2005 to 2012. Journal of Dairy Science 99, 1632-1648.
Laxminarayan, R., Duse, A., Wattal, C., Zaidi, A.K., Wertheim, H.F., Sumpradit, N., Vlieghe, E., Hara, G.L., Gould, I.M., Goossens, H., Greko, C., So, A.D., Bigdeli, M., Tomson, G., Woodhouse, W., Ombaka, E., Peralta, A. Q., Qamar, F.N., Mir, F., Kariuki, S., Bhutta, Z.A., Coates, A., Bergstrom, R., Wright, G.D., Brown, E.D. and Cars, O. (2013). Antibiotic resistance-the need for global solutions. Lancet Infectious Diseases 13, 1057-1098.
Mather, A.E., Reid, S.W.J., Maskell, D.J., Parkhill, J., Fookes, M.C., Harris, S.R., Brown, D.J., Coia, J.E., Mulvey, M.R., Gilmour, M.W., Petrovska, L., De Pinna, E., Kuroda, M., Akiba, M., Izumiya, H., Connor, T.R., Suchard, M.A., Lemey, P., Mellor, D.J., Haydon, D.T. and Thomson, N.R. (2013). Distinguishable Epidemics of Multidrug-Resistant *Salmonella* Typhimurium DT104 in Different Hosts. Science 341, 1514-1517.
Mie, A., Andersen, H.R., Gunnarsson, S., Kahl, J., Kesse-Guyot, E., Rembiałkowska, E., Quaglio, G. and Grandjean, P. (2017). Human health implications of organic food and organic agriculture: a comprehensive review. Environmental Health 16, 111. 22 pp.
Osterberg, J., Wingstrand, A., Nygaard Jensen, A., Kerouanton, A., Cibin, V., Barco, L., Denis, M., Aabo, S. and Bengtsson, B. (2016). Antibiotic resistance in *Escherichia coli* from pigs in organic and conventional farming in four European countries. PLoS ONE 11(6),:e0157049.
Sapkota, A.R., Kinney, E.L., George, A., Hulet, R.M., Cruz-Cano, R., Schwab, K.J., Zhang, G. and Joseph, S.W. (2014). Lower prevalence of antibiotic-resistant *Salmonella* on large-scale US conventional poultry farms that transitioned to organic practices. Science of the Total Environment 476, 387-392.
Smith-Spangler, C., Brandeau, M.L., Hunter, G.E., Bavinger, J.C., Pearson, M., Eschbach, P.J., Sundaram, V., Liu, H., Schirmer, P., Stave, C., Olkin, I. and Bravata, D.M. (2012). Are organic foods safer or healthier than conventional alternatives?: a systematic review. Annals of Internal Medicine 157, 348-366.
Snary, E., Pleydell, E. and Munday, D. (2006). Investigation of persistence of antimicrobial resistant organisms in broiler flocks: a mathematical model. Project OD2006. UK, Veterinary Laboratories Agency. Report March 2006, 70 pp.
WHO (2016). Strategic and technical advisory group on antimicrobial resistance (STAG-AMR): report of the fifth meeting, 23-24 November 2015, WHO Headquarters. Geneva World Health Organization. 9 pp.
Wierup, M., Wahlström, H., Lahti, E., Eriksson, H., Jansson, D.S., Odelros, Å. and Ernholm, L. (2017). Occurrence of *Salmonella* spp.: a comparison between indoor and outdoor housing of broilers and laying hens. Acta Veterinaria Scandinavica 59, 13.

13. Gene-edited organisms should be assessed for sustainability, ethics and societal impacts

A.I. Myhr[1] and B.K. Myskja[2]*
[1]Genøk-Centre for Biosafety, SIVA Innovation centre, PB 6418, 9294 Tromsø, Norway; [2]Department of Philosophy and Religious Studies, Faculty of Humanities, NTNU, 7491 Trondheim, Norway; anne.i.myhr@uit.no

Abstract

A number of new animal and plant breeding techniques (NBTs) has been developed, such as the genome editing technique CRISPR/Cas9. Internationally it is debated if gene-edited organisms should be subjected to the same risk assessment requirements as GMOs. Sweden has decided that gene-edited plants with no foreign genetic element should not count as GMOs, arguing that they are similar to plants altered through conventional breeding or plant mutagenesis. This regulatory discussion is not decided yet in the EU, but we will here argue that even if gene-edited plants and animals are exempt from GMO risk assessment, an approval procedure should include a broad assessment of such plants and animals. We take our point of departure from the Norwegian Gene Technology Act (GTA) which states that risk assessment of GMOs should be supplemented by an assessment of the sustainability, ethical and societal impact prior to regulatory approval of the novel products. Even if gene-edited organisms are considered comparable to non-GMOs in terms of risks, the technology has impacts that calls for an assessment of the kind required in the GTA. With NBTs it may possible to develop plants that have increased drought and saline tolerance relevant for the developing world. Such gene-edited plants can have positive, stable long-term effects on environment, economic and social conditions, and hence be argued to contribute to sustainability. Conversely, the same plants may also have adverse long-term environmental effects. Social benefits are such that are good, or at least not harmful, for small scale producers and consumers, not merely for patent holders and industrial farming. Possible examples are blight resistant potatoes and virus resistant pigs. The assessment of ethical impact becomes increasingly important when dealing with powerful technologies such as genome editing. One example is the potential for developing virus free pigs to be used growing human organs or other alterations that increase the usefulness of animals for industrial production. Questions related to welfare and protection of integrity needs to be evaluated. A wide assessment as required by the GTA will ensure that the NBTs will be beneficial for society in general.

Keywords: sustainability, social utility, GMO, ethical impacts, CRISPR, non-safety assessment

Introduction

Recent technical developments within biotechnology have opened the possibility for editing genetic information and expression in organisms in a faster and more targeted way than previously methods. These technical developments, termed new plant breeding techniques (NBTs) include genome editing techniques as CRISPR and TALEN, and has been adopted by academic and industrial research groups. With this new genome editing techniques it is possible to develop plants that can meet the societal challenges by climate change, such as plants that have increased drought and saline tolerance; to improve efficiency in animal husbandry, such as fish that are sterile, cows without horns and virus resistant pigs; and to combat diseases or pests with gene drives. The techniques also hold promises for animal breeding to improve the safety for growth of organs to be used in humans. The US Department of Agriculture has made the decision to exempt applications of genome editing from regulation. This includes mushroom that have prolonged shelf life (reduced browning process) and maize with changed production process for starch. In other parts of the world, as Europe, the issue of how to regulate is still open, and it is

EurSafe 2018 – DOI 10.3920/978-90-8686-869-8_13, © Wageningen Academic Publishers 2018

discussed whether or not genome edited organisms fall under the scope of existing definitions and legislations regulating GMOs. We will argue that independent of the regulatory frameworks covering gene-editing techniques and genome-edited organisms, an approval procedure for such plants and animals should include a broad assessment of certain 'non-safety' issues (Zetterberg and Björnberg, 2017). Such non-safety assessment should include an evaluation of the contribution to sustainability, the societal benefits and the ethical impacts.

How to regulate genome-edited organisms and what are the issues discussed

Many countries require regulatory approval before environmental release and use of GMOs in food, feed and fibre. The main element is an assessment of health and environmental risk. Some countries include an evaluation of socio-economic and ethical considerations, so-called non-safety assessments. The requirements were established after international, regional and national discussions during the 1990s and can be found in the Cartagena Protocol under the Convention of Biodiversity, the EU Directive 2001/18/EC (originally 90/220/EC) and regulation (EC) no. 1829/2003, and for example the Norwegian Gene Technology Act (NGTA). Other countries, for example the USA and Canada, have decided to regulate GMOs under existing frameworks, without specific regulatory requirements. Simplified, we can say that the European have a process-based system justified in the precautionary principle (PP), whereas the regulatory principle of the US and Canada is product-based, using substantial equivalence (Zetterberg and Björnberg, 2017).

The pressing question now is whether genome editing techniques create organisms that fall under or are exempted from current GMO regulation. Related questions concern whether the current regulatory frameworks must be revised and adapted to these new techniques, to ensure adequate handling of the techniques themselves as well as the resulting products. These issues are discussed at national level (as for example Canada, USA, Germany, Sweden, etc.), regional (European Union) and international levels (Convention on Biological Diversity and its protocols).

The arguments for exemption of genome edited organisms from regulation include: (1) the analogy to organisms found in nature; (2) the similarity to organisms originating from mutagenesis techniques; and (3) the excessive regulatory burden placed on GMOs that is stifling innovation as well as global trade of agriculture commodities (Jones, 2015). The argument with most impact is the second, stating that the products from gene-edited organisms are indistinguishable from products created by processes already excluded from legislation. The harm potential of gene-edited organisms is equal to these non-regulated organisms (Hartung and Schiemann, 2014; Nature Editorial, 2017).

At present, no definitive decision concerning gene-editing regulation has been made by parties to the Convention of Biodiversity or the Cartagena Protocol or member states of the EU. However, some countries have used their national legislation to make interpretation of how to regulate. For example, Sweden has decided that gene-edited organism where no recombinant DNA has been inserted, would be exempted from GMO regulation. Those that have novel genetic material inserted, will be regulated and labelled as GMOs. EU court adviser, Michael Bobek, recently suggested that the use of any technique for mutagenesis purposes in plants, including gene-editing, should be exempted from GM regulations, but he also argued that individual states can regulate their use. In the near future, The European Court of Justice will make a judgement on this question (Court of Justice of the European Union, 2018).

The inclusion of non-safety assessment in regulatory frameworks

The inclusion of non-safety assessment has been recognized in international frameworks (e.g. article 26 of the Cartagena Protocol on Biosafety), European and African fora (e.g. African Biosafety Model Law)

(Binimelis and Myhr, 2016; European Commission, 2014; Greiter *et al.*, 2011; Spök, 2010). In Europe, a new directive on GMOs was approved in 2015 (Directive (EU) 2015/412), allowing a Member State (or region) to adopt measures restricting or prohibiting the cultivation in all or part of its territory of a GMO, or of a group of GMOs defined by crop or trait, based on grounds such as those related to socio-economic impacts, avoidance of GMO presence in other products, agricultural policy objectives or public policy. At the national level, Norway is the country with most experience in assessments of non-safety requirements as part of legislation. The Norwegian Gene Technology Act (GTA) emphasizes the need to consider the social utility and contribution to sustainability of GMOs, as well as their direct and indirect impacts on agricultural practice.

The inclusion of such non-safety considerations is highly controversial (Zetterberg and Björnberg, 2017). There is also a lack of consensus on what aspects that should be taken into account in such assessments. Issues of debate include scope, methods and disciplines involved, timing of consideration, baselines and comparators, criteria and indicators, 'endpoints' or targets, the role of public participation, the relationship with other fields of knowledge and with other dimensions of risk assessment (Binimelis and Myhr, 2016; Falk-Zepeda and Zambrano, 2010; Spök, 2010). These are challenges that need further elaboration with the aim to establish methodologies to be used during the processes of framing, data gathering, assessment and decision-making related to broader evaluation (Catacora-Vargas *et al.*, 2017).

Norwegian Gene Technology Act: sustainability, benefit to society and ethics

The GTA of 1993 regulates the production and use of GMOs. For a GMO to be approved in Norway, the Act requires that it must not be harmful to health or the environment. Norwegian authorities must also consider whether the production and use of the GMO contributes to sustainable development, is of benefit to society and is ethically justifiable.

The Biotechnology Advisory Board is responsible for making a broad assessment of GMOs, and has a special responsibility for assessing sustainability, social benefit and ethical factors. In 2000, the Board published a report on how to operationalise the concepts of sustainable development, social benefit and ethical and social considerations in the GTA (This report was revised in 2006 and 2009, see Norwegian Biotechnology Advisory Board, 2010). The Board has also carried out projects aimed at operationalise the concepts of sustainable development, social benefit and ethics in the GTA. In addition, reports on insect-resistant genetically modified plants (2011) and herbicide-resistant plants (2013) has been published. In both projects, scientists from different scientific disciplines and institutions in Norway contributed as ad-hoc experts. The parameters that were elaborated by the experts in the projects included environmental, societal and economic aspects. Recently, the Board presented a report discussing a revision of the GTA, discussing a level-based approval system for the release of GMOs based on the degree of genetic change. The majority of the board supported such a system, simplifying the approval process for gene-edited organisms, while retaining non-safety assessments for all organisms the are subject to the Act (Norwegian Biotechnology Advisory Board, 2018).

The justification for inclusion of non-safety assessments

An early driver for precaution, in addition to safety concerns, was the fear of GM technology as a means for transferral of power over food production from democratic institutions to 'science, technology and the industries that increasingly control them' (Tait 2001, 185). This warrants making stronger demands on issues of public concern. These non-safety concerns include ethical issues (e.g. animal welfare and animal integrity), societal distributions of benefits and sustainability. When a basic concern in the debate is the distribution of benefits – given a shared risk – a demand for an assessment of societal benefits is reasonable. Likewise, when the concern is with the long-term effects on the environment due to altered agricultural practices, a short-term risk assessment is not adequate alone. The issue of sustainability must therefore also be addressed. These concerns are clearly thematised in the ethical debate, the first as social justice and the second as matter of responsibility.

Given the arguments above, it follows that if one takes a precautionary approach, a broader assessment of non-safety issues is justified (Myhr and Myskja, 2011). What are the arguments for *not* taking such an approach? The most important is that these criteria are vague, arbitrary and subjective (Zetterberg and Björnberg, 2017). They are political and not scientific terms (such as risk) and cannot become the subjects of scientific, objective assessments. This gives wide room for abuse, for irrelevant, subjective rejections contrary to fairness in treatment. In addition, the basis for the assessments are shifting over time and depend on agriculture systems applied as well as geography. For example, a sustainability assessment of an herbicide-tolerant GM soy plant at a given time may find that the variety does not reduce herbicide use, does not increase yields per acre, the long-term effects of the herbicide on farmworkers and affected eco-systems are unknown, etc. Sustainability assessment on a later stage, however, could show that the variety actually lead to significant reduction in tillage, which reduced CO_2 emissions and soil erosion, and there is registered reduction in herbicide use as farmers gather experience in effective use. This would have altered the sustainability assessment significantly, but this information was not available at the time of assessment. How can one weigh these very different sustainability factors, and how can one decide the correct time for assessment? Another example is gene drives to eradicate malaria. How can one objectively weigh the societal benefits of saving numerous lives against the uncertainty regarding ecosystem effects and eradicating of a species, which can be classified as sustainability and ethics assessments? This seems to place us in a choice between incommensurable values, where the final decision is arbitrary and subjective. We can quantify lives saved and the risk of gene edited mosquitos spreading beyond a certain area, and likewise, model sustainability, but these quantifications are always limited by available experience and knowledge. They are only complete with some element of conjecture.

However, arguments like these assume that avoiding assessment with no clear scientific answer is adequate, but that is morally wrong. We should strive for the best approximation, even if there are no precise, quantified basis for the decision. Ethical decisions are a matter of making the best arguments, and even if we cannot agree or achieve consensus, we can make a sound decision based on the right procedures where all relevant arguments are heard, and trusted, competent people make the final decision. Many have pointed out that risk analysis and management are also value based and 'subjective', showing that there is no direct way from fact to decision (Zetterberg and Björnberg, 2017). Judgement and deliberation is needed to achieve a socially robust result.

Concluding remarks

Exemption of plants and other organisms produced by gene-editing techniques from GMO regulatory framework would mean that there is no specific requirement to assess any potential effects on food, feed safety, health or environmental safety caused by the use of these techniques. It would also mean exemption from GMO labelling requirements, restricting consumer choice. However, even if gene-edited organisms are considered comparable to non-GMOs in terms of risks, and therefore exempt from safety assessments, we hold that the technology should be subject to an assessment of the non-safety concerns for two reasons. First, the ownership issue remains the same: this is a patentable technology (Ledford, 2017) although it is as yet not clear how that right is affected if the resulting organism also could have been produced by non-patentable methods. Second, the addition of foreign material is irrelevant for the potential for altering characteristics with impact on sustainability, societal issues and ethics.

References

Binimelis, R. and Myhr, A. I. (2016). Inclusion and implementation of socio-economic considerations in GMO regulations: opportunities and challenges. Sustainability 8 (1), 62.

Catacora-Vargas, G., Binimelis, R., Myhr, A. I. and Wynne, B. (2017). Socio-economic research on genetically modified crops: a study of the literature. Agriculture and Human Values, published online December 2017.

Court of Justice of the European Union (2018). According to Advocate General Bobek, organisms obtained by mutagenesis are, in principle, exempted from the obligations in the Genetically Modified Organisms Directive, press release 18.01.18. Available at: https://www.politico.eu/wp-content/uploads/2018/01/CP180004EN.pdf?utm_source=POLITICO.EU&utm_campaign=8ee908b06c-EMAIL_CAMPAIGN_2018_01_18&utm_medium=email&utm_term=0_10959edeb5-8ee908b06c-189810753. Accessed 22 January 2018.

European Commission. (2014). New EU approach on GMOs. Available at: http://ec.europa.eu/food/plant/gmo/legislation/future_rules_en.htm. Accessed 9 May 2015.

Falck-Zepeda, J. B. and Zambrano P. (2011). Socio-economic Considerations in Biosafety and Biotechnology Decision Making: The Cartagena Protocol and National Biosafety Frameworks. Review of Policy Research 28(2): 171-195.

Greiter, A., Miklau, M., Heissenberger, A. and Gaugistsch, H. (2011). Socio-economic aspects in the assessment of GMOs – Options for action. Report 0345. Environment Agency Austria (Umweltbundesamt), Vienna, Austria, 48 pp.

Hartung, F. and Schiemann, J. (2014). Precise plant breeding using new genome editing tehniques: opportunities, safety and regulation in the EU. Plant Journal 78: 742-752.

Jones, H. D. (2015). Future of breeding by genome editing is in the hands of regulators. GM Crops & Foods, 6 (4), published online March 2016.

Ledford, H. (2017). Bitter CRISPR patent war intensifies. Nature News, 26 October 2017. Available at: https://www.nature.com/news/bitter-crispr-patent-war-intensifies-1.22892. Accessed 24 January 2018.

Myhr, A.I. and Myskja, B.K. (2011). Precaution or Integrated Responsibility Approach to Nanovaccines in Fish Farming? A Critical Appraisal of the UNESCO Precautionary Principle. Nanoethics 5 (1): 73-86.

Nature Editorial (2017). Seeds of change: The European Union faces a fresh battle over next-generation plant-breeding techniques. Nature 520: 131-132.

Norwegian Biotechnology Advisory Board (2000). Sustainability, benefit to the community and ethics in the assessment of genetically modified organisms: Implementation of the concepts set out in Section 1 and 10 of the Norwegian Gene Technology Act. Available (in Norwegian) at: http://www.bioteknologiradet.no/filarkiv/2010/07/1999_04_11_baerekraft_samfunnsnytte_og_etikk_temahefte.pdf. Accessed 22 January 2018.

Norwegian Biotechnology Advisory Board (2011). Insect-resistant genetically modified plants and sustainability. Available (in Norwegian) at: http://www.bioteknologiradet.no/filarkiv/2011/06/rapport_baerekraft_110627_web.pdf. Accessed 21 January 2018.

Norwegian Biotechnology Advisory Board (2013). Herbicide-resistant genetically modified plants and sustainability. Available at: http://www.bioteknologiradet.no/filarkiv/2014/09/Herbicide-resistant_genetically_modified_plants_and_sustainability_NBAB.pdf. Accessed 21 January 2018.

Norwegian Biotechnology Advisory Board (2018). The Gene Technology Act – Invitation to public debate. Available at: http://www.bioteknologiradet.no/uttalelser. Accessed 24 January 2018.

Spök, A. (2010). Assessing Socio-Economic Impacts of GMOs. Issues to Consider for Policy Development. Lebensministerium/Bundensministerium für Gesundheit, Vienna, Austria. 123 pp.

Tait, J., (2001) More Faust than Frankenstein: the European Debate about Risk Regulation for Genetically Modified Crops. Journal of Risk Research 4(2): 175-189.

Zetterberg, C. and Edvardsson Björnberg, K. (*2017*). Time for a new EU regulatory framework for GM crops? Journal of Agricultural and Environmental Ethics 30(3): 325-347.

14. Representing non-human animals: committee composition and agenda

M. Vinnari[*] *and E. Vinnari*
University of Tampere, Faculty of Management, Kanslerinrinne 1, 33014 University of Tampere, Finland; markus.vinnari@uta.fi

Abstract

As a result of the 'political turn' in animal ethics and the broader 'animal turn' in the social sciences, scholarly work has begun to emerge that contemplates whether it is feasible to extend traditional concepts of political theory, such as citizenship and political agency, to non-human animals. In this largely conceptual discussion, the actual mechanism whereby non-human animals' interests could be taken into account in political decision-making has received little attention. Thus, the purpose of the present paper is to consider the practical procedures related to the composition and operation of an advocacy committee where non-human animals would be represented by human proxies. We propose a model in which various scientific and lay experts elect the members of a 'committee for non-human animals' that not only produces statements on bills that coercively affect sentient non-human animals, but also initiates political discussion on critical issues. We also provide an illustrative example of how such a committee could deal with issues related to a proposed sustainability transition from animal-derived to plant-based foodstuffs.

Keywords: decision-making, non-human animals, political agency, representation

Introduction

The treatment of non-human animals has become one of the key policy areas that modern societies need to tackle (Buller and Morris 2003). While animal ethicists have thus far focused on defending the claim that animals are worthy of our moral concern, they have only recently begun to consider the possibilities for enfolding animals within democratic processes. This phenomenon has been labelled the 'political turn' in animal ethics (Garner and Sullivan, 2016). A parallel, somewhat associated development has been the 'animal turn' in the social sciences, where it has been recognized that human beings and non-human animals are involved in a variety of social relations at home, in the workplace, and in such public spaces as parks. In contrast to other social scientists, however, political science scholars have been reluctant to consider the inclusion of non-human animals into the political sphere.

As a result of these respective 'turns' in animal ethics and the social sciences, scholarly work has begun to emerge that contemplates whether and how non-human animals could be enfolded within democratic processes (e.g. Cochrane, 2010, 2012). Within this literature, a debate has emerged concerning whether and how it would possible to extend traditional concepts of political theory, such as citizenship and political agency, to non-human animals (e.g. Donaldson and Kymlicka, 2011; 2015; Hinchcliffe, 2015; Hooley, in press; Ladwig, 2015). As this discussion and debate has to a great extent focused on conceptual issues, even those advocating the extension of democracy to non-human animals have only briefly touched upon the actual mechanisms whereby non-human animals' interests could be taken into account in political decisions. In doing so, most of these scholars have referred to models of advocacy or enfranchisement (Garner, 2016, 2017; Hooley, in press) in which non-human animals would be represented by human proxies. However, they have largely refrained from elaborating on how such representatives could be selected and what their work would entail in practice.

Svenja Springer and Herwig Grimm (eds.) ***Professionals in food chains***
EurSafe 2018 – DOI 10.3920/978-90-8686-869-8_14,

We acknowledge that a topic as novel and contested as the political representation of animals requires and deserves thorough philosophical and conceptual discussion. Yet, we are also eager to go beyond the yes/no debate since it is only possible to answer these questions by examining what an affirmative or negative answer might entail in practice. Taking the feasibility of animals' political representation as a given, we attempt to anticipate the kinds of arrangements that could be implemented to guarantee that animal interests are taken into account in political decision-making. Thus, the purpose of the present paper is to consider the practical implementation of an advocacy model of representing non-human animals. In particular, we will seek to answer the following research questions: Which items would be on the agenda of a committee for non-human animals? How could the members of such a committee be selected?

We consider these issues both on a general level and, for illustrative purposes, in the specific case of the food system. It has been suggested that, at least in affluent Western nations, ethical, environmental and public health related reasons warrant a sustainability transition from animal-derived to plant-based foodstuffs (Vinnari and Vinnari, 2014). Political decision-making associated with the implementation of such a transition offers a good opportunity for us to demonstrate how such questions could be processed by a committee for non-human animals.

The paper is organized as follows. In the second section, we review prior research at the intersection of political philosophy and animal ethics, focusing especially on citizenship and political representation. In the third section, we propose a procedure for selecting human beings to a governmental body representing the interests of non-human animals and a tentative agenda for organizing the workings of such a body. In the fourth section, we discuss the conclusions and implications of the study.

Literature review: political representation of non-human animals

Previous research situated at the nexus of animal ethics and political philosophy has investigated the applicability of concepts derived from political theory to non-human animals. In a much discussed and debated work, Donaldson and Kymlicka (2011) outline a political theory of animal rights based on an abolitionist approach, which denies all exploitation of non-human animals by human beings (Francione, 1996). More specifically, Donaldson and Kymlicka (2011) divide non-human animals into three groups – wild, liminal and domesticated animals[1]; based on their relations with human beings. The authors propose that domesticated animals can be conceived of as political agents and, consequently, human beings' co-citizens, whose interests regarding issues that affect them should be actively solicited and taken into consideration in political decision-making. They further argue that wild animals ought to be considered sovereign beings whose lives should not be infringed upon by human beings. Donaldson and Kymlicka's (2011) application of the concepts of citizenship and sovereignty to non-human animals has since been debated (see e.g. Donaldson and Kymlicka, 2015; Hinchcliffe, 2015; Ladwig, 2015).

One contested aspect in Donaldson and Kymlicka's (2011) proposition is their starting position, namely their subscription to animal rights theory. To circumvent such moral philosophical debates, some political scientists have tried to derive the necessity of considering animals in political decision-making by utilizing conceptual resources from political philosophy only. Garner (2016) for instance has problematized Donaldson and Kymlicka's (2011) citizenship model specifically on the grounds that it is based on contentious claims regarding the capacities of non-human animals. Garner's (2016) preferred alternative relies instead on the all-affected principle, according to which all those affected by a decision should be heard when making that decision (Eckersley, 2000; Goodin, 2007; see also Latour,

[1] We acknowledge that this division has its problems and animals can for example belong to multiple groups (Herzog, 2010). This can also lead to the exclusion of some animals, e.g. invertebrates from the considerations.

2004). In a similar vein, Hooley (in press) emphasizes that regardless of whether or not they fulfill the criteria of full-fledged political agents, non-human animals can be considered our fellow citizens whose good matters to the public good and whose preferences should therefore be solicited and responded to.

As most of this emerging literature has focused on macro-level abstractions, very few scholars have thus far considered how non-human animals' political agency or the all-affected principle could be translated into practice. Garner (2016, 2017) proposes an enfranchisement model whereby non-human animals' preferences are taken into account through human proxies elected to represent their interests in state legislature. It appears, however, that in his view the agenda of such a committee would be limited to reporting on bills that coercively affect non-human animals. Moreover, his only advice regarding the implementation of the enfranchisement model is that the human representatives could be selected by 'a constituency made up of organisations concerned about the wellbeing of animals' (Garner, 2017: 467). As a step towards the practical implementation of an enfranchisement model, in the following section we propose an agenda for a committee for non-human animals and then elaborate on the principles for selecting the human beings who would serve on that committee.

Hierarchy of needs and selecting representatives of non-human animals

In our proposed model, the agenda of a committee for non-human animals would be based on a hierarchy of needs type of approach. In the case of human beings, it is relatively widely accepted that individuals' needs can be placed in a hierarchical order, such as in Maslow's hierarchy of needs or the Max-Neef model of human-scale development (Marshall and Toffel, 2005). We combine these two approaches in our model in order to present a development-oriented agenda for the committee for non-human animals to pursue (Table 1). To elaborate on the various levels of the development hierarchy in the case of human beings, the baseline (zero) level target is for all human groups to be acknowledged. One does not need to go very far back in history to find places where not all human groups or their needs were considered to be relevant in policymaking.[2] Acknowledgement of existence is followed by what we call first-level targets, in other words fulfilling the basic needs of food and shelter as well as freedom from torture. The second-level targets in turn relate to basic rights, such as freedom from slavery and the right to a fair trial, while the third-level targets of equality of self-actualisation comprise political agency as well as freedom of thought and speech.

If we accept the above targets in the case of human development[3], in the name of consistency we can apply a similar rationale to non-human animals (Bruers 2013; Allievi *et al.*, 2015). Doing so would imply that animals would first need to be acknowledged as actors (Level 0), after which they should be allowed to fulfill their basic needs (Level 1). In the case of domesticated animals, the level of welfare has been defined in the form of the Five Freedoms (freedom from thirst and hunger, discomfort, pain, fear and stress as well as the freedom to express their normal behavior). Correspondingly, wild animals should be allowed freedom from excessive harm inflicted by humans.

From the perspective of basic rights (Level 2), the target for domesticated animals would be the right not to be killed by human beings. Because wild animals can be likened to inhabitants of foreign nations (Donaldson and Kymlicka, 2011) since they do not live in such close relationships with human beings

[2] In some cultures, this can be still the case for women, certain societal classes (e.g. untouchables in India) or sexual minorities.

[3] We acknowledge that our line of thought can be argued to resemble the human rights tradition of debating which rights are more important than others. Although that discussion appears to have abandoned hierarchies over the view that rights need to be seen as inalienable and indivisible, we argue that a hierarchical view is necessary in the case of non-human animals' rights as the latter are at the moment far from being universally accepted.

Table 1. Hierarchy of needs for human beings and non-human animals.

	Humans	Domesticated animals	Wild animals
Level 3	Freedom from conditions of compulsion Equality and self-actualisation • Political agency • Freedom of thought • Freedom of speech	Self-actualisation • Citizenship	Self-actualisation • Sovereignty
Level 2	Rights Basic rights • Freedom from slavery • Right to fair trial	Right to live • Right not to be killed	Species survival • Right not to go extinct/not to be killed
Level 1	Basic needs Basic survival • Fulfilment of basic needs (food, shelter) • Freedom from torture	Basic welfare • Five Freedoms (hunger, discomfort, pain, fear, express behaviour)	Harm avoidance • Freedom from excessive harm inflicted by humans
Level 0	Acknowledgement • To acknowledge different groups (such as women, various ethnicities, etc.) as relevant in policy-making	• To be considered as actors with interests relevant to policy-making	• Identify the species involved with interests relevant to policy-making

as domestic animals do, it might be necessary to define wild animals' rights in relation to species instead of individuals. A suggested target could be for instance that wild animal species would be secured the right not to become extinct because of human actions or, more extremely, the right not to be killed.

The Level 3 target for domesticated animals would be the possibility for self-actualization, meaning that these animals would possess some degree of citizenship, while for wild animals, self-actualisation would imply sovereignty (compare Donaldson and Kymlicka, 2011).

As concerns the selection of non-human animals' representatives to serve on the committee, we propose that these advocates would comprise experts, understood in a wide sense to include both scientists and laypersons[4]. These experts could be selected by an initial group of experts, who in turn would be solicited by government civil officials[5] from universities, research organizations and NGOs. This process can be facilitated by first defining the parameters related to the sought-after expertise in the form of questions that the committee is envisaged to consider in its meetings. The hierarchy of needs presented above can be of help in the selection of the experts and also in dividing them to sub-groups. For example, one sub-committee could discuss the animal welfare questions related to Level 0 and Level 1 development targets, and another could discuss the animal rights questions associated with Level 2 and Level 3 targets (Table 2). The necessity of having such sub-committees derives from the different moral philosophical

[4] The experts' work is expected to concern animal rights. Scientist, for example, are expected to have published on animal rights. Likewise, the NGO representatives should be from animal rights organizations.

[5] To guarantee its independence, the committee would not be positioned under any individual ministry but would address its reports directly to the Parliament, somewhat similar to a parliamentary audit committee.

Table 2. Example questions to be deliberated by a committee for non-human animals.

	Domesticated animals	Wild animals
Level 3	• How to secure citizenship/sovereignty for non-human animals? • How to achieve non-human citizenship/sovereignty in a way that does not increase inequality among humans? • How to utilise economic methods to guide the process? • How to guide the change at a feasible pace?	
Level 2	• Do domesticated animals have inherent value? • Should the killing of domesticated animals be prohibited?	• Do wild animals have inherent value? • Should the killing of wild animals be prohibited? • What kind of limitations to land use or consumption are needed to secure wild animals' rights?
Level 1	• Do non-human animals feel pain, sorrow, etc.? • To what degree should the suffering of domesticated animals be minimised?	• Do non-human animals feel pain, sorrow, etc.? • Should some type of hunting or fishing methods be banned to prevent animal suffering?
Level 0	• Are humans animals?	• What kind of life exists?

standpoints of animal welfarists, who accept human use of animals on condition that the harm experience by the latter is minimized, and animals rights advocates, who do not accept human use of non-human animals. Thus, it would be relatively difficult to organise joint discussions related to animal welfare and animal rights in one committee as these targets do not necessarily go in hand in hand, but can even in some cases be counterproductive (Francione, 1996).

On Levels 0 and 1, the representatives of animals could be biologists and philosophers. On Level 1, domesticated animals could additionally be represented by veterinarians and experts from animal welfare organizations. Wild animals, on the other hand, could be presented by biologists and experts from environmental NGOs. Related to Level 2 and 3 targets, domesticated animals could be represented by animal ethicists and experts from animal rights NGOs. On Level 3, more emphasis could be placed on sociologists and economists to guide the development process, as on that level the discussions of the committee would related mainly to the distribution of resources among relatively equal actors.

Conclusions

The non-human animal issue has become one of the areas where new ways to organize decision-making are needed. In this paper, we have argued that one way to recognize non-human animals as politically relevant actors would be to use scientific and lay experts as their representatives in political decision-making processes. In order for these processes to propel a transition towards the attainment of rights, and even citizenship/sovereignty, for non-human animals, we can conceptualize the agenda of a committee for non-human to as a hierarchy of needs. The starting-point for development is to acknowledge the different individuals/species as actors. The committee's subsequent target would be to secure at least some degree of welfare for the non-human animals, after which the committee's focus would shift to the issue of securing basic rights. The committee's ultimate target would be to secure non-human animals' freedom from conditions of compulsion, in other words making them equal to human beings and providing them with opportunities for self-actualisation. The selection of experts to serve on the committee for non-human animals could follow a similar hierarchical outline, utilizing different experts in different stages of the development process. The selection of these experts should be left to an initial group of experts.

References

Allievi, F., Vinnari, M. and Luukkanen, J. (2015). Meat consumption and production – analysis of efficiency, sufficiency and consistency of global trends. Journal of Cleaner Production 92: 142-151.

Bruers, S. (2013). The Ethical Consistency of Animal Equality. Sept 2013, DRAFT. Available online (Cited 18.12.2017). https://stijnbruers.files.wordpress.com/2013/05/the-ethical-consistency-of-animal-equality5.pdf.

Buller, H. and Morris, C. (2003). Farm animal welfare: a new repertoire of nature-society relations or modernism re-embedded? Sociologia Ruralis 43: 216-237.

Cochrane, A. (2010). An introduction to animals and political theory. London, Palgrave.

Cochrane, A. (2012). Animal rights without liberation. Applied ethics and human obligations. New York, Columbia University Press.

Donaldson, S. and Kymlicka, W. (2011). Zoopolis: A Political Theory of Animal Rights. Oxford, Oxford University Press.

Donaldson, S. and Kymlicka, W. (2015). Interspecies Politics: Reply to Hinchcliffe and Ladwig. The Journal of Political Philosophy 23: 321-344.

Eckersley, R. (2000). Deliberative democracy, ecological representation and risk: Towards a democracyof the affected. In M. Saward (Ed.), Democratic Innovation: Deliberation, Representation and Association (pp. 117-132). London, Routledge.

Francione, G. (1996). Rain without thunder: The ideology of the animal rights movement. Philadelphia, Temple University Press, 269 pp.

Garner, R. (2016). Animals, Politics, and Democracy. In Robert Garner and Siobhan O'Sullivan (eds.) The Political Turn in Animal Ethics (Rowman and Littlefield, New York, chapter 7).

Garner, R. (2017). Animals and democratic theory: Beyond an anthropocentric account. Contemporary Political Theory 16: 459-477.

Goodin, R. (2007). Enfranchising all affected interests and its alternatives. Philosophy & Public Affairs 35: 40-68.

Herzog, H. (2010). Some we love, some we hate, some we eat. HarperCollins, New York, 326 pp.

Hinchcliffe, C. (2015). Animals and the Limits of Citizenship: Zoopolis and the Concept of Citizenship. Journal of Political Philosophy 23: 302-320.

Hooley, D. (in press). Political Agency, Citizenship, and Non-human Animals. Res Publica. DOI 10.1007/s11158-017-9374-1.

Ladwig, B. (2015). Against Wild Animal Sovereignty: An Interest-based Critique of Zoopolis, Journal of Political Philosophy 23: 282-301.

Latour, B. (2004). Politics of Nature: How to Bring Sciences back into Democracy. Cambridge, Harvard University Press, 307 pp.

Marshall, J. and Toffel, M. (2005). Framing the Elusive Concept of Sustainability: A Sustainability Hierarchy. Environmental Science & Technology 39: 673-682.

Vinnari, M. and Vinnari, E. (2014). A framework for sustainability transition: The case of plant-based diets. Journal of Agricultural and Environmental Ethics 27: 369-396.

15. The challenge of including biodiversity in certification standards of food supply chains

S. Stirn[*] *and J. Oldeland*
Bodiversity, Evolution and Ecology, Institute for Plant Science and Microbiology, University of Hamburg, Ohnhorststr. 18, 22609 Hamburg, Germany; susanne.stirn@uni-hamburg.de

Abstract

Worldwide, agricultural production, food producers and retailers have a strong impact on biodiversity. Agriculture is the cause of up to 70% of the projected loss of biodiversity, mainly due to the expansion of cropland into forests, grasslands and savannahs. In the European agricultural landscape, intensive agriculture with its dependencies on the use of pesticides and synthetic fertilizers is another main cause of biodiversity loss. Therefore, the agricultural sector is asked to develop management options towards a more sustainable production and consumption pattern as set down in SDG 12 (Responsible Consumption and Production) and simultaneously to improve biodiversity indicators for SDG 15 (Life on Land). The Baseline Report of the EU Life-project 'Biodiversity in standards and labels for the food sector' analysed current certification standards and pointed out several gaps concerning the protection of biodiversity. In sum, effectively reducing the negative impacts of food production on biodiversity remains a big challenge due to the complexity of the matter. Although there exists a global agreement on the definition of 'biodiversity' (*sensu* UN Convention on Biological Diversity), the application of operable biodiversity indicators and procedures for monitoring trends of these indicators are still lacking behind. Finally, the allocation of responsibility within the supply chain with regard to the assessment and protection of biodiversity remains an open issue. We analyse the current practise of certification standards with regard to: (1) the definitions of biodiversity used; (2) the indicators used for documentation of compliance; and (3) the allocation of responsibilities within the supply chain.

Keywords: allocation of responsibilities, biodiversity indicators, certification standards, SDGs

Background

In 2015, the necessary shift to more sustainable consumption and production patterns was again stressed by the release of the 17 Sustainable Development Goals (SDGs). It was explicitly mentioned that 'consumption patterns need to be made sustainable, particularly lifestyles in industrialized societies, and reduce their ecological footprint to allow for the regeneration of natural resources on which human life and biodiversity depend.' The special role of businesses has been emphasised: 'It is in the interest of business to find new solutions that enable sustainable consumption and production patterns. A better understanding of environmental and social impacts of products and services is needed [...]. Identifying 'hot spots' within the value chain where interventions have the greatest potential to improve the environmental and social impact of the system as a whole is a crucial first step' (United Nations, 2015).

However, the relation between the food industry and natural resources is ambiguous: On the one hand side food production is strongly depending on healthy ecosystems providing valuable natural resources. Healthy ecosystems are characterized by a stable biodiversity which does not only provide raw materials (food, fibre, ingredients) but also clean water and healthy soils. At the same time, the agricultural sector is one of the main factors contributing to the loss of biodiversity globally. In terrestrial ecosystems, food production is estimated to contribute to 60-70% of total biodiversity loss in terms of animal and plant species (Kok and Alkemade, 2014). The main impact of food production on terrestrial biodiversity is through land use change, mainly the conversion of natural habitats into agricultural lands (Kok and

Svenja Springer and Herwig Grimm (eds.) **Professionals in food chains**
EurSafe 2018 – DOI 10.3920/978-90-8686-869-8_15, © Wageningen Academic Publishers 2018

Alkemade, 2014). Thus, there is a growing awareness of the need to pay more attention to biodiversity aspects. The loss of biodiversity or transformed ecosystems can have dramatic economic effects for the food industry such as the depletion of natural resources leading to supply chain disruptions, loss of reputation due to negative effects of economic activities on biodiversity or failure to comply with legal requirements.

In a previous analysis we looked at the respective data provided by the top German retailers in their publicly available reports. Although the quantity and quality of the information provided was quite different it became obvious that concerning their environmental impacts the retailers were predominantly disclosing data on CO_2-emissions and their efforts to reduce these. This is a comparatively simple task and makes good economic sense (win-win-solutions). In the case of the direct impacts of their food supply chains on natural resources, the retailers are aware of the cultivation phase of the raw materials being a hotspot of possible negative environmental impacts (e.g. the destruction of rain forests or the environmental damage caused by pesticides and fertilizers) (Rewe Group, 2016). In these cases the retailers are mainly relying on certification schemes and their labels to cover possible environmental impacts and to inform the consumers (Stirn *et al.*, 2016). However, it remains unclear to what extent and in which form these certification schemes cover biodiversity.

In the current analysis, we will therefore take a closer look at the topic 'biodiversity' within the task of measuring and improving the environmental impact along the food supply chain with the help of certification systems. We will take a closer look at: (1) the definition of biodiversity used; (2) the indicators of biodiversity reported; and (3) the balance of responsibility between producers, retailers, and consumers.

What is biodiversity?

The term 'biodiversity' is very much ambiguous and vague. In ecology, biodiversity comprises the richness, abundance, and composition of genes, species and ecosystems (United Nations, 1992). A useful definition of biodiversity was provided by Beierkuhnlein (2003) who stated that 'Biodiversity is the expression of the quantitative, qualitative, or functional diversity of biotic objects at different organisational levels [genes, species, ecosystems] within concrete or abstract spatial and/or temporal scales'. This definition is still not complete as new concepts such as phylogenetic diversity (Magurran and McGill, 2011) appear. Hence, 'biodiversity' itself is very diverse and thus becomes a complex topic, making it way more complicated to communicate than climate change.

This complexity makes it challenging to measure 'biodiversity' in a useful and practical way (Purvis and Hector, 2000). For example, there is no generally accepted framework for measuring biodiversity as this strongly differs between taxonomic groups, such as insects, mammals or plants. However, for each taxonomic group standardized approaches do exist. Policymakers have formulated national targets for the protection of biodiversity (National Strategy on Biological Diversity), however, only overriding and less specific targets have been set probably given the difficulty of measuring biodiversity (e.g. 'Biodiversity and landscape quality' is measured as the population development of 51 bird species in Germany). One of the few regulatory frameworks in Europe is EMAS (Eco-Management and Audit Scheme) which is an environmental management scheme for businesses. Within this management scheme members of the Advisory Boards are continuously working on improving core indicators for environmental reporting. Until today, the only indicator for biodiversity required by EMAS is land use expressed in 'm^2 of built-up area' (Office of the German EMAS Advisory Board, 2010). Besides these legislative frameworks, a number of international and national initiatives are currently developing biodiversity indicators, e.g. the Global Reporting Initiative (Global Reporting Initiative, 2016) which was the first to develop global standards for sustainability reporting of businesses. Thus, the development of new biodiversity indicators is ongoing and is urgently needed. How far biodiversity is implemented already in certification schemes which are of interest for retailers and consumers will be investigated more closely in the next chapter.

Certification schemes

Along the food supply chain the most crucial challenges are related to the assessment of the impact on biodiversity during the cultivation phase, especially for bulk commodities like palm oil, cacao and soybeans. Since retailers are not able to assess the impacts of the multitude of products offered, they mainly rely on certification schemes. This dependency makes the latter an important factor when it comes to the effectiveness in safeguarding from biodiversity loss. Accordingly, we looked at three different certification schemes and the indicators they are using to assess impacts on biodiversity. We chose UTZ (UTZ, 2016), Roundtable on Sustainable Palm Oil (RSPO, 2013) and Roundtable on Responsible Soy (RTRS) as representative examples for the crucial commodities coffee, cacao, tea (UTZ), palm oil (RSPO) and soybeans (RTRS).

All certification schemes apply different environmental indicators or standards for reporting: for example UTZ and the RSPO standards explicitly prohibit the deforestation of primary forests, whereas the RTRS standards, in more general terms, ask for the responsible expansion of soy cultivation provided that off-site environmental impacts have been assessed and minimized. Further, the term biodiversity and its indicators are not well defined and therefore are prone to different interpretations. We have summarized how our three aspects of biodiversity in the food supply chain are covered in the different certification schemes (Table 1).

Definitions of biodiversity and indicators used

First of all, it is visible from Table 1 that all certifiers apply terminology actually used in conservation science. For example, all certification schemes analysed require information on rare or endangered species. However, it is not said, which taxonomic group (e.g. plants, insects, birds, mammals, microbes) is of importance, how rarity is measured and at which spatial scale an endangered species is in fact 'endangered'. In particular the concept of spatial scale is tremendously important for measuring biodiversity, as biodiversity is scale dependent (Dengler and Oldeland, 2010): While rainforests in Ecuador are most species rich in plants on one hectare (942 species), mountain grasslands in Argentina lead the world records for plants on 1 m^2 (89 species) (Wilson *et al.*, 2012). Furthermore, only certain

Table 1. Overview on biodiversity related aspects covered by three certification schemes.

Aspect	UTZ	RSPO	RTRS
Definition of biodiversity used	• Threatened and endangered species • Ecological diversity • Habitats • Ecosystems	• Habitats of High Conservation Value • Rare, threatened or endangered species	• Habitats (native forests, wetlands) • Rare, threatened or endangered species
Indicators used for documentation of compliance and monitoring	• Presence of species • Presence and status of habitats • Protected areas	• Status of habitats • Status of rare species • Integrated Pest Management and invasive species	• Maps of habitats including type, extent, area • Monitoring of rare species presence • Biodiversity Action Plan

aspects of biodiversity are assessed (threatened species, primary forests) while others are not covered at all. For example, all certification schemes use the status or presence of a species as indicators. However, the important aspect of species abundance is neglected. As many biodiversity indices are abundance based, it is especially important to include abundance for monitoring change. There could be hundreds of individuals or only one individual of a bird species; without accounting for abundance, it would be simply noted as occurring in one year, but occurrence alone does not allow to properly infer on the condition of that bird population. Also, all certification schemes are focused too much on single species, while the species composition of an ecosystem is not considered, although the composition is most important for the functioning of an ecosystem. UTZ lists 'ecological diversity' but it is left unclear what this term means. Finally, although RTRS requires that at least 10% of native vegetation remains intact on a farm, it is not specified in which spatial arrangement this is effective for conserving biodiversity. This is in contrast to the well-known negative effects of habitat fragmentation (Fahrig, 2003). Therefore, the biodiversity indicators listed by the certification schemes could be valuable but until now they are not specific enough.

Global environmental NGOs like Greenpeace or Conservation International are continuously criticizing certification schemes to further the development of biodiversity indicators and to include other aspects of biodiversity (e.g. species abundance, composition). Examples of a further development of biodiversity indicators are the EU published Sectoral Reference Documents (SRDs) on the voluntary participation in a community eco-management and audit scheme (EMAS). The SRDs on Best Environmental Management Practice provide guidance and inspiration to organisations in specific sectors on how to further improve environmental performance. The first Sectoral Reference Document is related to the retail trade sector and was published in May 2015 (European Commission, 2015). For other sectors 'Best Environmental Management Practices' have been published, e.g. for the agricultural sector which is also of great importance for the food supply chain (Antonopoulos *et al.*, in preparation). Here certain key environmental performance indicators are listed (Table 2), each with specific measurements.

As an improvement on biodiversity indicators, for example, abundance metrics of key species are now included. Fertilizer application or stocking rates are adjusted to reach the so-called 'benchmark of excellence' which aims at maintaining and enhancing local biodiversity (Table 2). At present, these indicators are on a voluntary basis and are incentives for establishing best practice examples and guidelines. These improvements and the experiences made with the use of the new indicators should be incorporated into existing certification schemes.

Table 2. Excerpt of best environmental management practices for the agricultural sector as envisioned by the EU commission.

Best environmental management practices	Benchmark of excellence	Key environmental performance indicators
3.4. Landscape scale biodiversity management	A biodiversity action plan established with local biodiversity experts is implemented on the farm…	• N application rate (kg/ha/year) • Key species abundance metrics (no./m^2)
7.2. Managing high nature value grassland	… to maintain and enhance the number and abundance of locally important species	• Species frequency and diversity (no. and no./m^2) • Average stocking rate (livestock units/ utilized agricultural area) • Match stocking rate to biodiversity needs

Allocation of responsibilities

The third point we looked at was the allocation of responsibility for more environmental sustainability in the food supply chain by using certification schemes. Here the situation is equally complex, because with the usage of certification schemes the responsibility is firstly with the producers. They have to comply with the standards, e.g. they have to ensure that targeted biodiversity indicators are met. This makes it more complicated to be listed as a supplier of certified products. Extensive literature exists on the hurdle certification schemes may pose especially for smallholders (e.g. Brandi *et al.*, 2015). Thus, certification agencies should take over responsibility to enable producers to comply with the higher environmental standards, e.g. by providing access to relevant information and supporting training of the producers. Because of their market power retailers are no longer solely distributers of food. Instead, they are influencing patterns of production and consumption. Therefore, they can use their active role to nudge consumers to choose the more sustainable product, e.g. by expanding their range of certified products or using marketing communications and awareness raising campaigns (Lehner *et al.*, 2016; Schubert, 2017; Thaler and Sunstein, 2008). Finally, the responsibility of the consumers is to inform themselves about certification schemes and to preferably select the labelled products, thereby increasing the demand for more sustainable products. There is much discussion about the respective share of responsibility lying with each actor along the food supply chain (see e.g. Akenji, 2014) but it is widely accepted that coordinated action is necessary to achieve the necessary shift to more sustainable production and consumption pattern

Conclusion

Certification schemes seem to be a good starting point for companies in the food supply chain to deal with environmental aspects. Progress has been made with regard to biodiversity indicators by including conservation relevant aspects. The experiences with and the discussion on the certification standards will give opportunities to constantly improve the respective standards. By using these certification standards, we have shown that all actors along the food supply chain bear certain responsibilities. Further discussions on the balance of the responsibility between, producers, certification agencies, retailers and consumers are required for reaching the SDG 12, responsible production and consumption.

References

Akenji, L. (2014). Consumer scapegoatism and limits to green consumerism. Journal of Cleaner Production 63: 13-23.

Antonopoulos, I.S., Canfora, P., Dri, M., Gaudillat, P., Styles, D., Williamson, J., Jewer, A., Haddaway, N. and Price, M. (in preparation). Best environmental management practice for the agriculture sector – crop and animal production. Available at: http://susproc.jrc.ec.europa.eu/activities/emas/documents/AgricultureBEMP.pdf. Accessed 9 January 2018.

Beierkuhnlein, C. (2003). Der Begriff der Biodiversität. Nova Acta Leopoldina 87(328): 51-72.

Brandi, C., Cabani, T., Hosang, C. *et al.* (2015). Sustainability Standards for Palm Oil: Challenges for Smallholder Certification Under the RSPO. Journal of the Environment & Development 24(3): 292-314.

Dengler, J. and Oldeland, J. (2010). Effects of sampling protocol on the shapes of species richness curves. Journal of Biogeography 37: 1698-1705.

European Commission (2015). Commission decision (EU) 2015/801 of 20 May 2015 on reference document on best environmental management practice, sector environmental performance indicators and benchmarks of excellence for the retail trade sector under Regulation (EC) No 1221/2009 of the European Parliament and of the Council on the voluntary participation by organisations in a Community eco-management and audit scheme (EMAS). Official Journal of the European Union, L 127/25.

Fahrig, L. (2003). Effects of habitat fragmentation on biodiversity. Annual review of ecology, evolution, and systematics 34(1): 487-515.

Global Reporting Iniative (2016). GRI Standards. Available at: https://www.globalreporting.org/standards. Accessed 9 Januar 2018.

Kok, M. and Alkemade, R. (2014). How Sectors Can Contribute to Sustainable Use and Conservation of Biodiversity. CBD Technical series 79.

Lehner, M., Mont, O. and Heiskanen, E. (2016). Nudging – A promising tool for sustainable consumption behaviour? Journal of Cleaner Production (134, Part A): 166-177.

Magurran, A.E., and McGill, B.J. (Eds.). (2011). Biological diversity: frontiers in measurement and assessment. Oxford University Press.

Office of the German EMAS Advisory Board (2010). The new core indicators of EMAS III. Available at:http://www.emas.de/fileadmin/user_upload/06_service/PDF-Dateien/UGA_Infosheet_Indicators.pdf. Accessed 9 January 2018.

Purvis, A. and Hector, A. (2000). Getting the measure of biodiversity. Nature 405: 212-219.

Rewe Group (2016) Rewe Group Sustainability Report. Available at: http://rewe-group-nachhaltigkeitsbericht.de/2016/en/supply-chain. Accessed 9 January 2018.

Round Table on Responsible Soy (RTRS) (2017). RTRS Standards for Responsible Soy Production. Version 3.1, 01 of June, 2017. Available at: http://www.responsiblesoy.org/certification/nuestra-certificacion/?lang=en. Accessed 9 January 2018.

Round Table on Sustainable Palm Oil (RSPO) (2013). Available at: https://rspo.org/key-documents/certification/rspo-principles-and-criteria. Accessed 9 January 2018.

Schubert, C. (2017). Green nudges: Do they work? Are they ethical? Ecological Economics 132: 329-342.

Stirn, S., Martens, M. and Beusmann, V. (2016). EU-sustainability reporting requirements – an incentive for more sustainable retailers? In: Olsson, Araujo and Viera (eds.) Food Futures: ethics, science and culture. Wageningen Academic Publishers, 484-490.

Thaler, R.H. and Sunstein, C. (2008). Nudge: Improving decisions about health, wealth, and happiness. Yale University Press, New Haven, Conneticut.

United Nations (1992). Convention on Biological Diversity. Available at: https://www.cbd.int/doc/legal/cbd-en.pdf. Accessed 9 January 2018.

United Nations (2015). Integrated and coordinated implementation and follow-up to the outcomes of the major United Nations conferences and summits in the economic, social and related fields. Follow-up to the outcome of the Millenium Summit. United Nations General Assembly, A/69/L.85.

UTZ (2016) UTZ Guidance Document Nature Protection (Version 1.0, August 2016). Available at: https://utz.org/?attachment_id=4201. Accessed 9 January 2018.

Wilson, J.B. Peet, R.K., Dengler, J. and Pärtel, M. (2012). Plant species richness: the world records. Journal of Vegetation Science 23: 796-802.

16. Ranging in free-range laying hens: animal welfare and other considerations

J.-L. Rault
Institute of Animal Husbandry and Animal Welfare, University of Veterinary Medicine, Vienna, Veterinärplatz 1, 1210 Vienna, Austria; jean-loup.rault@vetmeduni.ac.at

Abstract

The demand for products from free-range farm housing systems has increased. Despite this, there is little scientific knowledge about how much laying hens go outside (i.e. 'range') and its implications for animal welfare. This presentation covers a combination of studies on the topic of free-range egg farming from an animal welfare point of view. The aim is to illustrate through the case study of Australian free-range eggs the challenge of aligning farming practices, consumers' expectation and other market forces. Contrary to previous studies that reported low range use (5 to 30%) based on the proportion of the flock outside at one time, recent studies that followed individual hens with radio frequency identification systems revealed that most hens (85 to 95%) accessed the range. Hence, better methods are required to assess range use, and more hens range that commonly believed. Online forum discussions were run as a virtual version of focus group between stakeholders. The results highlighted that the general public and animal advocacy groups emphasised the psychological and natural aspects of poultry welfare, whereas industry-related members emphasised its health aspects. Video observations on commercial farms revealed that the main behaviours displayed by hens while on the range did not differ in terms of nature, but rather in terms of frequency from the behaviours shown inside the shed. This only partly addresses the consumer belief that free-range allows for more 'natural' behaviours. From the hens' point of view, the welfare implications of ranging appeared minimal, from the measures collected to date (fear, stress, fitness). Nevertheless, hens varied substantially in their range use, leaving the question of 'why do hens range?'. The scientific evidence suggests that free-range offers the choice to the hens, and the market suggests that it also offers a choice to the consumers. However, consumers' expectation that laying hens should spent most of their time outside may not necessarily match the laying hens' wants. Whether animal welfare science, through its current tools and approaches, is able to inform all societal questions on animal use and on a timely manner is a topic for discussion.

Keywords: behaviour, farming, natural, outdoor, stakeholders

Introduction

The demand for free-range products has increased internationally, especially in regards to poultry eggs and meat. Nevertheless, the drives for such changes are unclear. The best way to address the demand for different farming practices remains the source of debate in society. I will use the example of the Australian free-range egg industry as a case study to illustrate the relationship between various actors, with a focus on animal welfare.

When I started this work, there had been a recent and fast growth in free-range egg production in Australia, with an increase of 64% from 2006 to 2011 and 24% in the year 2011 alone, with free-range eggs representing 34% of the retail egg market at that time. Free-range eggs in Australia now represents 41.3% of the retail egg market in volume and 51.4% in value (Australian Eggs Limited, 2017).

For the farmers, there was little scientific knowledge available at that time about the factors modulating access to the outdoor (i.e. ranging behaviour) in laying hens, and its implications for animal welfare. This

Svenja Springer and Herwig Grimm (eds.) ***Professionals in food chains***
EurSafe 2018 – DOI 10.3920/978-90-8686-869-8_16, © Wageningen Academic Publishers 2018

motivated investment by the egg industry in research on free-range egg housing systems and practices, with the aim to better understand the factors modulating hen ranging behaviour.

This presentation covers a combination of experimental studies on the topic of free-range egg farming from an animal welfare point of view. The aim of this paper is to use this case study to illustrate the complexity of the food production topic and the various facets of the animal welfare debate.

Scientific assessment of hen ranging behaviour

Traditionally, the amount of ranging by free-range laying hens was estimated by observing the proportion of the flock seen outside at one point in time or at different times during the day. These studies reported relatively low ranging, with a maximum of between 5 and 46% of the flock observed on the range at any one time, and large variations between times of the day, flocks and studies (Bubier and Bradshaw, 1998; Hegelund *et al.*, 2005, 2006; Nicol *et al.*, 2003; Zeltner and Hirt, 2003).

However, more recently, researchers have been able to use technologies such as radio-frequency identification (RFID) systems to overcome the technical difficulty of following individual birds in large flocks over extended periods of time. Using RFID tags on individual laying hens on experimental and commercial farms, a large majority of hens were found to access the range on a regular basis, with relatively consistent results across studies revealing different ranging patterns: 10% of heavy user hens (every day and long or far ranging), 80% of regular users (at least every other day), and 10% of indoor hens (never observed ranging over the study period) (Gebhardt-Henrich *et al.*, 2014; Larsen *et al.*, 2017a; Richards *et al.*, 2011). Hence, better methods are required to assess range use, and more hens range that commonly believed.

In comparison to data on the number of hens ranging, few studies have observed the behaviours of laying hens while ranging (Campbell *et al.*, 2017; Chielo *et al.*, 2016; Rodriguez-Aurrekoetxea and Estevez, 2016; Thuy Diep *et al.*, in press). Ranging could in fact means anything from laying down inactive on the range to foraging for food and walking long distances. Overall, the main behaviours display on the range does not appear to differ in terms of nature, but rather in terms of frequency from the behaviours shown inside the shed (Chielo *et al.*, 2016; Thuy Diep *et al.*, in press), where the hens are more active while ranging. These findings only partly address the belief that free-range allows hens to perform more 'natural' behaviours not possible indoor.

Scientific assessment of the implications of ranging for hen welfare

While several recent studies have focused on more accurate assessments of ranging behaviour by laying hens, only a handful of studies have investigated the implications of ranging on laying hen welfare.

The outdoor range offers a wide range of environmental stimuli, opportunities for exercise, dust bathing substrates, and foraging opportunities with a diversity of food items (seeds, insects). All these attributes could benefit hen welfare. Yet, the outdoor range also presents a risk of predation, imbalanced diet, increased exposure to pathogens and inclement weather. These risks could seriously compromise the welfare of free-range hens.

The literature on the welfare outcomes of free-range housing systems is inconsistent. For instance, several studies have reported that greater outdoor range use is inversely related to the prevalence of feather pecking (e.g. Bestman and Wagenaar, 2003; Nicol *et al.*, 2003). Mahboub *et al.* (2004) found that, on an individual basis, hens that spent more time outside had less feather damage but Hegelund *et al.* (2006) reported that plumage condition was not correlated with range use. Indeed, this is an example of a possible non-causal relationship, where plumage condition and ranging behaviour could both be related to a third, unidentified variable. Furthermore, there is a large heterogeneity in range use between hens, with different sub-populations as outlined before, which may result in different welfare implications according to individual ranging patterns.

The focus on inter-individual variation in range use is also more closely aligned with the fact that animal welfare should be assessed at the level of the individual animal, not at group level. Based on the few evidence to date, the welfare implications of ranging appeared minimal, from measures collected at individual level (Campbell *et al.*, 2017; Larsen *et al.*, in press). Nevertheless, hens varied substantially in their range use, leaving unanswered the question of why hens access the range. The scientific evidence suggests that free-range offers the choice to the hens between various environments (Larsen *et al.*, in press), and the market suggests that it also offers a choice to the consumers.

Animal welfare consideration from the human perspective

Free-range products are often perceived as healthier or safer by consumers (Harper and Makatouni, 2002). However, the underlying reasons for this position and the consumer construct of free-range remain poorly understood.

Online forum discussions between stakeholders (members of the general public, animal advocacy groups, and poultry farming industry) have been used to explore human attitudes and beliefs relevant to poultry welfare (Howell *et al.*, 2016). Across six online chats, participants (general public, n=8; animal advocacy group, n=11, chicken industry, n=3; research or veterinary practice who had experience with poultry but no declared industry affiliation, n=3) discussed poultry welfare and completed pre- and post-chat surveys gauging perceptions and objective knowledge about poultry management and welfare. The online surveys and forums shown to be an effective way of gathering information about perceptions and reasons for approval or disapproval of specific poultry management practices, as well as gauging objective knowledge of those practices. The results highlighted once again that the general public and animal advocacy members emphasised the psychological and natural aspects of poultry welfare, whereas industry-related members emphasised its health aspects. In regards to free-range, most participants rated the welfare of poultry housed in free-range systems to be higher than poultry housed indoor (Howell *et al.*, unpublished data). This study also highlighted that misunderstandings about current farming practices could be clarified in forums which contained industry representation (Howell *et al.*, 2016), suggesting that industry engagement can help to inform consumers on current farming practices. Participants also agreed on the need for enforceable standards and industry transparency.

The overarching debate and its implications for farming practices

The paper so far covered the scientific evidence relating to hen ranging behaviour and hen welfare. However, free-range production has increased because of consumer demands, rather than based on scientific evidence that it is better for the hens or that it results in a better product. Nevertheless, the Australian consumer's expectation that free-range laying hens should spent most of their time outside (according to personal conversations with policy makers), which do not match the laying hens' wants, who spend in average 3 to 4 hours ranging daily.

In Australia, there was no clear legislative definition for free-range egg production, apart from that hens should be allowed outdoor access for at least 8 hours a day. Subsequently, the Australian government decided to develop a definition for free-range egg production (but only after it initiated numerous lawsuits for two years against individual farmers for sub-optimal ranging on those farms). This resulted in a new definition, which states that 'free range eggs must come from hens that have meaningful and regular access to the outdoors, with a stocking density not exceeding 10,000 birds per hectare'. Unfortunately, the first part of this definition can be considered rather loose and subjective, while the evidence to support the second part is questionable.

As an animal welfare scientist working closely on the topic, a frustrating aspect has been the lack of focus on the laying hens themselves as the main concerned party. The debate became oversimplified to the point of a highly mediatised debate over range stocking density (as highlighted in the above definition), in hens per hectare of outdoor space. There was no evidence at the time that this was a pressing concern for the hens, or at least the most pressing concern to hen ranging behaviour or hen welfare. The intention of supermarkets and egg resellers to add range stocking density as a label on egg cartons fuelled the debate, as it could be interpreted by consumers as stocking density being a meaningful aspect of free-range production, which pre-empted later scientific evidence which showed that outdoor stocking density may only moderately impact on ranging behaviour and welfare (Campbell *et al.*, 2017). This debate between animal welfare organisations and the poultry farming industry on range stocking density ignored issues such as uneven hen distribution on the range (which would not be solved with limiting range stocking density) and aspects that appear more relevant for the hens from scientific findings, such as means to improve the quality of the range by offering attractive features on the range such as shelters (Larsen *et al.*, 2017b). There may also be problems induced by human perception. For instance, the provision of covered shelters in the outdoor area (called wintergarden or verandah) is a highly preferred area by laying hens (Chielo *et al.*, 2016, Thuy Diep *et al.*, in press). However, from a legislative point of view, these types of areas do not count as outdoor area in the UK for instance, which led to poor uptake of this type of structures by the farming industry there. This conundrum illustrates the complexity of aligning consumer perception, laying hen preferences which often relate to hen welfare, economical implications for farmers (in an often unstable market) and the establishment of standards that satisfy all parties and result in the desired outcome.

Conclusions

The case study of free-range eggs in Australia represents a relevant example of discrepancies between farming practices, consumers' expectation and animal wants, influenced by economic pressure and information (or lack thereof).

Animal welfare science, through its current tools and approaches, is able to provide data to inform debates on societal questions in regards to our care of animals. The scientific approach could be improved by defining research questions precisely addressing societal needs or concerns, and in transferring the scientific knowledge back to society or the influential actors at the right time.

However, the slow pace of research and legislation seems to have increasing difficulties to cope with the increasing pace of societal changes fuelled by social media and retailer competition. Supermarkets and other retailers, in between farmers and consumers, play an increasing role in changes in the demand for food products, such as animal welfare friendly or environmentally sustainable products. However, these key players have only recently started be more active and engaged with other parties such as scientists. It is also worth investigating whether animal and consumer interests are being faithfully reflected by the main actors influencing food production, and accurately addressed by scientists delivering data to inform the debate.

Acknowledgements

This research would not have been possible without the support of my PhD student, Hannah Larsen, collaborators and principally Prof. Paul Hemsworth, and research grant support such as from the Australian Egg Corporation Limited, now Australian Eggs Limited.

References

Australian Eggs Limited (2017). Annual report 2016/2017. Available at: https://www.australianeggs.org.au/dmsdocument/818-annual-report-2017. Accessed 2 January 2018.

Bestman, M.W.P. and Wagenaar, J.P. (2003). Farm level factors associated with feather pecking in organic laying hens. Livestock Production Science 80: 133.

Bubier, N.E. and Bradshaw, R.H. (1998). Movement of flocks of laying hens in and out of the hen house in four free range systems. British Poultry Science 39: 5.

Campbell, D., Hinch, G., Downing, J. and Lee, C. (2017). Outdoor stocking density in free-range laying hens: effects on behaviour and welfare. Animal 11: 1036-1045.

Chielo, L., Pike, T. and Cooper, J. (2016). Ranging behaviour of commercial free-range laying hens. Animal 6: 1-13.

Gebhardt-Henrich, S.G., Toscano, M.J. and Frohlich, E.K.F. (2014). Use of outdoor ranges by laying hens in different sized flocks. Applied Animal Behaviour Science 155, 74-81.

Harper, G.C. and Makatouni, A. (2002). Consumer perception of organic food production and farm animal welfare. British Food Journal 104: 287.

Hegelund, I., Sorensen, J.T., Kjaer, J.B. and Kristensen, I.S. (2005). Use of the range area in organic egg production systems: effect of climatic factors, flock size, age and artificial cover. British Poultry Science 46: 1.

Hegelund, I., Sorensen, J.T. and Hermansen, J.E. (2006). Welfare and productivity of laying hens in commercial organic egg production systems in Denmark. NJAS – Wageningen Journal of Life Sciences 54: 147.

Howell, T., Rohlf, V., Coleman, G. and Rault, J.-L. (2016). Online chats to assess stakeholder perceptions of meat chicken intensification and welfare. Animal 6: 67.

Larsen, H., Cronin, G., Smith, C., Hemsworth, P. and Rault, J.-L. (2017a). Individual ranging behaviour patterns in commercial free-range layers as observed through RFID tracking. Animal 7: 21.

Larsen, H., Cronin, G., Smith, C., Hemsworth, P. and Rault, J.-L. (2017b). Behaviour of free-range laying hens in distinct outdoor environments. Animal Welfare 26: 255-264.

Larsen, H., Hemsworth, P., Cronin, G., Smith, C. and Rault, J.-L. (in press). Relationship between welfare and individual ranging behaviour in commercial free-range laying hens. Animal, in press. doi: https://doi.org/10.1017/S1751731118000022.

Mahboub, H.D.H., Muller, J. and von Borell, E. (2004). Outdoor use, tonic immobility, heterophil/lymphocyte ratio and feather condition in free-range laying hens of different genotype. British Poultry Science 45: 738.

Nicol, C.J., Potzsch, C., Lewis, K. and Green, L.E. (2003). Matched concurrent case-control study of risk factors for feather pecking in hens on free-range commercial farms in the UK. British Poultry Science 44: 515.

Richards, G.J., Wilkins, L.J., Knowles, T.G., Booth, F., Toscano, M.J., Nicol, C.J. and Brown, S.N. (2011). Continuous monitoring of pop hole usage by commercially housed free-range hens throughout the production cycle. Veterinary Record 169: 338.

Rodriguez-Aurrekoetxea, A. and Estevez, I. (2016). Use of space and its impact on the welfare of laying hens in a commercial free-range system. Poultry Science 95, 2503-2513.

Thuy Diep, A., Larsen, H. and Rault, J.-L. (in press). Behavioural repertoire of free-range laying hens indoors and outdoors, and in relation to distance from the shed. Australian Veterinary Journal, accepted 31-07-17.

Zeltner, E. and Hirt, H. (2003). Effect of artificial structuring on the use of laying hen runs in a free-range system. British Poultry Science 44: 533.

17. Effect of farm size and abattoir capacity on carcass and meat quality of slaughter pigs

N. Čobanović, D. Vasilev, M. Dimitrijević, V. Teodorović and N. Karabasil*
University of Belgrade, Faculty of Veterinary Medicine, Department of Food Hygiene and Technology, Bulevar oslobodjenja 18, 11000 Belgrade, Serbia; cobanovic.nikola@vet.bg.ac.rs

Abstract

The study was designed to determine the effect of farm size and abattoir capacity and their interaction on carcass and meat quality of slaughter pigs. Pigs from small scale farm slaughtered at low-input production system had the highest backfat thickness, skin lesion score percentage of the human-inflicted type of skin lesions and the lowest lean meat content. Conversely, pigs from large scale farm slaughtered at high-input abattoir had the lowest backfat thickness, the skin lesion score, the percentage of the human-inflicted type of skin lesions and the highest lean meat content. In addition, pigs from small scale farm slaughtered at low-input abattoir had the highest pH_{45} value and the incidence of dark, firm and dry meat, while the highest percentage of normal and pale, soft and exudative pork was obtained from pigs from large scale farm slaughtered at the high-input abattoir. In conclusion, rearing conditions at farm level and handling procedure during the pre-slaughter period at small and large pork production systems negatively affected carcass and meat quality, compromised animal welfare and, thus, caused serious financial loss and ethical concern.

Keywords: high and low-input abattoir, large and small scale farm, PSE meat, DFD meat, skin lesions

Introduction

Pig production aimed at providing animals with high meatiness and good pork quality traits at the same time. Meat quality is not a predetermined and during a pig's life various factors can affect its characteristics – from the gene, health, feeding, slaughter weight and gender to pre-slaughter and slaughter conditions (Čobanović *et al.*, 2016a, 2016b; Karabasil *et al.*, 2017). Although a number of studies have been conducted to assess and improve carcass and meat quality, most have been carried out in pigs from large-scale pig farms and high-input abattoirs, which highlights the need to evaluate carcass and meat quality of pigs from small-scale farms and low-input production systems. The Republic of Serbia is a relatively small country in southeastern Europe, with a longstanding tradition of pig and pork production. Of almost 460 establishments that are approved to produce and place meat and meat products on the local market, most are low production systems with local significance. Furthermore, among Serbia's 149,000 agricultural holdings, around 76% are considered backyard farms, with fewer than 100 fattening pigs, around 23% are small-scale, so-called family farms, which are slightly bigger than backyard farms and produce for the local market, while only a small number of farms (around 400) are large production systems specialised in pig production (Bussel-Van Lierop *et al.*, 2015). Backyard and small scale farms deliver their fattening pigs to small local abattoirs, with a weekly slaughter rate of approximately 100 pigs, while large-scale farms usually deliver their fattening pigs to high-input abattoirs (Bussel-Van Lierop *et al.*, 2015). Unfortunately, there is a lack of papers focusing on the comparison of slaughter pigs from different production size farms and abattoirs capacity. This information could assist local farmers and pork producers to improve their production systems and obtain better carcass and pork quality. Therefore, the aim of this study was to determine the effect of farm size and abattoir capacity and their interaction on carcass and meat quality of slaughter pigs.

EurSafe 2018 – DOI 10.3920/978-90-8686-869-8_17, © Wageningen Academic Publishers 2018

Materials and methods

The study was conducted in 2016 on 240 slaughter pigs, about six months old, with a live weight (LW) of approximately 110 kg. All animals were of the same breed (Yorkshire × Landrace crossbreeds) and originated from different production size farms. The small scale farm produced only fatteners and consisted of one fattening unit with the capacity to finish a maximum of 100 pigs per cycle (180 days). Fattening unit consisted of two pens containing 50 pigs each. The pigs entered the small scale farm with an average weight of 30 kg where they were kept until reaching about 110 kg LW. The large scale farm was a conventional farrow-to-finish herd with free-range (i.e. outdoor) sows and confined (i.e. indoor) weaners and fattening pigs. The farm consisted of five fattening units and capacity to finish at least 2,000 pigs per cycle (180 days). Each unit consisted of 20 pens containing 20 pigs each. The pigs were fattened until they reach about 110 kg LW. The pigs were transported for slaughter for about one hour (stocking density of about 0.5 m^2/100 kg pig) in the same commercial transporter by the same driver. The pigs were held in a lairage for about 3 hours at a space allotment of about 0.65 m^2 per pig. Pigs were slaughtered in two abattoirs of different capacity. Low-input slaughter facility was defined as abattoir with a daily slaughter rate below 35 pigs, while high-input slaughter facility was defined as abattoir with a daily slaughter rate over 300 pigs. In both abattoirs, pigs were head-only electrically stunned, exsanguinated, and further processed using conventional industrial practice.

The carcasses were weighed immediately after splitting and final washing to obtain the hot carcass weight (HCW), and re-weighed 24 hours after chilling to determine the cold carcass weight (CCW). Backfat (BFT) and loin muscle thickness (LT) were measured with a metal ruler with an accuracy of 1.0 mm on the midline of the split carcass in millimeters: a fat measurement taken as the minimum fat thickness of the visible fat including rind covering the *M. gluteus medius* and a muscle measure taken at the shortest connection between the front (cranial) end of the *M. gluteus medius* and the upper (dorsal) edge of the vertebral canal. The lean meat content (LMC) (%) was calculated using ZP (*Zwei-Punke Messverfahren*) method (Commission Regulation (EC) No 1249/2008) based on the thickness of the backfat and loin depth according to the following formula: $y = 65.93356 - 0.17759 \times x_1 + 0.00579 \times x_1 - 52.54737 \times x_1/x_2$, with y = estimated lean meat content of the carcass (kg); x_1 = backfat depth (mm) and x_2 = loin muscle depth (mm). This formula is valid for carcasses weighing between 60 kg and 120 kg (HCW). Skin lesions (SLC) were visually appraised on the left side of the carcasses 45 minutes after slaughter based on the Welfare Quality® protocol (2009). The carcasses were divided into the following regions: (1) ears; (2) front part of the carcass (from the head to the end of the shoulder); (3) middle part of the carcass (from the end of the shoulder to the rear part of the carcass); (4) rear part of the carcass; and (5) limbs (from the accessory digit upwards). Each region of the carcass was scored based on a three-point scale: (0) no visible skin lesions or only one skin lesion bigger than 2 cm or skin blemishes smaller than 1 cm; (1) between two and 10 skin blemishes bigger than 2 cm; and (2) any wound penetrated into muscles or more than 10 skin blemishes larger than 2 cm. The scoring of the five regions of the carcass was combined into the following score from 0 to 2: (0) all carcass regions with a score of 0; (1) at least one carcass region with a score 1; and (2) at least one carcass region with a score 2. Carcass skin lesions were also classified as human-inflicted type bruises, fighting-type bruises and mounting-type bruises by visual assessment of shape and size to recognize their origin (Faucitano, 2001). Lesions due to biting are recognized as being about 5-10 cm in length, comma shaped and concentrated in high number in the anterior (head and shoulders) and posterior parts of the carcass (ham). Long (10 to 15 cm), thin (0.5 to 1 cm wide) comma shaped bruises densely concentrated on the back of pigs typically caused by the fore claws were classified as mounting-type bruises. Human-inflicted bruises as a consequence of the use of sticks leave large dark brown rectangular marks usually in the posterior part of the carcass (ham).

The pH and temperature of the loin muscle were measured 45 minutes after slaughter on the left half of the carcass at the level of the 10th and 11th ribs using a pH-meter 'Testo 205' (Testo AG, Lenzkirch,

Germany). Meat quality traits were both measured in duplicate, and the average of the two measurements was taken as a final result. Pork quality classes (pale, soft and exudative PSE, normal meat, and dark, firm and dry – DFD) meat were determined according to Čobanović *et al.* (2016a) using pH_{45} value. The carcasses showing pH_{45} values lower than 6.0 were classified as PSE meat, while the carcasses showing pH_{45} values higher than 6.4 were classified as DFD meat. The carcasses with pH_{45} between 6.0 and 6.4 were classified as normal pork.

Statistical analysis of the results was conducted using software SPSS version 23.00 for Windows (SPSS, 2015). Two-way ANOVA with Tukey's multiple comparison test was performed to test the effect of farm size and abattoir capacity, and their interaction on the carcass and meat quality traits. Accordingly, pigs were assigned to one of four groups arranged in a 2×2 factorial design based on the farm size (large and small scale farm) and abattoir capacity (high- and low-input production system). Data were described by descriptive statistical parameters as the mean value and pooled standard error of means (SEM). The Chi-square test was used to determine the incidence of pork quality classes and different types of skin lesions with respect to the farm size and abattoir capacity. Values of $P<0.05$ were considered significant.

Results and discussion

The effects of farm size and abattoir capacity on the carcass quality traits are shown in Table 1. LW, HCW and CCW were not influenced ($P>0.05$) by farm size and abattoir capacity. Pigs from small scale farm slaughtered at low-input production system had the highest ($P<0.05$) BFT and the lowest LMC. In contrast, pigs from large scale farm slaughtered at high-input abattoir had the lowest ($P<0.05$) BFT and the highest LMC. In the present study, large scale farm produced more uniform pigs, whose lean meat content is around the national average of ~50-52% (Petrović *et al.*, 2009). On the contrary, small scale farm produced more diverse pigs with lean meat content lower than the national average results. This can be attributed to the facts that small pig producers have a low level of economic organization of production, usually lack access to adequate veterinary extension services and also lack the resources to invest in necessary facilities, hygiene, genetics, and improvement of feeding for the production of high quality pork (Food and Agriculture Organization, 2007). Furthermore, high-input abattoirs most often deliver pigs from large scale farms but also maintain their own farms, where pigs are produced from the good genetic material under the supervision of veterinarians (Knecht *et al.*, 2016). In contrast, the large majority of low-input abattoirs producing for the local market. Their production is based on slaughter pigs sourced from small individual farms, while only a smaller number of pigs, usually of lower quality, are supplied from large farms (Bussel-Van Lierop *et al.*, 2015). This could explain the differences in carcass quality between pigs from small scale farm slaughtered at low-input production system and pigs from large scale farm slaughtered at a high-input abattoir.

The effects of farm size and abattoir capacity on the meat quality traits are shown in Table 1. Pigs from small scale farm slaughtered at low-input abattoir had the highest ($P<0.05$) SLC, pH_{45} value and the incidence of DFD meat, while the highest ($P<0.05$) percentage of normal meat quality was obtained from pigs from large scale farm slaughtered at a high-input abattoir (Table 1). In addition, the highest ($P<0.05$) percentage of the human-inflicted type of skin lesions was recorded in pigs from small scale farm slaughtered at low-input production system (Table 2), indicating rough handling during the pre-slaughter period. Rough handling on the day of slaughter increases the severity of skin lesions on a pig carcass. Moreover, if the severe skin lesions are provoked by rough handling at an early time during the pre-slaughter period, muscle glycogen stores are depleted, leading to lower production of lactic acid, and resulting in the occurrence of DFD meat (Čobanović *et al.*, 2016b).

Despite the fact that pigs from large scale farms slaughtered at high-input slaughter facilities produced the highest proportion of normal pork quality, the same pigs also had the lowest pH_{45} value and highest

Table 1. Mean values (±SEM) of carcass and meat quality in slaughter pigs according to farm size and abattoir capacity (n=240).[1]

	Small scale farm		Large scale farm		SEM	Significance[2]		
Abattoir capacity	Low-input	High-input	Low-input	High-input		FS	AC	FS×AC
Number of pigs	60	60	60	60				
Carcass quality traits								
LW (kg)	111.70	111.70	110.80	106.4	2.925	NS	NS	NS
HCW (kg)	87.93	87.94	87.76	83.82	2.360	NS	NS	NS
CCW (kg)	86.19	86.62	85.89	82.56	2.311	NS	NS	NS
BFT (mm)	24.79[a]	16.80[b]	21.57[c]	13.87[d]	0.191	*	*	*
LT (mm)	65.65	66.83	66.58	66.30	2.159	NS	NS	NS
LMC (%)	43.18[a]	49.90[b]	47.07[c]	52.46[d]	2.040	*	*	*
SLC	1.80[a]	1.05[b]	1.45[c]	0.65[d]	1.539	*	*	*
Meat quality traits								
pH_{45}	6.42[a]	5.98[b]	6.11[c]	5.86[d]	0.066	*	*	*
T_{45} (°C)	37.11[a]	37.24[a]	37.57[a]	38.75[b]	0.910	NS	*	*
Pork quality classes								
PSE (%)	8.33[a]	6.67[a]	8.33[a]	25.00[c]	-	-	-	-
Normal (%)	33.34[a]	56.67[b]	53.34[b]	75.00[c]	-	-	-	-
DFD (%)	58.33[a]	36.66[b]	38.33[b]	0.00[c]	-	-	-	-

[1] Different letters in the same row indicate a significant difference at $P<0.05$ ([a-d])

[2] AC: significance of abattoir capacity; FS: significance of farm size; AC×FS: significance of the interaction between farm size and abattoir capacity; *: statistical significance at ($P<0.05$); NS: not significant ($P>0.05$).

Table 2. Percentage of different types of skin lesions on pig carcasses according to abattoir capacity and farm size (n=240).[1]

	Small scale farm		Large scale farm	
Abattoir capacity	Low-input	High-input	Low-input	High-input
Number of pigs	60	60	60	60
Human-inflicted type skin lesions (%)	63.33[a]	23.33[b]	40.00[c]	10.00[d]
Fighting-type skin lesions (%)	13.33	23.33	21.67	20.00
Mounting-type skin lesions (%)	13.33	18.33	21.66	20.00

[1] Different letters in the same row indicate a significant difference at $P<0.05$ ([a-d]).

T_{45} values as well as the incidence of PSE meat (Table 2). It has been reported that pig's previous experiences with handling at the farm can affect how a specific slaughter pig will react to being handled in the future (Grandin, 2017a). Pigs raised on large scale farms are less accustomed to contact with people, and are more difficult to load and unload from a lorry if their first experience with people in their barns occurs on the day of slaughter (Grandin, 1988, 2017a). Hence, highly excitable pigs are difficult to drive, especially at high-input slaughter facilities. This leads to rougher handling at loading, unloading and in the stunning chute due to excessive and/or improper use of electric prods. Use of electric prods do not normally leave any trace on the carcass (unless applied violently), but negatively affects pork quality (Grandin, 1988, 2017b; Faucitano, 2001). This could probably explain why pigs from large individual farms slaughtered at high-input production systems had the lowest SLC ($P<0.05$, Table 1) and proportion of human-inflicted skin lesions ($P<0.05$, Table 2), but the highest occurrence of PSE meat ($P<0.05$, Table 1). Therefore, it can be presumed that the combined effects of high excitability of pigs and overuse and/or improper use of electric prods due to high slaughter rate caused a severe acute stressful reaction, resulting in an excessively fast drop in pH value in the first 45 minutes *post-mortem*, which, in combination with high muscle temperature, results in PSE meat (Grandin, 2017b). Producers should walk quietly through the finishing barns throughout the feeding period to reduce excitability problems and improve ease of movement at the abattoir (Grandin, 2017a,b). Likewise, playing a radio in the finishing barn, with a variety of music and talk, will greatly reduce animal excitability when a person enters the room (Grandin, 2017b). Furthermore, electric prods should be a moving tool of last resort and used only when absolutely necessary to the back of the pig behind the shoulder whereby the duration of the shock should not exceed one second (National Pork Board, 2014). Moreover, it has been reported that elimination of electric prod use during pre-slaughter handling will reduce the occurrence of PSE and the skin lesion score by 50% (Faucitano, 2000). Accordingly, pigs accustomed to contact with people will be calmer and easier to drive, and, therefore, meat obtained from those animals will have better quality, because they will be less likely to become excited even on a high speed slaughter line (Grandin, 1988).

Conclusion

The results showed that poor rearing conditions at farm level and inadequate handling procedure during the pre-slaughter period at small and large pork production systems caused serious financial losses, arising from the reduced carcass and meat quality. In addition, inadequate preparation of pigs to contact with people at farm level and use of electric prods during pre-slaughter handling became a direct cause of suffering, which compromised animal welfare and, thus, represented a serious ethical concern. Accordingly, it is strongly recommended for farmers and pork producers to use flags, rattle paddles, and/or plastic boards instead of electric prods and/or sticks as an effective prevention of needless suffering of animals during the pre-slaughter period. Education of the farm and abattoir personnel on pig behaviour and handling practices at farm level and during the pre-slaughter period and detailed verification of suppliers are of paramount importance to minimise pre-slaughter stress and to optimise carcass and pork quality. In addition, small scale farms can improve the final meat quality through investment in necessary facilities, hygiene, genetic selection programs and feeding strategy.

Acknowledgements

This paper was supported by Ministry of Education, Science and Technological development, Republic of Serbia, Project No. 31034 and III46009.

References

Bussel-Van Lierop, A., Van Wagenberg, C. and Somers, D. (2015). Serbian pig sector: an overview. Fact finding mission: opportunities for collaboration between Serbia and The Netherlands. Wageningen UR Livestock Research.

Čobanović, N., Bošković, M., Vasilev, D., Dimitrijević, M., Parunović, N., Djordjević, J. and Karabasil, N. (2016a). Effects of various pre-slaughter conditions on pig carcasses and meat quality in a low-input slaughter facility. South African Journal of Animal Science 46: 380-390.

Čobanović, N., Karabasil, N., Stajković, S., Ilić, N., Suvajdžić, B., Petrović, M., and Teodorović, V. (2016b): The influence of pre-mortem conditions on pale, soft and exudative (PSE) and dark, firm and dry (DFD) pork meat. Acta Veterinaria-Beograd 66(2): 176-182.

European Commission. (2008). EC Regulation No. 1249/2008 of 10 December 2008 laying down detailed rules on the implementation of the Community scales for the classification of beef, pig and sheep carcases and the reporting of prices thereof. ECOJ No. L337, 3-30.

Faucitano, L. (2001). Causes of skin damage to pig carcasses. Canadian Journal of Animal Science 81(1): 39-45.

Food and Agriculture Organization (2007). A systematic analysis of the agribusiness sector in transition economies: The Serbian meat value-chain.

Grandin, T. (1988). Environmental enrichment for confinement pigs.

Grandin, T. (2017a). On-farm conditions that compromise animal welfare that can be monitored at the slaughter plant. Meat Science 132: 52-58.

Grandin, T. (2017b). Recommended animal handling guidelines & audit guide: a systematic approach to animal welfare. AMI Foundation.

Karabasil, N., Čobanović, N., Vučićević, I., Stajković, S., Becskei, Z., Forgách, P. and Aleksić-Kovačević, S. (2017). Association of the severity of lung lesions with carcass and meat quality in slaughter pigs. Acta Veterinaria Hungarica 65(3): 354-365.

Knecht, D., Jankowska-Mąkosa, A. and Duziński, K. (2016). The Effect of Production Size and Pre-Slaughter Time on the Carcass Parameters and Meat Quality of Slaughtered Finisher Pigs. Journal of Food Quality 39(6): 757-765.

National Pork Board (2014). Transport Quality Assurance® Handbook. Available at: http://porkcdn.s3.amazonaws.com/sites/all/files/documents/Resources/04113.pdf.

Petrović, L., Tomović, V., Džinić, N., Tasić, T. and Ikonić, P. (2009). Parametri i kriterijumi za ocenu kvaliteta polutki i mesa svinja. Meat technology 50(1): 121-139 (in Serbian).

Welfare Quality® (2009). Welfare Quality® assessment protocol for pigs (sow and piglets growing and finishing pigs). Welfare Quality® Consortium, L., the Netherlands.

Section 3.
Ethics of production and consumption

18. Animals as objects: defining what it means to 'professionally' treat animals in meat production[1]

*M.-T. Schlemmer**, *H. Grimm and J. Benz-Schwarzburg*
Unit of Ethics and Human-Animal-Studies, Messerli Research Institute, Veterinaerplatz 1, 1210 Vienna, Austria; resi@schlema.com

Abstract

The fact that the human-animal relationship is ambivalent becomes especially apparent when looking at the issue of meat. The meat paradox describes the tension of loving animals and at the same time loving meat. It concerns consumers and producers alike (Loughnan *et al.*, 2012). Psychologically, it can be described as a cognitive dissonance often resulting in an unease about eating former living beings one usually cares about. One possibility for overcoming this is the objectification of animals, i.e. turning subjects into objects by denying them certain characteristics. This alleviates the dissonance, as the animals no longer have to be ethically considered. The paper sets out a philosophical conception of the objectification of animals used for meat production. After generally illuminating the meat paradox and the issue of objectification, Nussbaum's (1995) seven signposts of human objectification in pornography are transferred to the animal context resulting in four distinct indicators of objectifying treatment of animals: denial of being an end in itself, denial of preference autonomy, denial of individuality, and denial of sentience. Examples of pork production in Western industrialised countries show that objectification of animals is a socially established practice that is directly and indirectly requested from professionals working in the agricultural context. Finally, a definition of objectification of animals used for meat production will be provided, representing a starting point for answering ethical questions about the treatment of animals (as objects) and the associated responsibilities of consumers and producers.

Keywords: objectification, meat-paradox, cognitive dissonance

Introduction

The interaction of humans and animals has a long history and involves not only live but also dead animals. This is especially the case when it comes to the issue of meat, which is to be understood as the complex whole of production and consumption of animal meat. Humans, however, have a quite ambivalent relationship with animals which is captured by the meat paradox. While people on the one hand love animals and are concerned about their well-being, they at the same time use them for the production of meat and regularly consume them. This can give way to a tension they experience as a certain kind of discomfort (Loughnan *et al.*, 2012). Looking at this unease from a psychological point of view, the phenomenon of the meat paradox can be described as a form of cognitive dissonance, which is defined as an incongruity of different cognitions (e.g. attitude, knowledge). If cognitions are not in line with each other or with a certain behaviour, this results in discomfort (Festinger, 1957), e.g. the belief that animals have to be cared for and the fact that they are regularly consumed. One way to deal with this dissonance is to change the belief about their moral claim. Studies have shown that consumers of animal meat are eager to deny animals those characteristics which are thought of as being crucial for their moral status (e.g. mind, sentience, emotions; Bastian *et al.*, 2012; Bilewicz *et al.*, 2011; Loughnan *et al.*, 2010). Further, the psychological theory of Melanie Joy (2011) about the 'culture of meat' present in modern

[1] The content of this paper represents one aspect of the unpublished Master's thesis 'Denying animals to be subjects: the case of objectification of animals used for meat production' by M.-T. Schlemmer (University of Veterinary Medicine, Vienna).

EurSafe 2018 – DOI 10.3920/978-90-8686-869-8_18, © Wageningen Academic Publishers 2018

societies proposes that objectification, i.e. turning a subject into an object, is one option for dealing with the cognitive dissonance accompanying the issue of meat. Commonly, a subject can be understood as an entity that has to be considered morally, while an object is something that does not in itself have to be taken into account from a moral standpoint. This often means that as soon as a subject is denied its subject-status and turned into an object, any moral responsibility towards it becomes obsolete and it can be treated in a carefree way.[2]

A philosophical definition of animal objectification

While this psychological sight provides arguments for how meat eaters actually deal with the meat paradox, it does not exactly define what objectification means. Since there is a lack of a definition of animal objectification, the effort is made to transfer Martha Nussbaum's (1995) theoretical philosophical concept of objectification of women in pornography to the animal realm. She herself regards objectification as 'not only a slippery, but also a multiple, concept' (Nussbaum, 1995: 251) and is aware that each of the used key terms (e.g. agency) would need further elaboration and investigation. In her theory she states seven signposts that she thinks of as being indicative of objectification (Nussbaum, 1995). As the psychological findings indicate that forms of denial lie at the core of objectifying an entity, all features proposed by Nussbaum (1995) are formulated as a denial. To underline the thoughts of this transformation, practical examples from modern pig production as it commonly takes place in Western industrialised countries are given. By explaining objectification as 'speaking, thinking, and acting' (Nussbaum, 1995: 249) in a way that treats 'one thing as another' (Nussbaum, 1995: 256), Nussbaum gives a multi-layered definition of objectification, allowing to view consumers as well as producers as potential objectifiers of animals.

Denial of being an end in itself

Nussbaum's (1995) first signpost of objectification is instrumentality and focuses on the fact that objectifiers commonly treat subjects as mere instruments for their purposes, thereby according to them only instrumental value. Looking at modern pig production, this is certainly the case as these animals are only kept for the purpose of producing meat. They are bred, reproduced, fattened, and slaughtered only in order to provide people with meat and thereby are used as a means to an end of somebody else. Each of these animals is therefore denied status as an end in itself.

Denial of preference autonomy

Denying a subject autonomy and self-determination, in Nussbaum's (1995) opinion, also marks objectifying treatment. Although the classical interpretation of autonomy according to Immanuel Kant (1968[1785]: 434-435) cannot easily be applied to animals, it is possible to use Tom Regan's (2004[1983]) concept of preference autonomy. Preference autonomy is to be understood as the ability to initiate action in order to fulfil one's own preferences (Regan, 2004[1983]) and hence can be found in animals as well. Severely restricting pigs in their possibilities to follow these preferences (e.g. searching for food, engaging with social partners), as is the case in pig farming in Western industrialised countries, marks a clear denial of preference autonomy (Steiger, 2002). Since animals kept for meat production are furthermore owned by people who determine just about every aspect of their life and fate (Schrecker *et al.*, 1997), they are even more limited in their options to fulfil preference autonomy, which is why the signpost of ownership proposed by Nussbaum (1995) is subsumed under this feature as well.

---

[2] If there are moral claims addressing this object, they will not come from an intrinsic value of the object itself but arise from an external context (e.g. the owner's interests); such moral claims are usually regarded as weaker.

Inertness

Treating a subject as something that is lacking in agency and possibly also in activity, i.e. as being inert, is also an indicator of objectification, according to Nussbaum (1995). Unfortunately, however, Nussbaum (1995) does not give an intelligible definition of the terms involved. Therefore, a very basal definition of activity that scales it down to mere physiological activity, like protein synthesis and proliferation of muscular tissue, is assumed. This, however, cannot be denied to animals, as it is exactly this feature people make use of when producing and fattening animals for the production of meat. This aspect hence cannot be transferred to the animal context. Starting from such a basal definition it is unclear how a denial of agency and activity applies to the pornographic context Nussbaum (1995) writes about, as women also show bodily activity and are more than just dolls. Due to Nussbaum's (1995) lack of clarity regarding the terms involved and her use of ambiguous examples, however, the question remains whether this notion is useful in the analysis of objectification at all.

Denial of individuality

Nussbaum (1995) sees fungibility, i.e. treating a subject as something that is interchangeable with a subject of the same kind or another kind, as a signpost of objectification. Fungibility can be seen as having two aspects: the first refers to the function of the entity (as the common Latin origin fungi of fungibility and function already suggests) and the second one refers to its individual character. As the first aspect is already covered by the signpost of instrumentality, the second one relates to what Joy (2011) calls deindividualisation. It describes the fact that subjects who are regarded as being fungible can easily be replaced by other subjects, thereby neglecting their individuality (Regan, 2004[1983]). Keeping pigs in large groups only identifying them by numbers and refusing to recognise them individually, is an example for such a denial (Joy, 2011). In this sense, a denial of individuality is common in animals used for meat production.

Denial of sentience

Whereas the signpost of violability as suggested by Nussbaum (1995) focuses on the bodily integrity of a subject, the signpost of denial of subjectivity concentrates on its subjective experience. Neglecting one of these aspects or both clearly points into the direction of objectification as Nussbaum (1995) understands it. Since the physical and psychological health of animals is heavily interwoven and their potential to experience both is subsumed under their characteristic of being sentient (Passantino, 2008), it seems only logical to combine these signposts into a denial of sentience. The practical examples of this denial are obvious; not only are several (legally permitted) painful interventions routinely carried out by farmers (e.g. surgical castration without anaesthesia), but also do intensive management systems elicit enormous stress in the animals, e.g. because of regular mixing of sows in dynamic groups (Gregory and Grandin, 2007).

Transferring Nussbaum's (1995) signposts of objectification of humans to the animal context hence results in four distinct features of animal objectification (Figure 1). Animal objectification can therefore be defined as the treatment of animals – in form of speaking, acting, thinking, and/or actually treating them – as objects by denying them either one or more of the following characteristics: being end in themselves, having preference autonomy, having individuality, and being sentient. In the context of animals used for meat production in Western industrialised countries, such objectification is carried out by both consumers and producers, although perhaps especially so by producers, since objectification is an entrenched part of established management and production systems.

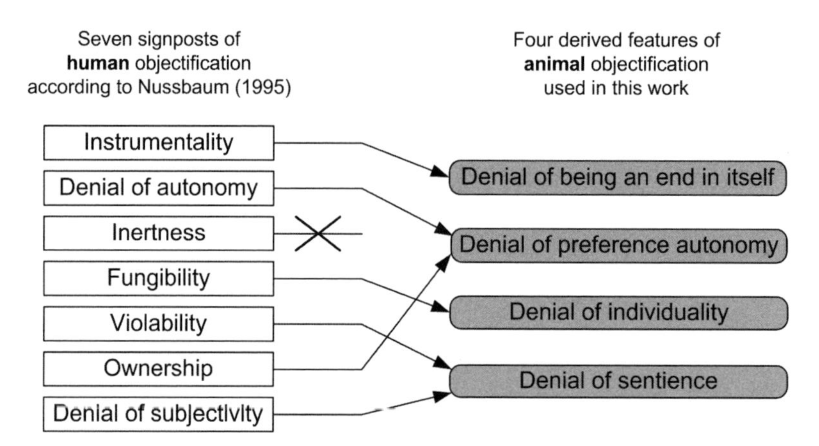

Figure 1. Derivation of the four features of animal objectification from Nussbaum's (1995) seven signposts of human objectification.

Discussion and outlook

The definition of animal objectification implies that the presence of only one denial suffices. If this is the case, then meat animals are clearly objectified. From Nussbaum's (1995) signposts of objectification, we can derive four distinct features of animal objectification (via specific forms of denial) that meat animals face on a regular basis. Still, it is an open question how such objectification can be evaluated morally. Applying different ethical theories (e.g. the animal rights theory) to the derived understanding of objectification could further clarify whether objectification is morally problematic or unproblematic. Such further analysis could help to reflect upon and reconsider our relationship with meat animals by way of questioning our established views and socially accepted habits which are ultimately of legal and political relevance.

References

Bastian, B., Loughnan, S., Haslam, N., and Radke, H.R.M. (2012). Don't mind meat? The denial of mind to animals used for human consumption. Personality and Social Psychology Bulletin, 38 (2): 247-256.

Bilewicz, M., Imhoff, R., and Drogosz, M. (2011). The humanity of what we eat: conceptions of human uniqueness among vegetarians. European Journal of Social Psychology, 41 (2): 201-209.

Festinger, L. (1957). A theory of cognitive dissonance. Row, Peterson and Company, Evanston, White Plains, 304 pp.

Gregory, N.G. and Grandin, T. (2007). Animal welfare and meat production (2nd ed.). CABI, Wallingford, 400 pp.

Joy, M. (2011). Why we love dogs, eat pigs and wear cows. An introduction to carnism (2nd ed.). Conari Press, San Francisco, 208 pp.

Kant, I. (1968[1785]). Grundlegung zur Metaphysik der Sitten. In Preußische Akademie der Wissenschaften (Ed.), Gesammelte Schriften (Vol. 4, pp. 385-464). De Gruyter, Berlin.

Loughnan, S., Haslam, N., and Bastian, B. (2010). The role of meat consumption in the denial of moral status and mind to meat animals. Appetite, 55 (1): 156-159.

Loughnan, S., Bratanova, B., and Puvia, E. (2012). The meat paradox: how are we able to love animals and love eating animals? In-Mind, 1: 15-18.

Nussbaum, M.C. (1995). Objectification. Philosophy and Public Affairs, 24 (4): 249-291.

Passantino, A. (2008). Companion animals: an examination of their legal classification in Italy and the impact on their welfare. Journal of Animal Law, 4: 59-92.

Regan, T. (2004[1983]). The case for animal rights (2nd ed.). University of California Press, Berkeley, Los Angeles, 425 pp.
Schrecker, T., Elliott, C., Hoffmaster, C.B., Keyserlingk, E.W., and Somerville, M.A. (1997). Ethical issues associated with the patenting of higher life forms. Westminster Institute for Ethics and Human Values, McGill Centre for Medicine, Ethics and Law, Canada. http://www.iatp.org/files/Ethical_Issues_Associated_with_the_Patenting_o.pdf. Accessed 9 February 9 2016.
Steiger, A. (2002). Die Würde des Nutztieres – Nutztierhaltung zwischen Ethik und Profit. In: M. Liechti (ed.) Die Würde des Tieres. Harald Fischer Verlag GmbH, Erlangen, pp.221-231.

19. Breeding Blues: an ethical evaluation of the plan to reduce calving difficulties in Danish Blue cattle

P. Sandøe[*], L.F. Theut and M. Denwood
University of Copenhagen, Department of Veterinary and Animal Sciences, Grønnegårdsvej 15, 1870 Frederiksberg C., Denmark; pes@sund.ku.dk

Abstract

Danish Blue Cattle is a breed of cattle originating from the Belgian White Blue Cattle (BWB), a breed that is characterized by double muscling, which in turn may lead to difficult calving. In Belgium, difficult calving in this breed is typically pre-empted by means of planned caesarean sections (CS). The breed association, under some pressure from its parent organisation the National Cattle Committee, first implemented an action plan to reduce the rate of CS following a media event and subsequent reactions from politicians in Denmark in 1998. Later, the breed association renamed the breed from Belgian White Blue to Danish Blue Cattle in an effort to distance the breed from its Belgian origin. The aim of this paper is to undertake an ethical evaluation of how this issue was handled, with specific focus on the actions of the breed association and the National Cattle Committee. This evaluation involves an objective assessment of the outcome of the action plan as well as a wider ethical assessment of the rationale and actions of the professional organisations in charge. We begin by describing the controversy in 1998, which led to the aforementioned action plan. Then, we evaluate how successful the action plan has been in achieving its stated goals. These results show that the action plan has achieved a decrease in the rate of CS from over 50% between 1990-1998 to below 10% between 2000-2013. There has also been a significant decrease in the rate of other types of difficult births in the breed. Finally, we evaluate the implementation of the action plan from the perspective of professional ethics, where the aim is to handle public controversies in such a way as to maintain acceptance from the surrounding society. This has clearly been a success. However, viewed from a consequentialist perspective with a focus on animal welfare, the outcome is more ambiguous. The welfare for Danish Blue cattle has improved, but on the other hand the breed still has a much higher level of CS than comparable breeds of cattle in Denmark.

Keywords: professional ethics, animal welfare, caesarean section, Belgian White Blue Cattle

Background

The Belgian White Blue (BWB), also known as Belgian Blue, is a modern cattle breed originating from Belgium, where it was formally established in 1973. The characteristic double muscling of the breed arose as a natural recessive mutation in the myostatin gene. Double-muscled individuals of local Belgian breeds attracted high slaughter prices in the 1950s. With the introduction of cattle caesarean section (CS) around the same time, it became possible to routinely deliver double muscled calves. This motivated breeders to select bulls that were guaranteed to pass on the double muscling gene mutation. These bulls were used for artificial insemination so that over time more and more calves were double muscled; and eventually a separate breed, BWB, was established with double muscling as one of the breed's main characteristics (Lips *et al.*, 2001).

BWB was introduced in Denmark in 1972, and the breeding of BWB quickly increased among a small group of devoted breeders. In 1978 there were 297 purebred BWB registered in Denmark, and in 1979 a breed organisation for Danish BWB was established (Stendal, 2004). The main role of BWB in Denmark has been, and still is, to produce semen that can be used for inseminating dairy cows so that offspring not planned be used as replacement heifers can be used for more efficient meat production.

EurSafe 2018 – DOI 10.3920/978-90-8686-869-8_19, © Wageningen Academic Publishers 2018

When used for this purpose, the offspring will not be double muscled as they possess only a single copy of the myostatin gene.

Danish breeders of BWB have also successfully engaged in export of bulls and semen to other countries and were part of international collaboration led by Belgian breeders. Production of beef cattle is not a big business in Denmark; and it is fair to say that breeding of BWB in Denmark was considered partly as a hobby for those involved (Stendal, 2004).

In the 1990's concerns were raised within the breed organisation about the high frequency of CS, which was regularly above 50%; and in the mid-1990's there was a controversy about the breed in Denmark's neighbouring country, Sweden. The Swedish government wanted to ban the breed, but the European Court ruled that such a ban would not be in accordance with EU law (the aim of which is to secure the free movement of goods, including recognized cattle breeds, across the borders of the countries within the EU) (Stendal, 2004).

In May 1998 the issue became a public controversy in Denmark due to a television documentary where BWB was used as an example of extreme breeding. There was a reaction from the minister of animal welfare legislation in Denmark, and pressure was put on the cattle organisations to deal with the issues. Therefore, the issue became a matter of professional ethics. The aim of this paper is to document and discuss how the Danish BWB association together with its parent organisation, the National Cattle Committee, handled the issue.

The public controversy and how it was handled by the breed organisation

In the aforementioned TV documentary, two veterinarians were shown performing a preventive CS on a BWB cow; after which another cow was shown with a series of visible CS scars. The owner of the cow on which the CS was performed was subsequently reported to the police for breaking the Danish Animal Welfare Law (Stendal, 2004). The case was heard by the Veterinary Health Council, which concluded that the law had *not* been violated.

However, the Veterinary Health Council also submitted a letter to the Ministry of Justice, which at that time was responsible for the Danish Animal Welfare Law, and stated that it was problematic 'from the point of view of animal ethics to continue the breeding of animals where a significant frequency of birth difficulties can be foreseen'. In the letter the council also asked that the Danish Council for Animal Ethics, an advisory board to the minister, should be consulted regarding the ethical acceptability of maintaining breeds 'where birth difficulties often occur' (Det Dyreetiske Råd, 1998).

Meanwhile, the cattle organisations acted very quickly on their own. Even before the above statement was made public, the Danish BWB association had presented an action plan for BWB in Denmark. According to sources with close ties to the organisation (Hansen, 2004; Stendal, 2004) the action plan came into existence after strong pressure from the National Cattle Committee.

The action plan was finalized and sent to the minister on the 15th of June 1998 – less than three weeks after the broadcast of the documentary. The stated goal of the plan was to bring down the level of CS in the Danish BWB from about 50% to approximately 10% within a timeframe of 5-10 years. The main purpose of the plan was to keep double muscled BWB but to breed the animals in a way so that CS would be required less frequently. The breeding efforts included selection of bulls which were known not to cause birth difficulties in their offspring, and a focus on managing the diet of the in-calf animals in such a way as to be leaner at parturition and therefore better able to give birth. A ban on routine preventive CS was also part of the plan (Det Dyreetiske Råd, 1998, Bilag 2).

Subsequently, the minister asked the Danish Council for Animal Ethics to make a statement on the issue. In this statement the Council endorsed the idea of an action plan that could serve as a basis for a voluntary agreement between the professional organisations and the Danish society represented by the minister.

A meeting was held between the minister, the Danish BWB association, various other cattle and agricultural organisations, the Veterinary Health Council, and the Danish Council for Animal Ethics. The conclusion of that meeting was that the Danish BWB breeding organisation should report back to the Danish Council for Animal Ethics on a yearly basis, and that the Council should follow the development and report back to the minister.

The only subsequent main addition to the original action plan was in 2005, when it was decided that BWB heifers should not be inseminated with semen from BWB bulls so that double muscled calves would be restricted to second (or later) calvings when the cow is fully mature.

Outcome of the action plan

A complete set of calving records for all purebred BWB animals registered with the breed association in Denmark during the period January 1985 – October 2013 was received from the Knowledge Centre for Agriculture, Denmark. Due to comparatively small numbers of recordings, observations prior to 1st January 1990 were discarded. This dataset theoretically reflects the population of purebred BWB cattle in Denmark during the period, although we note that registration with the breed association is not mandatory so an element of self-selection bias is likely to be present. We also note a comparatively high rate of incompletely recorded data: out of a total of 5,169 birth observations, 906 were discarded due to missing dam identification and 734 were discarded due to one or more implausible calving interval recorded for the associated 121 dams. This left a dataset comprising 3,529 birth observations of 1,791 male, 1,660 female, and 78 missing-sex calves from a total of 1,353 unique dams and 335 unique sires.

An overview of the distribution of 2,951 recorded live births and 316 recorded still births over time is shown in Figure 1 (excluding the 262 recordings that were missing birth status). The proportion of stillbirths over time appears to be relatively consistent, however the recorded birth rate increases dramatically between 1997-1999, followed by a decline to pre-1997 levels. This most likely represents a varying voluntary reporting rate to the breed association rather than a true change in birth rate over time. Note that the final year of recordings (2013) is incomplete.

Over the entire time period, there were a total of 2868 births with recorded calving score within the following categories: 1123 No Assistance, 508 Easy Assisted, 170 Difficult, 227 Vet Assisted, 840 CS. The relative distribution of these within the total number observed for each year is shown in Figure 2. There is a noticeable change in the relative rate of these recordings over time, particularly in terms of the proportion recorded as CS, which suddenly decreases around the broadcast in 1998 and continues to decline towards 2013.

Statistical analysis of the data presented was done using three generalised linear mixed models, as follows. Model 1 was a mixed effects logistic regression with recorded birth status (live or dead; n=3,220) as the response, model 2 was a mixed effects logistic regression with recorded CS (no or yes; n=2,855) as the response, and model 3 was a proportional odds mixed effects logistic regression with the ordinal variable of calving score (no assistance, easy assisted, difficult and vet assisted; n=2,017) as the response and excluding observations of CS. Fixed effects of calf sex and time interval (which was divided into a pre-phase before the TV documentary (1990-1997), an intermediate phase (1998-1999), and the post phase (2000-2013)) were used for all models along with a random effect of dam ID. The models were

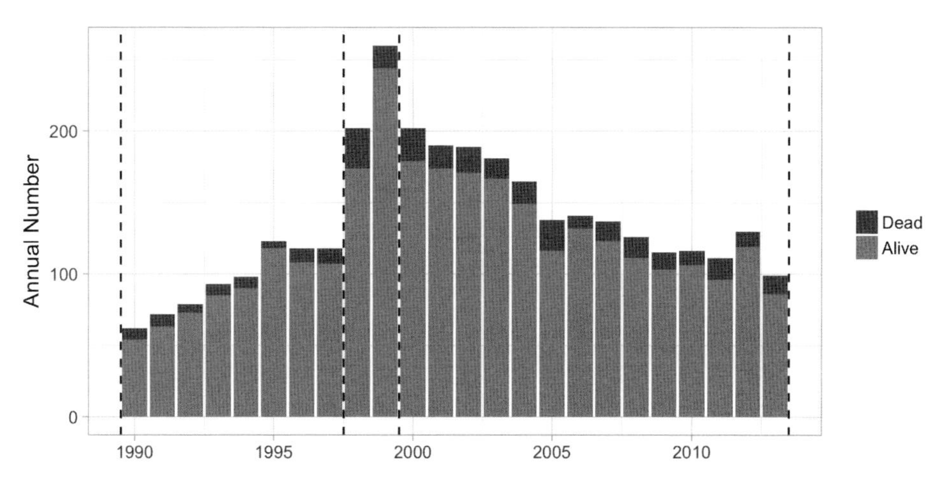

Figure 1. The recorded number of live and stillbirths of purebred calves during the time period January 1990 – October 2013, stratified by year. The dotted lines demarcate the periods of interest as follows: the pre-phase before the action plan, a two-year intermediate phase, and a subsequent post-plan phase.

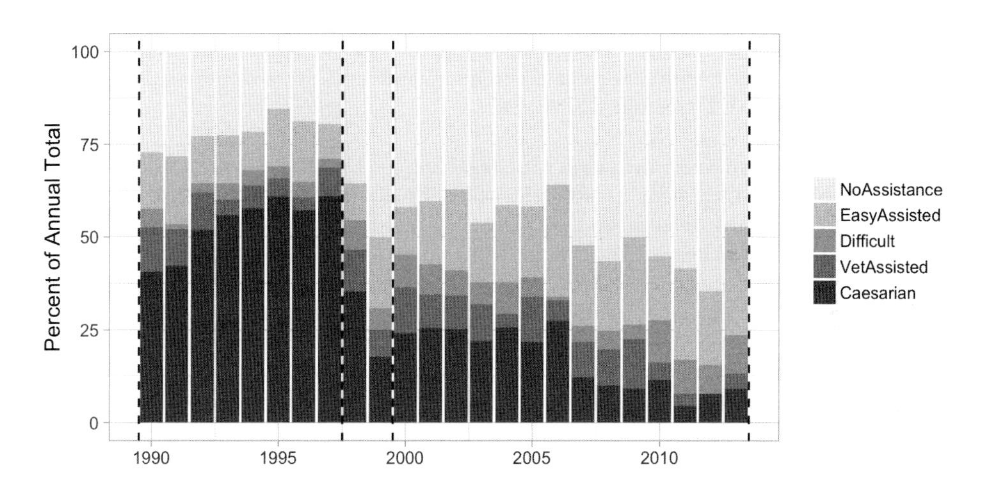

Figure 2. The recorded relative frequency of each calving score category for purebred Danish BWB cattle during the time period January 1990 – October 2013, stratified by year. The dotted lines demarcate the periods of interest as follows: the pre-phase before the action plan, a two-year intermediate phase, and a subsequent post-plan phase.

implemented using lme4 version 1.1-14 (Bates *et., al* 2015) and ordinal version 2015.6-28 (Christensen, 2015) packages for R version 3.4.2 (R Core Team, 2017).

The results indicate that there is no significant association between time period and stillbirths (model 1). However, the results for model 2 indicate significantly higher odds of CS in the time period 1990-1997 relative to 2000-2013 (odds ratio (95% CI): 24.74 (15.35-42.92) $P<0.001$) as well as the time period 1998-1998 relative to 2000-2013 (odds ratio (95% CI): 1.96 (1.27-3.06) $P=0.003$). Similarly, the results for model 3 indicate significantly higher odds of 'worse' calving scores (excluding CS) in the time period 1990-1997 relative to 2000-2013 (odds ratio (95% CI): 1.80 (1.28-2.54) $P<0.001$) although not for 1998-1998 relative to 2000-2013 (odds ratio (95% CI): 1.08 (0.78-1.48) $P=0.646$). The interpretation of this is a large and significant reduction in the reported rate of CS following the intervention, as well as a significant (although smaller) improvement in the reported non-CS calving scores over the same time period.

The overall conclusion is that the action plan has successfully achieved its main purpose of reducing the level of CS in BWB in Denmark, and has also concurrently reduced the reported rate of difficult calving. However, the changing number of reported births during the period (with a peak in 1999 that is approximately 4 times the smallest reported number in 1990) is a potential for concern, as this changing number of reported births may be due to a reporting bias within the community. It is conceivable that such a reporting bias could have changed over the time period, and that this may partly explain the significant associations between time period and rate of CS and difficult births. Further work is therefore required to investigate the cause of the observed variation in reported birth rate over time, as well as to investigate the reasons for the data quality issues that resulted in almost 32% of the data being discarded, before the results presented here can be regarded as conclusive.

Assessment of the development from two ethical perspectives

The Danish BWB association with support from its parent organisation kept sending reports of progress regarding the action plan at regular intervals until 2013 to the Danish Council for Animal Ethics. The council followed up on these with reports to the minister, and meetings with the organisations were held in 2005 and 2008 (Det Dyreetiske Råd, 2013).

Meanwhile, an interesting development took place regarding the Danish BWB association. It changed the name of the breed to Danish Blue Cattle, the name of the association to the Danish Blue Cattle association, and it broke its formal ties to the Belgian organisation. The background for this decision was a confrontation with the Belgian breeders at a conference in July 2011, where representatives from all countries present apart from the Belgians reportedly expressed concerns about the calving issue. In contrast, the Belgian representatives saw no reason for concern (Landsforeningen for Belgisk Blåhvidt Kvæg i Danmark 2011), which prompted the Danish organisation to decide to change the name of the organisation and breed, and distance itselves from the Belgian breeders, as the British breed association had already done in 2007.

It was also part of the stated background for the decision of the Danish BWB association that the company in charge of dairy cattle breeding in Denmark, VikingGenetics, had expressed concerns about delivering semen that could be associated with the Belgian breeding practice, where it is claimed that around 90% of BWB calving take place by means of CS (Landsforeningen for Belgisk Blåhvidt Kvæg i Danmark 2011).

It is noteworthy that a significant shift in mentality occured among the Danish breeders between the start of the process and 2011: Initially, the concern about reducing the level of CS was often described as sensationalist overreaction from the surroundings (Hansen, 2004; Stendal, 2004), whereas in 2011 the Danish Blue breeders seem to share this concern.

The Danish authorities' monitoring of the breeders of Danish Blue Cattle came to an end in 2013. In a letter from the Danish Knowledge Centre for Agriculture dated 29 November 2013, it is reported that the frequency of CS for Danish Blue Cattle for three subsequent years, 2010-2012, was below 10%

(Videncentret for Landbrug 2013). In light of this information the Danish Council for Animal Ethics decided to recommend the ministry to close the case.

In the letter from the Knowledge Centre, there is a table, unfortunately not backed by any reference to published studies, comparing calving rates of Danish Blue Cattle with those found in four other common beef breeds: Simmental, Blonde, Charolais, and Limousine. Interestingly, for all these breeds in the three mentioned years the frequency of CS is below 1%, whereas Danish Blue ranges between 6 and 9% (Videncentret for Landbrug, 2013). The Danish Council for Animal Ethics notes that it would be good for the Danish Blue Cattle to reach the same low level of CS as the other breeds of beef cattle, but that this most likely is not a realistic goal.

Viewed from the perspective of professional ethics, the main purpose of an action plan as described in this paper would be for professional organisations to handle public controversies in such a way as to maintain acceptance from the surrounding society. From this perspective, the development has clearly been a success.

However, viewed from a consequentialist perspective with a focus on animal welfare, the outcome is more ambiguous. On the one hand, the welfare of Danish Blue Cattle appears to have substantially improved based on the reported data – both by limiting the reported rate of CS and the potential negative consequences following this, and by simultaneously achieving a reduction in the reported rate of difficult calvings. But on the other hand, the breed still seems to have a much higher level of CS calving than comparable breeds of beef cattle; and from this perspective it could be argued that instead of continuing to breed Danish Blue Cattle there should instead be shift to a breed with lower levels of CS. Of course, it is debateable to which extent a CS gives rise to welfare problems (Vandenheede *et al.*, 2001).

Acknowledgements

Thanks are due to Anders Fogh, Videncentret for Landbrug, Kvæg, Mogens Stendal, advisor to Landsforeningen for Belgisk Blåhvidt Kvæg i Danmark, and Stine B. Christiansen, Det Dyreetiske Råd, for providing data and other information. We also want to thank Sara V. Kondrup for editorial assistance. Last but not least we are very grateful to Thomas Mark who contributed a lot in the early phases of the project here described but who did not have the time to engage as a co-author of this paper.

References

Bates, D., Maechler, M., Bolker, B. and Walker, S. (2015). Fitting Linear Mixed-Effects Models Using lme4. Journal of Statistical Software 67: 1-48.

Christensen, R.H.B. (2015). Ordinal – Regression Models for Ordinal Data. R package version 2015.6-28. Available at: http:/CRAN.r-project.org/package=ordinal. Accessed 25 January 2018.

Det Dyreetiske Råd (1998). Udtalelse om avl af dyreracer, hvor fødselsvanskeligheder vil forekomme hyppigt. Available at: http://detdyreetiskeraad.dk/udtalelser/udtalelse/pub/hent-fil/publication/udtalelse-om-avl-af-dyreracer-hvor-foedselsvanskeligheder-vil-forekomme-hyppigt-1998. Accessed 25 January 2018.

Det Dyreetiske Råd (2013). Opfølgning vedrørende Belgisk Blåhvidt Kvæg. 20 December 2013. København: Det Dyreetiske Råd.

Hansen, M. (2004). Et moderne TV-eventyr. In: Stendal, M. (ed.) Jubilæumsbog 1979-2004, 25 året for stiftelsen af Landsforeningen for Belgisk Blåhvidt Kvæg i Danmark. Available at: https://sw11536.smartweb-static.com/upload_dir/docs/Jubilaeumsbog-2004/Jubilaeumdsskrift%2C-netversion.pdf, pp. 15-18. Accessed 25 January 2018.

Landsforeningen for Belgisk Blåhvidt Kvæg i Danmark (2011). Forslag fra bestyrelsen og Avlsudvalget til behandling og vedtagelse på ekstraordinær generalforsamling den 3. december 2011. 15 November 2011. Available at: https://sw11536.smartweb-static.com/upload_dir/docs/Artikler%2C-taler/11-12-03-Bestyrelsens-begrundelse-for-navneaendringerne.pdf. Accessed 25 January 2018.

Lips, D., De Tavernier, J., Decuypere, E. and Van Outryve, J. (2001). Ethical Objections to Caesareans: Implications on the Future of the Belgian White Blue. Preprints of EurSafe 2001. Food Safety, Food Quality, Food Ethics, Florence. University of Milan: 291-294.

R Core Team (2017). R: A language and environment for statistical computing. R Foundation for Statistical Computing, Vienna, Austria. Available at: URL https://www.R-project.org.

Stendal, M. (2004). Belgisk Blåhvidt kvæg i Danmark. In Stendal, M. (ed.) Jubilæumsbog 1979-2004, 25 året for stiftelsen af Landsforeningen for Belgisk Blåhvidt Kvæg i Danmark. Available at: https://sw11536.smartweb-static.com/upload_dir/docs/Jubilaeumsbog-2004/Jubilaeumdsskrift%2C-netversion.pdf, pp. 7-14. Accessed 25 January 2018.

Vandenheede, M., Nicks, B., Désiron, A. and Canart, B. (2001). Mother-young relationships in Belgian Blue cattle after a Caesarean section: characterisation and effects of parity. Applied Animal Behaviour Science 72: 281-292.

Videncentret for Landbrug (2013). Status på frekvensen af kejsersnit hos kødkvægracen Dansk Blåkvæg. 29. November 2013. Skejby: Videncentret for Landbrug, Kvæg.

20. Dual-purpose chickens as alternative to the culling of day-old chicks – the ethical perspective

N. Brümmer, I. Christoph-Schulz and A. Rovers*
Thünen Institute of Market Analysis, Bundesallee 63, 38116 Braunschweig, Germany; nanke.bruemmer@thuenen.de

Abstract

This paper considers the ethical evaluation of dual-purpose chickens as an important prerequisite for societal acceptance. According to several studies, chicken farming is seen as one of the most controversial forms of livestock farming due to e.g. the common practice of culling day-old chicks. In 2016, an administrative court in Germany decided that the killing of the male chicks does not violate animal welfare legislation and is permitted for economic reasons. Within the society, this ruling was broadly discussed and largely rejected for ethical reasons. One alternative to the culling of day-old chicks are dual-purpose chicken breeds. These breeds can do both: the hens lay eggs and the cockerels put on meat. But the hens lay fewer and smaller eggs and the cockerels need more time and feed to grow. The ethical evaluation of the topic for this paper has been conducted in six focus groups. In addition, important stakeholders have been identified through a comprehensive literature study. The ethical matrix according to Mepham is a tool of applied ethics for the interpretation of the interests of stakeholders with regard to ethical principles. The matrix is based on the three ethical principles of well-being, autonomy and justice. The aim is to present a well-balanced consideration from different angles and thus to reduce the complexity of the topic. Usually, an innovation is compared with the status quo. In this case: the keeping of dual-purpose chickens with the current practice (culling of day-old chicks). We applied the matrix to five interest groups: dual-purpose chickens, consumers, egg industry, farmers and environment. The results show that the topic dual-purpose chickens as alternative to the killing of day-old chicks is very complex and therefore requires a differentiated consideration of ethical aspects.

Keywords: ethics, ethical matrix, laying hens, day-old chicks, dual-purpose chickens

Introduction

Poultry farming is one of the most controversial discussed farming practices in livestock farming, besides criticized stocking densities and the use of antibiotics, also because of the common practice of day-old male chicks of laying breeds being culled directly after hatch (Verbeke *et al.*, 2000; Leenstra *et al.*, 2011). In 2016, the killing of day-old chicks was subject of a court ruling of the Higher Administrative Court of North Rhine-Westphalia (Federal State in Germany). The ruling states that killing day-old chicks does not violate the animal welfare legislation and is allowed for economic reasons because technical solutions are not ready for use yet (Beckmann, 2016; Oberverwaltungsgericht NRW, 20 A 488/15). The court decision started a public debate about the ethical aspects of this common practice in layer hens breeding. Currently, some alternatives to this practice are discussed. One alternative is the keeping of dual-purpose chickens. This is a chicken breed that is suitable for both, meat and egg production (Rautenschlein, 2016). Because of the negative correlation between meat growth and laying performance which engendered the specialization in chicken breeding, the dual-purpose hens and cockerels cannot compete with the performance of hybrid chickens (Grashorn, 2013). The hens of dual-purpose breeds lay fewer and smaller eggs and the cockerels put on less meat and need more time and feed to grow than conventional laying hens and broiler chickens (Damme, 2003; Koenig *et al.*, 2011). As a consequence, meat and eggs from dual-purpose chickens would be correspondingly more expensive (Leenstra *et al.*, 2010; Diekmann *et al.*, 2017). These aspects have implications not only for consumers but also for other

EurSafe 2018 – DOI 10.3920/978-90-8686-869-8_20, © Wageningen Academic Publishers 2018

stakeholder 'groups' such as the environment. Therefore, this article will focus on the ethical evaluation of dual-purpose chickens as an alternative to the culling of day-old male layer-chicks. According to Marie (2006), the consideration of ethical aspects in livestock farming is an important prerequisite for social acceptance and consequently also for the acceptance of animal products.

Methods

Focus groups

The ethical evaluation of the topic is based on six focus groups with the focus on dual-purpose chickens that were conducted with 6-8 participants each in June 2016 in three German places (Berlin, Cloppenburg and Munich). The participants were recruited regarding quotas by a market research agency and were consumers of chicken meat and eggs. The aim of focus groups is to create an atmosphere that fosters an almost natural conversation setting with diverse opinions and statements (Lamnek, 2005). Discussed topics were amongst others the killing of day-old chicks, potential alternatives and the advantages and disadvantages of dual-purpose chickens. The approximately 90-minute discussions were recorded, transcribed and then subjected to a qualitative content analysis. As ethical aspects were also discussed sufficiently in this context, the findings from the focus groups give a valuable input to the ethical evaluation of dual-purpose chickens.

Ethical matrix

The ethical matrix according to Mepham (2000a) is a tool of applied ethics for explaining the interests of affected groups in relation to ethical principles. The aim is to present a well-founded consideration from different perspectives and thus to reduce the complexity and to provide guidance in the evaluation of a topic (Zichy *et al.*, 2014). The objective of an ethical matrix is not to provide decisions but to facilitate the ethical assessment. Participatory methods can encourage this evaluation in terms of fact finding (Kaiser and Forsberg, 2010; Schroder and Palmer, 2003). The ethical principles of well-being, autonomy and justice were identified by Beauchamp and Childress (1994) and are guided by the three main theories of ethics, namely utilitarianism (well-being), Kantianism (autonomy) and the contractualist theory of Rawls (justice). Usually, the ethical matrix compares an innovation with the status quo (Mepham, 2000a). In Table 1 the ethical matrix contains five stakeholder groups. These were identified by a comprehensive literature review. The stakeholder groups include: the chickens, the consumers as key actors when it comes to the market success of the products from the dual-purpose chickens, the egg and meat industry which has to deal with the changing product properties, the farmers who would have to adapt new husbandry forms and the environment which is affected by the higher resource consumption of dual-purpose cockerels (Bruijnis *et al.*, 2015; Grashorn, 2013; Leenstra *et al.*, 2010, 2011).

Results

Chickens

For the chickens, the usage of dual-purpose chicken breeds could imply wellbeing and autonomy if the husbandry is adapted to the needs and natural behaviour of the animals and if animal welfare measurements are taken. It could be argued as well that a longer fattening period of the cockerels has a positive effect on the wellbeing. The point that was mostly stressed in the focus groups was the fairness aspect. If the males could live as well and are not killed as day-old chicks, equality between both sexes is given.

Table 1. Ethical matrix applied to the usage of dual-purpose chickens (adapted from Mepham, 2000b: 612).

	Wellbeing	Autonomy	Justice
Chickens	Wellbeing of the animals (presumes a good animal husbandry)	Freedom of behaviour (presumes a good animal husbandry)	Gender equality (rearing both sexes)
Consumers	Affordable and high quality food	Freedom of choice between products (informed purchase-decisions)	Availability of affordable food
Egg and meat industry	Efficiency	Freedom of choice to market dual-purpose chickens	Fair trade and laws
Farmers	Good working conditions and satisfying income	Freedom of choice to keep dual-purpose chickens	Fair trade and laws
Environment	Nature conservation	Biodiversity	Sustainability and responsible use of resources

Consumers

The usage of dual-purpose chickens could affect consumers regarding all three ethical principles. If eggs and chicken meat become more expensive, the affordability of food could be affected. These assumptions depend on the question if dual-purpose chickens are implied nationwide or only exist in a niche market. If this is the case the consumers would have the autonomy to choose between different products provided that there is a clear labelling. Dual-purpose chickens could also have an effect on the consumption behaviour of the consumers because the appearance of meat and eggs is different to those from hybrid chickens.

Egg and meat industry

The transition to eggs and meat from dual-purpose chickens could have far reaching implications for the whole supply chain which has to be adapted to the differing properties of dual-purpose chickens starting from the slaughtering process (conventional slaughterhouses for broiler chickens turned out to be not suitable for dual-purpose cockerels) to the processing and marketing of the eggs and meat. The satisfaction of the principles wellbeing and autonomy strongly depends on the voluntary nature of the implication of dual-purpose chicken breeds.

Farmers

For the farmers who keep layer hens or broiler chickens the changeover to dual-purpose chickens would imply some changes in the husbandry system and management. If freedom of choice is granted depends on the legal ban of culling day-old chicks and the practicability of alternatives. Premise for the wellbeing of the farmers is that they generate enough income. Therefore, it would be important for the farmers that the dual-purpose chickens are economically profitable and fair trade laws exist.

Environment

The environment is also affected by the usage of dual-purpose chickens. Due to the fact that the cockerels are fattened almost twice as long as conventional broiler chickens until they reach a sufficient weight, more resources such as water and land for feed are needed. Additionally, more manure occurs which has also an environmental impact. Therefore, the principles wellbeing and justice could be violated because nature conservation and sustainability are negatively affected.

Conclusions

The ethical matrix applied to the usage of dual-purpose chickens as alternative to the killing of day-old male chicks demonstrates that the issue has many different ethical facets and illustrates that there is a moral problem that cannot be easily solved. The implementation of dual-purpose chicken breeds could have on the one hand positive implications for the chickens itself if the husbandry conditions improve but on the other hand might negatively affect the environment and also endanger the availability of affordable food for consumers or a satisfying income for farmers. To finally assess the usage of dual-purpose chickens from an ethical perspective it is necessary to weight the single aspects. Therefore, the ethical matrix could serve as a basis for discussions between the different stakeholders. It is not intended to solve the ethical problem (Mepham, 2000a). A weighting between the different dimensions is necessary in the first place, e.g. the issue whether animal welfare should have priority. For further discussion, the outcome for chickens and the environment could be presented by the according animal rights groups and environmental organizations. This procedure enables a comprehensive fact finding and can be used by policy makers for a detailed consideration of the topic dual-purpose chickens as alternative to the culling of day-old chicks.

References

Beauchamp, T.L. and J.F. Childress (1994). Principles of Biomedical Ethics. Oxford University Press, New York and Oxford.

Beckmann, M. (2016). Über den vernünftigen Grund im Sinne des §1 S. 2 TierSchG bei der Tötung von männlichen Eintagsküken. Natur und Recht 38 (6), 384-390.

Bruijnis, M.R.N., Blok, V., Stassen, E.N. and Gremmen, H.G.J. (2015). Moral 'Lock-In' in Responsible Innovation: The Ethical and Social Aspects of Killing Day-Old Chicks and Its Alternatives. In: Journal of Agricultural and Environmental Ethics 28(5): 939-960.

Damme K. (2003). Fattening performance, meat yield and economic aspects of meat and layer type hybrids. World`s Poultry Science Journal 59: 50-53.

Diekmann, J., Hermann, D. and Mußhoff, O. (2017). Wie hoch ist der Preis auf Kükentöten zu verzichten? Bewertung des Zweinutzungshuhn- und Bruderhahnkonzept als wirtschaftliche Alternative zu Mast- und Legehybriden. In: Berichte über Landwirtschaft 95(1): 1-22.

Grashorn, M. (2013). Verwendung der männlichen Küken der Legeherkünfte., in: WING-Themen in der Geflügelhaltung, 15 Mai 2013, S.1-4.

Kaiser, M. and Forsberg, E.M. (2001). Assessing Fisheries – Using an Ethical Matrix in a Participatory Process. In: Journal of Agricultural and Environmental Ethics 14: 191-200.

Koenig, M., Hahn, G., Damme, K. and Schmutz, M. (2011). Utilization of laying-type cockerels as 'coquelets': Influence of genotype and diet characteristics on growth performance and carcass composition. In: Arch.Geflügelk. 76(3): 197- 202.

Lamnek, S. (2005). Gruppendiskussion. Theorie und Praxis. Weinheim: UTB.

Leenstra, F.R.; Munnichs, G.; Beekman, V.; Heuvel-Vromans, E. van den; Aramyan, L.H.; Woelders, H. (2011). Killing day old chicks? Public opinion regarding potential alternatives. In: Animal Welfare 20: 37-45.

Leenstra, F., van Horne, P. and Van Krimpen, M. (2010). Dual purpose chickens, exploration of technical, environmental and economical feasibility. XIIIth European Poultry Conference, Tours, France.

Marie, M. (2006). Ethics: The new challenge for animal agriculture. In: Livestock Science 103: 203-207.

Mepham, T.B. (2000a). A framework for the ethical analysis of novel foods: The ethical matrix. In: Journal of Agricultural and Environmental Ethics 12: 165-176.

Mepham, T.B. (2000b): The role of food ethics in food policy. In: Proceedings of the Nutrition Society 59 (4): 609-618.

Rautenschlein, S. (2016). Einsatz des Zweinutzungshuhns in Mast und Eierproduktion: Ansätze für ein integriertes Haltungskonzept. In: Rundschau für Fleischhygiene und Lebensmittelüberwachung (RFL) 68(8): 276-278.

Schroeder, D. and Palmer, C. (2003). Technology assessment and the 'ethical matrix'. In: Poiesis & Praxis 1(4): 295-307.

Verbeke, W.A.J. and J. Viaene (2000). Ethical challenges for livestock production: Meeting consumer concerns about meat safety and animal welfare. In: Journal of Agricultural & Envrionmental Ethics 12(2): 141-151.

Zichy, M., Dürnberger, C., Formowitz, B., Uhl, A. (2014). Energie aus Biomasse – ein ethisches Diskussionsmodell. Springer Vieweg, Wiesbaden.

21. On-farm slaughter – ethical implications and prospects

J. Hultgren[1], C. Berg[1], A.H. Karlsson[1], K.J. Schiffer[2] and B. Algers[1]*
[1]Department of Animal Environment and Health, Swedish University of Agricultural Sciences, P.O. Box 234, 53223 Skara, Sweden; [2]Eldrimner – Swedish resource centre for artisan food production, Ösavägen 30, 83694 Ås, Sweden; jan.hultgren@slu.se

Abstract

The slaughter of farm animals to produce meat for human consumption involves a number of ethical dimensions. The animals are subjected to considerable welfare risks; pre-slaughter handling can be one of the most stressful events in the life of an animal. Stress at slaughter may lead to deteriorated meat quality through increased breakdown of glycogen in the muscles, resulting in abnormally lower or higher pH values which makes the fresh meat pale, soft and exudative or dark, firm and dry, respectively. In most developed countries farm animal production is undergoing structural changes resulting in fewer but larger herds with less time for management of individual animals, making them less tolerant to handling at the time of slaughter. Despite comprehensive legal restrictions and official control of commercial abattoirs, farm animal welfare outcomes at slaughter vary considerably and are in some cases unacceptably poor. The high line speed in large-scale slaughter results in demanding working conditions, making it difficult for stockpersons to deal with hassle and balking. A considerable proportion of the animals transported to slaughter spend one night at the abattoir before being slaughtered, which may increase stress levels if lairage conditions are poor. On-farm slaughter may be conducted at a stationary plant at the farm, in a mobile unit temporarily placed at or near the farm, or by stunning and bleeding on farm followed by carcass transport to a nearby plant for further processing. On-farm slaughter may have a potential to reduce pre-slaughter animal stress by shorter or eliminated transports, minimised exposure to unfamiliar environments, animals and persons, less time in lairage and a reduced slaughter line speed, which would be in line with the increased awareness of ethical issues related to the slaughter of animals. On the other hand, on-farm slaughter involves challenges regarding food and occupational safety, waste management and public health. Requirements regarding adequate bleeding, freedom from carcass contamination, veterinary inspections and carcass refrigeration must be met. The working conditions for farm and slaughterhouse personnel must be acceptable, thus avoiding unacceptable physical and psychosocial risks.

Keywords: abattoir, animal welfare, gunshot method, mobile slaughter, pre-slaughter stress

Introduction

Slaughter is the process of killing animals to produce meat for human consumption. Mainly farmed animals such as livestock, poultry or fish, but also companion or sports animals like horses, are subject to this process. Farmers sometimes look for alternatives to industrialised slaughter, because they aim at a high level of control over the entire production chain to achieve market benefits, or because they want to improve animal welfare or product quality. We have chosen to restrict the discussion to on-farm slaughter of cattle, sheep, pigs and broilers in the European Union, regardless of the final destination and possible marketing of the meat. By on-farm slaughter we mean stunning and killing of animals in the place where they were raised during at least the last part of their lives.

European Union (EU) legislation regarding animal protection and food safety in connection with slaughter mainly concerns animals meant for commercial meat production. Domestic ungulates (e.g. bovines and porcines) and domestic solipeds (equines) should be slaughtered in a slaughterhouse, following *ante-mortem* inspection by an official veterinarian.

Slaughter in practice

In Europe slaughter transports of live animals are mainly done by road. The information about travel distances and duration of transports is limited; slaughter transports above 8 hours occur both within national territories and between countries, but the majority of transports are likely to be considerably shorter (Dalla Villa *et al.*, 2009).

In large-scale commercial slaughter, cattle are stunned mechanically while restrained in a stun box, either by a cartridge-driven or pneumatic bolt gun which drives a metal rod into the brain or by a bullet weapon or shotgun. Regardless of the weapon the shot must be placed within a well-defined small area of the forehead to produce the desired effect. Pigs are usually stunned either with carbon dioxide gas or, in less industrial settings, by passing an electric current through the head or captive bolt. The current induces an epileptic seizure which lasts for less than 1 minute. Chickens are stunned by either carbon dioxide or argon gas or by head-to-body or head-only electrical stunning.

Killing aims to irreversibly cut the blood supply to the brain and stop the heart, by exsanguination. Large cattle, pigs and horses, are bled by severing the large blood vessels near to the heart in a thoracic cut or stick, while chickens and usually sheep are bled by cutting the main arteries near to the head, a neck cut.

Animal welfare risks and other risks

Even if much has been done to reduce animal suffering in connection with transport and slaughter, most farm animals still experience considerable stress shortly before and during the slaughter (Cockram and Corley, 1991; European Food Safety Authority, 2004; Warriss, 1990). The process around slaughter may be the most stressful single event in the animals' lives.

Since long, farm animal production in most industrialised countries undergoes a massive restructuration. Average herd sizes increase and work is automated, which decreases the time the animals are exposed to human contact and may lower the tolerance to pre-slaughter handling (Bunzel-Drüke *et al.*, 2009; Rushen *et al.*, 1999).

Studies have shown that animal protection at commercial abattoirs varies considerably and is sometimes weak (Atkinson *et al.*, 2013; Von Wenzlawowicz *et al.*, 2012). Examples of potentially stressful situations are handling practised by operators who are unfamiliar to the animals, waiting and possible overnight stay in lairage at the slaughter plant, inadequate access to feed and water, mixing with unfamiliar animals, unexpected visual impressions, odours, noise and temperature changes. The degree of influence on the animals is likely to depend on the nature, intensity and duration of different negative stimuli in combination with the sensitivity and susceptibility of the animals to such stimuli (European Food Safety Authority, 2004).

Cattle that have been reared outdoors under extensive conditions, with very limited human handling, can become heavily stressed when crowded, transported and restrained at the time of slaughter (Le Neindre *et al.*, 1996; Waiblinger *et al.*, 2006). Gregory *et al.* (2007) noted that inadequate stunning was more common in stressed (19%) than in calm cattle (8%). Certain cattle breeds are more difficult to handle than others. Some abattoirs do not accept certain breeds because of the risks involved.

Apart from reducing animal welfare, pre-slaughter stress can impair the quality of meat (Ferguson and Warner 2008; Friedrich *et al.* 2015). The probably most important meat quality problem in ruminants is 'Dark-Firm-Dry' (DFD) or 'dark-cutting beef', which causes the meat industry considerable losses (Scanga *et al.*, 1998; Shen *et al.*, 2009; Warren *et al.*, 2010). DFD occurs when the glycogen reserves of the muscles are depleted prior to slaughter, resulting in a high ultimate meat pH and a dark lean color and a dry, often-sticky lean surface. The high ultimate pH reduces shelf life of DFD meat. Animal injuries and stress related to rough handling can also cause meat rejection at slaughter and low carcass weights (Huertas *et al.*, 2010; Jarvis *et al.*, 1996).

Effective ways to reduce animal stress at slaughter is to ensure that the premises and equipment are designed to allow for a smooth and efficient propulsion of the animals, and that the stockpersons understand the principles of the work and have a positive attitude towards the animals and work (Coleman *et al.*, 2003; Hemsworth, 2003). Abattoir stockpersons may feel pressured by a high line speed and have difficulties to do their job properly due to inadequate premises, and may therefore perceive a loss of control, which may induce forceful driving and animal stress (Coleman *et al.*, 2012). Stressed animals tend to display behaviours which complicate driving, such as balking, backing or struggling, thus increasing stockperson frustration even more. Investments in equipment for automated driving and adequate head restraint during cattle stunning have been shown to decrease electric prodding and stress behaviours radically (Atkinson, 2009).

On-farm slaughter

On-farm slaughter may be conducted at a small stationary plant located on farm, in a mobile unit temporarily placed at or near the farm, or by stunning and bleeding on farm followed by carcass transport to a nearby plant for further processing.

Stationary on-farm slaughter

A stationary on-farm abattoir may serve the host farm only, or receive animals from several farms. In the latter case it resembles an industrial slaughterhouse, although the throughput may be relatively low; transported animals from other farms are basically exposed to the same stressful events and stimuli as in large-scale slaughter. If the abattoir only receives animals from the host farm, however, the road transport is eliminated. Permanent driveways and routines for transfer of animals from housing facilities to the plant can be developed and the stockpersons can be properly trained. However, small-scale slaughter may imply animal welfare and food safety risks due to shortcomings or deficiencies in equipment or routines for handling of animals and carcasses (Alban *et al.*, 2011).

Mobile slaughter

A mobile abattoir consists of a complete, self-contained plant which can be moved between farms that deliver slaughter animals. Prototypes of such plants have been developed in several European countries and commercial units are operating in e.g. Norway for sheep and reindeer. Since 2014 mobile slaughter of large cattle is conducted in Sweden by a unit contained in two trucks with trailers. One of the vehicles contains a cooling unit in which the carcasses are transported to a central tendering and cutting plant. The trucks and a portable inspection pen are parked as near to the animal premises as possible.

Compared to on-farm stationary slaughter, mobile slaughter offers additional flexibility. Slaughter can take place on any farm of sufficient size, where the plant can be set up. Additional animals can be delivered from neighbour farms without sufficient parking space. Skog Eriksen *et al.* (2013) found lowered blood glucose levels in lambs slaughtered in a mobile plant, compared to a large-scale industrial slaughterhouse.

From an animal-welfare perspective, a critical point in mobile slaughter is when the animals are taken the short distance from the housing facilities to the plant, if these are not adequately designed and the stockpersons are insufficiently trained. Hence there is a need to design systems for adequate animal handling when delivering animals to mobile slaughter. Furthermore, the mobile plant must meet the legal requirements in all parts of the slaughter process, including stunning, bleeding, veterinary inspection and carcass refrigeration, to secure animal welfare and food safety. A completely self-contained mobile plant places high demands on reliable electricity and water supply and waste disposal systems, but also on the skills and flexibility of the staff.

Stunning and killing before transport to the abattoir

One way to improve animal welfare, staff and food safety at slaughter of animals that are difficult to handle can be to stun them with rifle at a distance while kept in a familiar home enclosure together with herd mates. After bleeding the bodies can be transported to a nearby or on-farm abattoir for processing. No active cooling is required if the time between shooting and evisceration at the plant is less than 2 hours or the climate is cold. However, current EU legislation allows domestic ungulates intended for commercial slaughter only to be brought live into the slaughter premises, with the exception of slaughter for emergency purposes (EU Regulations 853/2004 and 854/2004). Despite these limitations, on-farm stunning and bleeding of cattle with the so-called 'gunshot method' is practiced widely in for example Germany. Shooting on farm is also used widely for fenced wild game in accordance with EU regulations.

Schiffer *et al.* (2017) studied free-range cattle that were stunned by a rifle shot in the forehead at 2.5-20 m distance. The authors concluded that the method with great certainly leads to immediate unconsciousness and death, provided that the shot is accurate. The project also revealed that the resulting meat was more tender compared to meat from animals in conventional slaughter (Friedrich *et al.*, 2015). The blood lactate levels of herd mates in the same enclosure as the animal being shot did not reveal any stress reaction (Schiffer, 2015). The results show that the 'gunshot method' has potential to improve animal welfare at slaughter, especially for animals that are difficult to handle. Appropriate methods for rifle stunning of cattle (the 'gunshot method') under different conditions require development. A highly qualified and well-trained shooter and calm shooting conditions are necessary for an adequate stun quality. It must be possible to bleed the animal without delay and take care of the carcass without exposing it to food-safety hazards.

Further research in this area may provide more definite proofs of ethical, animal welfare and meat quality benefits and thus pave the road to a slightly more liberal view on slaughter methods in the EU.

Conclusions

On-farm slaughter may have the potential to reduce pre-slaughter animal stress and improve meat quality by shorter or eliminated transport, minimal exposure to unfamiliar environments, animals and persons, less time in lairage and reduced slaughter line speed. This would be perceived as an improvement in relation to the ethical issues linked to the slaughter of animals. Issues related to food and occupational safety, waste management and public health should be considered. However, development is needed for wider application of mobile slaughter or on-farm rifle stunning before transport to an abattoir.

References

Alban, L., Steenberg, B., Stephensen, F.T., Olsen, A.-M. and Petersen, J.V. (2011). Overview on current practices of meat inspection in the EU. Danish Agriculture & Food Council, Copenhagen. Available at: https://www.efsa.europa.eu/en/supporting/pub/en-190. Accessed 5 January 2018.

Atkinson, S. (2009). Assessing cattle welfare at stunning. In: Proceedings of 43rd Congress of the International Society for Applied Ethology. July 6-10, 2009. Cairns, Australia, p. 79.

Atkinson S., Velarde A. and Algers B. (2013). Assessment of stun quality at commercial slaughter in cattle shot with captive bolt. Animal Welfare 22: 473-481.

Bunzel-Drüke, M., Böhm, C., Finck, P., Kämmer, G., Luick, R., Reisinger, E., Riecken, U., Riedl, J., Scharf, M. and Zimball, O. (2009). Praxisleitfaden fuer Ganzjahresbeweidung in Naturschutz und Landschaftsentwicklung – 'Wilde Weiden' [Guidelines for all-season outdoor husbandry in nature protection and landscape development]. Arbeitsgemeinschaft Biologischer Umweltschutz in Kreis Soest eV, Bad Sassendorf-Lohne, Germany. Available at: http://www.abu-naturschutz.de/projekte/abgeschlossene-projekte/qwilde-weidenq.html. Accessed 5 January 2018.

Cockram, M. and Corley, K.T.T. (1991). Effect of pre-slaughter handling on the behaviour and blood composition of beef cattle. British Veterinary Journal 147: 444-454.

Coleman, G.J., McGregor, M., Hemsworth, P.H., Boyce, J. and Dowling, S. (2003). The relationship between beliefs, attitudes and observed behaviours of abattoir personnel in the pig industry. Applied Animal Behaviour Science 82: 189-200.

Coleman, G.J., Rice, M. and Hemsworth, P.H. (2012). Human-animal relationships at sheep and cattle abattoirs. Animal Welfare 21, Suppl. 2: 15-21.

Dalla Villa, P., Marahrens, M., Velarde Calvo, A., Di Nardo, A., Kleinschmidt, N., Fuentes Alvarez, C., Truar, A., Di Fede, E., Otero, J.L. and Müller-Graf, C. (2009). Technical report submitted to EFSA. Project to develop Animal Welfare Risk Assessment Guidelines on Transport. Project developed on the proposal CFP/EFSA/AHAW/2008/02. IZSAM G. Caporale Collaborating Centre for Veterinary Training, Epidemiology, Food Safety and Animal Welfare, Teramo, Italy and World Organisation for Animal Health. Available at: www.efsa.europa.eu/en/supporting/pub/21e.htm. Accessed 5 January 2018.

European Food Safety Authority (2004). Opinion of the Scientific Panel on Animal Health and Welfare (AHAW) on a request from the Commission related to welfare aspects of the main systems of stunning and killing the main commercial species of animals (Question no. EFSA-Q-2003-093). The EFSA Journal 45: 1-29. Available at: www.efsa.europa.eu/en/efsajournal/pub/45.htm. Accessed 5 January 2018.

Ferguson, D.M. and Warner, R.D. (2008). Have we underestimated the impact of pre-slaughter stress on meat quality in ruminants? Meat Science 80: 12-19.

Friedrich, M.S., Schiffer, K.J., Retz, S., Stehling, C., Seuß-Baum, I. and Hensel, O. (2015). The effect of on-farm slaughter via gunshot and conventional slaughter on sensory and objective measures of beef quality parameters. Journal of Food Research 4: 27-35.

Gregory, N.G., Lee, C.J. and Widdicombe, J.P. (2007). Depth of concussion in cattle shot by penetrating captive bolt. Meat Science 77, 499-503.

Hemsworth, P.H. (2003). Human-animal interactions in livestock production. Applied Animal Behaviour Science 81: 185-198.

Huertas, S.M., Gil, A.D., Piaggio, J.M. and van Eerdenburg, F.J.C.M. (2010). Transportation of beef cattle to slaughterhouses and how this relates to animal welfare and carcass bruising in an extensive production system. Animal Welfare 19: 281-285.

Jarvis, A.M., Messer, C.D.A. and Cockram, M.S. (1996). Handling, bruising and dehydration of cattle at the time of slaughter. Animal Welfare 5: 259-270.

Le Neindre, P., Boivin, X. and Boissy, A. (1996). Handling of extensively kept animals. Applied Animal Behaviour Science 49: 73-81.

Rushen, J., Taylor, A.A. and De Passille, A.M. (1999). Domestic animals' fear of humans and its effects on welfare. Applied Animal Behaviour Science 65: 285-303.

Scanga, J.A., Belk, K.E., Tatum, J.D., Grandin, T. and Smith, G.C. (1998). Factors contributing to the incidence of dark cutting beef. Journal of Animal Science 76: 2040-2047.

Schiffer, K.J. (2015). On-farm slaughter of cattle via gunshot method. Shaker Verlag, Aachen. Doctoral Dissertation, University of Kassel.

Schiffer, K.J., Retz, S.K., Algers, B. and Hensel, O. (2017). Assessment of stun quality after gunshot used on cattle: a pilot study on effects of diverse ammunition on physical signs displayed after the shot, brain tissue damage and brain haemorrhages. Animal Welfare 26: 95-109.

Shen, Q.W., Du, M. and Means, W.J. (2009). Regulation of postmortem glycolysis and meat quality. In: Du, M. and McCormick, R.J. (eds.) Applied Muscle Biology and Meat Science. CRC Press, Boca Raton, USA, pp. 175-194.

Skog Eriksen, M., Rødbotten, R., Grøndahl, A.M., Friestad, M., Andersen, I.L. and Mejdell, C.M. (2013). Mobile abattoir versus conventional slaughterhouse – Impact on stress parameters and meat quality characteristics in Norwegian lambs. Applied Animal Behaviour Science 149: 21-29.

Von Wenzlawowicz, M., von Holleben, K. and Eser, E. (2012). Identifying reasons for stun failures in slaughterhouses for cattle and pigs: a field study. Animal Welfare 21, Suppl. 2: 51-60.

Waiblinger, S., Boivin, X., Pedersen, V., Tosi, M., Janczak, A.M., Visser, E.K. and Jones, R.B. (2006). Assessing the human-animal relationship in farmed species: A critical review. Applied Animal Behaviour Science 101: 185-242.

Warren, L.A., Mandell, I.B. and Bateman, K.G. (2010). Road transport conditions of slaughter cattle: Effects on the prevalence of dark, firm and dry beef. Canadian Journal of Animal Science 90: 471-482.

Warriss, P.D. (1990). The handling of cattle pre-slaughter and its effects on carcass and meat quality. Applied Animal Behaviour Science 28: 171-186.

22. Toward the research and development of cultured meat for captive carnivorous animals

B. Kristensen
Oregon State University, School of History, Philosophy, & Religion, 2520 SW Campus Way, Corvallis, OR 97331, USA; kristenb@oregonstate.edu

Abstract

In this paper, I respond to a major oversight in contemporary animal ethics discussions: the moral problem of captive predation. While attention has been given to wild animal predation, most notably by Jeff McMahan, the suffering caused in sustaining captive carnivorous animals is absent from the discourse. McMahan argues that humans have an obligation to reduce suffering by aiding prey animals in the wild. I present his arguments in favour of interfering with predation as well as arguments against such obligations. McMahan lays the moral groundwork for a wide range of interference; however, interfering to the extent he suggests is currently infeasible. His arguments become stronger when applied to captive animals. Captive predation presents a situation in which humans have already interfered, and where further interference to alleviate suffering caused in sustaining captive carnivorous animals is practical and morally necessary. The carnivorous animals I consider are those kept as pets, in zoos and aquariums, and in rehabilitation centres. Cultured meat is the ideal candidate to replace the meat being fed to these animals. Current research is predominantly directed toward supplementing meat consumption by humans. I argue that it makes sense to direct research and development of cultured meat toward supplementing the diets of captive carnivorous animals. Feeding these animals cultured meat would result in an immense reduction in suffering and death for animals who would otherwise be killed to sustain them.

Keywords: animal ethics, *in vitro* meat, cellular agriculture, captivity, carnivory

The case for interfering in predation

Human intervention in wild animal predation is addressed by Jeff McMahan in 'The Moral Problem of Predation' (2016). McMahan takes an individualistic view of nonhuman animals as carriers of utility. He argues that predation must be taken seriously and addressed when and if doing so is possible. Understanding the motivations for interfering with animals in the wild will clarify the case when considering those in captivity. I maintain that we have been conditioned to assume an idealistic picture of the nonhuman world in which, left alone from human intrusion, it always already is a utility maximizing enterprise. We often assume that we must let nature take its course when it comes to wild animal suffering. McMahan's work questions such presumptions in terms of wild animal predation.

A presupposition espoused by McMahan and myself is that the suffering perpetuated by humans consuming animals outweighs the benefits of such consumption, and therefore we are ourselves first obligated to stop. Some may argue that to prey is a different act than to simply eat, and also that a lack of moral reflection in nonhuman animals is a relevant distinction. However, McMahan argues that regardless, humans have a responsibility to alleviate suffering for animals that are victims of predation. He states:

> The case in favor of intervening against predation is quite simple. It is that predation causes incalculably vast suffering among its innumerable victims, and to deprive those victims of the good experiences they might have had were they not killed. Suffering is

Svenja Springer and Herwig Grimm (eds.) ***Professionals in food chains***
EurSafe 2018 – DOI 10.3920/978-90-8686-869-8_22,

> intrinsically bad for those who experience it and there seems always to be a reason, though not necessarily a decisive one, to prevent it – a reason that applies to any moral agent who is capable of preventing it. (McMahan, 2016: 271)

We may have a stronger moral reason to stop suffering resulting from our own actions, but should we not intervene in other cases when possible? McMahan maintains that we should see all suffering, even that not directly causally connected to human agency, as similarly impacting an individual animal's experience of suffering. Prey are often 'hopeless in the absence of intervention.' (McMahan, 2016: 286) If we see these animals as being able to experience better or worse states of being, and if avoiding suffering and promoting welfare is of paramount importance, then we should do what we can to assist them. If one takes seriously his presuppositions, then his argument is convincing, although largely inapplicable at this time. There are opportunities for intervening in some capacity. Reducing support for conserving certain carnivorous animals could potentially alleviate the suffering of prey animals. However, this brings along a new set of ethical problems in creating a hierarchy of species based on their inability to survive without consuming flesh. Further, potential ecological consequences could actually compound suffering. The advent of cultured meat may resolve this problem in captive situations, as I discuss later. First, I consider objections to intervening into the affairs of predators and prey in the wild.

For many, the suggestion that we should interfere with 'natural' behaviours of wild animals is upsetting (Michael, 2002). Naturalness should not, however, be understood to be a moral guide. Such a position also seems rooted in human-nature dualism. The naturalness of suffering that impacts humans matters little in considerations of whether to alleviate it. I see no reason why we should take a different approach toward nonhuman animals, given an understanding that they likely often suffer in comparable ways. Intervention has unforeseen implications. Death and suffering are essential to evolution, and ecological wellbeing depends on the presence of predator species. Again, this line of thinking becomes problematic in claiming that ecological processes are something we should look to in articulating ethical behaviour. It is difficult to conceive of how to alleviate suffering for individual animals under such an approach. Tom Regan (2004) also argues against interference. Under his rights approach, in which most animals are considered a subject of life, it seems that he would maintain that we have an obligation to come to the aid of prey animals. However, he states that 'as a general rule, they do not need help from us in the struggle for survival, and we do not fail to discharge our duty when we choose not to lend our assistance.' (Regan, 2004: xxxviii) When Regan states that these animals do not need help from us in the struggle for survival he seems to be appealing to a notion of benefit played out through evolution (McMahan, 2016: 285). While species will evolve features suited for their survival as a result of such a struggle, this is troubling from an individualistic rights approach such as Regan's since there does not seem to be any benefit to the many individual animals who suffer and die in the course of this struggle for survival.

Considering captive predation

Next, I discuss captive predation, the indirect predation carried out by carnivorous animals in captivity. First, I discuss companion animals with a focus on cats. I then shift to examples from zoos and aquariums. I conclude this section with a look at a rehabilitation centre for wild animals.

Pet food has a profound impact on animal suffering, as well as on the environment. If we considered the dogs and cats in the United States as their own country, that country would rank 5th in worldwide meat consumption (Okin, 2017). Since dogs are not obligate carnivores, I focus specifically on the impact of cats' diets (Knight and Leitsberger, 2016). There are an estimated 88.3 million house cats in the United States (Frischmann, 2009). A large portion of the ingredients in cat food are of animal origin. Many foods manufactured for cats contain by-products, or the remains of carcasses processed for human consumption, including meat from 4D animals (dead, dying, diseased, or disabled). There has also been a move toward using human-grade meats in premium pet foods (Okin, 2017). It takes approximately eight house cats to equal the intake of a human consuming 2,000 calories a day. This means that the house cats in the United States are consuming the same caloric intake as approximately 11 million humans.

A 2008 study found that more than 2.9 million tons of raw fish are fed to house cats globally (De Silva *et al.*, 2008). It was predicted that by 2010 the amount would surge to 4.9 million tons. Cats actually consume more fish than humans do in some industrialized nations. This same study found that in Australia the average house cat consumes approximately 13.7 kg of fish per year, whereas the average human consumes approximately 11 kg per year. Most of these fish are forage fish, which include sardines, herring, anchovy, and capelin. The suffering resulting from the killing of these fish has a cascading effect up the trophic levels since they sustain many animals up the food chain including marine mammals and birds (Pikich *et al.*, 2014).

Keeping wild animals in captivity is problematic and my discussion should not be interpreted as an endorsement of such practices. Apex predators such as lions and orcas are often the most desired animals by visitors to zoos and aquariums. Beyond the direct harm caused by keeping the animals captive is the inevitable killing of significant numbers of other animals to sustain them.

In the wild, lions gorge on up to 43 kg of meat a day (Packer, 2017). According to the Smithsonian National Zoo, its lions consume about 227 kg of ground beef per week (Pressler, 2011). This meat is produced to satisfy their dietary needs, it is not a byproduct as is the case with a significant amount of pet food. Though this is a snapshot of one particular zoo, it is common practice to feed big cats meat from mainstream animal agriculture.

Adult orcas eat 68 to 136 kg of food per day. In captivity this is mainly thawed herring (Anderson, 2006). With this consideration, the 22 orcas being kept by the three Sea World parks consume at least 1,500 kg of fish, or at least 2,750 individual fish a day (Whale and Dolphin Conservation, 2017). That is 547,000 kg of fish a year, or more than 1,000,000 individual fish. In the case of both lions and orcas, as well as most other captive carnivores in zoos and aquariums, the animals they eat are already dead.

Lastly, I consider the Marine Mammal Center in Sausalito, California which specializes in the care and rehabilitation of pinnipeds, predominantly California sea lions and harbor seals. Between 2014 and 2016, they rehabilitated 1,534 pinnipeds (Blacow, 2017). The Marine Mammal Center feeds their patients an estimated 37,000 kg of herring every year. To put these numbers in perspective, the largest herring weigh around 0.5 kg. Under a conservative estimate, that 37,000 kg of herring adds up to at least 68,000 individual fish killed solely to feed the approximately 500 pinniped patients.

Toward the development of cultured meat for captive carnivorous animals

McMahan's arguments become stronger when applied to captive predation. He indicates that we may have more of an obligation to step in and stop suffering in cases which are a direct result of human activity. Captive predation presents a situation in which we can both care for captive carnivorous animals and alleviate the suffering caused by their current consumption. Further, I maintain that those who oppose intervention into wild animal affairs would have little issue in the case of captive animals. If there was a way to reduce suffering for other animals with minimal or no change to their captive situations, it would seem that on this McMahan and Regan would agree. I conclude by considering a way to implement such a change.

Cultured meat is a promising candidate for supplementing the diets of captive carnivorous animals. Cellular agriculture is a quickly advancing technology. JUST Inc., formerly Hampton Creek Foods, promised to have cultured meat products on store shelves by 2018 (Garfield, 2017). Memphis Meats and Finless Foods in the United States, and SuperMeat in Israel, are also headed toward having cultured meat on the market within several years. In the near future cultured meat will be a viable option to feed captive carnivorous animals as well.

Arguments for cultured meat focus on its reduced environmental impact, as well as its implications for animal welfare (Tuomisto *et al.*, 2011). Cultured meat also eliminates many safety concerns from contamination and food-borne illnesses (Zaraska, 2013). Arguments against its production critique how cultured meat would impact our relationship with animals, as well as aspects of its development that often depend on continuing to use and kill animals. However, it is undeniable that fewer animals would suffer and die in a future in which cultured meat replaces conventional animal agriculture. Some companies have already committed to not using animals in the process of research and production.[1] The case in favor of research and development of cultured meat is usually made for the purpose of supplementing current consumption of meat by humans. Since humans do not have a nutritional need for meat, this focus seems to have more to do with a refusal to change dietary habits, than it does with actually meeting a nutritional requirement in a less harmful way. I argue that it is practical and morally imperative to also direct the research and development of cultured meat in the direction of supplementing the diets of captive carnivorous animals who must consume meat to survive. Notable exceptions to the current paradigm of research are Wild Earth, and Bond Pet Foods, two startups that are working to develop cultured meat pet foods.

It is likely that cultured meat will first be marketed for humans, then make its way to the market for animal food. A challenge to the introduction of cultured meat is consumer acceptance, and although animals fed cultured meat may not know the difference, it is possible that introducing it as animal food first may carry the stigma that cultured meat is akin to pet food. The biggest challenge, however, is the current cost of production (Kadim *et al.*, 2015). Once cultured meat is introduced to the market, it will likely be at a premium price, which will exclude many humans from consuming it, and also may make it economically infeasible for introduction to captive carnivores for some time. However, Wild Earth plans to unveil prototypes of cat food containing cultured mouse meat in late 2018 (Fox, 2018).

Once cultured meat is able to satisfy the nutritional needs of captive carnivores and the price of production drops, I believe it will be a viable candidate to replace the flesh from animals who are caught, raised, and killed. Immeasurable suffering could be eliminated. This poses no risks of ecosystem disruption. Humans have already intervened in a large way by either domesticating, catching and keeping these animals captive, or by raising wild animals in captivity.[2] The experience that animals have consuming flesh in captivity would likely be no different if that flesh came from a lab rather than an actual animal. An exception are captive carnivorous animals who are fed live animals. However, most captive carnivorous animals consume flesh from animals who are already dead. It will take longer to develop unprocessed meat, which is a requirement for some captive carnivorous animals. However, within the next ten years it should be practical to introduce cultured meat to animals who do not need unprocessed meats. Cultured meat realises the potential to eliminate the animal suffering and death that till this time has been a troublesome necessity in sustaining captive carnivorous animals.

[1] For example, San Francisco-based Finless Foods gets cells for research from fish that die of natural causes at The Aquarium of the Bay. See: https://leapsmag.com/forget-farm-table-lab-table-fresh-fish-making-waves.

[2] This does not justify captivity or suggest that a diet of cultured meat would necessarily improve the wellbeing of animals in captivity, but is merely a pragmatic approach to reducing suffering in the current operating paradigm.

References

Anderson, G. (2006). Killer Whales: Feeding. Course Materials from Marine Science Class at Santa Barbara City College. Available at: http://www.marinebio.net/marinescience/05nekton/KWfeeding.htm. Accessed 14 June 2017.

Blacow, A. (2017). Oceana. Behind the Scenes of Marine Mammal Rescue. Available at: http://usa.oceana.org/blog/behind-scenes-marine-mammal-rescue. Accessed: 7 June 2017.

De Silva, S.S. and Turchini, G.M. (2008). Towards Understanding the Impacts of the Pet Food Industry on World Fish and Seafood Supplies. Journal of Agricultural and Environmental Ethics 21, Issue 5: 459-467.

Fox, K. (2018). Why Wild Earth Cofounder Ryan Bethencourt Is Applying The Science Of Vegan Biohacking' To Pet Food. Forbes. Available at: https://www.forbes.com/sites/katrinafox/2018/03/15/this-vegan-biohacker-is-set-to-launch-the-first-cultured-protein-foods-for-pets.

Frischmann, C. (2009). Pets and the Planet: A Practical Guide to Sustainable Pet Care. Wiley Publishing, Hoboken, NJ: 2009. 312 pp.

Garfield, L. (2017). Hampton Creek says it's making lab-grown meat that will be in supermarkets by 2018. Business Insider. Available at: http://www.businessinsider.com/hampton-creek-lab-grown-meat-2017-6. Accessed 18 July 2017.

Kadim, *et al.* (2015). Cultured meat from muscle stem cells: A review of challenges and prospects. Journal of Integrative Agriculture 14, Issue 2: 222-233.

Knight, A. and Leitsberger, M. (2016). Vegetarian versus Meat-Based Diets for Companion Animals. Animals 6, Issue 9.

McMahan, J. (2016). The Moral Problem of Predation. In: Chignell, Andrew and Cuneo, Terence (eds.) Philosophy Comes to Dinner: Arguments About the Ethics of Eating. Routledge, New York, pp. 268-293.

Michael, M.A. (2002). Why Not Interfere With Nature? Ethical Theory and Moral Practice 5, Issue 5: 89-112.

Okin, G.S. (2017). Environmental impacts of food consumption by dogs and cats. PLoS ONE 12, Issue 8. Available at: https://doi.org/10.1371/journal.pone.0181301.

Packer, C. (2017) Lion Center Initiatives: Frequently Asked Questions. University of Minnesota: College of Biological Sciences. Available at: http://cbs.umn.edu/research/labs/lionresearch/faq. Accessed: 14 June 2017.

Pikich, *et al.* (2014). The global contribution of forage fish to marine fisheries and ecosystems. Fish and Fisheries 15, Issue 1: 43-64.

Pressler, M.W. (2011). Feeding animals at the National Zoo. The Washington Post. Available at: http://www.washingtonpost.com/lifestyle/kidspost/feeding-animals-at-the-national-zoo/2011/07/27/gIQAEOEABJ_story.html. Accessed: 28 June 2017.

Regan, T. (2004). The Case for Animal Rights. University of California Press, Berkeley. 474 pp.

Tuomisto, H.L. and Teixeira de Mattos, M.J. (2011). Environmental Impacts of Cultured Meat Production. Environmental Science & Technology 45, Issue 14: 6117-6123.

Whale & Dolphin Conservation (2017). The Fate of Captive Orcas. Available at: http://us.whales.org/wdc-in-action/fate-of-captive-orcas. Accessed 23 January 2018.

Zaraska, M. (2013). Is Lab-Grown Meat Good for Us? The Atlantic. Available at: http://www.theatlantic.com/health/archive/2013/08/is-lab-grown-meat-good-for-us/278778. Accessed 14 June 2017.

23. Don't be cruel: the significance of cruelty in the current meat-debate

P. Kaiser
Department of Philosophy, University of Vienna, Universitaetsstr. 7, 1010 Vienna, Austria; peterkaiser@aon.at

Abstract

The aim of this paper is to investigate the significant role of the concept of cruelty to animals with regard to its application in the current meat-debate. Even those philosophers who do not advocate vegetarianism/veganism are in almost unanimous agreement when it comes to the condemnation of factory farming: its practices cause massive, unnecessary suffering to animals, therefore they are cruel. Prohibition of cruelty to animals has been central to animal protection acts, and is still playing a pivotal role in the debate over determining moral status: If x can be wronged by cruelty, then x has at least some moral status. I will discuss Timothy Hsiao's (2017) recent attempt to defend factory farming by arguing that it is not cruel to animals since animals have no moral status. By focusing on his misconception of cruelty it will be shown that Hsiao's anthropocentric definition 'that an act is cruel if it reveals a corrupt character or if it corrupts one's character so as to make one more disposed to mistreating humans' is neither necessary nor sufficient. After a reconsideration of David DeGrazia's (2009) fundamental objections to indirect duty views I will consider four significant shortcommings of Hsiao's account: first, the dubious spillover aspect; second, his underdetermined conception of sadism. Third, it will be demonstrated that Hsiao's definition is too inclusive. Finally, it will be shown that Hsiao disregards different kinds of cruelty. With reference to Julia Tanner's (2015) distinction of four kinds of cruelty the many practices involved in factory farming ought to be analyzed as prime examples of indifferent cruelty.

Keywords: animal cruelty, moral status, factory farming, meat

Introduction

Whereas global meat production is still on the rise (Heinrich-Böll-Stiftung *et al.*, 2018), ethical defenses of meat are few and far between. Considering 'the moral complexities of eating meat' (Bramble and Fischer, 2016), even those philosophers who do not advocate vegetarianism/veganism are in almost unanimous agreement with the ones who do, when it comes to the condemnation of factory farming. Thus, without presupposing a particular ethical theory, a simple argument against meat consumption from factory farms briefly states: If it is wrong i.e. morally unjustified to knowingly cause unnecessary harm, pain, distress, fear, and (henceforth, in short) suffering to animals, then it is also wrong to support practices that do exactly that. Factory farming significantly causes massive, unnecessary suffering. To consume meat and other animal products from factory farms is to support practices that cause massive, unnecessary suffering. Therefore, it is wrong to consume factory farm products (DeGrazia, 2009; Hsiao, 2017; Rachels, 2004). Importantly, 'knowingly causing unnecessary pain and/or suffering' is a representative definition of cruelty (Tanner, 2015: 822). Now, who is in favor of cruelty? 'Nothing shocks our moral feelings so deeply as cruelty does' (Schopenhauer, 1998: 169). Legally, prohibition of cruelty to animals has been central to animal protection acts since the 17th century (*ibid.*: 818). For example, the current Austrian Animal Protection Act states: 'It is prohibited to inflict unjustified pain, suffering or injury on an animal or expose it to heavy fear' (TSchG §5.1). Ethically, the concept of cruelty is still playing a pivotal role in the debate over determining moral status: If x can be wronged by cruelty, then x has at least some moral status; i.e. moral significance in her/his/its own right. Or vice versa: If x has

EurSafe 2018 – DOI 10.3920/978-90-8686-869-8_23, © Wageningen Academic Publishers 2018

no moral status at all, then x cannot be wronged by cruelty. In what follows, I will focus on this aspect of the cruelty-debate.

No moral status

Timothy Hsiao (2017) has recently argued that factory farming (he consequently uses the euphemism 'industrial animal agriculture') is *not* cruel to animals, since animals have no moral status. In order to attribute moral status to humans only, i.e. deny animals any direct moral status, Hsiao needs to establish a morally significant anthropological difference between humans and animals. Unsurprisingly, neither sentience, consciousness, nor agency constitute exclusive human moral status. 'If the moral importance of our own conscious lives depends instead on some further fact that animals lack, then one can non-arbitrarily affirm that human lives matter while denying that animal lives matter' (*ibid*.: 43), so the prospect goes. Thus, the answer to the question of what it is that confers moral status is 'the traditional one', as Hsiao himself proudly presents: 'In order for a being to have moral status of any kind, it must have the capacity to reason.' (*ibid*.: 44) In other words, strictly anthropocentrically, Hsiao appears to reverse the order of Bentham's guiding questions: The morally significant question is not, Can they suffer? but, Can they reason?

Hsiao circumvents the common objection, the argument from marginal cases – infants, mentally disabled do have moral status despite their inability to reason – by, not so non-arbitrarily, claiming 'that all humans, regardless of age or disability, have moral status because they possess a rational nature. That is, all humans, in virtue of being the kind of organism they are, possess a basic or root capacity to reason' (*ibid*.: 47). However that may be; for a thorough critique of Hsiao's marked preference for 'metaphysically robust forms of essentialism' and 'neo-Aristotelianism' (*ibid*.: 48), see Puryear *et al.* (2017).

No status, no cruelty

No matter how far-fetched Hsiao's argument for exclusive human moral status may be, accordingly it follows, if animals have no moral status at all, then they cannot be wronged by cruelty. If it is denied that their sentience, precisely their pain and suffering, is morally significant in its own right, then this seems to imply 'that 'cruelty' to nonhumans is strictly impossible' (Pluhar, 1995: 320). No-status renders 'animal cruelty'-talk nonsensical. If a stone or a tree has no moral status, then the stone or the tree cannot be wronged by cruelty. If a cow is considered to be nothing more than a living 'thing' (Joy, 2011: 118), being objectified and commodified as mere production unit, then consequently she, respectively 'it', cannot be wronged by cruelty. If 'it' is being harmed then simply in the sense of being damaged, but not in the sense of being wronged. I term this straightforward conclusion the (Neo-)Cartesian 'no-status-no-cruelty view'.

As Descartes (in-)famously stated: 'My opinion is not so much cruel to animals as indulgent to men – at least to those who are not given to the superstitions of Pythagoras – since it absolves them from the suspicion of crime when they eat or kill animals' (Descartes, 1991: 365-366). Dear consumer, dear producer, don't worry, eat meat, and do not limit yourself 'in the pursuit of profit' (DeGrazia, 2009: 156). This practical advantage of human exceptionalism, the absolving consequence for the carnist's clear conscience – 'eating certain animals is considered ethical and appropriate' (Joy, 2011: 30) – is, what Hsiao is ultimately after (Hsiao, 2015). But precisely in times like these, when 'the superstitions of Pythagoras' not only affect ideas and ideals of 'animal lovers', but also prevail in animal ethics and animal cognition, let alone modern animal protection acts, it is hard to sustain this type of strict no-status view. This might explain, why, against all odds, no-status advocates accept the significant role of the concept of cruelty to animals. Hsiao echoes Peter Carruthers's (1992) Kantian-flavoured no-status-indirect-duty, respectively no-status-indirect-cruelty view:

> '[A]nimals lack moral status. ... However, this is not to say that we are permitted to do literally anything we want to animals, anymore than the fact that we can use our property for our own purposes implies the right to do whatever we want with it. We may not have duties directly to animals in respect of their welfare, but we do have duties to ourselves and each other that require us to respect animal welfare ... it is through harming animals that persons are wronged' (Hsiao, 2017: 48-49).

Despite the fact that in many cases it is perfectly legal to waste and destroy our very own property, let's keep those 'animal engines' running smoothly. A comprehensive critique of indirect-duty views is beyond the scope of this paper. But in the present context it is important to see that no-status views face the following dilemma: Either they are committed to the no-status-no-cruelty view. Then they have to abandon the concept of animal cruelty as strictly impossible – no matter how counterintuitive this might appear. Or no-status views accept the challenge of accounting for the wrongness of animal cruelty (*ibid.*). Then they have to concede that animals have at least some moral status, i.e. their suffering matters morally in their own right – which contradicts their very own position. This second horn of the dilemma reflects David DeGrazia's (2009: 148-150) fundamental objection to indirect-duty views. As the following discussion of Hsiao's conception of cruelty will show, he is in no position to meet the challenge of giving a satisfactory account of the wrongness of animal cruelty. However, examining the inconsistencies will help to get a better grasp of what it is that makes cruelty to animals cruel.

A formal definition of cruelty

Hsiao's strict anthropocentric, character-specific 'formal definition of animal cruelty' states 'that an act is cruel if it [1.] reveals a corrupt character or [2.] if it corrupts one's character so as to make one more disposed to mistreating humans' (Hsiao, 2017: 49). As Stephen Puryear *et al.* have correctly noted, Hsiao's definition needs to be understood as biconditional: 'It follows that factory farming is not (inherently) cruel only if this definition states a necessary and not merely a sufficient condition for cruelty, since otherwise cruelty might also take some other form which *is* inherently present in factory farming' (Puryear *et al.*, 2017: 312). It can be demonstrated that Hsiao's definition is neither necessary nor sufficient. I will consider four significant shortcommings of his account.

Spillover

Hsiao himself appears to be ambivalent about condition [2.] of his definition, which can be termed the spillover aspect as necessary condition for cruelty. On the one hand he notes that the 'link between violence to animals and violence to people is well-documented' (Hsiao 2017: 49). On the other hand, while trying to dismiss DeGrazia's (2009: 149) objection that there are cases of animal cruelty without any spillover to people, Hsiao abandons this very condition in favour of condition [1.]: 'If humans have duties to themselves to develop and cultivate a certain character makeup, then we may plausibly say that every act of animal cruelty harms the person himself, independent of any causal spillover to third parties' (Hsiao 2017: 49). However, according to Hsiao's own account, this concession undermines the strength of his definition significantly. In DeGrazia's words, one ought not to 'try to base a very certain moral judgment – that cruelty to animals is wrong – on speculative empirical assumptions about an undesirable spillover effect on humans' (DeGrazia, 2002: 26). Moreover, the second horn of the dilemma, regarding Hsiao's definition, becomes relevant, since Hsiao (again, like Carruthers before him) 'leaves unexplained *why* cruelty to animals is a vice and compassion to them a virtue. Animals, on this view, lack moral status and cannot be directly wronged. So why should pulverizing cows for fun reveal a defective moral character any more than does tearing up a newspaper for fun? The only plausible account of why cruelty is a vice acknowledges the moral status of its victims' (*ibid.*). This insight will become especially relevant in understanding cruelty as sadism.

Sadism

Given the first conditional of his definition – an act is cruel if it reveals a corrupt character – it is striking that even by Hsiao's own definition, there actually are some cases of animal cruelty in the context of factory farming. However, 'what counts as a cruel practice towards animals will be person-specific' (Hsiao, 2017: 49). Practices involved in factory farming ought not to be considered as 'inherently' cruel, analogous to warfare and the 'fact that soldiers can fight virtuously, honorably' (*ibid.*: 51). But this seemingly paradoxical affirmation of Tolstoy's saying 'as long as there are slaughterhouses there will be battlefields' can be interpreted as simply pointing to the fact that it is difficult to find any inherently, non-context-specific cruel actions. 'Tightening the vise on my workbench is not inherently cruel. But if someone's hand is in that vise, and I delight in ... the suffering that tightening it would cause, then tightening it suddenly becomes an instance of cruelty' (Puryear *et al.*, 2017: 315).

'A practice that evinces a cruel character for one person may not be considered cruel for another. John Smith may be able to work in a slaughterhouse for his entire life without there being anything amiss about his character, while the same profession may allow Hannibal Lecter an outlet to cultivate his sadism' (Hsiao, 2017: 49).

For the prototypical cruel agent is the sadist, undisputedly revealing a corrupt character by taking delight in causing another to suffer for the sake of suffering itself. Consider Hannibal as stun operator in a slaughterhouse who intentionally stuns cows or lambs unsufficiently, so that they remain conscious. Transported to the so-called 'stickers', another sadistic character like Hannibal cuts their throats in order to bleed them out, knowing that they consciously experience this process alive, and ultimately takes pleasure in, literally, silencing the lambs by inflicting pain and suffering on them. This example of sadism betrays Hsiao's no-status-indirect-cruelty view in several ways. Hannibal's initial goal is neither to expose his fellow workers to heavy fear; nor does he attempt to cause company owners to suffer by 'damaging' their precious property. Hannibal is not threatening them to carry out the same violent actions on them. This example is all about inflicting suffering on the factory farm animals he is dealing with. Indeed, the slaughterhouse profession may allow Hannibal 'an outlet to cultivate his sadism'; thereby making him less likely to torture other humans (see DeGrazia, 2002: 18 for making this point against spillover on humans). Moreover, the typical sadist might most likely choose *weak* human or nonhuman victims (Pluhar, 1995: 91). Cultivating sadism can also be understood in terms of cultivating a certain kind of empathy, as the phenomenological debate has convincingly shown: 'Just think of ... the sadist. A high degree of empathic sensitivity might come in handy if one wants to manipulate and exploit people. It is also a precondition for cruelty, since cruelty requires an awareness of the pain and suffering of the other' (Zahavi, 2014: 116).

In the present context the most important aspect of sadism is that in order to reveal – or better: satisfy – his sadistically corrupt character, Hannibal himself must believe that animals are not mere morally insignificant living things. Otherwise he could not take delight in causing them to suffer for the sake of suffering. If Hannibal were committed to the no-status-no-cruelty view, if he considered animals as mere organic machines with no morally significant conscious suffering and pain experience, then he would have to quit his job and cultivate his sadism by restricting himself to human victims; since substitutes won't do it for him: neither destroying lifesize toy animals, attending cathartic performances of Camus's 'Caligula', Brecht's 'Saint Joan of the Stockyards', participating in Hermann Nitsch's 'Theatre of Orgies and Mysteries', nor movies with disclaimers like 'no animals were harmed', 'all the violence was safely simulated'. Hannibal needs the 'real thing'.

The upshot of this discussion is that Hsiao's definition 'fails to capture an essential feature of cruelty, namely, that of having an improper regard for the badness of the suffering that one causes another'

(Puryear *et al.*, 2017: 314). Contrary to this, Hannibal has a stunningly proper regard for the badness of the suffering that he causes another. To him, Bentham's question 'Can they suffer?' is of tremendous importance: maximize the others' suffering for your own pleasure. 'Pervertedly pathocentric', he acknowledges the moral status of his victims. However, as we shall see in objection 4, most instances of cruelty take significantly other forms than sadistic cruelty.

Vandalism, not cruelty

There is a third major problem with Hsiao's character-specific, agent-corruption definition. It is not sufficient, insofar it is too inclusive. Simply put, there are many actions that may have a character-corrupting tendency, yet involve no cruelty, respectively sadism, at all. Consider, for example, theft, plagiarism, offering and/or accepting a bribe, malicious mischief, and – eminently suitable for Neo-Cartesians – vandalism defined as knowingly causing unnecessary damage. Taking delight in destroying an-/organic things is not to be confused with taking delight in causing another to suffer for the sake of suffering; though the latter might include the former. The difference between vandalism and cruelty as sadism is well-established; both might be ultimately rooted in 'the anatomy of human destructiveness' (Fromm, 1997). Obliterating this very difference would amount to committing an equivocation fallacy.

Four kinds of cruelty: the way ahead

Suspiciously, Hsiao focuses on sadistic cruelty only. However, the majority of factory farm workers, like Hsiao's 'John Smith', evidently are no sadists. Even so, the many practices involved in the meat-production business are prime examples of another type of cruelty: indifferent cruelty, what Tom Regan has called 'brutal cruelty': 'Some cruel people do not feel pleasure in making others suffer. Indeed, they seem not to feel anything. Their cruelty is manifested by a lack of what is judged appropriate feeling, as pity or mercy, for the plight of the individual whose suffering they cause' (Regan, 2004: 197). Sadistic as well as indifferent cruelty can be manifested in two ways, actively and passively: 'Passive behavior includes acts of omission and negligence; active, acts of commission' (*ibid.*). Following Regan and Julia Tanner's recent reference to his classification, there are at least four possible kinds of cruelty: (1) active sadistic cruelty, (2) passive sadistic cruelty, (3) active indifferent cruelty, (4) passive indifferent cruelty (Tanner, 2015: 823).

All four kinds of cruelty are actually manifest in (factory) farming practices. Hannibal as stunner is a typical example of active sadistic cruelty. Hannibal as worker surveying 'broiler' chickens in high-density confinement, some with broken wings, and some with twisted and broken legs, unable to hold their weight, being deprived of their most basic welfare conditions, is an example of passive sadistic cruelty. The same scenarios, without someone taking pleasure in it, are examples of active or passive indifferent cruelty to animals.

Agreeing with Tanner's definition that cruelty is 'knowingly causing unnecessary pain and/or suffering', it has to be stressed that first, applying the concept of indifferent cruelty explicitly requires an awareness of the moral significance of the animals' suffering. Indifference is not ignorance regarding the moral status of animals. However, indifference 'may result from ... routinely performing an action until one becomes desensitized, or numbed, to it' (Joy, 2011: 82). The meat-production business significantly facilitates indifferent cruelty, literally, from (factory) farm to fork. Moreover, 'our practices and legislation regarding indifferent cruelty are inconsistent insofar as they are applied to farm animals in a way that they are not to pets' (Tanner, 2015: 825).

Second, focusing on the necessity-aspect, the charge of indifferent cruelty implies that no sufficient moral justification for knowingly causing suffering can be given. The suffering involved in food chains is unnecessary either to achieve the particular end, i.e. human nutrition including meat; or, making the stronger case, the very end is not morally justifiable, since our nutritional needs can be met without subjecting animals to so much suffering. However, we do have readily available, adequate, cruelty-free alternatives to animal products (Rachels, 2004: 71; DeGrazia, 2009: 153; Tanner, 2015: 828).

Space limitations do not permit me to elaborate this point, however, in terms of future prospects, the next step would amount to 'a systemic reassessment of all farming practices' in the light of the four types of cruelty (*ibid.*: 834). This is how the philosophical debate could and should have a societal impact: it could strengthen the legal case for the prohibition against farm animal cruelty significantly, pointing the way ahead to reduce 'the extensive amount of socially acceptable forms of violence against animals' (Flynn, 2012: 4) – singing along to that old Elvis tune: 'Don't be cruel'.

References

Bramble, B. and Fischer, B. (eds.) (2016). The complexities of eating meat, Oxford University Press, Oxford, 232 pp.

Carruthers, P. (1992). The animals issue. Cambridge University Press, Cambridge, 224 pp.

DeGrazia, D. (2002). Animal rights. Oxford University Press, Oxford, 144 pp.

DeGrazia, D. (2009). Moral vegetarianism from a very broad basis. Journal of Moral Philosophy 6 (2): 143-165.

Descartes, R. (1991). Letter to Henry More, February 5, 1649. In: Cottingham, J. (ed.) The Philosophical Writings of Descartes Volume III. Cambridge University Press, Cambridge, 432 pp.

Flynn, C.P. (2012). Understanding animal abuse. Lantern Books, New York, 133 pp.

Fromm, E. (1997). The anatomy of human destructiveness. Pimlico, London, 680 pp.

Heinrich-Böll-Stiftung, BUND, Le Monde Diplomatique (2018). Fleischatlas 2018. Available at: https://www.boell.de/de/der-fleischatlas-2018-bestellen. Accessed 26 January 2018.

Hsiao, T. (2015). In defense of eating meat. Journal of Agricultural and Environmental Ethics 28 (2): 277-291.

Hsiao, T. (2017). Industrial farming is not cruel to animals. Journal of Agricultural and Environmental Ethics 30 (1): 37-54.

Joy, M. (2011). Why we love dogs, eat pigs, and wear cows. Conari Press, San Francisco, California, 208 pp.

Pluhar, E. (1995). Beyond Prejudice. Duke University Press, Durham, North Carolina, 392 pp.

Puryear, S., Bruers, S. and Erdös, L. (2017). On a failed defense of factory farming. Journal of Agricultural and Environmental Ethics 30 (2): 311-323.

Rachels, J. (2004). The basic argument for vegetarianism. In: Sapontzis, S. (ed.) Food for thought. Prometheus, Amherst, New York, pp. 70-80.

Regan, T. (2004). The case for animal rights. University of California Press, Berkeley, California, 450 pp.

Schopenhauer, A. (1998). On the basis of morality. Hacket, Indianapolis, Indiana, 226 pp.

Tanner, J. (2015). Clarifying the concept of cruelty. What makes cruelty to animals cruel. Heythrop Journal 56 (5): 818-835.

Zahavi, D. (2014). Self and other. Oxford University Press, Oxford, 296 pp.

24. Understanding food markets and their dynamics of exchange

W. Leyk
Friedrich-Alexander-Universität Erlangen, Theology Department, Ethics (Systematic Theology II / Ethics), Kochstrasse 6, 91054 Erlangen, Germany; pfrwleyk@aol.com

Abstract

Food scandals increasingly draw attention to conditions of production and products themselves which are increasingly morally encoded and furthermore subject to societal and cultural dynamics. Georg Simmels theory of economy as exchange brings consumers, producers and products together in comprehensive exchange scenarios and dynamics of objectification. Increasing desire for ethical products and use of labels are critically reviewed for their output of information and are located within a system of knowledge, trust and diverse stakeholder interests on food markets. The survey of these products also includes remarks on products having effects and showing agency in a frame of network-theory. In exchange these objects are brought together with subjects in a frame of new consumers awareness for diverse and conflicting perspectives on food-consumption. Postmodern and hybrid consumers are dependent on information, but also are overstrained tending to irrational choices. Referring to the proposal of subject-object interaction in exchange-scenarios already existing concepts for ethical market governance like consumer ethics and paternalistic (ethical) guidance by nudges are critically related to markets dynamics. Ethics for exchange of foods should take into account product-agency as well as consumers interests. It is proposed to turn to participative concepts like institutional governance, ethical matrix and focus on relational approaches.

Keywords: Actor-Network-Theory, animal ethics, ethical food, paternalistic guidance, objectification

Understanding food markets and their challenges

A diversity of dynamics like market forces, politics, societal and cultural convictions shapes food markets and individuals often find themselves in a web of relations and interests distorting their once deliberate choices. In this situation the ethical challenge is rather a descriptive than a normative one. Better understanding of these comprehensive challenges will facilitate decisions. Methodological frame of this approach is Georg Simmels 'Philosophy of Money' which serves as a heuristic background for a more profound perception of food markets. His understanding of economy brings together consumers, producers and their products in a dynamic scenario of exchange. Exchange transforms things into objects of desire in a comprehensive process of objectification and cultification. These valuation dynamics are connectable to increasing concern for food ethics, sustainability or methods of food production. Other thematic subtexts frame this survey. The fact, that animals are increasingly considered morally relevant, also demands new awareness for products and their context. Exchange brings together consumers, producers and products in a surprising scenario of interaction which best can be understood within actor-network theory. Actor-Network-Theory (ANT) proves a good choice since it does not imply any predetermined market or actor theory because focuses on effects which do not need acting subjects in the sense of common subjectivism. Network theory is not considered as an alternative but a dynamic complement to other methodology helping toward better understanding of comprehensive dynamics in the exchange-encounter of products and consumers.

Products talking: 'ethical' food and labels

Food scandals are proof, that products cannot be separated from a context of interests, emotions or convictions and they result in either a crisis of products acceptance or urgent desire for better products.

EurSafe 2018 – DOI 10.3920/978-90-8686-869-8_24, © Wageningen Academic Publishers 2018

Better products often are labelled 'ethical' and connected to values like animal welfare, environmental protection, fair trade or healthy eating. More than any other products they are concerned with consequences of production or use. They contribute much more to markets than just saturation of demand. They encourage reflections, public discourse and even establish their own milieus which are characterized by a new kind of interaction between consumers and products. Therefore the relation of customers and products can be considered reciprocal. Products 'agency' should not to be understood in a frame of subjectivism but rather be assessed within a primarily relational concept like network theory. This reference to network understanding does not imply a normative methodological decision since ANT is focused on dynamics and relations and does not imply specific economics with predefined attributes for actors in the market (Callon, 1999; Mattson, 2003: 3). The network-idea of product agency implies interactions between human and non-human agents. It is deliberately chosen because it is particularly applicable to animals in economy who as 'living products' set in motion dynamics of objectification as well as they produce 'moral codes' and narratives (Priddat, 2007: 31.45).

This new set of reciprocal relations between market-subjects and market-objects inspires a look at product representation by use of labels. Labels are issued by companies or regulatory institutions (like the Minister of Health, Nordic Council 2004). While labels pretend to inform about the product, they often shed more light on their users intentions, expectations and production issues than on the product itself. Labels are considered facilitating good decisions since they have potential for 'tackling' comprehensive challenges' (Thaler and Sunstein, 2008: 205) by reduction of complex matters to simple messages or symbols (Thaler and Sunstein, 2008: 203). It is questionable if labels further information. They rather have two antagonistic effects. They generate a kind of knowledge as well as they encourage trust which is the opposite of knowledge. The antagonistic effect of knowledge and trust (which is a practices of ignorance) can be referred to a theory of social systems and then is plausible as a systemic code of knowledge and ignorance. These contradicting effects set in motion dynamics of communication, rating practices and discourse. One could reflect, if knowledge and ignorance play an autopoietic role in the food system, but this conclusion should be left to experts in system theory. Empirically it is uncertain, if labeling facilitates deliberate choices during everyday shopping (Galizzi 2012: 15). A study from Denmark (Christensen *et al.*, 2016) makes evident, that customers not only seek knowledge, but also trust in the context of products, that is production methods or attitudes of the producer. Another study suggests, that labels influence consumer behavior up to 7%. It is likely, that many consumers rather rely on emotional narratives and on ethical aspects, than on information about e.g. ingredients (Scott, 2001). Critically it must be observed that labeling often rather serves consumer manipulation up to deception than transparency and information. Many labels suggest a comprehensive strategy for animal welfare which is not substantiated if one digs deeper. Ethical customers of dairy products for example often are concerned, how long calves are kept with their mothers but if they check out a respected green label like german *Demeter* (2018) they will find no explicit arrangement on this, while *Naturland* (2018: 20) at least suggests keeping calves for some days with their mother. Motherbound dairy farming is not even mentioned as an option and the same applies to the problem of male chicks killed during production of eggs. Almost a third of *Demeters* guidelines is busy with the label itself and other parts focus the final production-process. Often labels neglect significant parts of the food chain with the apology, that due to comprehensive value chains fully ethical products are unattainable anyway (Crane, 2001: 370). Lately a practice of labeling has developed which even might be judged as absurd: Vegan labeling focuses avoidance of possible animal components like proteine during production of wine or possible 'skin' contact with animals on tea plantation. This information strategy probably tells more about prospective vegan users than about the product. In the end Crane's pragmatic conclusion is, that there is no such thing as a definitive 'ethical product', but only partial ethical attributes which the customer must recognize, believe, value and, ultimately buy into.

Sobering result of this survey is, that labels often rather serve consumer confusion than transparency. Caught between knowledge and ignorance labeling provides insights sometimes more into the dynamics of food markets than into products. It lacks transparence and is not standardized. A truly reliable system of labels should be kept apart from autopoietic interests of the food system. Its standards and control mechanisms would have to be administered by impartial and reliable institutions. Collective Action of producers, consumers and supervision by non-governmental organizations are a thinkable model for this. Another proposal for product communication is (Mepham, 2013), that stores should provide computerized ethical matrixes which are better at reflecting comprehensive challenges of ethical products.

Subjects: hybrid and myopic customers

Without interested subjects there is no value. The relation of products and their buyers has intensified by appearance of a new brand of consumers. Ethical, well-informed and selective consumers are shifting from a consumer to a citizen mind-set) (Food Citizenship, 2017). They complement the idea of utility with specific moral codes and considerations about species appropriate production, fair trade and ecology. Ethical consumers are concerned individuals, often empathetic, educated in questions of animal welfare, sustainability and health impacts of food and they have a variety of choice-options at hand. They are hybrid and postmodern in so far as they integrate choice factors which first modernity would have judged irrational. Their idea of utility is an integrated and holistic one. Postmodern consumers are inconsistent and sometimes are characterized by the formula 'Aldi et Audi', meaning that customers of around the corner discounters drive up in expensive cars. They have left the narrow path of consequentialism and are moving in a web of social possibilities. These new choice-options also affect market architecture as they complement classical understandings of market equilibrium and dynamics of demand with added values and increasing use of moral semantics like sustainability, fairness or animal welfare. These moral codes result in a new normativity which is subjected to public discourse or societal commitments on morality (like e.g. the ban on fur-products) (Stehr, 2007: 66-75, 302-307).

A significant idea of economics is, that consumers should be well informed for deliberate choice. Food exchange meanwhile takes place within a considerably widened frame of knowledge and trust, emotion and psychology, morality and societal discourse. While this development opens a door for ethics in economy, there is also a drawback. Information overload causes customers to struggle with choices. Postmodern customers find themselves in a dilemma typical for 2nd modernity. Abundancy of information undermines knowledge. Knowledge and ignorance (trust), security and risk are dwelling side by side. Deception and a state of exhaustion turn the well informed into a myopic Customer who seeks for guidance of any kind (Bröckling, 2017: 15-44, 188) like social networks, reputation management, labeling and rating by use of data. The case of *myopic* and clueless consumer makes evident, how choices are embedded in a comprehensive web of agencies and efficiencies. But abundant information contributes to consumers disorientation and is no substitute for lack of first-hand experience. It is striking how critical consumers often have naïve and unrealistic ideas about nature and food production.

Ethics for a changed market?

These dynamics of valuation explain changes in once predictable consumer behavior. They suggest differentiated heuristics beyond already existing concepts like consumer ethics or paternalistic guidance by nudging.

Consumer ethics are shaped by shared consumer and corporate social responsibility on moralized markets. These markets have supplemented demand and supply mechanism with a desire for moral codes and product-narratives. Consumers now are able to upgrade exchange by introduction of quality aspects like political or cultural ideas (Food Citizenship, 2017; Lamla, 2006). Consumer social responsibility (CONSR) (Ahaus, 2009; Brinkmann *et al.*, 2008; Lamla, 2006) and food-citizenship are analogical to Corporations social responsibility (CSR). Fair Trade is considered a good example for it (Brinkmann *et al.*, 2008). It is hard to evaluate though the importance of ethical consumption. It is quite difficult to interpret numbers like the 10% which consumers are prepared to add for ethical products (De Pelsmacker, 2005; Arnot, *et al.*, 2006 count 13-18%). On one hand we are left with a 90% majority of non-ethical products ruling the market. On the other hand, a 10% margin of increased profit by such products is interesting for any company. German sausage producer 'Ruegenwalder' was able to increase sales by 17% by introduction of vegetarian products which now make 20% of the product portfolio. Limiting factor of consumer ethics is, that they are confined to certain specific milieus, rely on voluntary self-commitment and are neither controlled nor enforced by any institutions. Ethical consumption is but one of many interventions possible. Other interventions can be restrictive laws, animal activism or voluntary restrictions. Food-citizenship though can be an important market-force if individual interests are summing up to affect markets. Lamla observes increasing participation by use of information and media. They open the scenario of individual preference and choice for comprehensive considerations, democratic experimentalism, governance by deliberate use of data and so on.

Nudging is a concept of behavioral as well as political economics. Paternalist nudging guides consumers without coercion. We all know nudges: Urinals with fake plastic flies, traffic obstructions near schools and so on. Nudging is an appealing concept, but it should be noted, that it is not based on knowledge or ethical convictions but rather on general ideas about prevention and health care (Thaler and Sunstein, 2008). Nudges act below the level of knowledge, consciousness or rationality. They simplify choice by use of biological or cultural default settings like use of green as a color on labels for good nudges (Thaler and Sunstein 2008: 89). Realistic observers will notice that nudging is connected rather to psychology and behavioral science than to deliberate choice or values. Nevertheless, nudging promotes ethical issues like sustainability, health protection, animal welfare and others. Ethical reflection of nudging though focuses issues like autonomy and manipulation, self-government and governance. Sunstein (2016) appropriately calls nudging 'The Ethics of Influence' and concentrates on questions of legitimation, mandates and transparency. Nudging has been successful and is used by governments for market governance, especially if combined with choice architecture. With nudging the model of economic relations expands: Non-economic players like government and institutions exert influence on choices by well-meant rules and choice architecture. Paternalist guidance contradicts the idea of food-citizenship which is built on awareness, knowledge and resulting choices of consumers. On the other hand, nudges solve the dilemma of myopic and overstrained consumers by reducing problems complexity.

Ethics for food markets

Exchange of food is subjected to comprehensive dynamics of valuation. A crisis of social acceptance e.g. is neither caused by flaws in demand and supply nor by lack of consumers interest, it is caused by misunderstanding objectification dynamics, the role of knowledge and trust and neglection of products importance. Problems on food markets often are consequence of limited market-heuristics which neglect exchange contexts.

What kind of ethics does this new food market need? It seems trivial to one-dimensionally invoke ethical standards like honesty for product labels, concern for consumer-health or minimum standards in animal keeping or responsibility. Ethical governance of food markets will only partially be provided by consumer ethics, paternalistic guidance, by ethical labels, nor by invoked normativity of any kind. Comprehensive market dynamics suggest a differentiated methodology like ethical matrix, a research of dynamics of valuation and their connectivity to ethical considerations. Methods bringing actors together are advisable like the idea of institutions as a model for shared convictions or coordination of interests, implementation of compliance or governance by impartial organizations. Concerning the animal case, it is advisable to target situational relation-based reasoning rather than capacities. Down-to-earth ethics on food markets will not have to surrender ideals but ethics on complex markets will only be achieved by joint efforts for learning and better understanding of markets topography and their relational architecture.

References

Arnot, Chr., Boxall, P.C. and Cash, S.B., (2006). Do Ethical Consumers Care About Price? A Revealed Preference Analysis of Fair Trade Coffee Purchases in: Canadian Journal of Agricultural Economics 54: 555-565.

Bröckling, U. (2017). Gute Hirten führen sanft – über Menschenregierungskünste, Frankfurt.

Callon, M. (1999). Actor-network theory – the market test. Sociological Review 47: 181-195.

Crane, A. (2001). Unpacking the ethical product. Journal of Business Ethics 30: 361-373.

De Pelsmacker, P., Driesen, L. and Rayp, G. (2005) Do Consumers Care about ethics? Willingness to pay for Fair-Trade Coffee. The Journal of Consumer Affairs 39: 363-385.

Demeter (2018). https://www.demeter.de/leistungen/zertifizierung.

Food Ethics Council (2017). Food Citizenship – Report 2017, Available at: https://www.foodethicscouncil.org/our-work/food-citizenship.html. Accesses 10 January 2018

Food Ethics Council (2015). Food choices, advertising and ethics. What role for advertising in moving towards an ethical food system? A report of the Business Forum meeting on Tuesday 24th March 2015 Available at: https://www.foodethicscouncil.org/uploads/publications/150324%20Food%20choices%20advertising%20ethics.pdf. Accessed 20 January 2018.

Galizzi, M. (2012). Label, nudge or tax? A review of health policies for risky behaviours in: Journal of Public Health Research 1 Available at: https://www.ncbi.nlm.nih.gov/pmc/articles/PMC4140317. Accessed 20 January 2018.

Heidbrink, L. and Schmidt I. (2011). Das Prinzip der Konsumentenverantwortung – Grundlagen, Bedingungen und Umsetzungen verantwortlichen Konsums. in: Heidbrink, L., Schmidt, I., Ahaus, Die Verantwortung des Konsumenten. Frankfurt pp. 25-56.

Heidbrink, L., Schmidt, I. and Ahaus, B. (2011). Der verantwortliche Konsument. Wie Verbraucher mehr Verantwortung für ihren Alltagskonsum übernehmen können. Working Papers CRR (Center for Responsibility Research) 10 Available at: http://www.responsibility-research.de/resources/WP_10_Der_verantwortliche_Konsument+_.pdf. Accessed 20 January 2018.

Lorenz, St. (2006). Biolebensmittel und die 'Politik mit dem Einkaufswagen' in: Lamla, J., Neckel Sgh. (Hrsg) Politisierter Konsum – konsumierte Politik, Wiesbaden, pp 91-112.

Mattsson. L.G. (2003). Understanding market dynamics potential, contributions to market(ing) studies from actor-network theory. Available at www.impgroup.org/uploads/papers/4380.pdf. Accessed 20 January 2018.

Mepham, B. (2013). Food Choice. Food Ethics: Spoilt for choice? The magazine of the Food Ethics Council 8/1, 11-15.

Naturland (2018). https://www.naturland.de/de/naturland/richtlinien.html.

Nordic Council of Ministers (2004). Ethical Labelling of Food, Copenhagen.

Priddat B. (2007). Moral als Indikator und Kontext von Ökonomie, Marburg.

Scott, J.V, Anusorn, S. and James, Th. (2001). Consumer ethics: an application and empirical testing of the Hunt Vitell theory of ethics, Journal of Consumer Marketing, 18. Available at https:// doi.org/10.1108/07363760110386018. Accessed at 20 January 2018 pp 153-178.

Stehr, N. (2007). Die Moralisierung der Märkte: Eine Gesellschaftstheorie, Suhrkamp, Frankfurt.

Sunstein, C.R. (2016). The Ethics of Influence – Government in the Age of Behavioral Science, Cambridge.

Thaler, R.H. and Sunstein, C.R. (2008). Nudge – Improving Decisions about health, wealth and happiness, Penguin, London.

25. Exploring young students attitudes towards a sustainable consumption behaviour

C.B. Pocol[1], D.E. Dumitraș[1] and C. Moldovan Teselios[2]*
[1]University of Agricultural Sciences and Veterinary Medicine Cluj-Napoca, Manastur St., no. 3-5, 400372 Cluj Napoca, Romania; [2]Babes-Bolyai University, B-dul 21 Decembrie 1989, no.128, 400604 Cluj-Napoca, Romania; ddumitras@usamvcluj.ro

Abstract

A comprehensive understanding of the sustainable consumer behaviour needs a complex analysis of influencing factors and actors involved in the food chain. Concerns about sustainability issues are nowadays present in the whole food chain, including professionals, practitioners and consumers. The current paper intends to assess the sustainable consumption behaviour of young generations in Romania with the aim to improve the education process related to sustainability at university level. The study was conducted on students enrolled in life sciences programs, using a face-to-face survey. Education for sustainable consumption behaviour, decision buying criteria, and awareness about the flow of products within the food value chain were the main topics included in the questionnaire. The approach was undertaken on individual and global level with the aim to emphasize the complexity of the phenomenon. Results indicate that students assign different importance to the production process, commercialization networks, environmental impact, and social aspects. Respondents claim that the benefits of food consumption on own health is more important than assuring the animal welfare within the production process. By grouping the main aspects that define sustainable consumption behaviour it was noticed that the individual concerns (health and safety) are more valuable than the global ones (socio-economic and environmental). Universities are ethically responsible for educating students towards sustainable consumption habits. Similarly, young people should be aware that individual actions have impact on the collective well-being, for the present and future generations.

Keywords: education, concerns, ethical responsibility, sustainability, universities

Introduction

Sustainable food consumption behaviour represents an important topic nowadays for both, scientists and citizens because of its impact 'on the environment, individual and public health, social cohesion, and the economy' (Reisch *et al.*, 2013: 1). The importance of this subject is also related to the great pressure of the world population increase (Govindan, 2018). A comprehensive understanding of the sustainable consumer behaviour needs a complex analysis of influencing factors and actors involved in the food chain from producer to the final consumer.

Studies emphasize the low level of awareness and/or understanding of sustainability aspects of the food chain (Vermeir and Verbeke, 2006, Grunert, 2011). Even a positive attitude towards sustainable characteristics of products, as ethical consumer, does not always ends into a sustainable behaviour (Grunert, 2011). At consumer level, there is a gap between the positive attitude and the actual purchase behaviour (Vermeir and Verbeke, 2006). Price, personal experience, ethical obligation, lack of information, quality perception, inertia in purchasing behaviour, cynicism, and guilt were identified as factors that may cause the gap between the intention to consume ethically and the ethical purchase decision (Bray *et al.*, 2010).

Svenja Springer and Herwig Grimm (eds.) ***Professionals in food chains***
EurSafe 2018 – DOI 10.3920/978-90-8686-869-8_25, © Wageningen Academic Publishers 2018

The role of education for ethical and sustainable consumption behaviour is imperative for young generations as they represent the main segments that could attenuate the gap between attitude and behaviour in the future. According to Barth *et al.* (2014), the universities play an important role in creating patterns of sustainable consumption, which include respect for ethical and humanistic values. Pocol *et al.* (2017) emphasize the importance of understanding the conceptual significance of 'sustainability' in the creation of such patterns, followed by an active and responsible engagement in sustainable practices.

In this context, the current paper intends to assess the sustainable consumption behaviour of young generations in Romania with the aim to improve the education process related to sustainability at university level.

Material and methods

The study used an exhaustive sampling approach, including all 517 students enrolled at the time of data collection in life sciences programs from the following faculties of the University of Agricultural Sciences and Veterinary Medicine of Cluj-Napoca, Romania: Faculty of Agriculture, Faculty of Horticulture, Faculty of Animal Husbandry and Faculty of Food Sciences. A total of 392 students filled the questionnaire (a response rate of 76%). Data were collected in 2015 using a face-to-face survey. Education for sustainable consumption behaviour, decision buying criteria, and awareness about the flow of products within the food value chain were the main topics included in the questionnaire. The sample was balanced based on socio-demographic variables: residence (54% from urban area, 46% from rural area) and gender (56% male, 44% female). The age of 94.64% of respondents ranged between 18 and 26 years. The 5-point Likert scale was used as mean of measurement to determine the sustainable consumer behaviour of young students. Data was analysed using frequencies.

Results and discussion

The assessment of students' daily behaviour (food purchasing and consumption, transportation means) indicates a relatively medium importance to the sustainable aspects (Figure 1). The majority of students frequently consume animal food products (82%), indicating a non-sustainable food choice. According to Pfeiffer *et al.* (2017) a healthy and environmental sustainable diet is based on low animal-origin food products and high plant-based ones. Almost half of students use modes of transportation below the

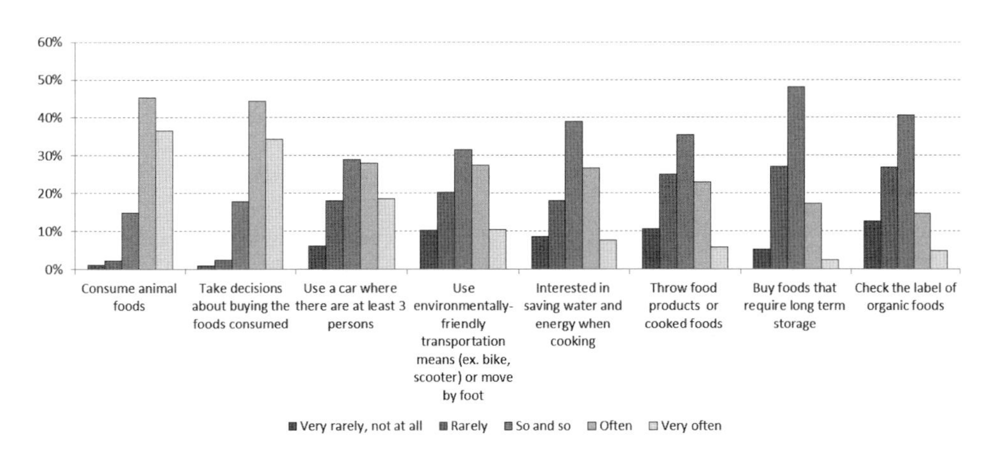

Figure 1. Daily behaviour of students related to sustainability.

average load level (3 persons) indicating a non-sustainable behaviour. The need of changing the behaviour of individual car users is emphasized by Steg (2006) who discusses four types of behavioural change that could contribute to achieve sustainable transportation, one of the behavioural changes referring to cycling and walking, alternatives used by 38% of respondents.

The buying purchasing criteria were analysed to understand the main factors that influence the decision about buying the food products, majority of respondents being the person mainly responsible with food purchasing in the household. The elements of marketing mix could have a critical role of changing purchasing patterns (OECD, 2008). The most important factors are quality and price of the product, much less importance being given to brand and environmental impact (Figure 2). In the case of Romanian consumers, a possible explanation of the ranking could be the low purchasing power due to the low development level of the country. However, a deeper investigation indicates that the choice is not always directly related to the purchasing power, as it is the case of German consumers who ranked the buying purchasing criteria in the same manner (OECD, 2008). The price was also considered as important criteria by Bray *et al.* (2010), who found that consumers are more concerned about the financial aspects than the ethical ones of the decision process.

Students assign different importance to the production process, commercialization networks, environmental impact, social and ethical aspects (Figure 3). As expected, the social and ethical aspects such as child labour, working conditions and decent wages are the most important in the students'

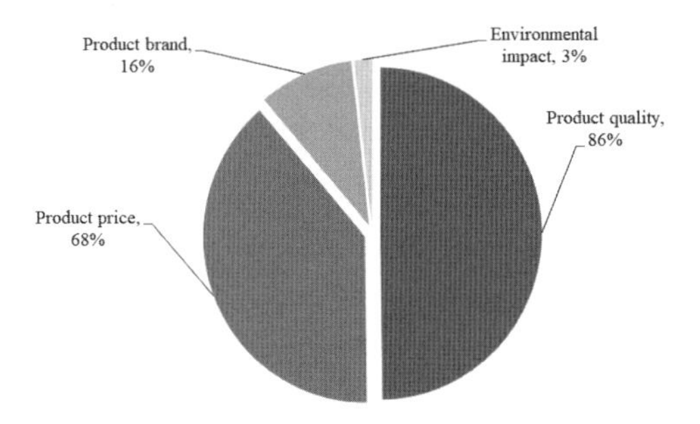

Figure 2. Buying purchasing criteria.

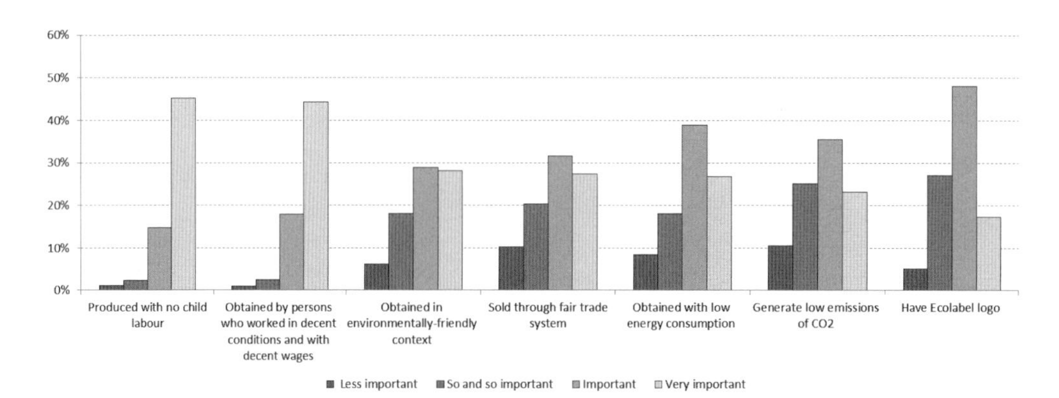

Figure 3. Importance of production and commercialization aspects related to sustainability.

opinion. On the other hand, the eco-production system and the low carbon emissions are the less important aspects in terms of food choice.

The sustainable diet is an important component of the sustainable consumption behaviour. Health benefits, pesticide use, animal ethics and welfare are among the issues considered when defining the sustainable diet (Garnett, 2014). All these issues were evaluated by students as important, but the highest scores were assigned to direct health benefits (Figure 4). The elements that could indirectly affect the consumer health (e.g. use of pesticides in the vegetal production systems and respecting animal welfare in farm animal breeding) are considered less important.

The EU citizens' concerns about sustainable consumption and production are constantly evaluated by the European Commission to improve the EU policies related to sustainable development. Knowledge and understanding of the Flower, the EU Ecolabel symbol, plays an important role in communication with the aim to increase the awareness of consumption sustainability (Grunert, 2011). The EU Ecolabel for consumers serves as an identity symbol to easy recognize two attributes of products: good quality and environmental friendly (EC, 2018). According to the Flash Eurobarometer, 45% of the Romanian respondents stated that the EU Ecolabel plays an important role in purchasing decision and 59% never heard or seen this label, percentages close to the EU average (Flash Eurobarometer, 2009). Moreover, only 15% of the Romanian respondents stated that they heard and bought products with the EU Ecolabel and 24% heard and did not bought products with this label (Flash Eurobarometer, 2009). Even if the food products group is still under discussion at EU level in terms of EU ecolabel criteria (EC, 2018), the recognition of the Flower was assessed in the current study as a symbol of sustainability, being correctly identified by only 25% young students (Figure 5).

By grouping the main aspects that define sustainable consumption behaviour it was noticed that the individual concerns (health and safety) are more valuable than the global ones (socio-economic and environmental) (Figure 6).

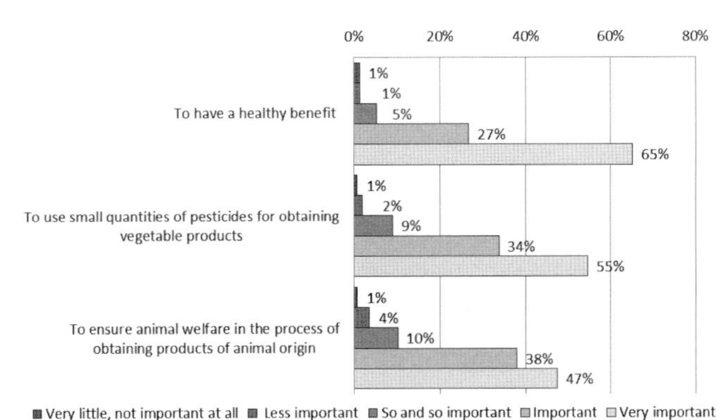

Figure 4. Importance of healthy issues related to sustainability.

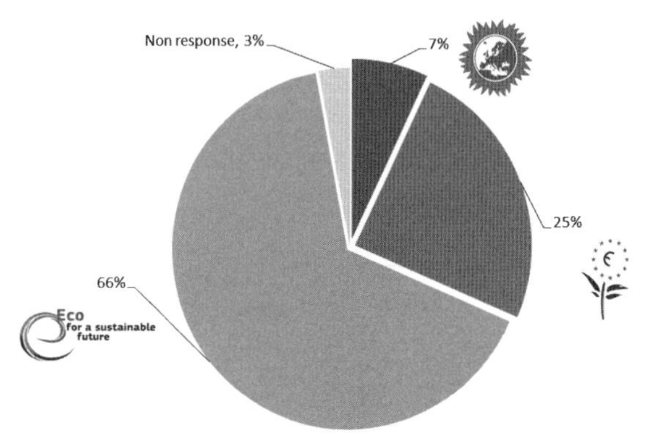

Figure 5. Knowledge of the EU Ecolabel.

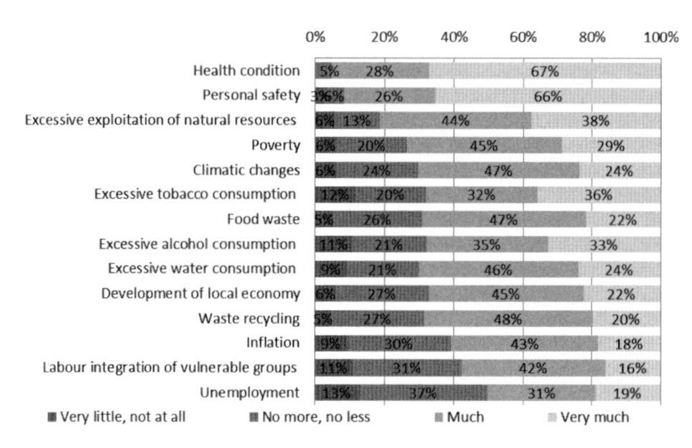

Figure 6. Concerns about aspects that define sustainable consumption behaviour.

Conclusions

The findings of this study suggests that understanding the consumption behaviour of young generations towards sustainability helps to explore how the current education programs may be improved to provide sufficient level of knowledge about sustainability. Universities are ethically responsible for educating students towards sustainable consumption habits. Similarly, young people should be aware that individual actions have impact on the collective well-being, for the present and future generations. After all, it is fundamental to understand that the entire process should be of common interest to increase awareness of the importance of ethical responsibility in consumption habits.

Acknowledgements

The study was conducted under a project entitled: European Master 'Green Food Industries', Project Number: 526585-LLP-1-2012-1-FR-ERASMUS-EMCR, Grant Agreement: 2012-2982/001-001, Sub-programme or KA: Erasmus Multilateral Projects: Support to the modernization agenda of higher education: curricular reform.

References

Barth, M., Adomβent, M., Fischer, D., Richter, S. and Rieckmann, M. (2014). Learning to change universities from within: a service-learning perspective on promoting sustainable consumption in higher education. Journal of Cleaner Production. 62:72-81.

Bray, J., Johns, N. and Kilburn, D. (2011). An exploratory study into the factors impeding ethical consumption. Journal of business ethics. 98(4):597-608.

European Commission (2018). Product groups and criteria. Available at: http://ec.europa.eu/environment/ecolabel/products-groups-and-criteria.html. Accessed 15 January 2018.

Flash Eurobarometer (2009). Europeans' attitudes towards the issue of sustainable consumption and production. Flash Eurobarometer, 256. Available at: http://ec.europa.eu/commfrontoffice/publicopinion/flash/fl_256_en.pdf. Accessed 16 December 2017.

Garnett T. (2014). What is a sustainable diet? A Discussion Paper. Oxford: Food & Climate Research Network, 31 pp.

Govindan, K. (2018). Sustainable consumption and production in the food supply chain: A conceptual framework. International Journal of Production Economics. 195:419-431.

Grunert, K.G. (2011). Sustainability in the food sector: A consumer behaviour perspective. International Journal on Food System Dynamics. 2(3):207-218.

OECD (2008). Promoting sustainable consumption. Good practice in OECD countries. Available at: https://www.oecd.org/greengrowth/40317373.pdf. Accessed 15 January 2018

Pfeiffer, C., Speck, M. and Strassner, C. (2017). What Leads to Lunch-How Social Practices Impact (Non-) Sustainable Food Consumption/ Eating Habits. Sustainability. 9(8):1437.

Reisch, L., Eberle, U. and Lorek, S. (2013). Sustainable food consumption: an overview of contemporary issues and policies. Sustainability: Science, Practice, & Policy. 9(2):1-19.

Steg, L. (2007). Sustainable transportation: A psychological perspective. IATSS research. 31(2):58-66

Vermeir, I. and Verbeke, W. (2006). Sustainable food consumption: Exploring the consumer 'attitude-behavioral intention' gap. Journal of Agricultural and Environmental ethics. 19(2):169-194.

Pocol, C.B., Arion, F.H., Dumitras, D.E, Jitea, I.M. and Muresan, I.C. (2017). The comprehension and study of the conceptual significance of 'sustainability' in agriculture and food production at university level. Journal of International Scientific Publications. Agriculture & Food. 5:587-597.

26. Ethical aspects of the utilization of wild game meat

R. Winkelmayer[1*] *and P. Paulsen*[2]
[1]*Pachfurth, Dorfstrasse 19, 2471 Rohrau, Austria;* [2]*Institute of Meat Hygiene, Vetmeduni Vienna, Veterinaerplatz 1, 1210 Vienna, Austria; tierarzt@winkelmayer.at*

Abstract

Wild game species in Middle Europe (esp. Germany, Austria, Switzerland) are utilized by hunting free-living specimens, or – to a small extent – by holding wild game in a similar way like domestic animals ('farmed game'), with meat or trophies as main products. Evidence exists that meat from wild game is favourable in various quality aspects over meat of comparable domestic species. As ethics is concerned, the production of game meat is often thought to be comparable to products from organic farming of domestic species. Currently, keeping wild animal in fenced areas for the purpose of hunting and hunting techniques which are based on mobilisation/disturbance of animals are under heavy debate as regards exposure of animals to unnecessary stress and pain, and related ethical concerns. Like in slaughter animals, indicators for *ante-mortem* distress and delayed onset of death exist for hunted game, e.g. insufficient *post-mortem* acidification of muscle tissue is associated with *ante-mortem* energy depletion (distress). The location of the wound channel allows assessing if sudden death occurred. Drive hunts on wild ruminants, suids and small game are associated with higher probabilities (compared to still hunts) for discomfort, distress and pain, due to chasing of animals and non-fatal wounds. Although field data demonstrate that it largely depends on the actual conditions if wound patterns and energy status of carcasses from one drive hunt differ from that from carcasses from still hunts, there is a higher probability of such adverse effects in sentient animals killed in drive hunts. This is an ethical issue, which needs to be dealt with in Codes of Practice, e.g. in guides to good practice in hunting. Such codes should preferably be issued by hunters' associations, or, if this is not feasible, codified in legislation on hunting.

Keywords: hunting, animal welfare, game meat quality, *ante-mortem* condition, pH-value

Introduction

Game meat available on the market originates from wild animals which have undergone some examination of the animal before killing and inspection of carcass and inner organs after evisceration. In the European Union, respective responsibilities of hunters ('trained persons') and official veterinarians are laid down in regulations EC No. 853 and 854/2004 (EC 2004a,b). Adherence to this inspection scheme and other codified food hygiene requirements assure that this type of meat as a high degree of food safety (Laaksonen and Paulsen 2015). The definition of 'wild game' in Reg. (EC) No. 853/2004 (EC 2004a) however includes also wild animals permanently or temporarily living in enclosed areas under conditions similar to wild game. Thus, it is not necessarily clear to consumers buying 'wild game' in the EU, if the animals were: (1) free-living; (2) kept in a fenced hunting area ('Jagdgatter') throughout the year or only in the winter during the feeding period ('Wintergatter'); or (3) if the animals had been released from breeding farms to the wild shortly before the hunt – a practice which is not uncommon for pheasants (Winkelmayer *et al.* 2008). This is an ethical issue, as there is broad consensus that hunting of released animals or of animals in fenced hunting grounds is not easy to justify (Fiala-Köck 2015), since these animals are raised solely for the purpose of being killed by shooting. Thus, animal act as 'moving targets'. Consequently, some critics coined the term 'pseudo-hunt'(Winkelmayer, 2009; Winkelmayer and Hackländer, 2008) to allow distinction from a restrictive use of natural resources in a sustainable and animal-welfare based way, which is easier to justify from an ethical viewpoint, irrespective of legal aspects (Seltenhammer *et al.*, 2011; Winkelmayer, 2014a,b).

EurSafe 2018 – DOI 10.3920/978-90-8686-869-8_26, © Wageningen Academic Publishers 2018

Irrespective of the abovementioned considerations, killing is – seen from a pathocentric (sentientist) view – a moral criterion (Wolf, 2014). There are differences if a gradual or egalitarian position is taken (Ott, 2014). Slaughter of farmed animals has to be done in a way to minimize fear and pain, and killing itself takes place after stunning. In contrast, killing by hunting is not preceded by stunning and there are various modes of killing and handling of animals before killing, partly according to preferences of hunters, partly according to other necessities.

Still hunting (blind hunting) allows killing of wild game with minimizing fear and pain, since the probability for lethal wounds is highest, and the animal has usually not been disturbed or stressed before. Stalking – which is common in the alpine regions – bears the risk that the animal is disturbed and moves, which gives less time to safely place a lethal shot. In drive hunts and battues, animals are deliberately mobilized, by means of dogs or humans, and hunters are awaiting the animals on their stands. Ideally, the animals are gently disturbed and just move to restore the flight distance. Such hunts are very often a thoroughly-planned professional way to reduce wild game population in a very short period of time in forest areas or when touristic use of landscapes requires restrictions in hunting (Wölfel, 2003). The latter reasons may justify drive hunts as a means of reducing wild animal population, although the risk for non-lethal shots wounds and thus, pain, is higher than in still hunts. Drive hunts organized for other purposes than reducing wild animal populations are pure 'societal events', which cause pain to sentient animals and, thus, are not acceptable from an animal ethics viewpoint (Winkelmayer, 2009, 2014a,b).

Motivations of hunters can also be examined from an ethical viewpoint. These range from 'recreational hunting' (termed 'blood-sport' by critics) to 'ultima-ratio-hunting' (or 'therapeutic hunting' in the sense of Varner, 1995), i.e. reducing numbers of animals in cultivated areas due to ecological or other (e.g. disease control) reasons. Subsistence hunting is usually considered ethically acceptable, but in Middle Europe, there is currently no necessity for subsistence hunting (Grimm and Wild, 2016; Wild, 2015).

Currently, there still is a broad consensus that farming of animals is necessary for human nutrition, and thus, informed consumers will accept hunting as a branch of agriculture and forestry. However, hunting purely for trophies or recreation is much less accepted, although this is a legal activity in the EU (Forster *et al.* 2006, FUST-Tirol 2008). It would be helpful in this context, if hunters adhere to the ethical requirement of minimizing distress, pain and fear and to foster positive impressions of animals (e.g. allowing animal to roam unrestricted) (Kaplan, 2017).

Game meat for the consumer

Game meat is generally perceived as healthy food (Hoffman and Wiklund, 2006; Hoffman and Cawthorne, 2012); in Europe, however, the issue of radioactivity resulting from fallout of Tchernobyl remains a problem in several regions. For meat of several species, a nutritionally favourable fatty acid composition of lipid tissues has been demonstrated. These intrinsic properties and a certified health status cover, however, not all 'qualities' expected by consumers. Similar to farmed animals, consumers want the meat to originate from animals living under decent conditions, and exposed to no or minimum disturbance, fear and pain prior to and during killing. For meat from wild game placed on the market, it is usually not known if animals had been killed with the first shot, or if they had been chased before killing, etc. It is conceivable that in these cases, meat from wild game is not ethically superior to meat from farmed animals raised in organic or conventional farming.

A clear distinction/labelling of meat according to origin of animals (free-living; fenced; released) and their mode of killing would allow consumers concerned about animal ethics to make an informed decision when buying game meat. Such information is conveyed to consumers in form of quality labels, e.g. the 'Genussregion Weinviertler Wild' brand (Genussregion Wienviertler Wild, 2018), which guarantees that the meat originates from free-living game killed by lethal shots (no gut shots) with a minimum of *ante-mortem* distress.

Apart from hunting, game can be farmed and slaughtered like farmed domestic animals. Animals are raised in an extensive way, and slaughter can be performed on the farm, thus avoiding stress during transport and handling. Some traditional extensive farming models, e.g. reindeer herding in Fennoscandia are seen as sustainable use of otherwise unproductive areas in conformity with animal, welfare standards. In principle, this model could be adapted for deer in Austria, e.g. animal are free-roaming in summer and then corralled into fenced areas in the winter ('Wintergatter'), where a part is slaughtered and the reduced population is fed during winter season and releasing in the spring.

Can the 'ethical quality' of game meat be measured?

From the considerations above, a provisional definition of 'ethical quality' of game meat would include: living conditions with a minimum of restrictions; minimized stress, pain and fear during the hunting event (incl. killing). Preferably, these requirements can be translated into criteria that can be checked on the carcass. This allows assessment on an individual-animal as well as on a hunting-event basis. Currently, there are at least two criteria that can be checked on a routine basis on the carcass, viz. number and location of shot wounds and the degree of *post-mortem* acidification of the muscles. The latter is based on the observation, that under normal conditions, *post-mortem* glycolysis in muscles – under anaerobic conditions – will effectuate lactic acid accumulation in meat and thus lower pH from around 7.2 to typically 5.5-5.6 (Lawrie, 1998). Consumption of carbohydrate stores *ante-mortem* can occur due to physical exercise, but also due to emotional stress, with the outcome, that *post-mortem* pH will not drop under ca. 6.0, which has consequences for meat shelf life and technological use. Likewise, extreme *ante-mortem* exercise could result in lactic acid accumulation before onset of death, with unusual low pH values already early *post-mortem* (e.g. around 5.3). Such findings have been reported for deer in drive hunts (Deutz *et al.*, 2006). Whereas such pH-deviations can be measured also after deboning and cutting, information about the location of shot wounds is lost during the cutting process. It is generally accepted that damages to the brain or the major blood vessels in the anterior thorax will cause sudden death (Winkelmayer *et al.*, 2005), thus abdominal or leg wounds indicate that animals experienced pain before death. Notably, both criteria have also implications on hygienic condition of meat. The elaboration of codes of good practice for hunters is an on-going task (Deutz *et al.*, 2006). These criteria, however, do not cover all ethical issues. For example, an animal kept in a fenced area and killed at the beginning of a drive hunt with a lethal shot would experience no pain due to the hunting itself, but the setting as such (breeding of animal in enclosed areas for the purpose of killing by hunting) still would be questionable from an ethical perspective.

Conclusions

From an animal ethics viewpoint, 'quality' of meat from wild game will differ according to animals' living conditions (free-roaming or raised in captivity and released for killing, or kept in fenced areas), mode of hunting (mobilisation of game bear the risk of non-lethal shot wounds) and motivation of hunters (population reduction for ecology or disease control purposes or purely as a 'sport'). The ethical requirement of minimizing distress, pain and fear and to foster positive impressions should be prioritized by hunters, which means that certain animal keeping and hunting techniques should be critically reviewed, under scientific criteria and in relation to ethical standards. This relates, in particular, to raising pheasants or ducks in captivity and releasing them shortly before a hunt, where they are wounded by shot, which by nature bears the risk of non-lethal wounds and of pain. Likewise, drive hunts on large game deliberately raised for the purpose of being killed are a critical issue in terms of animal ethics. Such an approach would also add credibility to the game meat market.

A pragmatic approach to ethical quality of meat from game is to provide information to consumers on *ante-mortem* conditions and the type of hunting, e.g. in the framework of quality labels. In addition, hunting techniques which comply with the ecological need for population management and animal welfare requirements have to be developed.

References

Deutz, A., Völk, F., Pless, P., Fötschl, H. and Wagner, P. (2006). Game meat hygiene aspects of dogging red and roe deer. Archiv für Lebensmittelhygiene 57: 197-202.

EC (2004a). Regulation (EC) No 853/2004 of the European Parliament and of the Council of 29 April 2004 laying down specific hygiene rules for on the hygiene of foodstuffs. Official Journal of the European Union L139/55.

EC (2004b). Regulation (EC) No 854/2004 of the European Parliament and of the Council of 29 April 2004 laying down specific rules for the organisation of official controls on products of animal origin intended for human consumption. Official Journal of the European Union L155/206.

Fiala-Köck, B. (2015). Jagd und zeitgemäßes Tierschutzverständnis. Tagungsband Jagdtagung Stainz, November 1, 2015, 11 pp.

Forstner, M., Reimoser, F., Lexer, W., Heckl, F. and Hackl, J. (2006). Nachhaltigkeit der Jagd. Prinzipien, Kriterien und Indikatoren. avBUCH, Vienna, Austria, 126 pp.

FUST-Tirol (2008). Jagdgatter' und Aussetzung von Wildtieren zum Abschuss. – FUST-Position 7; Forschungs und Versuchsprojekt Alpine Umweltgestaltung' des Förderungsvereins für Umweltstudien (FUST-Tirol, Achenkirch). Available at: http://www.fust.at/wp-content/uploads/Positionen_07-–-FUST-Tirol.pdf. Accessed 17 March 2018.

Genussregion Weinviertler Wild (2018). Available at: http://www.genuss-region.at/genussregionen/niederoesterreich/weinviertler-wild/index.html. Accessed 17 March 2018.

Grimm, H. and Wild, M. (2016). Tierethik zur Einführung. Junius Verlag GmbH, Hamburg, Germany, 252 pp.

Hoffman, L. C. and Cawthorn, D.-M. (2012). What is the role and contribution of meat from wildlife in providing high quality protein for consumption? Animal Frontiers 2: 40-53.

Hoffman, L. C. and Wiklund, E. (2006). Game and venison – meat for the modern consumer. Meat Science 74: 197-208.

Kaplan, H. F. (2017). Tierrechte – Das Ende einer Illusion? Books on Demand, Norderstedt, Germany, 108 pp.

Laaksonen, S. and Paulsen, P. (2015). Hunting hygiene. Wageningen Academic Publishers, Wageningen, The Netherlands, 304 pp.

Lawrie, R. A. (1998). Lawrie´s Meat Science. 6th ed. Woodhead Publishing Limited, Cambridge, UK, 336 pp.

Ott, K. (2014). Jagd aus naturethischer Sicht. Tagungsband zum 7. Rotwildsymposium der Deutschen Wildtierstiftung, Hamburg, Germany. Available at: http://rothirsch.org/tagungsband-zur-jagdethik-veroeffentlicht. Accessed 17 March 2018.

Seltenhammer, E., Hackländer, K., Reimoser, F., Völk, F., Weiß, P. and Winkelmayer, R. (2011). Zum ethischen Selbstverständnis der Jagd. Österreichs Weidwerk 4: 8-12.

Varner, G. E. (1995). Can animal rights Activists Be Environmentalists? In: Pierce, C. and VanDeVeer, D. (eds.). People, Penguins, and Plastic Trees. Basic Issues in Environmental Ethics. Wadsworth Publ., Belmont (CA), USA, pp. 254-273

Wild, M. (2015). Wem wird die Waidgerechtigkeit gerecht? Schweizer Tierschutz STS, 3. Wildtiertagung. Braucht es die Jagd? Olten, February 12, 2015.

Winkelmayer, R. (2009). Animal welfare during hunting: the ethical perspective. In: Smulders, F.J.M. and Algers, B. (eds.) Food safety assurance and veterinary public health, vol.5: Welfare of production animals: assessment and management of risks. Wageningen Academic Publishers, Wageningen, the Netherlands, pp. 205-220.

Winkelmayer, R. (2014a). A note on game meat, animal welfare and ethics. In: Paulsen, P., Bauer, A. and Smulders, F.J.M. (eds.). Trends in game meat hygiene: From forest to fork. Wageningen Academic Publishers, Wageningen, the Netherlands, pp. 373-376.

Winkelmayer, R. (2014b). Ein Beitrag zur Jagdethik. Österreichischer Jagd- und Fischerei-Verlag, Wien.

Winkelmayer, R. and Hackländer, K. (2008). Der Begriff 'Jagd' – eine Differenzierung. Österreichs Weidwerk, 9, 10, 11/2008.

Winkelmayer, R., Malleczek, D., Paulsen, P. and Vodnansky, M. (2005). A note on radiological examination of the thoracal cavity of roe deer to identify the optimum aiming point with respect to animal welfare and meat hygiene. Wiener Tierärztliche Monatsschrift – Veterinary Medicine Austria 92: 40-45.
Wolf, J.-C. (2014). Das Beraubungsargument gegen die Tötung von Tieren. TIERethik 1(8): 7-13.
Wölfel, H. (2003). Bewegungsjagden – Planung-Auswertung-Hundewesen. Leopold Stocker, Graz-Stuttgart, 190 pp.

Section 4.
Food ethics

27. Questioning long-term global food futures studies: a systematic, empirical, and normative approach

Y. Saghai[1], M. Van Dijk[2,3], T. Morley[2] and M.L. Rau[2]
[1]Johns Hopkins University, Berman Institute of Bioethics, 1809 Ashland Avenue, Baltimore, MD 21205, USA; [2]Wageningen Economic Research, P.O. Box 29703, 2502 LS the Hague, the Netherlands; [3]International Institute for Applied Systems Analysis, Schlossplatz 1, 2361 Laxenburg, Austria; saghaiwork@gmail.com

Abstract

Studies of the futures of food answer questions such as 'do we need to increase global agricultural production to feed the world sustainably in 2050?' Conclusions vary dramatically. Similar variations and uncertainties are striking with respect to many other dimensions of food security and food systems. The sheer heterogeneity of methods used to explore the futures of food seems to undermine meaningful comparisons and aggregation between studies. These issues and others compromise responsible collective choices vital for humanity, nonhuman animals, and Earth systems. Disagreements on what policies and social actions we should adopt to shape the future of food depend on how we assess the evolution of food security and food systems over the long-term (at least 20 years into the future). Building upon foresight practice and theory, our team borrows tools from economics, STS, and philosophy to shed light on global food futures and food ethics. In this paper, we will introduce readers to some of our unpublished and provisional findings. We will cover two questions: (1) What does a systematic review of global food security modelling and projection studies reveal about predominant methods, food security indicators, drivers of change, and the range of future global food security projections? (2) Should the usual notion of a 'plausible' future explicitly or implicitly invoked in global food futures scenarios to delineate the range of futures worth exploring be modified or abandoned to set free our epistemic, ethical, and political imagination?

Keywords: foresight, plausibility, models, ethics

Introduction

Studies of the futures of food answer questions such as 'do we need to increase global agricultural production to feed the world sustainably in 2050?' Conclusions vary dramatically. Similar variations and uncertainties are striking with respect to many other dimensions of food security and food systems. The sheer heterogeneity of methods used to explore the futures of food seems to undermine meaningful comparisons and aggregation between studies. These issues and others compromise responsible collective choices vital for humanity, nonhuman animals, and Earth systems. Disagreements on what policies and social actions we should adopt to shape the future of food depend on how we assess the evolution of food security and food systems over the long-term (at least 20 years into the future). Building upon foresight practice and theory, our team borrows tools from economics, STS, and philosophy to shed light on global food futures and food ethics. In this paper, we will introduce readers to some of our unpublished and provisional findings. We will cover two questions: (1) What does a systematic review of global food security modelling and projection studies reveal about predominant methods, food security indicators, drivers of change, and the range of future global food security projections? (2) Should the usual notion of a 'plausible' future explicitly or implicitly invoked in global food futures scenarios to delineate the range of futures worth exploring be modified or abandoned to set free our epistemic, ethical, and political imagination?

EurSafe 2018 – DOI 10.3920/978-90-8686-869-8_27, © Wageningen Academic Publishers 2018

A glimpse at a systematic review of global food security modelling and projection studies

To address the major and complex issue of food security, policy makers currently making decisions need to have insights into the potential future pathways of global food security. A number of studies have assessed various aspects of future global food security (Baldos and Hertel; 2016; Hasegawa *et al.*, 2015). Often, the results of these studies vary widely and are difficult to compare because of differences in methodology (Godfray and Robinson, 2015), obstacles to structurally comparing the results of global economic and agricultural simulation models (Von Lampe *et al.*, 2014), heterogeneity in definitions of output indicators (Van Dijk and Meijerink, 2014) and divergence in the choice of projections and scenarios (Reilly and Willenbockel, 2010).

Primary and review studies often use the terms 'projections' and 'scenarios' interchangeably. However, projections are alternative quantitative results of running a model based on different assumptions or inputs. In contrast, a scenario refers to a 'plausible, comprehensive, integrated and consistent description of how the future might unfold' (Van Vuurenen *et al.*, 2014, 377). In the food futures modelling literature scenarios have a quantitative component (including projections) and a qualitative component, as well as a narrative storyline that links important statements about the future that may or may not be quantifiable. This review focuses on projections and, when relevant, related broader scenarios that global food futures.

Each of the review studies covers only a selection of models, projections and scenarios that have been used in the literature and focuses on a relatively narrow set of food security indicators. In contrast, the goal of our systematic literature review is to rigorously and transparently identify, evaluate, and summarize the results of all global food security modelling studies since 2000. All data will be stored in the Global Food Security Projections Database and can be used by the research community to benchmark the results of upcoming food security modelling studies.

Methods

We used the guidelines for the qualified application of systematic review by the Evidence for Policy and Practice Information and Co-ordinating Centre (University of London) and the Cochrane Handbook for Systematic Reviews of Interventions to organize the review. First, we searched for studies by using six broad exclusion/inclusion criteria: (1) Topic (focus on food security); (2) Global coverage; (3) Projections and scenarios; (4) Quantification (the results of the study are quantified using some type of model, such as econometric, computable general equilibrium or partial equilibrium); (5) Time horizon (the study presents projections for the year 2030 or beyond); and (6) Year of publication (2000-2018). To find the relevant literature, we used global repositories of scientific literature and searched the grey literature. Figure 1 depicts a PRISMA (preferred reporting items for systematic reviews and meta-analyses) diagram that summarizes process and results of our search strategy. The query of the repositories resulted in 3647 studies; 45 studies were included in the database.

One of the aims of this review is to show the range and uncertainty of global food security projections. Unfortunately, the results of the models are not always easily comparable because of differences in projections and variable definitions, methodology and time horizon. In order to make the data comparable, we mapped all scenarios to the Shared Socio-economic Pathways (SSPs) scenarios. The SSPs (Van Vuuren *et al.*, 2017; O'Neill *et al.*, 2017) are a set of five storylines that describe potential but 'plausible' global futures: inclusive and sustainable growth (SSP1), business as usual (SSP2), fragmentation through regional rivalry (SSP3), increasing inequality (SSP4) and resource intensive high growth (SSP5).

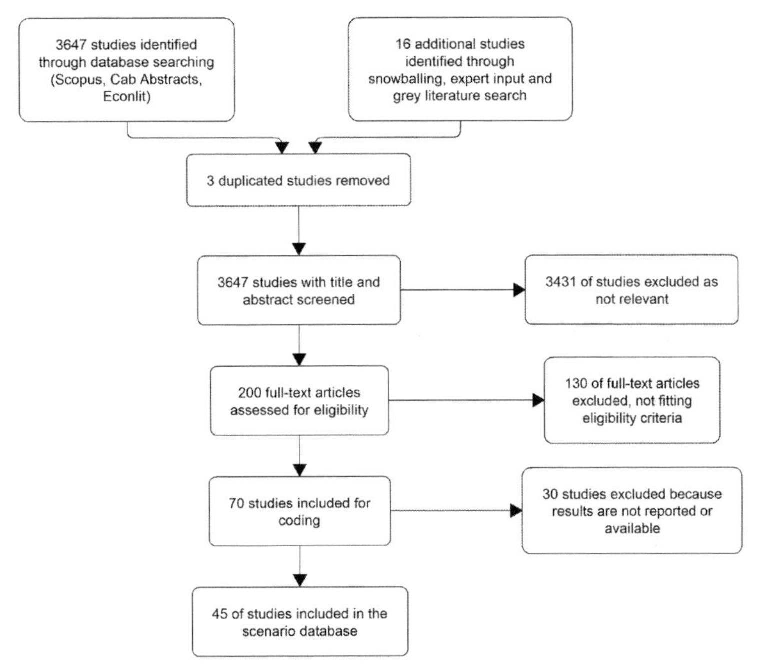

Figure 1. Prisma diagram of the study selection.

Results

The number of global food security projection studies has substantially increased over the last two decades from 1 in 2003 to 70 by 2018. The global food price crisis (2007-2008) resulted in renewed attention to global food and nutrition research including global food security projection studies. There is a transition from studies that present the results of one model to multi-model comparisons that present and discuss the results of an ensemble of models. Many of these studies are produced as part of the Agricultural Model Intercomparison and Improvement Project (AgMIP) that was initiated in 2010.

The majority of studies employ simulation models to project food prices, food demand or people at risk of hunger to the year 2050 (e.g. IMPACT, GLOBIOM, MAGNET). The second most commonly used approaches are statistical models that use regression techniques to estimate future food security (Tilman *et al.*, 2011; Bodirsky *et al.*, 2015).

Comprehensive assessments of future food security should, at least, present a set of indicators that cover all of the four dimensions of food security: availability, access, stability and utilization. The FAO has prepared a comprehensive list of around 20 national-level indicators that capture the four dimensions of food (in)security (FAO, 2018). Comparing this list with the indicators produced by the selected studies suggests that the model studies present a very narrow view of food security. None of the studies presents an indicator on stability and only a few studies address the utilization dimension.

Three factors can explain the large range of food security outcomes. First, in contrast to the AgMIP exercises, where all models use harmonized assumptions on drivers and attempt to align the implementation of qualitative scenario assumptions, the input data of the studies in our review differs considerably. All these differences and variations result in a large bandwidth of results. Second, differences in methodologies to model long-term global food security can strongly influence the results. Systematic model comparisons in AgMIP showed that structural differences between computable general equilibrium and partial equilibrium, assumptions on technological change, and the way food demand is modelled are important factors which explain differences in model outcomes. Hertel and Baldos (2016) found that apart from technological change, income, capital, labour and land elasticities are critical determining factors of model output, although they only have received very limited attention in the literature. Finally, differences in reporting of results (e.g. definition of variables, units and aggregation) are also an important factor that explains the wide range of outcomes.

The provisional results of our systematic review offer a synthesis of the methods and results of recent global food security assessments. The final set of selected studies shows a variety in terms of methodologies, modelling of driving forces, presentation of food security indicators and range of outcomes. Since the first studies in the beginning of the millennium, modelling approaches have advanced and the latest studies capture a wide number of food security drivers, including, among others, population and income growth, technological change, trade and shifts in diets. It is therefore striking that many drivers are still neglected, such as multiple dimensions of inequalities (beyond income growth), aquaculture, land policy and use, or post-harvest wastage. Indicators that are presented by the modelling studies overlook stability and utilization dimensions of food security. Finally, the range of outcomes is large, even after they are mapped to the same SSPs.

Plausible and implausible global food futures

A cursory look at the food security and food systems debate and a close reading of global food futures studies show that judgments about the plausibility or implausibility of selected futures abound and are rarely explicitly justified. Why is this the case? What is meant by plausibility? Are plausibility judgements indispensable and justifiable or would we be better off without them to set our epistemic, ethical, and political imagination free to explore a broader range of alternative food futures, where for instance, more radical, rapid, and global changes in food production, distribution of land, access to food, modes of consumption, and moral attitudes towards certain foodways would be deemed plausible?

To start answering these questions, we need to look at the function of plausibility in futures studies and its pragmatics, without benefiting from the insights of historians of futures studies who have not yet investigated the evolution of conceptions of 'plausible futures' in depth. Science, Technology and Society scholars, on the other hand, have shown that judgments of plausibility and conflicts about such judgements occur in general during foresight practice and are not reported or explicated in published documents (Van Asselt *et al.*, 2012).

Why is plausibility so important in futures studies, yet often confused with probability or feasibility (Ramirez and Selin, 2014)? In theory, foresight experts converge on the view that the exploration of the future should not be limited to probable futures because complex problems with a long-term temporal scale, such as the evolution of global food systems, generate many futures that it is impossible to forecast. Thus, the main solution is to explore a range of futures more broadly construed than quantifiable probable futures, and more narrowly construed than the set of all logically possible futures: plausible futures.

However, the criteria for characterizing plausibility remains either vague, objectionable, or implicit, ranging from evidence that a similar event has occurred in the past in sufficiently similar circumstances, to proof of concept that some event or process could occur in the future (Wiek *et al.*, 2013). The space between the probable and the possible needs to be charted to expand our imagination about the future, which tends to be anchored to a bias toward certain constructions of the past. But we also want to avoid idle speculations: credibility and evidentiary support matter. Plausibility judgements have thus played a major role in futures studies at the very least since the work of Herman Kahn on scenarios.

The approach to scenario construction that Kahn, its creator, advocated states that 'scenarios are hypothetical sequences of events constructed for the purpose of focusing attention on causal processes and decision-points' (Kahn and Wiener, 1967). Therefore, the job of scenario builders is not limited to presenting images of challenging and relevant alternative futures. Rather, the selection of a set of alternative futures has to be non-arbitrary, inform decisions, and have a heuristic value. To do this, scenario builders ought to expand their imagination and knowledge-base and *causally* explain how we could plausibly get from the present to a small set of challenging, relevant, and distinct futures.

Scenario theorists and practitioners have naturally criticized this influential approach, given well-known philosophical disputes about the meaning and demandingness of causal explanations and notorious practical difficulties in hunting causes in sciences and other social practices. Some foresight theorists and practitioners went beyond raising objections against the exclusive recourse to causal explanations to demonstrate plausibility: they have put forward several alternative approaches. The most influential approach advocates abandoning the requirement of plausibility altogether and argues for a radical pluralistic view according to which both present and future worlds are constructed and do not demand plausible explanations to support them (Vervoort *et al.*, 2015).

Conclusions

To sum up, the broader ethical and political debate about global food futures depends in part on making progress on methods used to explore global food futures, including a combination of modelling and scenarios, and in part on our willingness to discuss the meaning and value of plausibility judgments about considered futures. Should we expand the types of explanations required for providing support to plausibility judgments to include a wider variety of explanations (structural, narrative, etc.) or ought we abandon the concept altogether to avoid judgments that are either epistemically conservative or based on unrecognized ethical, social, and political values?

Acknowledgement

This research was jointly funded by a Stavros Niarchos Foundation grant to John Hopkins University and a grant from Wageningen University and Research.

References

Baldos, U.L.C. and Hertel, T.W. (2016). Debunking the 'new normal': Why world food prices are expected to resume their long run downward trend. Global Food Security 8: 27-38.

Bodirsky, B., Rolinski, S., Biewald, A., Weindl, I., Popp, A. and Lotze-Campen, H. (2015). Global Food Demand Scenarios for the 21st Century. In: Belgrano, A. (ed.) PLoS ONE 10 (11). FAO: e0139201.

Food and Agriculture Organization of the United Nations (2018). Food security statistics. Available at: http://www.fao.org/economic/ess/ess-fs/en. Accessed 17 March 2018.

Glenn, J. C. and The Futures Group International (2009). Scenarios. In Glenn, J. C. and Gordon, T. (eds). Futures research methodology-Version 3-0. T.J. Editorial desconocida. Washington, DC.

Godfray, H.C.J. and Robinson, S. (2015). Contrasting approaches to projecting long-run global food security. Oxford Review of Economic Policy 31 (1): 26-44.

Hasegawa, T., Fujimori, S., Takahashi, K. and Masui, T. (2015). Scenarios for the risk of hunger in the twenty-first century using Shared Socioeconomic Pathways. Environmental Research Letters 10:1. IOP Publishing: 014010.

Hertel, T.W. and Baldos, U.L.C. (2016). Attaining food and environmental security in an era of globalization. Global Environmental Change 41: 195-205.

Kahn, H. and Wiener, A. J. (1967). The year 2000: A framework for speculation on the next 33 years. Macmillan.

Le Mouël, C., and Forslund, A. (2017). How can we feed the world in 2050? A review of the responses from global scenario studies. European Review of Agricultural Economics 10 (8): 1-51.

O'Neill, B.C., Kriegler, E., Ebi, K.L., Kemp-Benedict, E., Riahi, K, Rothman, Van Ruijven, B.J., Van Vuuren, D.P., Birkmann, J., Kok, K., Levy, M. and Solecki, W. (2017). The roads ahead: Narratives for shared socioeconomic pathways describing world futures in the 21st century. Global Environmental Change 42: 169-80.

Ramírez, R. and Selin, C. (2014). Plausibility and probability in scenario planning. Foresight, 16 (1): 54-74.

Robinson, S., Van Meijl, H., Willenbockel, D., Valin, H., Fujimori, S., Masui, T., Sands, R., *et al.* (2014). Comparing supply-side specifications in models of global agriculture and the food system. Agricultural Economics 45 (1): 21-35.

Tilman, D., Balzer, C., Hill, J. and Befort, B.L. (2011). Global food demand and the sustainable intensification of agriculture. Proceedings of the National Academy of Sciences of the United States of America 108 (5): 20260-4.

Valin, H., Sands, R.D., Van Der Mensbrugghe D., Nelson, G.C., Ahammad, H., Blanc, E., Bodirsky, B., *et al.* (2014). The future of food demand: understanding differences in global economic models. Agricultural Economics 45 (1): 51-67.

Van Asselt, M., Van 't Klooster, S., Van Notten, P.W.F. and Smits, L. (2012). Foresight in action: Developing policy-oriented scenarios. Routledge.

Van Dijk, M. and Meijerink, G.W.W. (2014). A review of global food security scenario and assessment studies: Results, gaps and research priorities. Global Food Security 3 (3-4): 227-38.

Van Vuuren, D.P., Kriegler, E., O'Neill, B.C., Ebi, K.L., Riahi, K., Carter, T.R., Edmonds, J., *et al.* (2014). A new scenario framework for climate change research: scenario matrix architecture. Climatic Change 122 (3): 373-386.

Van Vuuren, D.P., Riahi, K., Calvin, K., Dellink, R., Emmerling, J., Fujimori, S., Samir K.C., Kriegler, E.and O'Neill, B. (2017). The Shared Socio-economic Pathways: Trajectories for human development and global environmental change. Global Environmental Change 42 (January): 148-52.

Vervoort, J. M., Bendor, R., Kelliher, A., Strik, O. and Helfgott, A.E. (2015). Scenarios and the art of worldmaking. Futures, 74: 62-70.

Von Lampe, M., Willenbockel, D., Ahammad, H., Blanc, E., Cai, Y., Calvin, K., Fujimori, S., *et al.* (2014). Why do global long-term scenarios for agriculture differ? An overview of the AgMIP Global Economic Model Intercomparison. Agricultural Economics 45 (1): 3-20.

Wiek, A., Withycombe Keeler, L., Schweizer, V. and Lang, D.J. (2013). Plausibility indications in future scenarios. International Journal of Foresight and Innovation Policy, 9 (2-3-4): 133-147.

28. Four sociotechnical imaginaries for future food systems

P.B. Thompson
Michigan State University, Departments of Philososophy and Community Sustainability, 503 S. Kedzie Hall, East Lansing, MI 48824, USA; thomp649@msu.edu

Abstract

This paper sketches four archetypal characterizations of how food will be produced, processed, distributed and consumed over the coming half century – a time in which all manner of social association will be influenced by climate change, growing scarcity of resources relative to human population and climate change. Each archetype reflects what Sheila Jasanoff has called 'a sociotechnical imaginary' that generates scenarios or visions of the future that are richly dependent on a technical infrastructure and on a pattern of future technical innovation. In this paper, 4 such imaginaries are sketched briefly: technological modernization (a continuation of food system innovations that began in the 20th century); sustainable intensification (a model emphasizing more efficient use of ecosystem services; 'extensification' (a return to less intensive land use) and urban agriculture (a model that is driven by traditions of urban activism, planning and information technology).

Keywords: agriculture, urban food systems, sustainability, food sovereignty

It will be obvious to everyone who has spent more than a moment reflecting on the current state of affairs in food production, distribution and consumption that there are competing evaluations of how the food system is performing, and how it should (or could) change. In this paper I sketch 4 sociotechnical imaginaries as competing visions. As Sheila Jasanoff argues, it is in the comparison of sociotechnical imaginaries that their normative commitments and implications become clear. Although technological modernization will certainly continue to influence agricultural production as well as food processing, awareness of key failures in existing food systems are becoming evident, especially in cities where neighborhoods may not possess an economic or technical infrastructure for fresh or healthy food access, or may restrict local control of food systems. Each of the three alternatives is, in effect, responding to a different ethical interpretation of why and in what respect technological modernization has failed, and why urban planners and social activists must take a more proactive role in developing the food systems of the future. After an introductory discussion of sociotechnical imaginaries, the paper proceeds by discussing each of the four archetypal imaginaries in turn.

Sociotechnical imaginaries

Jasanoff's recent work relies on the concept of an 'imaginary' to characterize the way that innovators, engineers and developers of technology envision the transformation of existing realities into a future shaped by the realization of mere possibilities. This concept goes considerably beyond conventional adjectival meanings of imaginary, and draws heavily on the philosophical characterization of the social imaginary as it has been developed in the writings of Cornelius Castoriadis and Charles Taylor. What Jasanoff adds is a keen appreciation of the way in which institutional practices are shaped and reshaped both through material infrastructure. Technical systems play many roles, one of which is simply the infrastructure for human interaction: roads, bridges and communication systems having dramatic impact on the pace and scale at which institutional practices are reproduced. In addition, technical innovations can challenge fundamental institutions such as markets or property rights. One recent transformation in food systems has been a shift in the carrier for an organism's phenotypic traits – the characteristics of keen interest to food producers. While the social imaginary has understood that plant features such as height, maturation time, seed shattering and the like are biologically grounded in genes, they have

EurSafe 2018 – DOI 10.3920/978-90-8686-869-8_28, © Wageningen Academic Publishers 2018

been functionally tied to seeds for all practical purposes. Now they are tied to sequences that can be transferred from plant to plant, and this technical innovation has led to alterations in the institutional practices of farming and plant variety development. Jasanoff and her students have explored a number of cases in which R&D has been shaped by existing and envisioned imaginaries that include both social and technical mechanisms for reproduction and transformation (Jasanoff and Kim, 2015). Hence the term sociotechnical imaginaries.

Technological modernization

According to this vision, agriculture is modernizing, and all segments of the food system are following a path that has been blazed by industrialization in other sectors of the modern economy. The most visible element of this path is the transformation of the labor process through mechanization and other technologies that replace both onerous and repetitive tasks and also skilled workmanship. Thus the crop production of the early 19th century has been entirely changed through the use of tractors, specialized harvesting equipment and chemicals the eliminate or reduce the amount of time and skill needed for weeding and controlling plant pests. Food processing has moved out of the household almost entirely, while rail and other transport infrastructure has created global markets. Livestock husbandry and production of meat, milk and eggs have been the last part of the food system to undergo radical change. This is, of course, the story of industrial agriculture and further description would rapidly become tedious.

The industrialization of the food system is often glamorized through highlighting the reduction of backbreaking labor, global increases in total food production and gains in reduction of hunger. Whether these claims are true or not, these are not the actual philosophical commitments that undergird technological modernization. In fact, the industrialization of the food system has occurred in response to market opportunities, on the one hand, and the discipline of price competition, on the other. This archetype imagines that food systems are composed almost entirely of profit-seeking firms, whether they be farms, trading companies, input suppliers or food retailers. This structure leads right to the doorstep of the consuming household, where non-market values may or may not exert sovereignty. Consumer values may call out for preserving biodiversity or humane treatment of livestock and if so, markets will respond. But firms within the system do not have the luxury of ignoring market forces.

Sustainable intensification

In contrast to the status quo, this sociotechnical imaginary is in fact driven by value commitments, along with a robust and scientifically informed set of assumptions about the future. These assumptions include projected growth in the global population and climate change. The other key value commitment is 'feeding the world.' This is implicitly understood as a goal for the food sector viewed in aggregate terms, rather than as an entitlement that could be exercised by individual human beings. At a minimum, it implies that the sector can produce agricultural commodities in amounts sufficient to satisfy nutritional demands of the entire global population.

Sustainable intensification is the sociotechnical imaginary of scientific and policy elites. It is based on persuasive studies indicating that if this package of value commitments is to be realized in over the next half century, food production methods will need to become much more efficient in their use of land and water, and in minimizing loss due to spoilage or waste. Unlike the other imaginaries I will discuss, sustainable intensification has been articulated and promoted in fairly explicit terms. This terminology was prominent in a study from the United Kingdom Office of Science (2011), and was only slightly less emphasized in a report on future needs in food systems from the U.S. Association of Public and Land Grant Universities (2017). The terminology has been adopted by major agricultural firms such as

Monsanto (Monsanto, 2016). It has become a target of explicit critique from food systems advocates who do not see it as marking a significant difference from status quo practice in industrial agriculture. That is: technological modernization (Thompson, forthcoming). Although advocates for sustainable intensification do undertake significant criticism of mainstream industrial agricultural technologies, it is perhaps fair to say that social institutions and even firms at the retail end of the food system are not prominent in this imaginary. The focus is on agriculture.

Extensification

Specialists in international agriculture adopted the term 'intensification' as a contrast to increasing the food supply by expanding the ecological footprint, a strategy they sometimes refer to as 'extensification.' This is not what I have in mind here, and perhaps there is a better name. As a sociotechnical imaginary extensification shares two key features with sustainable intensification. One is a reaction to the drift of technological modernization, and its accompanying problems of environmental damage and social dislocation. The other is a central focus on agriculture. However, unlike sustainable intensification, this sociotechnical imaginary is heavily focused on the social institutionalization of the food system. Whatever we decide to call it, there is a sociotechnical imaginary that is more or less shared by many people, including many self-described food activists, farmers, chefs, planners and a good deal of the general population. One element is an assumption that most food will be produced on farms in the future, coupled with the presumption that these farms will conform roughly to the owner-operated household enterprise stereotype that has typified the majority of food systems since the invention of agriculture. However, this archetype envisions a food system with social institutions that differ markedly from those that have evolved through the market-oriented thrust of technological modernization. Active work to realize this imaginary is often committed to the development of alternative institutions such as reinvigorated farmers' markets, food hubs, local sourcing initiatives and programs that connect farmers to institutional kitchens in schools and hospitals. Innovations in distribution and processing are thus much more fundamental to this sociotechnical imaginary than they are in sustainable intensification. Indeed, mainstream agricultural elites may have difficulty in imagining such innovations as a form of technology. This illustrates a philosophically important feature of sociotechnical imaginaries: implicit assumptions determine what counts as a technical innovation and subsequently influence the extent to which alternative configurations of the food system are even imaginable.

There is, however, quite a bit of diversity about the rationale for extensification, and an accompanying assortment of strategies on how to pursue it. For some exponents, the focus is on socio-economic institutions that favor household farmers. For others the underlying issue is health, and the rationale for localizing food systems resides in their presumed impact on diets. Food sovereignty stresses community control over food systems. For organizations such as Via Campesina that represent rural communities whose identity and social forms revolve around food production, food sovereignty translates into an implicit commitment to extensification, though it may place more emphasis on preserving existing social institutions, as opposed to innovating new ones. There are also environmental rationales in which the presumptively smaller scale of household farms is more consistent with norms of ecological stewardship. Bio-dynamic versions of organic agriculture turn upon the ideal of the farm as a self-contained regenerative ecological system. This vision gets linked to broader conceptions of rural communities as exhibiting forms of social sustainability that have been advocated by Wendell Berry (Thompson, 2017). The mixture of ethical values driving extensification creates internal tensions within the community of people pursuing this archetype. It can so be difficult to determine how well certain strategies conform to the overarching paradigm of extensification. As I have framed it, the sociotechnical imaginary I am calling extensification is shaped by its commitment to the continuation of household farming, on the one hand, and to technological and institutional innovations that reconfigure the relationships between household food producers and consumers of food. It is distinguishable from sustainable intensification via that sociotechnical imaginary's failure to emphasize social or distributive technologies as central to its mission, and distinguishable from urban agriculture in ways that are yet to be articulated.

Urban agriculture

A review of the extensive literature associated with the term 'urban agriculture' would suggest quite a bit of overlap (if not virtual identity) with what I have been calling extensification. Like extensification, it places significant emphasis on social relationships among various actors in the food system, but unlike it, urban agriculture has no deep commitment to the household farm. In this sociotechnical imaginary, food sovereignty's emphasis on controlling one's food destiny comes to the forefront, but unlike Via Campesina, the participants in this vision of the food system do not imagine themselves to be farmers, and the communities in which they live are not organized or recognized through rituals of planting, harvest and a shared destiny based on agriculture. Thus to the extent that urban agriculture overlaps extensification, it has some of the same weaknesses. But this is a sociotechnical imaginary that is emerging in large cities that have relied on non-agricultural sectors – manufacturing, energy, finance, government, healthcare and even the arts – as their primary generator of employment and economic activity. This vision of the food system draws less upon prior experience with the production, processing, distribution and consumption of food than any of the other three, and more on the imperatives of planning for the needs of large, complex populations centers. Thus one additional point of vulnerability is simply the fact that this imaginary has much more to prove.

Of course many of the initiatives now associated with urban agriculture have already been driven by the evident failure of market institutions, even in cities with considerably less than 10 million people. Neighborhood based urban food production in American cities has been motivated by the perception that local communities must take control of the food system (White, 2010; Block *et al.*, 2012). Initiatives such as Growing Power in Milwaukee or D-Town Farms in Detroit have frequently been conceptualized as if they were instances of the extensification sociotechnical imaginary. They are described along with localizing food initiatives that connect restaurants with farmers or the emphasis on nutrition and access to vegetables is placed at the forefront (DeMattia and Lee Denny, 2008). But other elements of existing urban agriculture stress employment or educational opportunity – features more typically associated with the goals of urban planning (Steele, 2008).

Other initiatives that combine urban planning's traditional focus on jobs and economic development with urban control of a food system are less compatible with the vision of extensification. Controlled-environment production is, in one sense, an extension of long recognized plant growth technology deployed in greenhouses. As these systems have achieved greater and greater efficiencies, they are now being proposed in space-saving vertical applications (e.g. in urban structures) as a way to address access and control-of-production issues that are seen to be the key challenges of food security in existing inner cities, as well as the megacities of the future (Despommier, 2011). This sociotechnical imaginary is under serious development at technology centers that are not traditionally associated with the food system. Patrick Brown's Impossible Foods project at Stanford provides one striking example (Kolodny, 2017), while Caleb Harper's Open Agriculture project at MIT is another (Sargianis, 2018).

Conclusion

The sketch that I have provided of four sociotechnical imaginaries is, of course, radically underspecified. More careful and extensive discussion of them could form the basis for a more robust theoretical treatment, but the notion of a sociotechnical imaginary attains is power more through its capacity for promoting critical reflection and dialog than as a device for predicting the future. Respondents may justly reject the boundaries that I am suggesting, and they may do so because they think things will turn out differently. My four archetypes might also be challenged on normative grounds: isn't there some way to envision a future that combines the good in each of these possible future food systems? Perhaps there is, and if so, that is exactly the kind of reflection I hope to provoke.

References

Association of Public and Land Grant Universities (APLU) (2017). The Challenge of Change: Harnessing University Discovery, Engagement and Learning to Achieve Food and Nutrition Security. APLU, Washington, DC. Available at: http://www.aplu.org/library/the-challenge-of-change/File. Accessed 13 January 2018.

Block, D. R., Chávez, N., Allen, E., and Ramirez, D. (2012). Food sovereignty, urban food access, and food activism: contemplating the connections through examples from Chicago. Agriculture and Human Values, 29(2), 203-215.

DeMattia, L., and Lee Denney, S. (2008). Childhood obesity prevention: successful community-based efforts. The ANNALS of the American Academy of Political and Social Science, 615(1), 83-99.

Despommier, D. (2011). The vertical farm: controlled environment agriculture carried out in tall buildings would create greater food safety and security for large urban populations. Journal für Verbraucherschutz und Lebensmittelsicherheit, 6(2), 233-236.

Jasanoff, S., and Kim S.-H. (2015). Dreamscapes of Modernity: Sociotechnical Imaginaries and the Fabrication of Power, University of Chicago Press, Chicago.

Kolodny, L. (2017). Impossible Foods CEO Pat Brown says VCs need to ask harder scientific questions, techcruch.com Available at: https://techcrunch.com/2017/05/22/impossible-foods-ceo-pat-brown-says-vcs-need-to-ask-harder-scientific-questions. Accessed 18 January 2018.

Monsanto Corp. (2016). Growing Better Together: Monsanto 2016 Sustainability Report. Accessed Jan. 13, 2018 at https://monsanto.com/app/uploads/2017/05/2016-sustainability-report-2.pdf.

Sargianis, K. (2016). In the field: Caleb Harper, Director, MIT Open Agriculture Initiative, cook's science. Available at: http://www.cooksscience.com/articles/interview/in-the-field-caleb-harper-director-mit-open-agriculture-initiative. Accessed 14 January 2018.

Steele, D. (2008). Beyond the backyard garden: urban agriculture in Milwaukee. Next American City. at:http://c.ymcdn.com/sites/www.fuelmilwaukee.org/resource/resmgr/regional_news_pdfs/6.10.08_urban_agriculture.pdf. Accessed 14 January 2018.

Thompson, P.B. (2017). The Spirit of the Soil: Agriculture and Environmental Ethics, 2nd Ed. Routledge, London and New York.

Thompson, P.B. (Forthcoming). Sustainable intensification as a social imaginary, In Contested Sustainability Discourses: From Food Sovereignty to Sustainable Intensification, D. Constance, Ed.

UK Office of Science (2011). Foresight, The Future of Food and Farming: Challenges and Choices for Global Sustainability. The Government Office for Science, London.

White, M.M. (2010). Shouldering responsibility for the delivery of human rights: a case study of the D-town Farmers of Detroit. Race/Ethnicity: Multidisciplinary Global Contexts, 3(2), 189-211.

29. Ethical perspectives on molecular gastronomy: food for tomorrow or just a food fad?

G. Precup[1], A.M. Păcurar[1], L. Călinoiu[1], L. Mitrea[1], B. Rusu[1], K. Szabo[1], M. Bindea[1], B.E. Ştefănescu[2] and D.C. Vodnar[1]*
[1]Department of Food Science. University of Agricultural Sciences and Veterinary Medicine, Cluj-Napoca, Romania; [2]Department of Pharmaceutical Botany, Iuliu Haţieganu University of Medicine and Pharmacy, Cluj-Napoca Romania; dan.vodnar@usamvcluj.ro

Abstract

This paper is arguing the concept of molecular gastronomy and its derived culinary trend, note-by-note cooking, as a possible scenario for the 'food of the future' and sheds light on the controversial principles of molecular cooking. In the context of world hunger, poverty and rapid population growth, forecasted to 10 billion people by 2050, molecular gastronomy`s next trend, 'Note by Note cooking', come up with the solution of using pure chemical compounds rather than animal or plant tissues for designing new foods, but still trying to create the same taste, consistency, colour or trigeminal sensation. The approach is still in its infancy, but it gives specific examples of its sustainability. However, this new way of designing foods faces the 'food neophobia' of people worried about the safety aspects, nutrition, toxicology and economics. Therefore, this paper will focus on ethical questions that should be addressed: What compounds can be used? Is it sure that they aren`t harmful? What are the implications for traditional cooking? Will note-by-note cooking be more expensive than current methods of obtaining food?

Keywords: pure compounds, food neophobia, alternative

Introduction

The global challenges that humankind is facing are jeopardizing the future of our planet and of the next generations, if current trends continue and new sustainable visions will delay to appear. There is an ever-growing demand especially for food, energy and water, as the world`s population is exploding and expected to reach 10 billion people by 2050 (FAO, 2017). This staggering fact emphasizes the huge challenge that the agro-food industry faces to sustainably feed and ensure nutrition security for these people, the majority of whom live in underdeveloped regions. An innovative and revolutionary solution came from a famous French chemist, Hervé This who believes to have found the key solution for the eradication of world hunger and proposed the concept of 'Note by Note cooking' in 1994, in the magazine 'Scientific American'. It is the newest trend of molecular gastronomy and stands for obtaining food and dishes by using pure compounds or mixtures instead of meat, fish, vegetable or fruits tissues (This, 2013).

The concept sounds revolutionary and the conservators in matters of food could perceive it as blasphemous, but its inventor asserts that this way of obtaining food is 'more energy-efficient and environmentally friendly than with all-natural ingredients.' The scientist also states that more nutritious and healthy food could be obtained, by using 'Note by Note cooking' (This, 2013). However, people are concerned about ethical issues in terms of nutrition, safety, economy, feasibility or losing food identity, by extinction of traditional food methods.

Furthermore, when talking about molecular gastronomy, a lot of confusion and controversy still raises. This term was coined in 1988 by the Hungarian Oxford physicist Nicholas Kurti and the French INRA chemist Hervé This and refers to the scientific discipline that uses methods to better understand

EurSafe 2018 – DOI 10.3920/978-90-8686-869-8_29, © Wageningen Academic Publishers 2018

and control the molecular, physicochemical, and structural changes that occur in foods during their preparation and consumption (This, 2006). Molecular gastronomy is often misused in the media with molecular cooking, which is an application of molecular gastronomy and refers merely to the way that chefs are preparing the dish, by adopting the tools and techniques developed by scientists to their own style of cooking (This, 2006). This was also the first to incorporate the introduction of new tools and methods from scientific laboratories such as rotary evaporators, liquid nitrogen, sintered glass filters, ultrasound probes, or ingredients like sodium alginate, agar-agar, calcium lactate into the kitchen (Precup, 2017; Blanck, 2007).

Therefore, in the *Haute Cuisine* field today, it is of vanguard to surprise the guests with fancy dishes that contain fake caviar made from sodium alginate and calcium, spaghetti from fruits, vegetable or chocolate or instant ice creams obtained by using liquid nitrogen. Famous chefs like Ferran Adrià (elBulli, Spain), known as the founder of the molecular cuisine trend, but also Heston Blumenthal (The Fat Duck, Bray, Berkshire, UK), or Thomas Keller (Per Se, New York City), have been labeled as molecular gastronomists. Being disturbed about this misconception, they have written a public paper entitled 'Statement on the "New Cookery"' in 2006, to clarify the confusion between their real intentions and the connotations of molecular gastronomy (Ruy, 2009).

Thus, molecular gastronomy does not refer to a fancy trend for privileged people who can afford to spend a small fortune on luxury food, but chemistry and physics behind the preparation of any dish. For example, why steak cooked in a pan turns brown and chemists know that the browning is due to Maillard reaction, between amino acids and carbohydrates or a mayonnaise becomes firm because it contains proteins and lecithin in the egg yolk, that serve as emulsifiers.

Having some chemistry knowledge, foodies who do not afford to pay a fortune in a top restaurant could also experiment in their own kitchen, if they can overcome the fear to use chemical substances as ingredients. For instance, if they cannot afford a good olive oil, they could add to their oil a flavor with a freshly cut grass scent, named hexanal. Likewise, 1-octen-3-ol or benzyl trans-2-methylbutenoate can offer a mushroom taste in a dish (This, 2002).

The ethical questions regarding the safety and nutrition aspects of using these 'chemical compounds' in foods could arise, as the food 'neophobia' determined people to place the trust in foods eaten at young ages and fear new foods.

Ethical challenges of note-by-note cooking

In order to be able to find an answer if molecular gastronomy, in particular its new trend, 'Note by Note cooking', could represent a sustainable solution for the future of food, several ethical questions should be addressed: What compounds can be used? Is it sure that they aren`t harmful? What are the implications for traditional cooking? Will note-by-note cooking be more expensive than current methods of obtaining food?

What compounds can be used?

This affirms in his book that note-by-note cooking is not 'chemistry' and the compounds used should not be confused with molecules, chemical products or synthetic compounds (This, 2014). People often confuse the relation between a whole and its parts are probably unaware that the food industry already use pure compounds such as water, sodium chloride (or salt), sucrose (sugar), gelatin. A vast range of compounds, from saccharides to aminoacids could be used in note-by-note cooking and the laboratory separation processes like filtration, direct or reverse osmosis, and vacuum distillation could be used to obtain the pure fractions. To create a food product, it is mandatory to first understand what chemicals give foods their taste, structure and aroma, using techniques such as mass spectrometry to identify the constituent parts. These building blocks can be extracted from animal or plant tissues or made artificially. Agro-food waste, as it represents a big environmental challenge, could also represent a rich source of valuable compounds like phenolic compounds, antioxidants or fatty acids, which can be useful for designing new foods, and their recovery could be economically attractive (Baiano, 2014; Rudra *et al.*, 2015; Vodnar *et al.*, 2017).

Is it sure that they aren`t harmful?

As note-by-note cooking stands for obtaining foods by using pure compounds, toxicity could be avoided by simply not using the toxic compounds. For instance, the meat cooked on the fire using a barbecue, releases carcinogenic compounds called benzo[α]pyrenes (Farhadian *et al.*, 2011) or the potatoes we eat contain glycoalkaloids, a-solanine and a-chaconine, responsible for the greening of the potatoes peels; toxic myristicin (6-allyl-4-methoxy-1,3-benzodioxole) from nutmeg; estragole; some glucosinolates from cabbages (This, 2013). Conversely, the question of toxicology still requires more studies to investigate the effect of consuming low doses of compounds for a long time.

What are the implications for traditional cooking?

People could refuse to adopt note-by-note cooking as they fear that traditional cooking would disappear and they would lose their food identity. Food neophobia is generally regarded as the reluctance to eat, or the avoidance of new foods, whereas familiar foods that people used to eat are usually correlated with good and healthy food. But the argument that traditional foods are safer to eat only because they are old is not sufficient (Guiné, 2016). There is a lot of scientific evidence and arguments that refute some food products legitimized by old food habits, like smoked products or red meat, which could increase the risk of obesity and developing several diseases like colon cancer (Bouvard *et al.*, 2015; Canella *et al.*, 2014; Zur Hausen, 2012).

Will note-by-note cooking be more expensive than current methods of obtaining food?

To answer this question, the cost of investment in new equipment has to be taken into consideration, as well as the increasing cost of energy and the diminishing resources. In today`s world, more than 90% of the energy used for kitchen stoves is wasted (Sovacool, 2013). This affirms that compounds can be obtained at lower costs in terms of energy, if using note by note cooking. For example, when cooking a meat, in order to obtain the flavour, it has to achieve approximately 200 °C, whereas the compounds responsible for odour and taste can be faster obtained by using note by note cooking. Moreover, a sauce from wine could be obtained only by using 0.417 kWh with a cost of 0.05 euros per sauce (This, 2012, 2013). Compounds obtained from plants will be less expensive and more efficient, as food spoilage could be avoided by extracting the water from fresh fruits and vegetables (This, 2014).

Conclusion

Nevertheless, this revolutionary concept of note by note cooking is still in its infancy, but goes beyond artistic endeavor and could represent a valuable alternative in the attempt to overcome the ethical, social, environmental and economic problems that burden our food chain.

References

Baiano, A. (2014). Recovery of biomolecules from food wastes-a review. Molecules 19:14821-14842.

Blanck, J. F. (2007). Molecular gastronomy: Overview of a controversial food science discipline. Journal of Agricultural & Food Information, 8(3), 77-85.

Bouvard, V., Loomis, D., Guyton, K. Z., Grosse, Y., El Ghissassi, F., Benbrahim-Tallaa, L. and Straif, K. (2015). Carcinogenicity of consumption of red and processed meat. The Lancet Oncology, 16(16), 1599-1600.

Canella, D. S., Levy, R. B., Martins, A. P. B., Claro, R. M., Moubarac, J. C., Baraldi, L. G. and Monteiro, C. A. (2014). Ultra-processed food products and obesity in Brazilian households (2008-2009). PloS one, 9(3), e92752.

FAO (2017). The future of food and agriculture – Trends and challenges. Rome.

Farhadian, A., Jinap, S., Hanifah, H. N. and Zaidul, I. S. (2011). Effects of meat preheating and wrapping on the levels of polycyclic aromatic hydrocarbons in charcoal-grilled meat. Food chemistry, 124(1), 141-146.

Guiné, P.F., Ramalhosa, R., & Paula Valente, L. (2016). New foods, new consumers: innovation in food product development. Current Nutrition & Food Science, 12(3), 175-189.

Precup, G., Calinoiu, L. F., Mitrea, L., Bindea, M., Rusu, B., Stefanescu, B. E. and Vodnar, D. C. (2017). The Molecular Restructuring of Classical Desserts by Using Food Industry By-Products. Bulletin of University of Agricultural Sciences and Veterinary Medicine Cluj-Napoca-Food Science and Technology, 74(2), 58-64.

Rudra, S.G., Nishad, J., Jakhar, N. and Kaur, C. (2015). Food industry waste: mine of nutraceuticals. International Journal of Scientific Environmental Technology 4:205-229

Ruy, D. (2009). Lessons from Molecular Gastronomy. Log, (17), 27-40.

Sovacool, B. (2013). Energy and ethics: Justice and the global energy challenge. Springer.

This, H. (2002). Molecular gastronomy. Angewandte Chemie International Edition, 41(1): 83-88.

This, H. (2005). Molecular gastronomy. Nature materials 4(1): 5-7.

This, H. (2006). Food for tomorrow? EMBO reports 7 (11): 1062-1066.

This, H. and Rutledge, D. (2009). Analytical methods for molecular gastronomy. Analytical and bioanalytical chemistry 394: 659-66.

This, H. (2012). Solutions are solutions, and gels are almost solutions. Pure Applied Chemistry.

This, H. (2013). Molecular gastronomy is a scientific discipline, and note by note cuisine is the next culinary trend. Flavour 2:1.

This, H. (2014). *Note-by-note Cooking: The Future of Food.* Columbia University Press, USA, 255 pp.

Vodnar, D.C., Călinoiu, L.F., Dulf, F.V., Ştefănescu, B.E., Crişan, G. and Socaciu, C., (2017). Identification of the bioactive compounds and antioxidant, antimutagenic and antimicrobial activities of thermally processed agro-industrial waste. Food Chemistry 231, pp.131-140.

Zur Hausen, H. (2012). Red meat consumption and cancer: reasons to suspect involvement of bovine infectious factors in colorectal cancer. International journal of cancer, 130(11), 2475-2483.

30. Identity or solidarity food – *ex-ante* responsibility as a fair culture approach

C. Moyano Fernández
Department of Philosophy, Autonomous University of Barcelona (UAB), C/ de la Fortuna, Bellaterra, 08193 Barcelona, Spain; cristian.moyano@uab.cat

Abstract

We associate culture with the historical tradition, those events and past behaviors that set a trend in our identity. However, culture not only keeps looking back, but also turns to the future. Culture can be projected as an emerging possibility. Thus, the movements of power (political, economic and social) generated around the culture, find a different sense if they are not encouraged by the same conventional force, from the same customs. Politicians or capitalism are often pointed out as the cause of most of the unfairness associated like for example malnutrition. However, they are treated as inseparable institutional pieces, which go hand in hand, instead of understanding that capitalism is at its core a cultural rather than a political process. In this second it finds its support but its origin is sustained by the *ethos* of people, by our lifestyle which is difficult for us to give up deliberately. Because of habit and custom we have been anchored to the comforts afforded by the capital of the welfare society, saving the cognitive effort involved in exercising empathy. Often, acting with justice implies to act according to the moral code that displays hegemony in the society of our time. But if we expand our view and assume that globalization has allowed our sense of responsibility not to be marginalized at a local level, then the discourses of justice must be harmonized with a multicultural recognition. Moral pluralism finds its heterogeneity in several past customs, but a common fair culture can be built if it aims for *ex-ante* responsibility. In that case, is reasonable to promote solidarity as a political movement to protect the health of our neighbours and next generations while part of our identity is rejected? In the ethical theory proposed here, identity and solidarity rather than meeting in an exclusive dilemma are integrated to deal with food inequalities which raises one of the greatest injustices that we live nowadays and our descendants will inherit.

Keywords: equality, sufficiency, sustainability, global justice, human authenticity

Introduction

I am interested in thinking twice about the reason why we do not make some ethical decisions almost effortless that could substantially improve the quality of life of many people. Especially in food. What prevents us from sacrificing some of our expensive and dispensable diets, without renouncing adequate nutrition so we can give those goods and resources to those suffering from malnutrition? I have found that one of the main reasons, which inhibit our ethical response, is the risk of thinking that we can lose a trait of our sociocultural identity and consequently losing social recognition. This is so given that a hegemonic feature of occidental identity seems to be the capitalist *ethos*. Is our cultural identity an impediment to act with greater justice in the unequal world where we live? It will depend on how we philosophically understand this identity, whether we protect its capitalist cultural base or encourage it to be formed on a more fair culture.

First, I discuss two ways of understanding social justice based on the understanding of a moral person: in a more global, interpersonal and aggregative way, in Singer style and in a more personal, differentiated and non-transferable way, in Nussbaum style. Both approaches show their interest in pluralism as well as a commitment to social equality. However, outlining a fair culture of food, like the one we are trying to search, requires thoughts that investigate the principles of a fair policy beyond moral anthropology.

Svenja Springer and Herwig Grimm (eds.) ***Professionals in food chains***
EurSafe 2018 – DOI 10.3920/978-90-8686-869-8_30, © Wageningen Academic Publishers 2018

Then, I go on to argue that our consumption decisions are usually conditioned by a hegemonic cultural tendency, and that such domain manipulates our identity sense. Looking for social recognition from a special identity is what has led, in part, to sustaining and reproducing a specific diet such as the one rich in meat products.

Then, presuming there are unfair food inequalities due to mainly cultural reasons, I wonder what should change in our culture so that we all have the same opportunities to be well nurtured. Here I argue that the answer is to point out the aspiration of equality. An egalitarian theory of justice to be solid needs to be constrained by sufficiency. I justify the latter on the basis of the idea that equality must be an intergenerational right and therefore the sustainability feature is indissociable. In particular, as regards the food sector a responsible consumption from most developed countries consists in the first place of not exceeding the limit from which the food-choice capability of less developed countries is biased until the point of not choosing between some minimum foods. Secondly, this responsible consumption must assume that its elective limit guarantees that future generations can have sufficient options. Then I conclude that if our diets are unsustainable to the point of leaving alimentary capability below a critical threshold (Sen, 1999; Venkatapuram, 2011) to those who share this world with us or to whom will inherit this land, then our food culture can not be considered fair.

Finally, I conclude that before seeing identity as a burden for social justice or opposed to ethical behaviors such as solidarity, may well serve as a support, if for example it expands its identification with humanity.

Moral anthropology for social justice

First, to answer the question of whether an ethical action such as universal and truly disinterested solidarity necessarily requires the privation of a specific feature of our cultural identity, I show two ways of understanding social justice according to its type of moral anthropology. On the one hand, the conception of person supported by Peter Singer and on the other hand the one defended by Martha Nussbaum. I consider important to go into these two anthropological positions since they clarify a vision of human identity that includes a preeminent ethical sense.

Singer has always been characterized as an utilitarian philosopher, so his idea of moral action is based on reason rather than in emotion. In his defense of the figure of the effective altruist as the most ideal example of a moral person he maintains that although emotional motivation is important to undertake ethical action the spark that ignites and redirects passion can and should be rational (Singer, 2015). Here, Singer subscribes to the thought of the philosopher Richard Keshen, for whom the most essential of a person's ethical life is the recognition that others are like us and their lives and welfare matter as much as ours. This fits with his concept of reasonable self-esteem, preaching that the reasonable person cannot have self-esteem and at the same time ignore the interests of others and he recognizes that their welfare has the same importance as his own. Thus, the most solid foundation for self-esteem would be to live an ethical life, in which one contributes as much as possible to make the world a better place. Doing so would not be altruism in a sense that implies renouncing what one would prefer to do neither alienation nor loss of identity, as Bernard Williams stated (Williams, 2006). It would be, on the contrary, the expression of the core of one's own identity.

On the other hand, Martha Nussbaum is usually shown as the heir of a deontological and above all teleological philosophy. In her original project of the capabilities approach she overcomes the individualism that would characterize a fully liberal scheme, leaving behind Rawlsian conceptions, to commit to equal opportunities. While for Kant and John Rawls the freedom of oneself consists in being responsible of one's preferences (Puyol, 1999: 168-169), in Nussbaum we find the doubt about the formation of those preferences (Nussbaum, 2000). How autonomous are we in our choices?

According to her, we should all be able to form a conception of the good life and engage in critical thinking about the planning of one's own life and practical rationality. With this assertion she pursues to protect our autonomy from possible adaptive preferences (Dworkin, 2011), without ignoring the local specification of policies that follow local customs without allowing pluralism to either violate the general characteristics of moral equality that it is showed in her list of basic capabilities. Thus, the moral frontier between people must be built on equality of capabilities, since that one contains respect for difference. Finally, another positive aspect of Nussbaum's approach is that starting from the fragility rather than from the autonomous solidity that Kant or Rawls suppose as an intrinsic characteristic of the human being it is shown a greater critical sensitivity towards the social structures where we live. The assumption of human characteristics such as dependence or fragility avoids making a hasty abstraction of social inequalities while promoting cooperation and social connection (Young, 2011) in order to achieve greater individual autonomy.

The cultural turn of the identity sense

To some philosophers, promoting autonomy helps to build self-identity more authentically (Taylor, 1992). However, our fragility habitually leads us to dependence, fact that seems to distance us from authenticity (MacIntyre, 1999). But this does not imply that we are without identity, because we always have several coexisting. We assume to perform various roles that better adjust to the stereotypes and traditions that dominate our society (Taylor, 1995), rather than those consciously and deliberately wished. This is how we daily perpetuate and adapt ourselves to cultural identities more based on the economic or political motive of a hegemonic enterprise than genuinely human.

In the same way, it is quite indisputable that identity plays an important role in food preferences. Of course, this is not the only factor driving our particular wishes, but it is one of the main factors at a collective level. Consuming certain food in the past centuries, even millennia, has been a symbol of social status and power. Nowadays, if the meat consumption is increasing fastly in less developed countries, it is largely due to the search for an identity recognition policy. The logic of capitalism is written in a liberalist key and it scarcely tolerates an economy of the common good (Felber, 2015) that manages natural capital (Helm, 2015). This idea has been assumed and recreated by the majority of the population from the industrialized countries, in order to reach the welfare and happiness states promised by the cultural industry (Adorno and Horkheimer, 2002).

Since ancient times, the choice of certain foods over other has often been due to an almost political aspiration, in the most etymological sense of the term of being admitted to the community circle of the *polis*. This symbol of power has been pursued, by virtue of achieving a valid and worthy social recognition. Nevertheless, only those enjoying strong economic privileges could go for it. It was the nobility and the clergy who gained a respectable identity by showing their power by buying land, servants or various foods; while the other estates associated their poverty condition to the narrow range of consumption choices they could afford.

Nowadays, however, the opportunities for a varied consumption have become widespread, reaching all countries whose industrial development has rocketed since the twentieth century first. Unfortunately, among the less developed countries in this sense, freedom of choice remains unfairly constrained by economic disadvantages. But those of us who live surrounded by the privileges of the welfarist society can now decide what to consume with hardly any economic impediment. This fact has caused, among other effects, a huge increase in meat consumption (FAOSTAT digital source). However, at present, the culture (and not the economy) stimulates and encourages us to eat products of animal origin.

It is true that the economy, operating from the background, motivates some dominant spheres, companies and lobbies to culturally disseminate the advantages in consuming meat (Singer, 2006: 275). Thus, with the manipulative pressure of the mass media exercised on the population, they manage to alienate the population from themselves and generate a demand according to the capitalist offer from the dominant spheres. Economic inequalities are radicalized in an extreme way. But this aspiration to enrichment comes to us through a specific cultural proposal, stating that what moves our decision is not the economy but the mere tacit acceptance of the hegemonic culture.

Despite the fact that hundreds of scientific studies support the relegation of meat in our diets without undermining our nutritional health, this information continues to be eclipsed by reports and advertisements sponsored by meat companies (Campbell and Campbell, 2006). Knowing this, health is not the best argument for going without meat in our diets (FAO/WHO, 2014), not even for a gustatory pleasure since cooking with spices nowadays makes it easy to imitate almost any flavor. In any case, it is for other reasons, such as the identitarian ones anchored to an unfair cultural tradition as capitalism.

Losing an identity seems to mean being excluded from a collective as well as not being recognized, with the consequent discriminatory marginalization that it may imply (Honneth, 1995). And despite we do not want to live that possible inequality culture has decided to hatch it. But how should we understand equality? And should we protect it whatever it takes?

Food equality and sufficiency in an intergenerational time

If the prevailing culture of the most industrialized countries that determines our identity results in serious social inequalities, such as malnutrition suffered by almost one billion people while others suffer from obesity (FAO and WHO, 2014), how should a more fair culture be articulated? On what cultural conditions should equality be built, if this one was politically desirable?

I will start by defending the following premise: for a real equality and a solid social justice, it must commit to sustainability. Equality, as a value *per se*, lacks a commitment to sustainability unless it is introduced into the temporary vector. Everybody can enjoy the same food at a certain time, but if this situation lasted for years resources would be exhausted and the situation would turn unfeasible. So that 'everybody' is a very questionable group if we take into account only those who cohabit in the present and exclude those who will live in the next generation. For equality to be real and lasting, not a mere passing stage, we must aspire to guarantee it for those of us who are here and for our descendants. Otherwise, equality, rather than being esteemed a respectable value for humanity, would be an exploitable value subject to a kind of race to take possession of the benefits before those who will be born after us. Thus, not worrying about leaving a land on which the future humanity has the opportunity to live in dignified conditions is to fail in the project of solidly consolidating social and global justice.

There is another premise that I consider pertinent to embrace: guaranteeing that sustainable equality over time necessarily requires starting from a level of sufficiency. Regarding the relationship between sufficiency and equality, I contemplate five possible lines: (1) we all have insufficient, but some of us more than others; (2) we all have at least enough, but some of us more than others; (3) we all have equally insufficient; (4) we all have equally enough; (5) we all have more than equally enough.

Sufficiency is conceived here as a threshold or boundary line above which our standard of living is satisfactory and below which it is unsatisfactory (as regards the realization of our human dignity). Then, starting from these 5 possible ways of development, the reason that guides us towards our blossoming would welcome options 2, 4 and 5, discarding 1 and 3. In the first instance, it is assumed that any human reason before a scenario in which we did not know our fate would proceed, given that by betting on everyone to overcome the sufficiency threshold would be the way to always ensure a minimum development and wellbeing. Note that affirming this hierarchy, equality has been relegated to the background, while reason has prioritized the sufficiency of a *prima facie* right.

In the first instance, our rational argument seems to be motivated by the benefit of an optimal survival of our species (Dawkins, 1976). Now we need to settle among the remaining options to identify on which one a social and global justice must be focused. Once these needs or minimum capabilities that make up sufficiency met, our social conscience goes further and often tries to find equality, either for envy or for ethics and justice. But as it was already pointed out, real equality must be sustainable in order to be equally enjoyed by our future generations, so the option that *a priori* infringes the least this idea would be 4. With it, everybody would be over the threshold without which dignity or equality are not possible, and what is more we would not endanger our future due to an abusive consumption of goods.

Betting on a sufficiency that ensures our dignity seems to be the most reasonable approach given that the theoretical way is the one that best preserves the projection of a justice concerned about global health and the harmonious survival of our humanity. And, in fact, the proposal of a sufficiency that includes equality would be even more profuse. Therefore, as far as our food choices are concerned, we can discern what food we can consume globally (equality), so that we ensure a good nutrition that allows us to develop ourselves with dignity (sufficiency) and does not result in an ecological crisis (sustainability).

Being aware of the unfair long-term consequences generated by industrialized meat products (Moyano, 2017), the most reasonable option is to limit the promotion of their consumption as soon as possible. It is a promotion based on the strong cultural idea that we must consume the same to recognize ourselves as equals but also carefree to guarantee equal opportunities to our descendants. So it is necessary to take a turn to this culture that leaves relegated sustainability and sufficiency and only prioritizes the mere equality of the present.

Conclusion: responsible identities in a fair culture

At the beginning, two philosophical positions of social justice have been put on the table: on the one hand, the effective altruism of Peter Singer and, on the other hand, the capabilities approach of Martha Nussbaum. As it has been said, although both thoughts handle a different sense of a moral person, they complement each other in several aspects, for example in their attempt to link equality and social connection with a genuine sense of human identity. However, they barely include a moral consideration for future generations or for the sustainability of the natural environment. In order to join these positions by filling the emphatic vacuum left by an extension of ethical concern for humanity in general, a consequent reformulation of culture is necessary. In relation to the right to a fair diet, it should contemplate the thoughts on sufficiency, equality and sustainability. This kind of culture would embody a concept of a moral person different from those perpetuated by the capitalist impulse.

The idea of human identity can derive from an altruistic self-esteem in the sense of Keshen and Singer or from a non-instrumental interdependence in the sense of Nussbaum. Both philosophical anthropologies allow us to build an interesting identity for ethics. But for our daily act of eating to be coherently responsible, we should also make a commitment to the aforementioned triad (equality, sufficiency and sustainability) of the fair culture. Thus, to evoke a food behavior in this identity line is through an *ex-ante* responsibility.

In other words, preserving the right to a fair diet as it has been proposed implies a series of positive duties (such as preventive responsibility or solidarity) towards others: not so much because of a deontological or political obligation, but because of a coherent human good sense. And then, being solidarity consists in preserving our identity, understood as it resulting from the union of the anthropologies of Singer and Nussbaum with the aspiration to achieve an equal and sustainable sufficiency.

Acting by mere tradition can lead to a responsibility understood as guilt, *ex-post*, while acting according to a fair culture is to take decisions assuming a responsibility towards the future, *ex-ante*. When we eat we must consider this responsibility if we do not want to lose the truly human identity that integrates such a precious value as justice.

References

Adorno, T. and Horkheimer, M. (2002). Dialect Enlightenment. Stanford University Press, Stanford.

Campbell, T. C. and Campbell, T. M. (2006). The China Study. BenBella Books, Dallas.

Dawkins, R. (1976). The Selfish Gene. Oxford University Press, Oxford.

Dworkin, R. (2011). Justice for Hedgehogs. Harvard University Press, Harvard.

FAO and WHO. (2014). Framework for Action. In: Second International Conference on Nutrition. Rome.

FAOSTAT. Available at: http://www.fao.org/faostat/en.

Felber, C. (2015). Change Everything: Creating an Economy for the Common Good. Chicago University Press, Chicago.

Helm, D. (2015). Natural Capital: Valuing the Planet. Yale University Press, Yale.

Honneth, A. (1995). The Struggle for Recognition: The Moral Gramar of Social Conflicts. Cambridge Polity Press, Cambridge.

MacIntyre, A. (1999). Dependent Rational Animals. Open Court Publishing, Illinois.

Moyano, C. (2017). Fragmented food autonomy: meat industry as a social determinant of health and counter-capability factor. In: Justice in Health Care – Values in Conflict. EACME Conference Book, Barcelona.

Nussbaum, M. (2000). Women and Human Development: The Capabilities Approach. Cambridge University Press, New York.

Puyol, A. (1999). Justícia i salut. Universitat Autònoma de Barcelona, Bellaterra, pp.168-169.

Sen, A. (1999). Development as freedom. Oxford University Press, Oxford.

Singer, P. (2006). The Way We Eat: Why Our Food Choices Matter. Rodale, New York.

Singer, P. (2015). The most good you can do. Yale University Press, Yale.

Taylor, C. (1992). *The Ethics of Authenticity*, Harvard University Press, Harvard.

Taylor, C. (1995). Philosophical Arguments. Cambridge University Press, Harvard.

Venkatapuram, S. (2011). Health justice: an argument from the capabilities approach. Polity Press, Cambridge.

Williams, B. (2006). The Point of View of the Universe: Sidgwick and the Ambitions of Ethics. In: The Sense of the Past: Essays in the History of Philosophy. Myles Burnyeat (comp.). Princeton University Press, Princeton.

Young, I.M. (2011). Responsibility for justice. Oxford University Press, Oxford.

Section 5.
Food politics: policy and legislation

31. EU Welfare States, food poverty and current food waste policy: reproducing old, inefficient models?

L. Escajedo San-Epifanio[1*] *and A. Inza-Bartolomé*[2]
[1]*Faculty of Labour Relation and Social Work, University of the Basque Country, Department of Constitutional Law and History of Political Thought, Postbox 644, 48080 Bilbao, Spain;* [2]*Faculty of Labour Relation and Social Work, University of the Basque Country, Department of Sociology and Social Work, Calle Los Apraiz no. 2, 01006 Vitoria-Gasteiz, Spain; leire.escajedo@ehu.eus*

Abstract

Distribution of food aid in kind seemed to have been left behind in the 80s and 90s in the EU Countries as other social protection programmes developed. However, the European crisis and the austerity measures, together with external factors, such as the fight against food waste, exponentially increased food charity activities. This paper analyses the approach of the EU to food poverty as well as those measures to fight food waste that could be described as food donation (or redistribution). How Food Banks and Food Charity are understood nowadays is another key issue to be analysed, as well as the misfunctions that in some cases are created through them in the social protection systems. In addition to that, the steps towards a new model will be presented and discussed, from a point of view of human dignity and human rights.

Keywords: food waste, social emergencies, food charity, food aid, food banks

Public responses to food poverty in the EU: moving towards obsolete models?

In the last few decades, and following the devastating consequences of the economic crisis in the EU, models of food aid that were abandoned in the 80s and early 90s have emerged again in some Welfare States. Food operations and aid in kind had been identified as ambivalent – even controversial – in international cooperation some years ago (Pérez de Armiño, 2000) and had only a very residual presence in social intervention models that sought social inclusion and the promotion of social autonomy. Hence, it is striking that since 2008, food aid in the form of goods has increased exponentially in the EU and various Governance levels, including the European Commission, are designing policy measures to promote food donations to people in situation or risk of social exclusion.

It is convenient to offer here, in an introductory way, a series of data that will help us properly situate the context in which this situation is taking place. Then, and in three separate sections, we will break down what we understand are the keys or factors that have driven, and drive, this movement towards the donation of food in kind, addressing in the final thoughts the consequences of this trend and the need to rethink these strategies and redirect them in the medium and long term.

Perhaps the most striking aspect of the response of the EU to food poverty is that it is not an answer that originated (at least not primarily) from the observation of people in need. The fight against food waste is the political framework in which reference is made to the fact that millions of Europeans lack enough purchasing power to adequately access food.

Employment and current levels of social aid are not, as of today, a guarantee of access to a decent life in the EU (Caritas, 2014). According to the EU Commission 55 million Europeans (EU Com, 2016) and, according to Caritas 25% of the citizens, live below the threshold of minimum income (Caritas, 2015) unable to meet basic needs such as food, clothing or housing. In the absence of a political response

EurSafe 2018 – DOI 10.3920/978-90-8686-869-8_31, © Wageningen Academic Publishers 2018

that improves the attention to these people, different factors have led to a notable increase in the charitable function performed by food banks in the strict sense and the food aid distribution entities that collaborate with them (Escajedo San-Epifanio, 2018). According to the FEBA, 5.7 million Europeans are served by its network. And the document of the commission to promote the donation of surplus food, which has been transferred to the EU Member States for its consideration, proposes to increase the food that is distributed through this network to serve more people and do so to a greater extent.

The roll of social services is displaced as consequence of the policies against the squandering. And at the same time it contributes, as some experts have been denouncing for years (Perry *et al.*, 2016), to the institutionalization of a model that, in concept, arose as an emergency mechanism destined to 'disappear' with time. Ronson and Caraher (2016) remind the theories of Seibel affirming that food aid organizations are 'successful failures', that is, they seem to proliferate indefinitely without anybody pointing out that the results did not meet the needs of those to whom they were intended to help, because their voluntary or faith-based nature makes criticism socially unacceptable. There is a halo effect around such organizations that do good deeds. Meanwhile, charitable distribution, if not articulated appropriately, may help to stigmatize and perpetuate poverty, and leave in the hands of the third sector or even of corporations the responsibility which corresponds to public institutions (Escajedo San-Epifanio *et al.*, 2018). In the following sections we analyze the approach of the EU to food poverty from food waste, which in some way has determined the development and expansion of European Food Banks, to such a point that it becomes a way of establishing itself as a food distribution formula alternative to vulnerable sectors. All this is supported by public opinion and the political system that understands it as a way to avoid the exclusion of certain groups.

Food poverty and the fight against food waste in the EU

With few exceptions, food waste has not been the focus of attention in the social sciences until very recently (Evans *et al.*, 2013). Suddenly, however, it has become a notable and growing focus of attention, especially in light of the important deployment of strategies that are being promoted, among others, by the FAO and the EU institutions (Escajedo San-Epifanio, 2016). Among them the 'Roadmap to a Resource-Efficient Europe' (EU Commission, 2011) and the document 'Strategies for a more efficient food chain in the EU' (EU Parliament, 2012). In this second document, the European Parliament offers a series of data describing the reality of food waste in the EU and warns that, regrettably, among the food waste a significant amount of 'safe and perfectly edible food' has been found. The amount given in 2011 was 90 million tons (179 kgs per person) and the data published in 2016 pointed to 88 million tons (European Commission, 2016). In these figures, the unprocessed fruits and vegetables and the catches returned to the sea were not included (Escajedo San-Epifanio, 2016).

Three major groups of strategies are proposed to fight against food waste, which is rejected from the economic, environmental and ethical point of view. First, measures aimed at improving the efficiency of food chains. Secondly, and for those cases in which it is not possible to improve efficiency, we find recovery strategies for surplus food (for human or animal food purposes). And, finally, recycling strategies or obtaining new products from the transformation of food waste.

Food poverty appears tangentially when the second group of strategies (food recovery) is explained. The recovery of food for donation purposes, which in some cases is described as a redistribution of food, is explained as a strategy that responds to two different problems (Commission, 2016): on the one hand, it reduces food wastage; on the other hand, it implies a support in the fight against poverty. However, indirectly difficulty in the access to food is recognized, although it is not one of the objectives of the food waste policy.

Food Banks as a tool to fight social emergencies

Models of food distribution that had been abandoned in the decades of the 80s and 90s in the EU, remaining only in a very residual way, have resurfaced in the wake of the economic crisis (especially since 2008). Food is also a metaphor for the modern processes of exclusion in advanced societies. Even if there is overproduction and waste in the consumer society, market access to some products is almost impossible for people in a situation of economic vulnerability. However, some of the leftover products finally get to them through a charity system. That is something critical, as far as that people's position varies greatly from a person entitled to welfare, to a user who asks for (and somehow depends on) the discretionary generosity of other citizens.

That of the Food Banks is a more complex history than it seems at first glance. At the very beginning, their action was explained as a solution that linked, on the one hand, urgent needs of food aid and, on the other hand, leftover food. But nowadays we cannot say they are emergency food systems any more. Some food banks allow companies to reduce their stocks in a profitable way and citizens who lack sufficient purchase power are invited to take that leftover food through charitable entities.

Some consequences are: (1) Even if it is substituting the role that the welfare state should assume, society thinks highly of food charity. (2) People at risk of exclusion are being excluded from the consumer society, and only those with sufficient incomes could have autonomous access to food. (3) The problem of food overproduction and its environmental and economic consequences is being morally compensated with charitable actions, but that means ignoring the consequences that food charity systems have on the autonomy and rights of excluded people. (4) Food redistribution is presented as a good way to face food waste in the EU, but the people that suffer food exclusion do not appear to be considered part of the 'food waste issue'. (5) Very few voices are stressing that this way of functioning implies an involution of the welfare state.

Food Banks are non-profit organisations whose main objective is to combat food insecurity by means of the distribution of food via intermediary entities. In some cases, although not always, this is surplus food which, if not donated, would become food waste. Although there are numerous types of Food Banks, the functioning of many of them depends to a large extent upon the work of volunteers. With a few exceptions, these are entities that distribute and/or deliver the food as and when it is received; that is, without processing it (Escajedo San-Epifanio *et al.*, 2018). The first Food Bank was founded in the United States in the 60's of the 20th century, thanks to the initiative of John Von Hengel, who worked as a volunteer in a soup kitchen (Schneider, 2013: 756). On the other hand, the first European Bank was founded in Paris in 1984 and two years later the European Federation of Food Banks (FEBA), a non-profit organization composed of 264 Food Banks from twenty-two member countries in Europe that collects food, which distributes to charitable organizations and social centers was founded. After an exponential growth of poverty up to 2014, the volume of activity of banks continues to grow, although the number of users has slightly decreased. However, current users are in a situation of acute need. In high-income countries, the increase in the need for food aid will be significant in the future (Gentilini, 2013).

Non-critical political and social support to food charity: some reflections

Feeding the hungry is an ethical duty in almost all societies and religions of the world (Pérez de Armiño, 2000). There is a tendency to think that food is what people need in a situation of poverty and in the collective imagination, giving that food appears as a formula 'of pure solidarity'. It is forgotten, however, that in that donation, not everything is altruism. Charity can become a way to impose or maintain power relations and / or arouse dependency or passivity in the recipients, moving them away from formulas that lead them towards social inclusion. Just as in international cooperation food aid does not always generate positive impacts (due to its economic, social, political and cultural effects), aid in kind also has its risks in the context of social assistance. Some investigations reveal (Pérez de Armiño, 2014) that the appeal to the obligation of the State to comply with the right to food is not frequently found in the speech of Food Banks. Mentions to the right to adequate food are infrequent and the activity of food banks does not give rise to messages of protest or political criticism; as this author explains, Food Banks have become 'a form of uncritical solidarity'.

To a certain extent Food Banks are useful to minimize the dissonance, but it is essential for politicians to implement the potentially problematic policy and to see that despite the essential welfare reforms they support, Food Banks will never abandon those unable to feed themselves and their children, and are also essential for society in general, which can continue without problems despite hurtful disparities (Ronson and Caraher, 2016: 86). Thus, while the mirage that immediate food assistance is available to people who cannot feed themselves exists, there will be little encouragement from governments to review the adequacy of their welfare programs, and also, there will be little pressure on the part of the population to do so (Tarasuk *et al.*, 2014). Thus, the notion of hunger is reinforced as a matter of charity, not politics (Riches, 2011: 768).

Consequences of obsolete food altruism models and the human right to food: final reflections

Viewed from a different perspective, an extended definition of food insecurity would be the following: 'limited or intermittent access to a healthy, safe and nutritionally acceptable diet that is accessed in a socially acceptable manner' (Anderson, 1990). It implies, therefore, not to compensate the hunger at any price and in any way. Among the criticisms of the way of operation and the service offered by the Food Banks, the effects on health that users may suffer due to its continued use and the inadequacy of what is offered stand out: it is associated with a wide spectrum of negative repercussions in health (Silvasti and Riches, 2014: 197). In this sense, in the opinion of Tarasuk and MacLean (1990: 332), institutional dependence on charity food assistance should be recognized as a major obstacle to good health, and cannot be seen as effective or adequate responses to serious and widespread food problems. The current food aid distribution model reproduces a welfare logic that favors disempowerment, the loss of personal autonomy and stigmatization, as well as chronifying impoverishment and social exclusion (Gascón and Montagut, 2015: 74). People who come to them lose part of their autonomy and inherent human dignity, because they have to accept charity food regardless of their needs and preferences (Riches and Silvasti, 2014: 9). And they have to prove that they have the right to receive what are essentially the leftovers from society (Tarasuk and MacLean, 1990: 332).

The engine that has driven the development of current models of social intervention (according to a human rights approach, that is, promoters of the autonomy of the person) is overlooked. In addition, the transfer of a public benefit (right) to a private (graceful) benefit, among other things means the loss of standardized claims processes, and gives these organizations the ability to arbitrate and decide who receives and who does not receive certain grants (Mata and Pallarés, 2014: 17). The current food aid system, organized as if an emergency were to be faced, shows some root problems and misfunctions: inaccessibility, inadequacy, inappropriateness, indignity, inefficiency, insufficiency, and instability (Poppendieck, 1999), Later, some other problems, 'ins' were added to the former: ineffectiveness, inequality, institutionalization, invalidation of entitlements, and invisibility. Something is to be done while governments dwarf and depoliticize hunger and food poverty as a relevant social problem, the corporatization and establishment of Food Banks globally led by the European Federation of Food Banks and the Global Foodbanking Network is strengthened in the majority or rich OECD countries (Riches and Silvasti, 2014b: 9).

Under the title 'Dignity: end everyone with hunger in Scotland', the Independent Working Group on Poverty (IWGFP, 2016) expressed strongly that no citizen of a country as prosperous as Scotland should make its food dependent on charities. The same phrase can be projected to the whole of the EU, one of the regions with more economic resources of the whole Planet. The human right to food was recognized decades ago in record time, but at the same time is one of the most systematically violated rights of the entire Universal Declaration of Human Rights (Escajedo San-Epifanio, 2018). In the Global North the recognition and implementation of this right requires analysing, on the one hand, who suffers from the failures of the food system (hunger, malnutrition, inadequate diets) and why; and on the other hand, what structural reasons or what actions or punctual inactions have an impact on their difficulty in accessing food adequately.

Acknowledgements

This work has been funded by: the Project 'Poverty and Food Aid resources in Vitoria/ Gasteiz: moving towards a new social intervention model', Fundación Vital (P.R. Inza-Bartolomé); the Project 'JAKI-ZUBIAK. Building strategies against food waste and food poverty', US 17/22, granted by University of the Basque Country, 2018-2019 (P.R.s Escajedo San-Epifanio and Rebato Ochoa); the research group URBAN ELIKA, Univ. of the Basque Country (P.R. Escajedo San-Epifanio); 'Multilevel Constitutionalism' Research Group, University of the Basque Country (PR López Basaguren); MINECO DER-2017-86988-P (PR López Basaguren).

References

Anderson, S. (1990).' Core indicators of nutritional state for difficult to sample populations'. *Journal of Nutrition*, 102: 1559-1660.

Caritas Report, 2014. The European crisis and its human costs. A call for fair alternatives and solutions. Available at: http://www.caritas.eu/sites/default/files/caritascrisisreport_2014_en.pdf.

Caritas Report, 2015. *Poverty and inequalities* on the rise; Just social systems needed as the solution. Availabe at: http://www.caritas.eu/news/crisis-report-2015.

Escajedo San-Epifanio, L. (2018). 'El despilfarro de alimentos y el derecho humano a una alimentación adecuada en la UE'. In: Escajedo San-Epifanio/ Rebato Ochoa/ López Basaguren (ed.), *Derecho a una alimentación adecuada y despilfarro alimentario*, Tirant lo Blanch, Valencia, Spain.

Escajedo San-Epifanio, L., Inza Bartolomé, A. and De Renobales, M. (2018). 'Food banks'. In: *Encyclopedia of Food and Agriculture Ethics*, 2nd ed., Springer.

Escajedo San-Epifanio, L. (2016). 'The politics of Food Waste and Food poverty in the EU', in. Olsson *et al.* (eds.), *Food Futures: ethics, science and culture*. WAP, The Netherlands.

EU Commission (2011). *Roadmap to a resource efficient Europe*. Availabe at: http://www.europarl.europa.eu/meetdocs/2009_2014/documents/com/com_com(2011)0571_/com_com(2011)0571_en.pdf.

EU Commission (2016). *Working document to prepate guidelines on food donation*. DG SANCO, 22th June 2016.

EU Parliament (2012). *Resolution of 19 January 2012 on how to avoid food wastage: strategies for a more efficient food chain in the E*U. Available at: http://www.europarl.europa.eu.

Evans, B. M. and Shields, J. (2000). 'Neoliberal restructuring and the third sector', CVSS, Ryerson University Working Paper Series, no. 13. http://www.ryerson.ca/~cvss/WP13.pdf.

Gascón, J. and Montagut, X. (2015). *Bancos de alimentos ¿Combatir el hambre con las sobras?*. Barcelona: Icaria-Asaco.

Gentilini, U. (2013). Banking on food: The state of food banks in high-income countries. *IDS Working Papers,* 415, pp. 1-18. Availabe at: http://www.ids.ac.uk/files/dmfile/Wp415.pdf.

IWFPG. (2016). Dignity: Ending Hunger Together in Scotland. Available at: www.gov.scot/Publications/2016/06/8020.

Mata, A. and Pallarés, J. (2014). Del bienestar a la caridad. ¿Un viaje sin retorno?. *Aposta. Revista de Ciencias Sociales*, 62, pp. 1-23.

McIntyre, L., Tougas, D., Rondeau, K. and Mah, C. (2015). 'In'-sights about food banks from a critical interpretive synthesis of the academic literature'. *Agriculture and Human Values,* pp. 1-17. DOI: 10.1007/s10460-015-9674-z

Pérez de Armiño, K. (2000). *Ayuda alimentaria y Desarrollo. Modalidades, criterios y tendencias.* HEGOA, University of the Basque Country (Spain).

Pérez de Armiño, K. (2014). 'Erosion of rights, uncritical solidarity and food banks in Spain'. In: Riches / Silvasti (eds.), *First world hunger revisited. Food charity or the right to food?*Basingstoke: Palgrave Macmillan, pp. 131-145.

Perry, J., Williams, M., Seffton, T. and Haddad, M. (2016). Emergency Use Only. Understanding and reducing the use of Food Banks in the UK. London, CPAG – OXFAM.

Poppendieck, J. (1999). *Sweet Charity: Emergency Food and the End of Entitlement.* New York: Penguin Books.

Riches, G. (2011). Thinking and acting outside the charitable food box: Hunger and the right to food in rich societies. *Development in Practice, 21*(4-5), 768-775.

Riches, G. and Silvasti, T. (2014b). 'Hunger in the rich world: food aid and right to food perspectives'. En G.Riches y T. Silvasti (eds.), *First world hunger revisited. Food charity or the right to food?* Basingstoke: Palgrave Macmillan, pp. 1-14.

Ronson, D. and Caraher, M. (2016). 'Food banks: big society or shunting yards? Successful failures', en Caharer/Doveney (eds.). *Food poverty and insecurity: international food inequalities. Food Policy.* London: Springer, pp. 79-88.

Schneider, F. (2013).'The evolution of food donation with respect to waste prevention'. *Waste Management, 33*(3), pp.755-763.

Tarasuk, V., and MacLean, H. (1990). The institutionalization of Food Banks in Canada: a public health concern. *Canadian Journal of Public Health,* 81, 331-332.

Van der Horst, H., Pascucci, S. and Bol, W. (2014). The 'dark side' of food banks? Exploring emotional responses of food bank receivers in the Netherlands. *British Food Journal,* 116 (9): 1506-1520. doi:http://dx.doi.org/10.1108/BFJ-02-2014-0081.

32. How should people eat according to the United Nations' 2030 Agenda for Sustainable Development?

H. Siipi[1] and M. Ahteensuu[2]*
[1]Philosophy Unit, 20014 University of Turku, Finland; [2]Research Career Unit, 20014 University of Turku, Finland; helsii@utu.fi

Abstract

The United Nations' '2030 Agenda for Sustainable Development' sets a goal for ending hunger, achieving food security, improving nutrition and promoting sustainable agriculture. The goal has various subgoals. In this paper we discuss the strengths and limitations the goal and its subgoals have in guiding people's food choices. We state that the multitude of the goals makes practical choices complex. Yet, multitude of goals is needed in order for the agenda to have a motivating power. However, some factors relevant for food choice seem to be missing from the subgoals listed. Finally, we challenge the traditional forms of dietary guidance and suggest that guidance can only serve the goals if its focus is shifted from food groups and particular foodstuffs to the whole food chain of a specific food item.

Keywords: sustainable development, food security, dietary guidance, ethics

Introduction

The United Nations' '2030 Agenda for Sustainable Development sets 17 Sustainable Development Goals which cover various issues from ending poverty, enhancing gender equality, safe cities, combating climate change and halting biodiversity loss. The second sustainable development goal reads as follows: '[e]nd hunger, achieve food security and improved nutrition and promote sustainable agriculture'. The second goal has eight subgoals which concern, for example, access to safe and nutritious food, increasing agricultural productivity, maintaining ecosystems, adaptation to climate change, maintaining genetic diversity of crop plants and farm animals, and securing incomes of small-scale farmers. Related to the second goal and its subgoals crucial questions emerge: What should people eat? What should they omit to eat?

In this paper we analyse, what kinds of recommendations regarding these two questions the second sustainable development goal and its subgoals give and are able to give. What kinds of challenges need to be met in translating the goal into practical eating guidance? In which way the goal can be useful and helpful regarding food related decision-making?

The questions are relevant as sustainable development goals are normative. They are action-guiding, meaning that they state an aim, a state of affair to be sought after. This ideal state of affairs should be used to formulate different policies, dietary guidelines and recommendations. The approach we take is philosophical, specifically that of analytic practical ethics. We believe that this kind of an analysis can be fruitful and provide added value to policy making and operationalization. Evaluating policy options and making choices that promote the second goal requires making comparisons between food someone gets and food that is enough for him or her (Cadieux and Blumberg 2014). Assessing what is enough is based on complicated and partly normative views regarding normal growth and weight, what is safe enough, and what counts as hunger, for example (on critical views regarding criteria for hunger in the millennium development goals, see Pogge, 2013, 2015). Further questions arise concerning the very idea of sustainable diets. According to Johnston *et al.* (2014), '[t]here does not yet exist an agreed on approach or tool to determine the level of sustainability of a diet or the trade-offs associated with any attempts

EurSafe 2018 – DOI 10.3920/978-90-8686-869-8_32, © Wageningen Academic Publishers 2018

or recommendations to increase the sustainability of a diet'. Similarly, Auestad and Fulgoni's (2015) review-based analysis found that methods, indicators and impact categories to evaluate sustainability of dietary patterns vary significantly. This said there appears to be extensive consensus on certain issues, for example, that reducing beef and dairy consumption together with avoiding food waste contribute to sustainability.

Multitude of goals

The second sustainable development goal brings together various different kinds of subgoals. In order for a diet to be according to the second sustainable development goal and its subgoals, it should, among other things, increase the incomes of small-scale food producers, help to maintain ecosystems, strengthen the capacity to adopt climate change, maintain genetic diversity, correct distortions in agricultural markets, be safe, nutritious and sufficient to its eater, and contribute to achieving food security. As can be seen from the following standard definition, this subgoal of food security alone consists of numerous components, which relate to quantity and quality of food as well as eaters' preferences:

> Food security exists when all people, at all times, have physical and economic access to sufficient, safe and nutritious food to meet their dietary needs and food preferences for an active and healthy life. (FAO, 1996)

With respect to the presented question regarding what people should and should not eat, this approach of multiple subgoals has both strengths and challenges.

Although seemingly independent, some of the subgoals may be interconnected in practice. Excessive meat consumption, for example, may be harmful from the point of view of various components including its impact on ecosystems, human health and sufficiency of food. Impacts on ecosystems and sufficiency of food are intertwined as food production is dependent on ecosystem factors such as quality of soil and amount of pollutants and pests. In short, the low negative environmental impact of a diet may contribute to food availability in terms of preserving the soil and land, which enables food production. In similar lines, avoiding over-harvesting and pollution is necessary for the future usage of fisheries. As Macdiarmid (2013: 14) notes, 'health can depend on the affordability and access to good quality of food, and the quality of the food can depend on the land and soil in which it is produced (i.e. environment)'.

Despite these practical connections between the components, including so many desirable features as subgoals has the implication that a particular food item may fare very well in regard to one subgoal, while not meeting another. For example, the nutritional evidence[1] to eat fish may conflict with environmental evidence to not to eat certain (commonly eaten) at-risk species (Lang and Barling, 2012). In similar lines, low fat dairy products can provide essential nutrients. However, the ecological impacts of dairy products are not very good and removing cream from dairy products creates a dilemma of food waste. Garnett (2014: 5) summarizes the situation: 'it is very unclear what such a [sustainable] diet might look like on the plate. It [the goal] also suggests that these multiple 'goods' are synergistic, when inevitably there will be trade-offs'.

As Garnet suggests, it is probable that there is no single diet that is the best alternative regarding all subgoals. Thus, at a practical level some subgoals have to be prioritized over others or at least a way for balancing between different components must be found. The trade-off may be, for instance, between the amount of food and its safety. Maximal safety standards may lead into wasting edible food, whereas boosting the amount and availability of food may sometimes happen at expense of safety. Thus, the goals

[1] 'Evidence' is here understood broadly to refer, roughly, to a science-based recommendation.

must in operationalization be accompanied with guidelines on how to balance between different trade-offs. For example, how to choose between a maximally healthy diet, another that maximally protects ecosystems, and a third one that serves both goals to some but less than maximal extent?

There are also questions of how to interpret the goals, i.e. what do they mean in practice? For example, how should the expressions 'sustainable agriculture', 'strengthening capacity for adaptation to climate change' and 'correct and prevent trade restrictions and distortions in world agricultural markets' be understood? At the moment, there are no generally accepted criteria for an agricultural practice being sustainable. Even though everybody seems to favour economic fairness, there is a wide disagreement on what it actually means in practice. In similar lines, decision-makers are far from agreeing on right measures regarding climate change. Even if agreement about interpretation of the goals were reached, disagreement regarding means for attaining the (sub)goals remains. There is much disagreement, for example, regarding the best ways of changing the food styles of people (see e.g. Korthals, 2015) and making the food economically accessible for them (Dilley and Boudreau, 2001).

However, on the other hand, bringing multiple factors together is good from the point of view of actually bringing about a change in food systems and in food style choices of individuals. The issue of dietary choice is not simple. Rather the food-related decisions have to take into account various factors including the ones presented above. Guidelines that fail on this may also fail to motivate people. As Korthals (2017: 419) notes, people are not motivated to follow guidelines that only acknowledge one aspect of food (e.g. healthiness) and which disregard important information (e.g. environmental information). But are all factors relevant to food choice included in the second sustainable development goal and its subgoals?

What is left out?

Even though the second sustainable development goal covers a wide range of relevant issues, some things that may for some people be relevant for food choice are left out. Taste and other aesthetic factors are not mentioned beyond the food security definition's requirement of food being culturally acceptable. The cultural acceptability is usually interpreted as a requirement of individuals not being forced to eat foods that are against their cultural or religious views. A Muslim society, for example, should not be considered food secure if the only protein source available is pork (Pinstrup-Andersen, 2009: 5-6; Siipi, 2015: 195-196). However, the issue of cultural acceptability may also be understood in a broader sense as a requirement of not adopting morally problematic practices against food cultures of others. Moreover, it might be taken to include food sovereignty, which FAO defines as follows:

> Food sovereignty is the right of peoples to define their own food agriculture; to protect and regulate domestic agricultural production and trade in order to achieve sustainable development objectives; to determine the extent to which they want to be self-reliant. (FAO, 2005)

The second sustainable development goal mentions the need for extra attention regarding vulnerable groups (such as women and indigenous people) and takes a stand on equal access to land, resources and knowledge. The goal also calls for fair food markets. Yet, some issues related to justice are missing. These include some non-economic fairness aspects such as misrecognition (e.g. exclusion, cultural subordination and disrespect) and lack of participation possibilities. Misrecognition, which draws the attention to 'deeper' processes and institutionalized practices that bias distribution or outcomes, has often been neglected also in the traditional justice theorizing (Fraser, 2009). Are the food systems compatible with the second sustainable development goal fair regarding individuals and families' possibilities to retain their food culture and make food-related life style choices? As Sandler (2015) puts it, 'food policies should not presume one culture or religion'. As an example, he discusses the US GM-food labelling policy which is quite ignorant of the fact that some people want to avoid GM-food for reasons that are not health-related. Most consumers today rely on large food companies and supermarkets for their food. These companies as well as politicians may make food policy choices without hearing the marginalized groups who can sometimes be most seriously affected by the decisions (Thompson, 2015). Injustice can also take the form of using cultural knowledge without permission and/or appropriate compensation (Heldke, 2003; Sandler, 2015; see also Lang and Barling 2012).

The concerns about treatment of animals in food production are also missing from the second goal. At least in theory, a food system causing considerable suffering to animals might fulfil the goals. The issue of animal suffering is strikingly relevant, since animal suffering in food production often results from making the process more efficient. For example, fast growth of broiler chickens causes them painful deformations and growing huge amount of pigs in small space causes stress to the animals. Efficiency of animal production serves many of the presented subgoals such as food sufficiency as well as increasing productivity and incomes of small-scale food producers. The issue cannot at a practical level be handled merely by listing food items that cause and do not cause suffering. Eggs and dairy products, for example, are sometimes produced in ways which cause considerable suffering to animals, but it is also possible to produce them in more humane (but probably less efficient) ways (for critical view, see Višak, 2013).

How should the questions be answered? Should we talk about food stuffs and food products?

Generally speaking, the second sustainable development goal and its subgoals are rather abstract and do not directly tell consumers and decision-makers what should and should not be eaten. Yet, at the same time, they seem to be and seem to be meant to be relevant for food policies and dietary guidance. Traditionally, dietary guidance and dietary policies have been based on food groups (e.g. vegetables, cereals, meat, fish, dairy, etc.) (see e.g. Macdiarmid *et al.*, 2011). If the goals are to be met (even partly), major changes in dietary guidance may be needed (cf. Merrigan *et al.*, 2015). The focus has to be shifted from food groups (e.g. meat, cereals) and particular foodstuffs (e.g. chicken, beans) to the whole food chain (from farming, to transport, to wholesale and retail, to cooking, and to waste disposal) of a specific food item. Ileana F. Szymanski (2016: 12) goes as far as saying that the term 'food' should apply not to items but to networks that create food. Producing an item at different times, in different ways and at different places may make a difference to its environmental impact (see Macdiarmid, 2013 on greenhouse gas emissions) as well as to its healthiness (e.g. through pesticides used) and economic fairness. There may be a huge difference in these respects, for example, between tomatoes produced in a greenhouse in Finland in winter and the ones growing in the field in Spain in summer time. Similarly, ways of cocoa production (including the labour and transport issues) matter greatly to the question of whether a particular chocolate bar can be taken as being in accordance with the second sustainable development goal.

In order for consumers to be able to contribute to the realization of the sustainable development goals by their own daily dietary choices, impacts of different food choices must be somehow communicated to them. In similar lines, political decision-makers as well as the food industry needs information about the impacts different foods have on the sustainable development goals. Since the choice may not be made simply on the basis of food item types (e.g. tomato or chocolate), other solutions must be found. Baggage labels (such as fair trade and carbon footprint labels) may provide part of the solution. Yet, they often concern only one aspect mentioned in the goals. As pointed out by Korthals (2017) at its best food guidance might stimulate people to acquire critical capabilities for making their food choices. By these capabilities they could contribute to realization of sustainable development goals not only by following labels and recommendations but by critically reflecting the information at hand and making their own assessments of different kinds of influences of their food choices.

Conclusions and some suggestions

The presented challenges do not, of course, imply sustainable development goals being useless or insignificant. Rather we believe that they are important and have the potential to make the world a better place. Yet, as presented above, there are challenges regarding the implementation of the goals into practical choices regarding food. This is due to the multitude of subgoals some of which are quite abstract and leave room for different interpretations. Moreover, since few foods are likely to serve all the goals set, priorities and trade-offs between different subgoals should be further discussed and clarified. Finally, reaching the goals seems to necessitate changes in the ways the dietary guidelines and food policies are usually presented. Instead of focusing on types of foodstuffs (e.g. tomato, sugar), the focus should be more on the whole food chain from field to fork – and even beyond to food waste (Thompson, 2015).

We do not have a recipe to solve the challenges presented. However, we have a number of suggestions for further discussion. First, consumers and decision-makers should be informed about easy cases. For example, meat and dairy produced at factory farms is unlikely to contribute to the goals, whereas vegetables grown outdoors by a local small-scale farmer are likely to serve the goals. Second, people should become aware of the goals and they should be given concrete possibilities to act on the goals, not only by their purchasing choices but also in other ways. Could, for example, hobby gardeners contribute to maintaining diversity of seeds and plants (and maybe even some farmed animals)? Might it be possible to find ways in which consumers (both individuals and ones making purchases for schools and senior homes, for example) could easily support food producers of vulnerable groups?

References

Auestad, N. and Fulgoni III, V.L. (2015). What Current Literature Tells Us about Sustainable Diets: Emerging Research Linking Dietary Patterns, Environmental Sustainability, and Economics, Advances in Nutrition: An International Review Journal 6: 19-36.

Cadieux, K.V. and Blumber, R. (2014). Food Security on System Context. In: Thompson, P. B., Kaplan, D. (eds.) Encyclopedia of Food and Agricultural Ethics. Springer.

Dilley, M. and Boudreau, T.E. (2001). Coming to Terms with Vulnerability: A Critique of the Food Security Definition, Food Policy 26: 229-247.

Fraser, N. (2009). Scales of Justice: Reimagining Political Space in a Globalizing World. Columbia University Press.

Food and Agriculture Organization of United Nations (2005). The State of Food and Agriculture. Available at: http://www.fao.org/docrep/008/a0050e/a0050e00.htm. Accessed 26 April 2017.

Food and Agriculture Organization of United Nations (2012). Sustainable Diets and Biodiversity: Directions and Solutions for Policy, Research and Action (editors: Burlingame, B., Dernini, S., Nutrition and Consumer Protection Division & FAO). Available at: http://www.fao.org/docrep/016/i3004e/i3004e.pdf. Accessed 25 April 2017.

Food and Agriculture Organization of United Nations (1996). Rome Declaration on World Food Security and World Food Summit Plan for Action. Available at: http://www.fao.org/docrep/003/w3613e/w3613e00.HTM. Accessed 25 April 2017.

Garnett, T. (2014). What is a Sustainable Healthy Diet: A Discussion Paper. Food Climate Research Network.

Heldke, L. (2003). Exotic Appetites: Ruminations of a Food Adventurer. Routledge.

Johnston, J.L., Fanzo, J.C. and Cogill, B. (2014). Understanding Sustainable Diets: A Descriptive Analysis of the Determinants and Processes That Influence Diets and Their Impact on Health, Food Security, and Environmental Sustainability, Advances in Nutrition: An International Review Journal 5: 418-429.

Korthals, M. (2015). Which Forms of Nudging Food Decisions Are Ethically Acceptable?. In: Dumitras, D. E., Jitea, I. E. and Aerts, S. (eds.) Know Your Food: Food Ethics and Innovation. Wageningen Academic Publishers.

Korthals, M. (2017). Ethics of Dietary Guidelines: Nutrients, Processes and Meals. Journal of Agricultural and Environmental Ethics 30: 413-421.

Macdiarmid, J.I. (2013). Is a Healthy Diet and Environmentally Sustainable Diet?, Proceedings of the Nutrition Society 72: 13-20.

Macdiarmid, J.I., Kyle, J., Horgan, G., Loe, J., Fyfe, C., Johnstone, A. and McNeill, G. (2011). Liverwell: A Balance of Healthy and Sustainable Food Choices. Available at: http://assets.wwf.org.uk/downloads/livewell_report_jan11.pdf. Accessed 26 April 2017.

Merrigan, K. *et al.* (2015). Designing a Sustainable Diet: Sustainability as Dietary Guidance Created Political Debate, Science 350 (9th Oct.): 165-166.

Pinstrup-Andersen, P. (2009). Food Security: Definition and Measurement, Food Security 1: 5-7.

Pogge, T. (2013). Powerty, Hunger and Cosmetic Progress. In: M. Langford, Sumner, A. & Yamin, A. E. The Millenium Development Goals and Human Rights: Past, Present and Future. Oxford University Press.

Pogge, T. (2015). The Hunger Games (December 15, 2015). Available at: SSRN: https://ssrn.com/abstract=2823609 or http://dx.doi.org/10.2139/ssrn.2823609.

Sandler, R.L. (2015). Food Ethics: The Basics. Routledge.

Siipi, H. (2015). Requirements of Safety and Acceptability in Food Security Definitions. In: Dumitras, D. E., Jitea, I. E. and Aerts, S. (eds.) Know Your Food: Food Ethics and Innovation. Wageningen Academic Publishers.

Szymanski, I. (2017). What Is Food? Networks, Not Commodities. The Routledge Handbook of Food Ethics, edited by Rawlinson, M. C. and Ward, C., Routledge, London.

Thompson, P.B. (2015). From Field to Fork: Food Ethics for Everyone. Oxford.

Višak, T. (2013). Killing Happy Animals: Explorations in Utilitarian Ethics. Palgrave MacMillan.

33. Sustainability, ethics, and politics: NGOs' advocacy discourses on anti-GM food

Y.C. Chiu and F.Y. Li*
Department of Bio-industry Communication and Development, National Taiwan University, No. 1, Sec. 4, Roosevelt Rd., Taipei 10617, Taiwan (R.O.C.); ychiu@ntu.edu.tw

Abstract

The potential benefits and harms of genetically modified food constitute a topic that inspires debate. Controversies over GM food reflect an interrelationship between social, political, and corporate forces. Among the societal agents disseminating information on GM food, NGOs' discourses play a crucial role in constructing perspectives and building agendas. The present study explored NGOs' discourses to determine how GM food is represented in Taiwanese society. We analyzed the three major NGOs in this field: Homemakers United Foundation, Homemakers Union Consumers Co-op, and No-GMO-Lunch. By collecting their pamphlets, books, publications, and website posts related to GM food from 1998 through mid-2017, a total of 482 articles underwent qualitative inductive analysis. The results showed that in addition to addressing uncertainties about health risks, claims that GM foods harm society on a worldwide scale are related to sustainability, ethics, and politics. The analyzed NGOs have attempted to promote consciousness of environmental sustainability among farmers and the general public by emphasizing that planting GM food could destroy the ecological balance and be detrimental to environmental justice. Consumption ethics has also been emphasized by encouraging people to refuse to buy or eat GM food. More crucially, in terms of political concerns, the analyzed NGOs have framed the development and implications of GM food as results of collusion within the 'government-scientist-enterprise complex.' Furthermore, war metaphors have been used to describe how by importing GM food, the government had retreated from the defensive line of guarding public health. Discourse is a strategic resource by which NGOs can empower farmers and the public and contend with governments and enterprises. By using specific phrasing and metaphors and introducing scientific terminology, NGOs can act as meaning architects and knowledge brokers to potentially increase their influence over how topics are defined and reach the level of legitimacy required to further participate in the policymaking arena.

Keywords: genetically modified (GM) food, non-governmental organization (NGO), discourse, Taiwan

Introduction

How genetically modified organisms (GMOs) may benefit or damage society and individual health is a controversial topic. The controversy over GM food reflects an interrelationship between social, political and perhaps corporate forces embedded in society's reactions to GM food technology. Among all agents in society disseminating information on GM food, the discourse of non-governmental organizations (NGOs) plays a crucial role in constructing perspectives and building agendas (Gemmill and Bamidele-Izu, 2002). Using various strategies, NGOs may influence governmental policy and practice, and reveal corporate behaviour and their otherwise obscure relationship with governments (Finger and Princen, 1994).

In Europe, NGOs' viewpoints on issues regarding GMOs include sustainable development, consumer protection, farmer welfare, and ethical concerns (Finger and Princen, 1994). For example, NGOs in France challenge both the risk assessments and policies related to GMO ignoring uncertain impacts on

EurSafe 2018 – DOI 10.3920/978-90-8686-869-8_33,

ecology (Roy and Joly, 2000). Furthermore, NGOs in Holland have raised ethical concerns over GMO safety control (Purdue, 2000; Schurman, 2004). These examples indicate that NGOs can successfully redefine the value of technology and spotlight undesirable effects of GMOs in society (Bauer and Gaskell, 2002).

In addition to focusing on GMOs' negative impact on society, another major strategy adopted by anti-GM food NGOs is to campaign against manufacturers. Researchers have indicated that NGOs' advocacy campaigns have targeted several international companies such as Monsanto, Avrnyid, Syngenta, BSAF and Dupont, all primary producers of genetically engineered seeds (Finger and Princen, 1994). They delegitimize and reveal problems with these corporations. The influence of NGOs over firms is indirect, as they attempt to influence critical players who have some influence over the firm's economic outcomes, such as politicians, voters, consumers, suppliers, and shareholders. NGOs achieve this by calling upon their interests, roles and responsibilities, values and norms. Moreover, NGOs also attempt to set agendas in the media which have an indirect influence over critical players (Finger and Princen, 1994).

In Taiwan, GMOs are all imported because planting is forbidden (Council of Agriculture, 2015). Most of the GMOs imported to Taiwan are soybeans, which are major ingredients for many popular daily foods, such as soybean milk and Tofu (Food and Drug Administration, 2014). Therefore, it is interesting to investigate how NGOs construct the problems of GM food in Taiwan, where GMOs are major ingredients in daily food and must be imported from foreign countries.

In this study, we focused on analysing the discourses of NGOs to explore how they frame the disputes over GM food in Taiwan. We view discourse as a strategic resource and a social and linguistic construction (Hardy *et al.*, 2000). This approach considers that social reality is shaped by different groups who compete to have their perspectives heard and recognized in ways that serve their own interests or purposes. Strategy is a particular rhetoric that provides a common language to determine, justify, and give meaning to the actions that an organization performs (Eccles and Nohria, 1993; Hardy *et al.*, 2000).

The aims of this study were to analyse NGOs' discourse to explore how NGOs frame GM food and how these allegations reflect their values regarding GMOs and their manufacturers in Taiwan.

Method

We analysed three major anti-GM NGOs in Taiwan: (1) Homemakers United Foundation, (2) Homemakers Union Consumers Co-op, and (3) No-GMO-Lunch. Their pamphlets, publications, and website posts related to GM food were collected from 1998, when the first article related to GM food appeared, through to July 31, 2017. A total of 482 articles were analysed by a qualitative inductive approach to inspect the discourses (Patton, 2005). Two authors read all articles independently, with research questions in mind. We specifically scrutinized the language, words, and sentences used to delineate GM food (Gee, 2014). We discussed once a week what we have learned from the collected articles and collaborated on developing major themes grounded in the texts (Patton, 2005). Three major themes finalized were sustainability, ethics, and politics.

Results

In addition to addressing the uncertain health risks, three main themes that NGOs constructed concerning the harm to society caused by GM food and the wider world were related to sustainability, ethics, and politics. The following paragraphs demonstrate these three themes and their changes through the decades.

Sustainability: evoke farmers' consciousness

In the beginning of their anti-GMO discourses, NGOs attempted to evoke farmers' and people's consciousness about environmental sustainability by emphasizing that planting GM food will destroy the ecological balance and act against environmental justice. NGOs illustrated broad environmental sustainability concerns.

> Environmental and ecological impacts result from large-scale cultivation of GM crops. The risks include the following: (1) new substances in GM crops may kill the predators of pests and the parasites, ..., (2) GM crops have been shown to have toxic effects on animals... (3) GM crops may change the biological balance in the soil, ..., (4) GM crops may cause gene flow, ... (a2000090001)

Later, NGOs further indicated that the current research about the impact of GM crops on the environment still cannot be conclusive. In response to the uncertain effects of GM crops, NGOs promoted 'GMO-free zones' campaigns, and encouraged farmers to guard the sustainability of agriculture and environment.

> People who support GM said that there is no evidence to show that GM crops are harmful. However, the development of GM technology has spanned only 13 years, so it is difficult to provide any long-term research data to prove the impact of GM crops on people's health and environment. (a2008100001)

> GMO-free zones were established to safeguard food ecology, farmers' wisdom and rights, and protect sustainability in agricultural systems. (a2009040001)

Ethics: unveil 'playing God' of the technology

One of the aspects of GM food that NGOs focused on was the ethics of GM technology. They explicitly questioned the application of GM technology as just like 'playing God'. NGOs criticize scientists who support GM technology as only considering the development of science and technology, but ignoring issues related to value, justice, and ethics.

> Evolution is a lengthy process. Slight changes in the gene structure are the results of a creature's adaptation to its environment. However, scientists who conducted genetic engineering disturbed natural regulation. (a2000070001)

> With the support of technology, humans have controlled and wantonly changed nature. Only time will tell if human beings have to pay a price for GM technology application. (a2000110001)

> When you eat GM foods this means that you agree with the values related to herbicide overuse, environmental pollution, and transnational biotech corporations controlling the global food supply. However, experts who support GM technology have always said that discussion of the scientific truth is enough. Values and social justice are ignored. (c2015021002)

The above excerpts indicate that NGO discourses not only advocated that the possible impact of GM technology on ecology was unethical, but also condemned scientists' integrity. In addition, consumption ethics was also stressed by advocating that people 'refuse to buy' and 'refuse to eat' GM food. NGOs emphasized this consumption practice not only for health but also for recognizing the value of sustainability and ethics. They empowered people's rights to knowing and choosing their food.

> As a consumer, you shall begin to pay attention to food labels, select non-GMO food, and change the world with consumer's power. (a2009020003)

> No matter whether supporting GM food or not, consumers should have the right to choose food. (c2014121301)

Politics: the collusion between government and enterprises

More importantly, in terms of political issues, NGOs used discourses to delegitimize international biotech corporations. NGOs allege that the development and implications of GM food in a society is the result of a collusion between governments and enterprises. Furthermore, the war metaphor was used to describe that in international trade, the government retreated from the defence line of guarding public health by importing GM food.

> The U.S. government often put pressure on European countries and Canada to ask the government for fewer controls and restrictions on GM crops.... However, it's not surprising that the government acts in concert with international biotech corporations because the latter always provide political donations to support presidential candidates. (a2000090001)

In recent years, NGO discourses have suggested that the issue of GM food import is not only related to the environment or health, but also to complicated political issues, such as international relations and trade exchanges. Given the difficulties owing to Taiwan's weak international status, to maintain its relationship with the United States, the government could only agree to import GM soybeans.

> To respond to the U.S. government's threat in the name of free trade, although Taiwan is often in a weaker position on the negotiating table, the government needs to take the unique eating habits of Taiwanese into trade negotiation and policy making consideration, and cannot just yield to another country's standards.... The next food security defence we are forced to abandon may be the regulation of GM food. (c2016050901)

NGO discourses also frame international biotech corporations as evil companies. The enterprises ignored the impact of GM crops on biodiversity and the environment, and used patent rights protection as an excuse to prohibit farmers from saving seeds. Moreover, the enterprises claimed that GM crops can feed and save the world from hunger. However, the NGOs uncovered their lies, and pointed out corporations were instead the culprits of food shortages.

> GM crops are not a sustainable way to feed the world. On the contrary, only few international GM enterprises are fed fully. (c2015081201)

In NGO discourses, scientists were framed as a player in the government-enterprise collusion. The discourse pointed out that what people believe to be objective scientific research may not be neutral but is rather manipulated. NGOs cited the research of independent scientists to reflect that the experimental results presented by the 'government-scientists-enterprises complex' were not credible.

> The belief that science is the top priority is problematic and extremely fragile. The information that we were brainwashed with is obviously the result of the collusion of

enterprises, governments, and the academic society. Whether GM crops and foods are big swindles or not? It seems highly possible now. (c2015081201)

Discussion

Discourse is a strategic resource for NGOs to empower farmers and the public, and to contend with government, and enterprises together. By using language and metaphors, and introducing scientific terminology, NGOs act as meaning architects and knowledge brokers, which may increase their influence to define the issue and gain legitimate power to further participate in the policymaking arena.

By analysing NGO discourses, we find that NGOs' major allegations on GM food relate to three main themes: sustainability, ethics, and politics. In addition, along with their own learning, experiences, and the changing social environment, NGO discourses against GM food have also transformed. In the beginning, NGOs were more focused on sustainability and increasing farmers' consciousness about the potential harm GMOs could cause. However, in recent years, they have focused on more ethical and political aspects. The discourse also empowers consumers and citizens to take political action and disclose the suspicious collusion between the government and industry.

By analysing NGO discourses, this study demonstrates that without traditional power resources, NGOs use discourse as a resource to attract media, set agendas, and more importantly to transfer scientific terminology into understandable lay language, describing the problematic issues of GM food for the public. Discourse is a strategic means that NGOs use to struggle with government, business, and scientists in the public communication arena.

This study indicates that, to some extent, discourse reflects the wider relations of power and that it is important in maintaining this power structure (Ebrahim, 2001). For the institutes that have less orthodox power and less political and economic resources, such as the anti-GMO NGOs considered in this study, discourse indeed is their significant resource and strategy to participate in this power battle.

References

Bauer, M. and Gaskell, G. (2002). Researching the public sphere of biotechnology. In: Gaskell, G. and Bauer, M. (eds.) Biotechnology: the making of a global controversy. Cambridge University Press, Cambridge, USA, pp. 1-17.

Council of Agriculture (2015). Taiwan has not yet approved for genetically modified crops planting. Available at: https://www.coa.gov.tw/theme_data.php?theme=news&sub_theme=agri&id=5134&RWD_mode=N. Accessed 5 January 2018.

Ebrahim, A. (2001). NGO behaviour and development discourse: cases from western India. Voluntas: International Journal of Voluntary and Nonprofit Organizations 12(2): 79-101.

Eccles, R. and Nohria, N. (1993). Beyond the hype. Harvard Business School Press, Boston, USA.

Finger, M. and Princen, T. (1994). Environmental NGOs in world politics: Linking the local and the global. Routledge, London, New York, 262pp.

Food and Drug Administration (2014). Genetically modified organisms approved for import in Taiwan. Available at: https://www.fda.gov.tw/tc/siteContent.aspx?sid=3976. Accessed 5 January 2018.

Gee, J. P. (2011). An introduction to discourse analysis: theory and method. Routledge, New York, USA, 218pp.

Gemmill, B. and Bamidele-Izu, A. (2002). The role of NGOs and civil society in global environmental governance. In: Esty, D.C. and Ivanova, M.H. (eds.) Global environmental governance: options and opportunities. Yale Center for Environmental Law and Policy, New Haven, USA, pp.77-100.

Hardy, C. and Palmer, I. and Phillips, N. (2000). Discourse as a strategic resource. Human relations 53(9): 1227-1248.

Patton, M.Q. (2005). Qualitative research. John Wiley & Sons, Ltd, New York, USA.

Purdue, D.A. (2000). Anti-genetix: the emergence of the anti-GM movement. Ashgate, Aldershot, UK, 161 pp.

Roy, A. and Joly, P.B. (2000). France: broadening precautionary expertise? Journal of Risk Research 3(3): 247-254.
Schurman, R. (2004). Fighting 'Frankenfoods': industry opportunity structures and the efficacy of the anti-biotech movement in Western Europe. Social Problems 51(2): 243-268.

34. Technology neutrality and regulation of agricultural biotechnology

P. Sandin[1], C. Munthe[2] and K. Edvardsson Björnberg[3]*
[1]Dept of Crop Production Ecology, Swedish University of Agricultural Sciences, P.O. Box 7043, 75007 Uppsala, Sweden; [2]Dept of Philosophy, Linguistics and Theory of Science, University of Gothenburg, P.O Box 200, 40530 Göteborg, Sweden; [3]Dept of Philosophy and History, KTH Royal Institute of Technology, Teknikringen 76, 100 44 Stockholm, Sweden; per.sandin@slu.se

Abstract

Agricultural biotechnology, in particular genetically modified organisms (GMOs), is subject to regulation in many areas of the world, not least in the European Union (EU). A number of authors have argued that those regulatory processes are unfair, costly, and slow and that regulation therefore should move in the direction of increased 'technology neutrality'. The issue is becoming more pressing, especially since new biotechnologies such as CRISPR increasingly blur the regulatory distinction between GMOs and non-GMOs. This paper offers a definition of technology neutrality, uses the EU GMO regulation as a starting point for exploring technology neutrality, and presents distinctions between variants of the call for technology neutral GMO regulation in the EU.

Keywords: GMO, EU, law, ethics

Introduction

Societies have a need to oversee, guide, and regulate the development, dissemination and use of technology for reasons of protection of cherished values. Such protective policies might be characterized as hard or soft. Soft ones include measures such as voluntary moratoria, as for instance implemented for transgenic organisms in the wake of the Asilomar conferences (Berg, 2008), various guidelines and industry codes of conduct, and the more recent calls for caution regarding human germ-line applications of gene editing and gene driving (Cicerone *et al.*, 2015). Hard ones include outright bans, obligatory pre-market approval processes and mandatory product standards. Many examples of the latter can be found in the European Union (EU) regulation on genetically modified organisms (GMO). In recent years, such policies have been attracting criticism of a certain kind, questioning a combination of inconsistency and inflexibility of common hard policy solutions. It has been pointed out that similar types of risks and uncertainties of different types of technology are often treated very differently, since different policy solutions are applied to different technology types, also within the same policy area, such as agriculture. It has also been pointed out that the hard policy solutions are often tardy to adapt to changes due to technological and scientific advances. This latter phenomenon moreover tends to increase the unequal policy treatment of different technologies. For this reason, calls to make hard protective technology policy 'technology neutral' have been heard, for instance in the areas of climate change technology and biotechnology, in turn provoking critical reactions regarding feasibility (Azar and Sandén, 2011, Munthe, 2017).

What is technology neutrality?

In this paper, we are referring to the notion of technology neutrality as a feature of regulatory structures. Such structures include actual legal statutes, case law, instructions and decrees for public agencies based in these statutes, routines designed within such agencies to comply with said instructions, and orders

EurSafe 2018 – DOI 10.3920/978-90-8686-869-8_34, © Wageningen Academic Publishers 2018

to parties given by agencies. Following Greenberg (2016), we accept the conceptual basis that such neutrality is seen as opposed to specificity.

It is never technologies *per se* that are regulated, but actions by agents that make some sort of use of a technology. The notion of technology neutrality and specificity of regulatory structures also has to be understood in scalar rather than binary terms: there may be more or less of it. Technology neutrality can thus be thought of as situated on a continuum with 'full technology neutrality' and 'full technology specificity' as the opposing (ideal) end points. Moreover, the degree of technology neutrality must be understood as relative. One aspect of this relativity regards *what part* of a regulatory structure we focus on (general statute, derived agency decrees, or case rulings, for instance). Another aspect has to do with how technologies are individuated, and how that relates to the current state of technological development. Take vehicles. There is no radical difference in function between a small pick-up truck and a large truck. However, the quantitative difference in size and weight may be perceived to matter from certain regulatory standpoints, so that in order to operate the larger, heavier vehicle, a different type of driver's license is required. At the same time, the bottom line of this regulatory framework will be common to all types of trucks: the rule that an appropriate license is required to drive them. Moreover, the rationale behind the regulation also applies equally, typically expressed in statute in terms of requirements for different types of licenses having to ensure some standard of reliable and safe driving ability. These latter aspects of the regulatory structure will be less specific (and more neutral) than each of the license requirements for the different type of trucks.

Based on these considerations, we will in the following apply the following working criterion of technology neutrality:

A regulatory structure, S, for a technology, T_1, is more technology neutral to the extent that

a. other technologies, T_2 ... T_n, that are similar to T_1 in terms of the rationale behind S, are also subjected to S, and
b. S applies more similar rules to T_1 and T_2 ... T_n when this similarity is greater

In addition, S for T_1, is more technology neutral to the extent that

c. other technologies, T_2-T_n, that are similar to T_1 in terms of the rationale behind S, are subjected to some other regulatory structure, S', that applies more similar rules to T_1 and T_2 ... T_n when this similarity is greater.

It is implied that technology neutrality decreases, and technology specificity increases, when these features are less present.

When we consider some regulatory structure aimed at protecting cherished values, its rationale will always consist of a combination of such values, and a normative idea of the importance of protecting them (both determined politically).

Case: The EU regulatory system on GMOs

The legislative work on biotechnology in plant breeding has developed along two parallel tracks (Zetterberg and Edvardsson Björnberg, 2017). One track is commonly referred to as 'process-based', which means that the technology of genetic modification is used as a trigger for regulatory oversight. This track is followed by the EU and its Member States. According to EU Directive 2001/18/EC on the deliberate release into the environment of genetically modified organisms and Regulation 2003/1829 on genetically modified food and feed, a genetically modified variety may be released into the environment or put on the European market only if it satisfies a set of licensing requirements. This is to be contrasted with 'product-based' legislation according to which the organism's traits, regardless of how they were obtained, in combination with the environment into which the organism will be introduced, determine which legal demands must be met in order for a release permit to be granted. This legislative track is followed by the United States and Canada, among other countries (McHughen and Smyth, 2008; Macdonald, 2014). The EU GMO legislation can be said to be more technology specific, since it singles out new crops developed by use of a certain technology (genetic modification) and subjects them to more stringent licensing requirements than other crops, while the US and Canadian legislations are in various degrees more technology neutral.

Proponents of agricultural biotechnology often argue that current EU regulatory processes are unfair, costly, and slow (Masip *et al.*, 2013). In a typical statement, Eriksson and Ammann (2017) talk about 'the regulatory discrepancy between the relatively unregulated so called conventional breeding techniques and the overregulated transgenic techniques' (Eriksson and Amman 2007, p. 1). They hereby imply a case for unfairness, due to a perceived violation of the basic requirement of justice that similar cases should be treated uniformly, and that differences in treatment must be justified. Critics of current regulation argue in this way by claiming that the distinction between genetically modified organisms and non-GM organisms is 'meaningless' (Ricroch *et al.*, 2016). One particularly fierce critic, Tagliabue (2016), calls GMO an 'inconsistent term', an 'inchoherent expression [which] is arbitrary' and a 'bogus concept', which is 'illogical' – all accusations fitting within one paragraph. However, also parties who are opposed to the extended use of industrial agricultural biotechnology have voiced dissatisfaction with the GMO regulation of the EU when realizing that it may not apply to most biotechnology as the methods used are moving from old school hybrid DNA technology to the latest 'gene editing' technology of CRISPR (Kim and Kim, 2016). At the time of writing, the issue remains undecided (Abbott, 2018). This criticism of the EU regulation of GMOs has a strong tendency to mix rather different messages, namely complaints about inconsistent standards and complaints about the level of regulatory requirements. A requirement of coherence in EU law (Winter, 2016) can be used as a basis for requiring that similar regulatory arrangements apply to relevantly similar technologies. But that does not answer either what makes different technologies relevantly similar or what level of regulatory force should be thus equally applied across some set of technologies. We note that the biotechnology advocates apparently would like to see regulatory force to be uniformly relaxed, while biotechnological sceptics desire the opposite move. We may thus distinguish the following variants of the call for technology neutral GMO regulation in Europe:

- TN1: Agricultural technology regulation in the EU should become more technology neutral.
- TN2: Non-GMO biotechnology should be subject to the same level of regulatory force as GMO technology in current regulation.
- TN3: GMO technology should be subject to the same level of regulatory force as non-GMO biotechnology in current regulation.
- TN4: Both GMO technology and non-GMO biotechnology should be subject to more forceful regulation than what is the case in current GMO regulation.
- TN5: Both GMO technology and non-GMO biotechnology should be subject to less forceful regulation than what is the case in current non-GMO biotechnology regulation

TN1 is consistent with each of TN2-TN5, but all of these are, in turn, mutually incompatible. It is obvious that the standard arguments summarized above do not settle the question to what extent T1 enjoys support, and if this support in that case speaks for any of TN2-TN5.

Discussions about technology neutral regulation often revolve around tensions between the effectiveness of regulatory structures and generic qualities thought to be required by legal systems to be justified. One central feature of a good legal system in both of these respects is the system's ability to regulate actions and behaviour over time and among agents in a consistent way. This feature is captured by the requirement of 'legal certainty'. Legal certainty means that the regulatory system is precise and understandable and that the implications of the system can be foreseen by those to whom it applies. As noted above, it has been argued that the present EU process-based regulatory system for GMOs fails to meet the requirement of legal certainty, since it is unclear whether plants that have been developed through genome editing fall under the regulation or not. More technology-specific regulatory systems are generally more vulnerable to this type of objection than more technology-neutral systems, especially in areas where there is rapid advancement in science and technology and where the boundaries of the technology are in need of continuous redefinition as a consequence of this. This means that the classification of technologies in relation to regulatory structures is critical for ascertaining its legal certainty. If technologies are classified as relevantly similar, they should be regulated similarly. If not, it is easier to justify differences in regulatory requirements. But what should be considered as relevant similarity? Here, it becomes necessary to go beyond the legal structure under consideration and look to underlying assertions thought to justify it. These may be more basic legal rules (such as principles found in constitutions or international conventions), but ultimately they will as a rule force us to consider the basic ethical rationale of law, legal systems and the considered areas of regulation, in our case, GMO. This brings us to moral philosophy (Gregorowius *et al.,* 2012).

Technology neutrality and ethics: a way forward?

We can broadly distinguish between 'consequentialist' and 'non-consequentialist' concerns that are invoked to justify different variants of technology neutrality or technology specificity, depending on what value the regulation intends to protect. Of course, a particular piece of regulation need not be justified solely with regard to one such concern. Multiple ways of justifying a given regulatory item are possible.

Both technology neutrality and technology specificity can be justified on straightforwardly consequentialist grounds depending on circumstances. A decidedly technology-specific policy (say, subsidies for vehicles using a particular kind of fuel) can be argued for on the basis that such a policy contributes to technological development that would otherwise never get off the ground due to high initial costs, low demand, and so on. However, many standard ways of arguing for technology neutrality are consequentialist. Typical examples include Tagliabue's (2018) call for 'product, not process'. Such arguments state that a particular technology ought to be regulated since doing so, or not doing so, will have certain desirable or undesirable consequences. Thus, if different technologies would have similar regulatory consequences, they should be treated similarly. If the (valid consequentialist) rationale of a regulatory structure for some specific technology is satisfied by some other technology, this would therefore seem to provide a *pro tanto* reason for subjecting the latter technology to similar regulation as the former. That is, they should be classified as relevantly similar from a legal standpoint. In terms of our criteria for technology neutrality, this, in turn, would seem to be a case of calling for more technology neutrality rather than less.

Another familiar ethical stance is that certain technologies are subject to strong non-consequentialist considerations. Ronald Sandler has argued that agricultural biotechnology ought to be resisted, since it is contrary to some virtues such as 'humility', which human beings ought to display in their relationship with nature (Sandler, 2004). In a similar vein, there are familiar objections to modern biotechnologies in the form of 'playing God' or 'unnaturalness' arguments (Siipi, 2015). They express the notion that certain qualitative features of a sub-class of technologies make them subject to some strong moral restrictions and that, therefore, there should be specific regulation for this sub-class, in spite of the fact that the technologies may be very similar to other technologies to which other regulatory solutions apply.

Autonomy and freedom provide a further ethical base that may be played out in either a consequentialist or a non-consequentialist version. It is interesting here because of its role to support regulation regarding very specific uses of biotechnology, but not others, namely commercial retail of goods containing some sort of biotechnologically manufactured element (such as a GMO). The idea is that producers and retailers should be legally forced to provide information to facilitate informed consumer choice on the basis of preferences and values related to GMO (positive or negative), usually in the form of mandatory labelling. Here, the reasons for the regulation will depend on the presence of differences of preference and values that are considered to be important enough to justify regulation from a perspective of autonomy and freedom.

A final ethical perspective concerns the regulatory handling of ignorance and uncertainty about facts of relevance from any of the other ethical perspectives, often expressed in terms of a precautionary principle (Munthe, 2011). In this area, Munthe (2017) has recently presented an idea for how biotechnologies may be classified as more or less similar in a number of ways of importance from a precautionary standpoint, and concluded that the current EU GMO regulation could and should probably be made more technology neutral from a precautionary ethical standpoint by subjecting more traditional agricultural practices and more modern biotechnology, such as GMO, to similar regulatory requirements. Hansson (2016) has argued that the current special regulatory arrangements for GMOs in Europe originally had a rationale based in lack of knowledge that no longer applies with equal force, likewise concluding that general agricultural technology regulation and that of traditional GMOs should be made more similar. However, both authors hold that new technological developments, for instance in the field of synthetic biology, present vast uncertainties and areas of ignorance, thereby possibly being apt for more specific regulatory attention. Again, a regulatory structure may very well be viewed as both quite technology specific from some standpoint, and at the same time more technology neutral in relation to a specific regulatory rationale (where the rationale might be, for instance, precaution).

References

Abbott, A. 2018. European court suggests relaxed gene-editing rules. Nature 19 January, https://www.nature.com/articles/d41586-018-01013-5.

Azar, C. and Sandén B. A. 2011. The elusive quest for technology-neutral policies. Environmental Innovation and Societal Transitions 1, 135-139.

Berg, P. 2008. Meetings that changed the world:Asilomar 1975: DNA modification secured. Nature 455: 290-291.

Cicerone, P.J., Dzau, V.J., Bai, C. and Ramakrishnan, V. 2015. International Summit on Human Gene Editing: A Global Discussion. Meeting in Brief. Washington: The National Academies Press.

Eriksson, D. and Amman, K. H. 2017. A universally acceptable view on the adoption of improved plant breeding techniques. Frontiers in Plant Science 7: 1999 doi: 10.3389/fpls.2016.01999

Greenberg, B. A. 2016. Rethinking technology neutrality. Minnesota Law Review 100: 1495-1562.

Gregorowius, D., Lindemann-Matthies, P. and Huppenbauer, M. 2012. Ethical discourse on the use of genetically modified crops: A review of academic publications in the fields of ecology and environmental ethics. Journal of Agricultural and Environmental Ethics 25: 265-293.

Hansson, S.O. 2016. How to be cautious but open to learning: time to update biotechnology and GMO legislation. Risk Analysis 36:1513-7.

Kim, J. and Kim, J.S. 2016. Bypassing GMO regulation with CRISPR gene editing. Nature Biotechnology, 34: 1014-1015.

Macdonald, P. 2014. Genetically modified organisms regulatory challenges and science: A Canadian perspective. Journal für Verbraucherschutz und Lebensmittelsicherheit 9 (Suppl 1): 59-64.

Masip, G., Sabalza, M., Pérez-Massot, E., Banakar, R., Cebrian, D., Twyman, R. M., Capell, T., Albajes, R. and Christou, P. 2013. Paradoxical EU agricultural policies on genetically engineered crops. Trends in Plant Science 18: 312-324.

McHughen, A. and Smyth, S. 2008. US regulatory system for genetically modified [genetically modified organism (GMO), rDNA or transgenic] crop cultivars. Plant Biotechnology Journal 6: 2-12.

Munthe, C. 2011. The Price of Precaution and the Ethics of Risk. Dordrecht: Springer.
Munthe, C. 2017. Precaution and ethics: handling risks, uncertainties and knowledge gaps in the regulation of new biotechnologies. Bern: FOBL.
Ricroch, A.E., Ammann, K. and Kuntz, M. 2016. Editing EU legislation to fit plant genome editing. EMBO Reports 17: 1375-1369.
Sandler, R. 2004. An aretaic objection to agricultural biotechnology. Journal of Agricultural and Environmental Ethics 17: 301-317.
Siipi, Helena. 2015. Is genetically modified food unnatural? Journal of Agricultural and Environmental Ethics 28: 807-816.
Tagliabue, G. 2016. The necessary 'GMO' denialism and scientific consensus. JCOM 15 (04): Y01.
Tagliabue, G. 2018. Reject the 'Gmo' fallacy, in terms of both safety concerns and socioeconomic issues. Geographical Review 108: e1-e5
Winter, Gerd. 2016. Cultivation restrictions for genetically modified plants. European Journal of Risk Regulation 1: 120-142.
Zetterberg, C. and K. Edvardsson Björnberg. 2017. Time for a new EU regulatory framework for GM crops? Journal of Agricultural and Environmental Ethics 30: 325-347.

35. The single story about the foodbank

L. Pijnenburg
Wageningen University and Research, CPT (philosophy), Hollandseweg 1, 6706 KN Wageningen, the Netherlands; leon.pijnenburg@wur.nl

Abstract

The dominant, academic story about the food bank says that it is as a de-politicising charity organisation (Dowler, 2016). There is, however, a danger to telling a single story (Adichie 2009). It consists of (1) mis-presenting the way in which food bank organisations understand their own activities and goals, and (2) not (fully) recognising the work they actually do, and the possible normative justification for it. To question a dominant story one needs an elaborated counter story which I cannot provide here and now. What I hope to do is more programmatic in nature. The plan here is to point at a questionable, basic element of the dominant story, that is: its critique on the charity approach, and then sketch a different story that also can be heard at the food bank. This will lead me to a reflection on the difference between the dominant 'rights based approach' and a 'dignity approach' that I reconstruct and derive from empirical research, and to an attempt to normatively justify the form of emergency food aid that the food bank provides. The distribution of food, donated by companies and private individuals and to be prepared at home, at some 530 food banks that are united in the 'Association of Dutch Food Banks' (ADFB), forms the empirical frame of reference in which I argue.

Keywords: food bank, charity, human dignity, human rights

One dominant story

The dominant story highlights at least four characteristics of the food bank as an institute or organisation: (1) it de-politicizes the problem of food-insecurity, (2) it functions as a smokescreen for government, (3) it subjects 'the poor' to a neoliberal welfare regime (in terms of privatising responsibility) and (4), it functions symbolically as a moral safety valve because of its 'ethos of charity' (Williams *et al.*, 2016: 2293-2294). These four story lines are intertwined and used mainly from a critical point of view that emphasizes social inequality, food insecurity, and the right to food. Riches (2011: 771, but also Riches *et al.*, 2014) argues for a human rights approach to adequate food and considers food banks to be 'symptoms and symbols not only of broken social safety nets but also of failing food and income redistribution policies', and he urges us to 'to think outside the charitable food-aid box'. The human right to food, says Riches, constitutes a better, alternative model. Many others, like Poppendieck, Dowler, and Lambie-Mumford, agree with Riches that the charity approach to food banks is, in a way, part of the whole problem because it diverts attention away from or, even worse, substitutes charity for the political responsibility that resides in governmental institutions. Poppendieck (1994) summarized the two kinds of societal response to hunger and poverty. The charity response is: (1) associated with voluntarism, neighbourliness, localism, spiritual good and personal involvement, it is (2) motivated by compassion and aims at alleviating hunger, and (3) its organising principles are: benevolence and care. The justice response is (1) associated with dignity, entitlement, accountability and equity, it is (2) motivated by (a feeling for) fairness and aims at limiting inequality, and (3) its organising principles are solidarity and the common interest. She notes that, although governments have become involved with the emergency food system in many ways, 'the phenomenon is overwhelmingly a project of the private, voluntary sector and of especially religious organizations' (Poppendieck, 1994: 69).

Descriptions of what it means for a food bank to be a charity organisation, however, are not always clear. Looking at more recent work it is striking that a clear description is completely lacking or given

EurSafe 2018 – DOI 10.3920/978-90-8686-869-8_35, © Wageningen Academic Publishers 2018

ex negativo: 'the food bank ..... has been characterised by its charitable nature – it is situated outside of the state' (Lambie-Mumford, 2017: 29). Insofar as the experience and views of volunteers and food bank receivers are discussed in the literature, the general picture is rather negative and depressing. Low quality of distributed food parcels has a negative impact on feelings of 'self-worth'; interactional experience at the food bank shows a mixture of (fear of) humiliation and blame, (compulsory) gratitude and shame; and being a receiver seems to be linked to a lower status (Van der Horst *et al.*, 2014; Smith-Carrier *et al.*, 2017). These views from the inside seem to confirm the expectations that many associate with the idea of the food bank as a traditional charity organisation that is voluntary and community/ faith based. The Trussell Trust Foodbank Network in the UK, with more than 400 projects, is a case in point, although some would not entirely agree with this. The professionalization, coordination and scale of these organisations has made them 'different from the historical responses to hunger', says Lambie-Mumford (2017: 9), 'like churches and other charitable initiatives'. However, to what extent this really has affected the nature of these organisations, that is: their own self-understanding in terms of the motives, reasons and expectations that volunteers and receivers have, is a question that has not been answered yet. Recent work seems to confirm the traditional image of community and faith-based volunteering at the food bank; so much so, that in an ethnography of the food bank the 'church door' functions repeatedly as a metaphor for the way into the food bank (Garthwaite, 2016: 83, 93, 148).

My proposition is that this is only one of the stories that can be heard in the daily practice of the food bank. Food banks in one specific socio-cultural, even national, environment show clear differences when compared with food banks in other environments; differences in terms of how volunteers and receivers understand their own organisation, role and functioning.

A different story

Research results sometimes show cracks in a story through which a different sound can be heard. Although volunteers (who distribute the parcels) and receivers sometimes think and act within the framework of a 'compulsory gratitude' that characterizes the traditional charity model, at other moments they clearly realise, in the words of some volunteers, that maybe people 'did not end up in this situation because of their own fault. (....) Receivers should not get the idea that you are the benevolent giver'. Food bank receivers also challenge the mark or stigma of a lower status: 'But I do not have to be ashamed. Somebody else put me in this position. ... I want to be treated as a human being of equal value', and a volunteer says that it's wrong to think that they should just handle their money better instead of holding up their hands: 'That reasoning is false' (Van der Horst *et al.*, 2014: 1513-1515).

Researchers emphasize that receivers and volunteers indicate that 'there should be no shame when there is no 'own fault' involved. In all interviews interviewees feel they have to address the issue of fault. In all but a few they claim that they are not to blame for their circumstances. Rather they blame "society", "bad luck" or "circumstances"' (Idem: 1515).

Although many responses are undeniably couched in a vocabulary of emotions and feelings, they nonetheless also carry with them a rational and justifiable argument. As can be learned from the interviews quoted above, they express a clear argument, given by both volunteers and receivers, that what is at stake at the food bank is an issue of justice and human dignity. Some human beings have become vulnerable and are hurt by conditions that are beyond their control, and other human beings, by making the statements they make and by rolling up their sleeves as volunteers, recognise and acknowledge this (Williams *et al.*, 2016: 2301).

People who give expression to the idea of emergency food organisations like the food bank as practices of human justice, cannot only be found on the distributing and receiving side of the counter at the foodbank. We (Hebinck and Pijnenburg, unpublished) also asked the volunteers who take care of the managerial tasks of the food banks in the Netherlands. After first inviting them to describe their own understanding of the food bank in institutional terms of goals, importance, and type, we confronted them with the dominant idea of the foodbank.

In their answers the local management described their own organisations in secular terms as voluntary organisations that contribute to reducing 'food poverty' by way of distributing free food parcels to people who, for whatever reasons and mostly not because of their own doings, do not have the financial resources (anymore) to provide for themselves (and their family) enough food to survive and function normally. Many also mentioned a secondary goal: to reduce food waste; and some also point at attempts and efforts to (re-) connect receivers (again) to governmental and/or professional help. Besides the official language used in the responses, which more or less copies the policy statements of the ADFB, at times a moral vocabulary surfaces when the *raison d'être* of the food bank is described: 'It is important that this work is done, to prevent that vulnerable people will fall between two stools'. With a moral-political twist: 'It is important to do this work because the government falls short on its task, and to our minds citizens should take care of fellow-citizens'.

Responses to the charity conception of the food bank were clearly more extreme and outspoken in that the expected receiver's attitude of 'humility and gratitude' was rejected almost entirely. Wordings used were: 'too extreme', 'ridiculous', 'insulting', 'nonsense'. Although some prefer a more 'neutral' language and speak about a professional organisation that offers help by way of delivering food parcels, a majority talks about mutual respect, equality, solidarity and moral responsibility. It seems that the hierarchical/paternalistic connotations of words and terms within the orbit of charity's referential field of meanings are unacceptable for the average moral consciousness of the food bank management. 'It's not about charity, it's about a fellow human being who's facing a rough time and needs some support. Some solidarity', a respondent says.

Human rights, human dignity and room for justification and critique

It is tempting to interpret this other story about the food bank, its volunteers and receivers, as a story about the food bank as a human rights practice in its own right. Although this line of thought should be investigated further, the talk about duties, rights and obligations at the food bank mainly points, not to a legal, also not to a charity, but to a moral issue. 'Solidarity and care for the vulnerable in society should be a core value for everybody, to avoid divisions in society', says a respondent (Hebinck and Pijnenburg, unpublished). In the practice of the food bank one can find a strong moral motivation and justification of that practice, constructed around the notion of human dignity. I consider human dignity to be the fundamental source of our indignation at inhumane situations and conditions (Habermas, 2010). As such, this indignation encourages us to say or do something, to revolt or to fight for rights, with the aim of preventing and/or eradicating these degrading circumstances. Compared to human rights, judgements about human dignity signify and highlight above all the moral character of an act, a situation or a condition. This places the assessed situation in a relational context of moral (joint) responsibilities, moral rights and duties (and not in the context of legal rights and duties).

Here, I can only touch upon this complicated relation between human rights and human dignity, but, in order to make a connection with our present debate I, inspired by works of Forst (2011) and Honneth (1996), would like to introduce the normative sociological concept of 'room for justification and critique' (R4J&C). It is a concept that applies to all societal levels and it contains the idea that people always experience more or less restrictions on their freedom and responsibility to question or justify acts (or results thereof) which, in the worst case scenario, violate their human dignity. The concept of R4J&C can replace the metaphor of 'the game' in Offe's analysis of 'the tripartition of the universe of life chances' (Offe, 1996). Instead of making a distinction between winners, losers and non-players in modern society's game, we can analytically distinguish people with proper access to institutionalised and established R's4J&C, which enables them to improve the quality of their life chances more or less, and people without proper access to these R's4J&C. These are the new 'outcasts' or 'excluded' that have been expelled by the process of globalisation since the last quarter of the previous century. The social consequences of this development are visible: 'The sources of solidarity are drying up, with the result that social conditions of the former Third World are becoming commonplace in the urban centres of the First World' (Habermas, 1998: 122). Instead of over-emphasizing a human being's liberal rights 'to have a choice' (Garthwaite, 2016; Lorenz, 2015), the concept of R4J&C ties together liberal and democratic responsibilities, rights and obligations. The main advantage of it lies not only in its potential to connect with a broader analysis of modern society, but also in the possibilities it offers to investigate the practice of the food bank as a R4J&C, a 'contested space' (Williams *et al.*, 2016), where people can find more or less support in their struggle for recognition and human dignity.

Towards a normative justification of emergency food aid: taking up the slack

By taking the basic moral indignation at violations of human dignity, which emanates from the practice of the food bank, as a point of departure for a moral/ethical justification of its *raison d'être*, a further bias of the dominant story emerges: it's state-centeredness.

But first the moral justification of the food bank as an organisation for emergency food aid. Faced with images of Bengalese people dying from hunger and malnutrition, Singer wrote in a 1972 seminal paper, we morally ought to prevent it; that is: if it is (1) within our power and if we (2) do not have to sacrifice something comparably morally significant (or, more moderate: something morally significant). Of course, one does not have to be a utilitarian like Singer to justify the statement that it is morally obligatory to prevent something bad from happening. In any event, Karnein stays agnostic about when duties like these arise and discusses Singers 'drowning child scenario' as an example for 'any situation in which, according to your preferred moral theory, a circumstance triggers a duty for all agents able to address it, and there happens to be more than one such agent' (Karnein, 2014: 593). She asks us to imagine a situation of three drowning children and three capable adults of whom two do not do their share (that is: to help a child). Does A have to 'take up the slack' left behind by those others? Karnein argues convincingly that A, although B and C are not doing their 'fair share', still has the responsibility to help all the children as much as possible. One should not mix up the two issues that are involved: on one hand 'what agents owe to third parties', and on the other hand 'how fellow duty bearers should relate to each other' (Karnein, 2014: 607). Karnein's reasoning gets an interesting twist when she observes that the relationship between duty bearers, and therefore also the 'slack-taking', can be horizontal and vertical. Within the present context this distinction can be used to make clear that the food bank is factually taking up the slack that the government, as the original duty bearer in the vertical dimension, has left behind.

Emergency food aid at the food bank is provided in countries where the state's capacities have been shrinking since the late 1970's. In the process of globalisation the state has lost much of its influence over the conditions for production that previously generated taxable incomes and profits. Further analysis should provide more insight here. For now, it raises sufficient doubts about the mantra of the single, dominant story: the state, the state, before it's too late!

References

Adichie, C. (2009). The danger of the single story, TED Talk.

Dowler, E.A, (2016). 'Just food': contemporary challenges for richer countries. In: I. Anna S. Olssen *et al.* (eds.), Food Futures; ethics, science and culture, Wageningen Academic Publishers 2016.

Forst, R. (2011). The ground of critique: on the concept on human dignity in social orders of justification. Philosophy and Social Criticism 37 (9).

Garthwaite, K. (2016). Hunger Pains: life inside foodbank Britain, Policy Press.

Habermas, J. (1909). The inclusion of the other, MIT, Cambridge/Massachusetts.

Habermas, J. (2010). The concept of human dignity and the realist utopia of human rights. Metaphilosophy, 41:4.

Honneth, A. (1996). The struggle for recognition, Cambridge, MIT

Karnein, A. (2014). Putting fairness in its place: why there is a duty to take up the slack. Journal of Philosophy, 111:11.

Lambie-Mumford, H. (2017). Hungry Britain, Policy Press.

Lorenz, S. (2015). Having no choice. Journal of Exclusion Studies, 5:1.

Offe, C. (1996). Modern 'Barbarity': a micro-state of nature? Constellations, 2:3.

Poppendieck, J. (1994). Dilemmas of emergency food. Agriculture and human values, Fall.

Riches, G. (2011). Thinking and acting outside the charitable food box; hunger and the right to food in rich societies. Development and Practice, 21:4-5.

Riches, G. *et al.* (eds.). (2014). First World Hunger Revisited: Food charity or the right to food?, London: MacMillan.

Singer, P. (1972). Famine, Affluence and Morality. Philosophy and Public Affairs, 1:3.

Smith-Carrier, T. *et al.* (2017). 'Food is a right ... Nobody should be starving on our streets'. Journal of Human Rights Practice, 9.

Van der Horst, H. *et al.* (2014). The 'dark side' of food banks? British Food Journal, 116:9.

Williams, A. *et al.* (2016). Contested space: the contradictory political dynamics of food banking in the UK. Environment and Planning, 48:11.

36. Things, patents, and genetically modified animals

M. Oksanen
Department of Social Sciences, University of Eastern Finland, P.O. Box 1627, 70211 Kuopio, Finland; markku.oksanen@uef.fi

Abstract

The development of genetically modified (GM) animals for food and biomedical experimentation and the claims of intellectual property rights (IPRs) on products, brands, and processes that involve these animals are inseparable. The paper examines philosophically IPRs employed as a legal instrument to protect the control of animal-related innovations and the creation of novel genetic material in animals. It focuses on what kind of moral relation to animals is being formed if patents and other forms of exclusive control are granted for animals used in food production and for their genetic material and how the resulting human-animal relationship has been criticized. It also explores the open-source approach, or Bio-Linux, in animal genetic modification. The paper argues that even if animal-related innovations were not granted patent protection, for example on the basis of morality exclusion, there are other forms of ownership that offer just as powerful control as patents over the maintenance of animal life and the direction of its evolution. These forms include the ownership of animals as things, or as tangible property, and trademarks. Together they enable the in-depth self-regulation of business in the development and maintenance of animal breeds and allow other less desirable breeds to disappear. Thus, the patents on animal genetic material do not imply a qualitatively novel moral relationship between humans and animals. Are patents then as morally acceptable as the ownership of animals as things? The answer to this question depends on what the production of patentable information involves, in particular, what kind of treatment of animals it allows and what kind of sentient beings it brings into existence.

Keywords: intellectual property, open-source, genetic material, ethics

Introduction

This paper examines philosophically intellectual property rights (IPRs) to animal-related innovations. It asks what kind of moral relation to animals is being formed if patents and other forms of exclusive control are granted for animals, for their genetic material and for the methods of modification. It also addresses the critique of the resulting human-animal relationship and explores the open-source approach, or Bio-Linux, in animal genetic modification.

Animals in property relations

Property rules concerning animals govern the acquisition, use and transfer of animals and provide the owners with certain responsibilities for their health and well-being and for the safety of the rest of society. Within property rules, animals are typically conceived of as things but due to the new methods of breeding that involve the use of genome technologies, IPRs to animals has become an issue of discussion. 'Thing' is a technical term for any tangible object of property other than land and intellectual property. Unlike tangible property, IPRs does not seem to imply thinghood, but at a closer look it becomes obvious that animal genetic material is subject to being both intellectual property and tangible property or things.

In what precise sense animals are the subject of IPRs depends on two elements. First, it depends on the economically valuable quality in them that has been created by some identifiable human actor and granted exclusive protection within the system of intellectual property. For example, transgenic animals could be used as bioreactors to produce some valuable substance, and the enhancement or

EurSafe 2018 – DOI 10.3920/978-90-8686-869-8_36, © Wageningen Academic Publishers 2018

engineering of this productive capacity in animals, sometimes located in its DNA, can be protected by means of intellectual property. Usually, patentable information must have characteristics of novelty, non-obviousness, disclosure and 'being capable of industrial application' or utility. What exists naturally, lacks an identifiable author, cannot be distinguished from other pieces of knowledge in a meaningful way, or otherwise fails to meet the criteria of patentability, belongs to the domain of public goods, and it can – within the bounds of other rules followed in a society – be used freely. The substance matter of patent is not limited to any single animal but is information that can materialize in many individuals and perhaps across species boundaries (Adams, 2003).

The second element is the genetic material in individual animals. Thus, intellectual property depends on the reproduction of genetic material in which case the right holder can determine the reproduction of genetic lineage. For example, the creators of the labradoodle, a 'designer dog' for dog-allergy families, could have asserted some rights to this variety, or a crossbreed; patenting was not a likely option but trademarking. The purpose of trademarks is to protect verbal and symbolic expressions so that consumers can become familiar with certain qualities in goods and services or in their providers. Having a trademark for a genetic lineage does not require genetic modification. There is at least one dog breed – ELO – the name of which is trademarked in Germany. As regards farm animals, the Black Hereford is a legally protected brand in the USA.

The maintenance of a breed is based on ethical and professional codes that the breeders have adopted. In addition, contracts are widely used to guide further breeding. A great deal of ethical discussion on patents seems to concentrate on regulation exercised through formal laws while – currently, at least – a much greater volume of breeding occurs in the world of self-regulated businesses. Of course, the world of self-regulation is shaped by human safety or animal welfare legislation to some extent. In the end, the actual control over flesh-and-blood animals is and will remain as crucial as ever and the proprietary activities are brand- or trademark-driven by individual breeders or their associations.

Critiquing the ownership of animals

The idea of animals as things, as objects of property, has been strongly criticized. The 'abolitionist approach' has crystallized its ultimate aim as follows: The basic right not to be treated as the property of others; rather (some) animals are persons, not mere things. Francione (2010) claims that when we recognize that animals have rights, we ought to recognize that they cannot be treated as property, as human belongings. Garner (2010) instead claims that ending animal's property status is neither a necessary nor a sufficient condition for realizing animal rights, because 'The abominable treatment of many supposedly free humans throughout the world is an important reminder that granting formal rights does not necessarily result in better treatment in reality'. Cochrane (2012) instead provides a conceptual argument stating that animal rights are not contradictory to the human ownership of animals since ownership as such does not entitle the owner to use animals in morally unacceptable ways. His critique stems from a qualified understanding of ownership and one could ask whether qualified ownership is proper ownership. The issue then converts into an issue of the real meaning of ownership (Oksanen, 1998).

It is quite impossible to undo property relations in regard to animals except by de-domesticating or re-wilding those animals that are currently domesticated. In the world of domesticated animals, there will necessarily be a human actor who claims to have stronger interests in the specific individual animal than any other human. Consequently, property in animals is put in place since having property is to have the strongest legitimate interest in that animal. In practice, even the most looked-after animals, companion animals, are the property of their custodians (cf. Fruh and Wirchnianski, 2017). The crucial issue here is ultimately what we mean by the concept of ownership and to what kinds of relations it gives rise. It is obvious that the owners of companion animals want to be kind and caring, although in reality their behavior can be inadequate, if not harmful.

Critiquing animal patents

Even if there is no inherently novel property relation to animals that patents and other forms of intellectual property exhibit, the following question arises: Is the nature of social power – that is, the power over other humans and not over animal subjects – different when there are patents for animals in addition to the conventional ownership of animals as things? Some scholars argue that the possibility of patenting animal genetic material is the nail in the coffin in the human domination of animals. For instance, Linzey (1993) claims that when we consider animals as 'patentable creations', we make 'the claim of absolute sovereignty over animals'.

Another critique of animal and plant patents comes from Shiva (1997) who claims that the patenting of living organism is violence in two different senses: 'First, life forms are treated as if they are mere machines, thus denying their self-organizing capacity. Second, by allowing the patenting of future generations of plants and animals, the self-reproducing capacity of living organisms is denied.' The first point is similar to that of Francione's critique of animal ownership, despite minor difference in wording. The latter point is about restricting the autonomy of animals, more specifically their autonomy to reproduce. This is, again, the whole point of human selection of biological varieties – in general, it would be rather difficult to choose which kind of human interference with the evolutionary process is 'violent' and which is benevolent.

What is common to Linzey's and Shiva's critiques is, I believe, an assumption that being the subject of patents and IPRs implies a somewhat qualitatively different control of animal lives than is discernible in the conception of living animals as mere property objects. My claim is that this change does not happen because it is impossible to see how animals could be even more than things or have more thing-qualities than they currently have. The thing is a separable entity and the possibility of patents for their generation does not change this legal fact. Furthermore, when we think of domination of individual animals, the policy of granting trademarks to animals can have as profound an impact on the evolution of dog and cattle breeds as genetic modification. Patenting animal inventions might, nevertheless, alter human relations in regard to animals (as property) and it might increase the power of the patent holder at the expense of others; the change of human relations to animals is of a lesser nature.

On the other hand, Linzey (1993) claims that, '[...] patenting represents the attempt to perpetuate, to institutionalize, and to commercialize suffering to animals.' It is undeniable that there is a lot of suffering in the breeding and use of production animals and therefore those who deem this suffering in animals as unjustified also require that there should be no production animals. When there are no production animals, there is no need for patenting those inventions that are based on the existence of the production animals. Since there are however animals bred by humans, we live in a non-ideal world from the perspective of ethical vegetarians and vegans and animal welfarists in general. Therefore, the question of the role of patents in the worsening of animals' condition is pertinent. The patents for animal genetic material can be bad for animals, if patents authorize the use of animals in ways that result in suffering.

Linzey's critique does not necessarily undermine the idea of animal patents. All patent applications are separate cases that require a specific decision, and if a patent application involves animal cruelty that is criminalized elsewhere in law, Linzey's critique falls short. To reach an informed judgement on animal patents, the larger normative landscape has to be recognized. Quite often, there are norms that promote human ingenuity and there are norms that curb it: thus, the overall regulatory system consists of contesting norms and the appropriate public authority must use discretionary power in weighing them.

Neutrality of patents?

Adams (2003) says that animal patenting is not ethically neutral. Ethical neutrality here refers to a position that animal patents do not imply a qualitatively novel relationship between humans and domestic animals. As Adams sees it, patents can be ethically neutral only if the basic right of the patent owner was to exclude others. Because patents are not merely about exclusion, the reasoning does not hold since having a patent implies having a right to utilize the patent and thus to impose some activities on animals. In case of animal patents, it provides 'an affirmative privilege to exploit commercially the object of the property right' (*ibid.*), i.e. both flesh-and-blood animals and the abstract right. Without the idea of animals as things, the idea of animal-related patents could not work. But again we can ask whether intellectual property in animals results in a deeper and qualitatively different control of animal lives.

Adams's argument against animal patents can be represented as follows:
1. Patents for animals results in the right to use them (as stated in the patent).
2. It is not morally acceptable to treat animals in ways that patents entitle.
3. Therefore, patents for animals are not morally acceptable.

As such, this line of argumentation is rather persuasive. It is the case, after all, that the patent decision is a very detailed and fully reflected-upon decision by the relevant public authority. As long as all the relevant laws and other legal material have been taken into account, the decision represents the official stance of the government and thus is enforceable (provided that all possible legal procedures have been accomplished). If the decision rests on the inadequate consideration of relevant laws, then it is questionable. Of course, the aim in legislation is not to produce inconsistent laws. In any case, there are ethical limits to the patentability of animal innovations and these are expressed, for instance, in the European Patent Convention. As Liddell (2012) has pointed out, such provisions require lawyers and other authorities in patent offices to take a stance on morality and this is 'the principal difficulty'.

If there is nothing more against animal patenting than the arguments against the ownership of animals as things, then the argumentation rests on rather a weak basis. But this apparent neutrality is the stance that Adams aims reject by emphasizing that patent decisions really affect the lives of individual animals.

If Adams's case does not hold, if the thinghood of animals produces equally uniform breeds, and if we cannot find any other arguments that show the intrinsic wrongness of patenting animal genetic resources, then we seem to be close to a situation where we have to ask whether they are acceptable. This is however not my preferable conclusion.

Open-source biology and animals

Let us consider the open-source biology (Bio-Linux, creative commons or copyleft) approach to the ownership of innovations in animal genetic resources. The idea of open-source is that while new high-tech uses of animals will be developed, the exclusive patents or the like and the payment of royalties would play no role in the mechanism of the acquisition of income from one's work on breeding. The results of breeding activities in the form of animal genetic material would remain as intellectual commons. The idea of Bio-Linux is to block the emergence of allegedly harmful monopolies over resources, over the animal genetic material in our case (Srinivas, 2002; Carlson, 2010). Would this approach provide us with a different moral relationship with animals than the standard IPRs approaches do? Would it differ from the prevailing practices? And could it be accepted by radical animal rightists? The answer, I think, to all of these questions is negative.

The open-source approach does not outright reject the genetic modification of production animals, and therefore, for those who consider that the source of wrongness lies in genomic technology and the human domination of animals it brings about, this offers no change in policy or mentality. The animal genetic material, independently of its production, is solely the property of the party who owns it concretely. Moreover, the idea of animal genetic resources as non-patentable material does not lead to the ideas of animal self-ownership or animal liberation. When we think of animal genetic material as commons, it is still open to various kinds of human use, including breeding in conventional or technologically advanced ways. In other words, the open-source approach only addresses interhuman relationships and is agnostic about the ethics of human-animal relationships.

There is one question to pose: If there are no exclusive animal patents, would such a state of regulation function as a disincentive to the development of animal genetic modification? There is an almost inseparable bond between IPRs and genetic engineering. So, a critic may reason that by getting rid of IPRs, we can also get rid of genetic modification. In light of the extensive expansion of Linux-based software in information technology, this may be rather wishful thinking. Of course, there are differences between information technology and biotechnology. Most importantly, the use of live animals in experimentation is strictly regulated and cannot be done at home, whereas there are no such limitations regarding the computers and software. Nevertheless, the notions of biohacking and do-it-yourself biology manifest trends that can have implications for the engineering of animal genomes, too.

Conclusion

The paper has argued that even if animal-related innovations were not granted patent protection, for example on the basis of morality exclusion, there are other forms of ownership that offer just as powerful control as patents over the maintenance of animal life and the direction of its evolution. These forms include the ownership of animals as things, or as tangible property, and trademarks. Together they enable the in-depth self-regulation of business in the development and maintenance of animal breeds and allow other less desirable breeds to disappear. Thus, the patents on animal genetic material do not imply a qualitatively novel moral relationship between humans and animals. Are patents then as morally acceptable as the ownership of animals as things? The answer to this question depends on what the production of patentable information involves, in particular, what kind of treatment of animals it allows and what kind of sentient beings it brings into existence. Patents for animal genetic material do not necessarily lead to new and additional suffering of animals; no more than do other forms of ownership and non-ownership of animals. We can treat animals badly and we need neither property rights nor patents to do so.

References

Adams, W.A. (2003). The Myth of Ethical Neutrality. Canadian Business Law Journal 39: 181-213.

Carlson, R.H. (2010). Biology is Technology. Cambridge, Mass.: Harvard University Press.

Cochrane, A. (2012). Animal Rights Without Liberation. New York: Columbia University Press.

Favre, D. (2000). Equitable Self-Ownership for Animals. Duke Law Journal 50: 473-502.

Francione, G.L. (2010). The Abolition of Animal Exploitation. In: Francione, G.L. and Garner, R. The Animal Rights Debate: Abolition or Regulation? New York: Columbia University Press, pp. 1-102.

Fruh, K. and Wirchnianski, W. (2017). Neither Owners Nor Guardians. Journal of Agricultural and Environmental Ethics 30: 55-66.

Garner, R. (2010). A Defense of a Broad Animal Protectionism. In: Francione, G.L. and Garner, R. The Animal Rights Debate: Abolition or Regulation? New York: Columbia University Press, pp. 103-174.

Liddell, K. (2012). Immorality and patents. In: Lever, A. (ed.). New Frontier in the Philosophy of Intellectual Property. Cambridge: Cambridge University Press, pp. 140-71.

Linzey, A. (1993). Created Not Invented: A Theological Critique of Patenting Animals. Crucible, April-June 1993: 60-67.
Oksanen, M. (1998). Environmental Ethics and Concepts of Private Ownership. In: Dallmeyer, D.G. and Ike, A.F. (eds.) Environmental Ethics and the Global Marketplace. Athens: University of Georgia Press, pp. 114-139.
Shiva, V. (1997). Biopiracy. Toronto: Between the Lines.
Srinivas, K.R. (2002). The Case for Biolinuxes: and Other Pro-Commons Innovations. In: Sarai Reader 2002: The Cities of Everyday Life. New Delhi: Center for the Study of Developing Societies.

Section 6.
Veterinary ethics: methods, concepts and theory

37. The recognition of animals as patients – the frames of veterinary medicine

M. Huth
Unit of Ethics and Human-Animal-Studies, Messerli Research Institute, Veterinaerplatz 1, 1210 Vienna, Austria; martin.huth@vetmeduni.ac.at

Abstract

The aim of this paper is to investigate in the social frames that determine veterinary practice in different ways. In veterinary ethics, the focus is often directed tripartite structure of veterinarian – owner – animal. But this triad turns out to be embedded in wider structures that build the implicit basis for a differential treatment of different animals. Hence, they pre-determine the triadic clinical encounters. Sophisticated technologies like blood donation, renal transplants, prostheses, etc. are provided to companion animals regularly. The individuals are considered patients with a normative demand for beneficence even similar to humans. Livestock is treated differently. A cow with acute renal failure would not get a donation but would be euthanized. In epidemics, economic constraints might widely overrule the individual's status as patients; e.g. in the food-and-mouth disease culling is performed because it is easier and cheaper than vaccination. The concern for the patient is limited to preferably painless killing. Hence, we can understand some animals (e.g. livestock) as 'liminal patients'. Consequently, these frames determine a 'differential allocation' of the recognition of animals as patient in different ways and degrees with crucial, however, often tacit impacts on concrete decision-making. Public expectations as well as the hidden curriculum of the veterinary profession determine how 'one should treat' a cow, a dog or a rat. However, these frames are not entirely determinant but are changeable.

Keywords: recognizability, vulnerability, beneficence, liminal patient

Introduction

This paper aims to deal with the question of the status of a patient *as patient*. Being a patient equals being an individual who merits care. However, not every living being is under any condition the source of obligations for medical treatment, more generally, of *beneficence* (Beauchamp and Childress, 2007). This applies to human beings – one can point at the fact that the status of being a patient is variable for foetuses (consider the controversies regarding abortion), severely impaired neonates or brain dead individuals – and *a forteriori* to animals. In some animals (mostly pets) we assume a moral obligation to provide the best available (or affordable) care. The animals' alleged health related needs should be pursued. In animals conceived of as livestock, the aim of treatment might be primarily (but not necessarily exclusively) efficiency for farmers. In other animals (e.g. pests, animals diseased from a contagious infection) even killing is a socially approved practice and expected from veterinarians and others. '[M]any people believe it morally right to kill vermin (such as rats), few have scruples over killing insects if they are 'nuisance', and virtually everybody considers we are morally obliged to kill bacteria that cause infectious diseases' (Mepham, 2016: 119).

However, if we consider morally relevant characteristics like socio-cognitive abilities as unconditioned rationale for obligations for veterinarians (take the example of rats), then at least veterinarians would face severe contradictions in their practice with significant consequences; moral conflicts easily turn into psychological burdens (Verweij and Meijboom, 2015).

EurSafe 2018 – DOI 10.3920/978-90-8686-869-8_37, © Wageningen Academic Publishers 2018

In contrast, the backdrop of the upcoming reflections is a relational approach to the status of the patient in veterinary medicine and its ethical implications (in contrast to moral individualism). First, the concept of the patient is analysed to show its practical significance. Drawing from Cora Diamond (1978) and Alice Crary (2010), it will be shown that the reduction of ethical considerations to allegedly morally relevant capacities of animals potentially fails to explain common moral convictions and practices and, hence, fails to deal with moral conflicts. On the contrary, they could even contribute to aggravating conflicts instead of mitigating them. Second, applying Charles Rosenberg's concept of disease as frame (1997) and Judith Butler's theory of recognizability (Butler, 2009) to veterinary medicine, an alternative approach will be put forward which focuses on the differential recognition of animals in society. Third, the critical power of this ethical approach will be made visible. Since one could easily come up with the objection that the status of the patient is arbitrary if it depends on recognition, it will be shown that the focus on the *recognition of vulnerability* particularly undermines arbitrariness.

What is a patient?

In the 'Principles of biomedical ethics' (2001), Tom Beauchamp and James Childress introduce the so-called 'principilism'. In a nutshell, they propose a theory for biomedical ethics basing on four fundamental principles that are to be balanced against each other: beneficence, non-maleficence, autonomy and justice (Beauchamp and Childress, 2001: 12). From my point of view, the *primus inter pares* is the principle of beneficence. If there is no initial obligation to take care for a patient, the principles of non-maleficence, autonomy and justice lack their significance; even the non-maleficence (historically denoted 'primum non nocere' and, hence, classified as primal, only makes sense if we already have a patient we owe – at least some kind of – respect). The initial hypothesis of this paper is that the status of being a patient and the moral duty of beneficence (for the patient's sake) are inextricably linked. Furthermore, the status of being a patient is not a matter of yes or no (like the inherent value in Tom Regan – either an animal conforms to the subject-of-a-life criterion and has inherent value or this is not the case, Regan, 2004: 264). Instead, it is suggested that in the context of veterinary medicine animals might appear *more or less* as patients and could in case be a 'liminal patient' who might even be killed but at least with care and as humane as possible.

In companion animals, veterinary medicine uses more and more 'high tech strategies' (Springer and Grimm, 2017) of medical treatments pursuing the supposed individual´s quality of life.[1] Sophisticated technologies like blood donations (Ashall and Hobson-West, 2017), renal transplants (Schmiedt *et al.*, 2008), up to four leg prostheses show an alleged obligation to treat cats and dogs with comparably high efforts and expenses (although these measures are not entirely uncontroversial). It is assumed that the (alleged) health related needs of an animal form the goal (and thus legitimation) of veterinary treatments (Grimm and Huth, 2017). In contrast, animals conceived of as livestock would hardly be receiver of donor organs or blood donation. Treatments are regularly performed to maintain productivity and efficiency (e.g. the cure fertility disorders in dairy cows; Jengaar, 2014) or to prevent epizootics, zoonosis and food-borne diseases (here 'treatments' could also comprise culling instead of vaccination for economic reasons; Hinchliffe, 2016). However, this does not lead to the consequence that there is absolutely no concern about these animals' welfare and health as empirical data e.g. regarding culling show (Stassen and Cohen, 2016).

Generally speaking, in animal ethics we find a strong paradigm that animals should be considered and treated according to their moral status derived from their mental capacities. In Singer, the equal capacity to suffer combined with the quality of interests suffices for an equal considerations of all relevant animals

[1] However, one could call this assumption into question since it is not entirely clear if the motivation to provide sophisticated health care could also be selfish – one just does not want to lose a best friend.

(Singer, 2011: 73-75). Similarly, in Regan, all animals that conform to the subject-of-a-life criterion have an inherent value that has to be respected (Regan, 2004: 264). These are but two protagonists of a bulk of authors proceeding from shared assumptions: The rationale for a moral status is provided by individual abilities or characteristics; hence, we can subsume these approaches under the category of 'moral individualism' (Crary, 2010).

This approach is in a significant tension to the widely acknowledged tripartite structure of veterinarian – animal – animal owner in veterinary medicine (Kimera and Mlangwa, 2016). Within moral individualistic theories, the obligation towards an animal should not be thwarted by the owner's notion of the decent treatment (dependent on own ideas about a liveable life, on the will – or the ability – to take more or less high economic efforts, on the particular relation between animal and owner, etc.). If the individual has a moral status, and this status constitutes obligations, the veterinarian should be focused only or mainly on the patient. Consequently, one could conclude that the mentioned triad is somewhat beside the ethical point. The differential treatment of animals with similar or equal cognitive abilities would then be an insurmountable problem. However, an exclusive focus on the animal risks losing the owner's compliance and, hence, a customer or, on the long run, the whole customer base.

However, things change if we take relationalism as the point of departure. The critique of moral individualism has entered the debates in animal ethics right after (or even during) the consolidation of the discipline. Already in 1978 Cora Diamond has published her paper 'Eating meat and eating people' in which she shows that there is a significant pitfall in moral individualism: The reduction of moral consideration to individual capacities could not cover all morally relevant convictions and practices or are even in tension to them. Using the thought experiment of consuming human victims of car accidents she points out that in a pathocentristic approach there would be no reason to criticize such a practice. No-one (or no-one's interest) is violated, no-one suffers; thus, not eating human dead seems bluntly squeamish. However, hardly anyone would adopt that 'conviction'. At the same time, moral individualism turns different treatment of animals into a logically and ethically inacceptable inconsistency. This also applies to veterinary medicine in which (in terms of cognitive abilities) similar animals are targeted differently in blood donations or kidney transplants (pets) or in practices like (prophylactic) mass culling (livestock).

In contrast, Diamond claims that a primal significance of beings in our lifeworld is normatively relevant: 'We can most naturally speak of a kind of action as morally wrong when we have some firm grasp of what kind of beings are involved.' (Diamond, 1978: 469) This applies to humans in contrast to animals, but also to the differential notion of animals (Huth, 2016: 125). In a similar vein, Alice Crary claims that 'we are necessarily guided by a conception of the kinds of things that matter in lives like ours' (Crary, 2010: 26). Hence, it is our pre-given relation to living beings that constitutes the particular sort of moral obligation.

This alternative approach to animal ethics has hardly been applied to the discourses of veterinary ethics. However, this seems a valuable source for a consideration of the status of an animal as patient. The most basic assumption would then be a twofold one: First, the triad veterinarian – animal – owner does make sense because relations are morally significant. And second, this triad is be conceived of as embedded in a broader normative infrastructure that sustains different legitimations of treatment. Diamond's and Crary's critique of moral individualism is based on a structure of significances that determines the range of moral considerations and approvable practices.. If we consider the triad of veterinary medicine as embedded in broader (normative) significations it follows that veterinary practice is not or not only reliant on a moral status derived from individual capacities but on relations and significations.

A similar structure of recognition is detectable in Chervenak's and McCullough's reflections regarding human medicine. According to them, a human being is recognizable as a patient only if (1) she takes part in a clinical encounter and (2) if there are possibilities for a proper treatment of symptoms acknowledged as a disease (Chervenak and McCullough, 1996). This is – although not seamlessly – transferable to the context of veterinary medicine (Grimm and Huth, 2017). Point (1) fully applies to veterinary medicine. An animal does not turn into a patient without any clinical encounter. Point (2) is applicable in a different manner than in human medicine. While in humans an acknowledged disease (as listed in the ICD-10) is a powerful rationale for treatment of virtually any human being, animals are considered differentially. The recognition of animals as patients is differential due to the infrastructure of (moral) significance of particular animals (as livestock, as pets, as vermin, as wild animals, etc.).

In a similar vein, Charles Rosenberg has put forward a theory of disease as frame: 'In some ways, disease does not exist until we have agreed that it does, by perceiving, naming and responding to it.' (Rosenberg, 1997: xiii) The same, so the argument goes, applies to the framing of animals as patients or in as contrast vermin or production unit. To take once more the example of the rats, we do not conceive of these animals as patients meriting beneficence unless one regards them as pets (which is occasionally the case). Only then we consider the alleged sufferer as 'fully' entitled to sympathy and care (*ibid.*: xvi). The selective recognition of a being as patient is invested with a unique configuration of social characteristics and triggers disease specific responses (Rosenberg, 1997: xviii). Hence, not any ill body is conceivable as a patient's body. The status itself becomes a crucial element in a complex network of obligations, practices, institutions, structures and social debates. Moreover, it is important to note that particularly in animals the status of being a patient is a matter of degrees. While a pet is regularly considered as a patient that merits the best available or affordable treatment, a cow could be treated – on the first glance – merely with regard to economic efficiency. However, 'even' the cow might not be treated merely like an object and anaesthesia or analgesia as well as adequate care are required in most of the interventions. The cow thus can be considered as a 'liminal patient'.

Selective frames – the differential recognizability of animals as patients

In what follows, Judith Butler's concept of recognizability (Butler, 2009: 2) will be explained and considered – together with Rosenberg's concept of the frame – as a valuable source for a theoretical analysis of veterinary practice and its immanent differential treatments of similar or even equal animals.

The respective animal that appears in perception and maybe as patient is manifest only within frames that are prior to individual decisions. According to Butler, there is a primal social ontology of (vulnerable) bodies that differentially determines our possibilities to be responsive to needs and suffering (Butler, 2009: 22). There is no animal per se that tells us in an unconditioned manner about our obligations towards her/him. Even the law conceives of animals according to their socio-cultural significance. Legal regulations in Austria, for example, proceed from conceptions of animals according to their use and significance as pets, livestock, etc.; practices are allowed or prohibited accordingly. Empirical findings that show a significant similarity of cognitive abilities in different animals have often little (although surely not no) impact on regulations and practices.

Hence, decisions in regard of treatments, omissions, euthanasia or culling are to be considered as initially socially framed and therefore veterinary practices are reliant on 'more general conditions that prepare and shape a subject for recognition' (*ibid.*: 5) and 'produce' a recognizable being (*ibid.*: 6). The embeddedness of veterinary medicine in socio-cultural frames is an inevitable fact that organizes and predetermines the practices that are seen as state-of-the-art, overblown or unnecessary. It is not a matter of arbitrary consideration or of personal decisions to declare four leg prostheses or kidney transplants as a decent clinical measure to treat dairy cows or rats. It is important to note that this recognizability of animals as patients is inevitably selective (*ibid.*: 24f.) and tacit (*ibid.*: 42). We are not always aware of the fact that we classify and are responsive to animals in these different ways. This is even visible in the critique of culling healthy animals as a control measure during an outbreak of an infectious disease. This kind of stamping out is not necessarily dismissed because of the unjustified killing of self-aware beings but because the 'natural function'[2] of the animal is not to be fulfilled (Cohen and Stassen, 2017: 140). Hence, a killing of these individuals is in common convictions morally acceptable only for adequate reasons and under decent circumstances. The presumed moral infrastructure and its structure of recognizability deeply determine these common convictions.

Not only the common sense but also the veterinary profession as embedded in society is influenced by these structures of (moral) significance. Tiffany Withcomb has famously pointed out a hidden curriculum in veterinary medicine (Withcomb, 2014) that shapes not only a certain habitus or language but also paradigmatic notions of animals and related 'normal' treatments. During training, a veterinarian will implicitly learn what kind of treatment is decent for a dairy cow, a beloved family dog or a rat (if met in a clinical encounter). This implicit attitude will pre-determine the scope of considerations in any clinical case. A veterinarian would usually not come up with the idea of kidney transplants e.g. in fattening pigs and would *a forteriori* prevent herself or himself from proposing such a treatment to a farmer.

Fragile frames – the matter of contingency

One last issue is significant regarding the frames of veterinary medicine. It is quite easy to see that the treatment of animals and the related ethical concerns are dependent upon particular socio-cultural contexts. However, a critique of ethical relationalism as mere reconstruction of a particular moral common sense does not touch this theory. These frames are not to be conceived of as fully determining. There is always wiggle room in particular situations since the status of being a patient is not a matter of yes or no. Situations as well as individuals are singular and require a sensitivity and responsiveness that might warp the structure of recognition we are familiar with. In any particular case we act within a general framework but a particular case is never a mere case of a rule (Huth, 2016: 128). Thus the critical potential of the veterinary profession is manifest in any clinical encounter in which common practices might be called into question.

Conclusion

This paper aimed to propose a relational approach to veterinary ethics. On the first glance, one could assume that the differential treatment of animals in veterinary practice is an ethical inconsistency. Is it justifiable that veterinarians more or less equally take part in animal experiments, contribute to efficiency in meat production and dairy industry and in clearly life-extending measures in pets – although these different animals are substantially equal in terms of cognitive abilities? A view like this could sharpen moral conflicts veterinarians experience on a regular basis anyway. Things change if we adopt a relational approach to veterinary ethics. In a relational approach to veterinary ethics we consider animals according to their significance in our life-world. Then a differential treatment of animals is potentially legitimized by the fact that veterinarians act within social frames that pre-determine which animals appear to what extent as patients – within the range from patients in the 'full sense' to liminal patients. However, any singular case might thwart this pre-determination and lead to a re-consideration of otherwise tacit structures and habits.

[2] Here, the word natural of course does not point at a natural predestination of the animal but at a social normality and value-laden expectations.

Acknowledgement

Gratitude is owed to Elena Thurner at the Messerli Research Institute who contributed to this paper with important suggestions.

References

Ashall, V. and Hobson-West, P. (2017). Doing good by proxy: Human-animal kinship and the 'donation' of canine blood. Sociology of Health & Illness 39 (9), pp. 908-922. Beauchamp, T.L. and Childress, J.F. (2001). Principles of Biomedical Ethics. Oxford University Press, Oxford, UK.

Butler, J. (2009). Frames of War. When is life grievable? New York, Verso.

Cohen, N.E. and Stassen, E. (2016). Public moral convictions about animals in the Netherlands: culling healthy animals as a moral problem. In: Meijboom, F.L.B. and Stassen E. (eds.) The end of animal life: a start for an ethical debate. Wageningen Academic Publishers, Wageningen, The Netherlands, pp. 137-148.

Crary, A. (2010). Minding what already matters. A critique of moral individualism. Philosophical Topics Vol. 38, Nr. 1, pp. 17-49.

Chervenak, F.A. and McCullough, L.B. (1996). The Fetus as a Patient: An Essential Ethical Concept for Maternal-Fetal Medicine. The Journal of Maternal-Fetal Medicine 5: pp. 115-119.

Diamond, C. (1978). Eating Meat and Eating People. *Philosophy* Vol. 53, Nr. 206, pp. 465-479.

Grimm, H. and Huth, M. (2017). One Health: Many Patients? A Short Theory on What Makes an Animal a Patient. In: Jensen-Jarolim, E. (ed.) Comparative Medicine. Disorders Linking Humans with Their Animals. Springer, Cham, Switzerland, pp. 219-230.

Hinchliffe, S. (2016). More than one world, more than one health: Re-configuring interspecies health. Social Science and Medicine, 129, pp. 28-35.

Huth, M. (2016). Humans, Animals and Aristotle. Aristotelian Traces in the Current Critique of Moral Individualism. In: Labyrinth. An International Journal for Philosophy, Value Theory and Sociocultural Hermeneutics, 18 (2), pp. 117-136.

Jeengar, K.M. (2014). Ovarian cyst in dairy cows: old and new concepts for definition, diagnosis and therapy. Animal Reproduction 11(2): pp. 63-73.

Kimera, S.I. and Mlangwa, J.E.D. (2016). Veterinary Ethics. In: ten Have, H. (ed.) Encyclopedia of Global Bioethics. Springer Reference, Cham, Switzerland, pp. 2937-2947.

Regan, T. (2004). The case for animal rights. University of California Press, Berkeley, USA.

Rosenberg, C. (1997). Introduction: Framing disease. Illness, society, and history. In: Rosenberg C. and Golden, J. (eds.) Framing Disease. Rutgers University Press, New Brunswick, USA.

Schmiedt, C.W., Holzman, G., Schwartz, T. and McAnulty, J.F. (2008). Survival, Complications, and Analysis of Risk Factors after Renal Transplantation in Cats. Veterinary Surgery, 37, pp. 683-695.

Singer, P. (2011). Practical Ethics. Cambridge University Press, New York, USA.

Springer, S. and Grimm, H. (2017). Hightech-Tiermedizin: Eine Herausforderung für die professionseigene Moral? Tierärztliche Umschau Vol. 72, pp. 280-286.

Verweij, M. and Meijboom, F.L.B. (2015). One Health as Collective Responsibility. In: Dumitras, D.E., Jitea, I.M., and Aerts, S.: Know your Food. Food Ethics and Innovation, Wageningen Academic Publichers, Wageningen, The Netherlands: pp. 144-149.

Withcomb, T. (2014). Raising the Awareness of the Hidden Curriculum. Veterinary Medical Education: A Review and Call for Research. In: JVME 41 (4), pp. 344-357.

38. Considering animal patients as subjects?

K. Weich
Unit of Ethics and Human-Animal-Studies, Messerli Research Institute, Vienna, Veterinaerplatz 1, 1210 Vienna; kerstin.weich@vetmeduni.ac.at

Abstract

The following paper will start from two recent developments in veterinary medicine: the prioritisation of animal welfare in the list of veterinarians' responsibilities, as well as the beginning of a discussion about the ethical dimensions of veterinary treatment and the resulting focus on a patient-centered ethics. The animal patient is being conceptualized in analogy to the patient of (human-)medical ethics: as some sort of person (rather than an object or good), whose individual perspective ought to be an active part in the making of decisions concerning its (medical) treatment. This readiness to accept wide-ranging similarities between humans and animals, however, ends at the principle of autonomy: since animals lack the capacity to speak like humans, their own perspective cannot be included in the process of deciding over their treatment. This is a crucial point, the paper will argue, since it threatens to undermine the entire patient-status of animals: they are incapacitated, attributed to the advocacy of the veterinarian (and/or owner), excluded from the ethical discourse on them. The present paper suggests a second look at this situation by questioning the human-animal-divide that structures it. Taking the ethical status of the animal as patient, whose dignity and welfare are the first responsibility of veterinarians, seriously, requires to work out alternative ways, in which animals could be articulating themselves as subjects.

Keywords: veterinary ethics, animal patient, autonomy

Introduction: vets up for animal patients!

The call for veterinarians to make animal welfare their normative priority has been renewed recently. In the UK, for example, the British Veterinary Association (BVA) is promoting the role of the profession as animal welfare champions by the publication of its 'Vet Futures' project (BVA, 2016). The central aim of this report is to 'support members to maximise their animal advocacy potential and achieve good welfare outcomes for animals.' (BVA, 2016: 3). In Germany, the Federal Chamber of Veterinarians (BTK) adopted a new Code of Ethics in 2015. The discussions that took place in its preparation focused on the relationship between veterinary medicine and animal welfare. Although the protection of animals was consistently accepted as an important value for the veterinary profession, the question of its prioritisation raised some controversy. Accordingly, a widespread veterinary journal paper was entitled: 'What comes first for the vet: the sausage or the pig?' (Tölle, 2015). The general introduction of the Code of Ethics, however, gives a clear, although implicit answer to this question, by ranking 'the respect for the dignity of animals' at the top of the veterinarians' responsibilities (BTK, 2015). In consequence, the discussion about how veterinarians ought to respond to this ethical demand has flourished.

The advancement of the animal to the top of the long list of duties and responsibilities that characterize the modern veterinary profession is accompanied by the inclusion of clinical issues into the profession's ethics. As Abigail Woods explains, to provide veterinary treatment to an animal is no longer automatically considered treating it ethically, but has itself become an ethical issue: 'Instead of assuming that the veterinary care of animals was, by definition, ethical, veterinarians now recognised the multiple ethical dilemmas it posed.' (Woods, 2011: 13) Ethical discussions of veterinary treatment form a fundamental part in veterinary ethics, including reflections on the differences and similarities between medical and veterinary ethics (Johnston, 2013; Whiting, 2011). Consequently, veterinary ethics becomes more and more patient-centered (Grimm and Huth, 2017; Mullan and Fawcett, 2016).

EurSafe 2018 – DOI 10.3920/978-90-8686-869-8_38, © Wageningen Academic Publishers 2018

Veterinary ethics: animal or client?

In this paper, I try to develop some preliminary thoughts about the impact of this development on veterinary ethics. I start with some reflections about its effects on the tripartite relationship between veterinarian-animal-owner, which is generally referred to as a unique feature and source of the ethical challenges veterinarians have to face. Thus, the so-called fundamental problem of veterinary ethics is often expressed as:

> [...] should the vet surgeon give primary consideration to the animal or the client? (Batchelor and McKeegan, 2012)

or

> ...to whom does the veterinary owe primary obligation: owner or animal? (Rollin, 2006)

Although these quotes refer to the concept of a 'primary' obligation, the binary structure of 'either client or animal' suggests that a firm decision between these two options is possible and would solve the dilemma. This impression becomes even stronger in the influential analogy of the paediatrician and the garage mechanic, drawn by Bernard Rollin (2006): Is the veterinarian fundamentally more like the paediatrician or the garage mechanic?

This analogy is of great heuristic value. It reflects the ambiguous legal and moral status of animals in our society. It therefore helps to read common problems and ethical challenges in veterinary practice in the light of the inclusion or exclusion of animals into the moral community. In this analogy, the option of the garage mechanic refers to the anthropocentric model, in which the exclusion of the animal of the moral community is based on the argument of a human-animal-divide. The status of the animal patient is that of a mere object, therefore neither the veterinarian nor the owner has obligations towards the patient, because it is an animal. As high-lighted above, this option of the garage mechanic has been rejected by recent developments in the field of veterinary medicine. Instead, the veterinarian is called to take the animal into account like a paediatrician their patient. In this perspective, the animal is included into the moral community. By analogizing the veterinarian to the paediatrician, the animal appears in similarity to a human patient. Simultaneously, by making the analogy, the moral obligations of the veterinarian become defined as the particular moral obligations of a physician. A big part of Medical Ethics is concerned with the normative regulation of the power and the authority that physicians have as professionals. Medical ethics is striving for the prevention of abuse of the 'aesculapian authority', by implementing a patient-centered ethics. Physicians should not take advantage of the structural inferiority of patients, but supply care and protection in line with their patients' interests, values and subjective ways of experiencing illness and therapy.

Principles of medical ethics

In this context, the concept of the four principles, formulated by Beauchamp and Childress (2009), has become a very prominent model in medical ethics. This model succeeds in grasping the particular responsibility of surgeons for their patients. Surgeons should conduct themselves not only by taking into account clinical data (harm and benefit) and social effects (justice), but also by acting in respect for the autonomy of their patients. This means that they have to strive for a decision, which does not only involve or treat the patient in a fair way, but also includes his personal and individual perspective as an important fact in the making of this decision. For example: a surgical intervention in the face can prevent a disease but will lead to a notable scar. A physician that follows a patient-centered ethics should respect that this harm weighs differently from person to person: for an elderly patient the scar might be less of a problem than for a young patient. The central importance of taking into account the patient's perspective is reflected by the widely accepted model of the informed consent, as a prerequisite for a legal and morally legitimate medical intervention (Ashall, *et al.*, 2017). Medicine should aim for the wellbeing of the patient (Main, 2006). By implementing 'respect of the autonomy of the patient' as a guiding principle for reaching this aim, it is indicated, that a disrespect for the patient's personal perspective could in itself restrict his wellbeing. More importantly, it indicates that the definition of 'wellbeing' in each specific case dependents on each patient's individual perspective. To switch from the garage mechanic to the paediatrician thus means to include the animal patient's perspective into its medical treatment. More specifically: to owe (primary) obligation to the animal by taking the wellbeing of the animal patient as a measure for the ethical quality of veterinary treatment. This seems only natural when it comes to exploring the moral obligations of a veterinarian, whose self-image as well as the social expectation in the veterinary profession, has clearly left behind the role-model of a garage mechanic in favour of that of a physician, who is dedicated to provide service for his patients (Mullan and Fawcett, 2016).

Respect for the patient's autonomy

The argument of this paper is based on the observation, that although the focus on the patient in veterinary ethics clearly aims on fostering the protection of animals under and by veterinary care, those animals remain excluded from these considerations. This hypothesis will be explained by examining the current debate on the tensions between the principle of autonomy and the concept of animals as patients. My argument is that, in this respect, the theoretical reflection often fails to meet its own normative requirement of including the animal patient in the picture of veterinary responsibility. As a source of this failure, the binary logic of the human-animal-divide can be identified. Finally, I argue, that, in order to proceed in the process of including the animal patient's subjective perspective, the logic of identities, similarities and differences could be shifted to the logic of relation and response.

It can easily be seen that the model of an autonomous patient is closely related to a certain understanding of subjectivity as a human feature. In a medical context, the notion of autonomy is closely linked to the construction of a person capable of giving consent, a capacity, which is usually attached to speech and free will. This idea refers to the concept of the Cartesian subject, which is defined as individual, autonomous and sovereign 'man'. This concept of a subject is closely linked to the idea of an unbridgeable human-animal divide, in fact, the Cartesian subject is rooted in an exclusion of the animal: humans are subjects, exactly because they have the capacity to speak and express their subjective perspective which animals lack (Oliver, 2009). The idea of a subject that goes in line with the idea of the exceptionality of being human forms the basis of the current concept of a patient, which is rendered practicable by legal stipulations as well as by ethical considerations, resulting in an effectual protection of patients as subjects. To act accordingly to the principle of 'respect for the autonomy' of the patient means to make the consent of the patient a prerequisite for medical interventions.

Animal patient advocacy

To include animals in this construction seems, then, as necessary as it seems senseless (Ashall *et al.*, 2017). The principles of (human) medical ethics are indeed being adopted more and more by veterinary ethics. As in animal welfare or in the discourse of animal rights, the application of human medical ethics to veterinary medicine is justified by the similarities between animals and humans, between a veterinarian and a paediatrician, between veterinary and human medicine, between an animal patient and a human (child) patient. This analogy, however, usually ends end at the principle of autonomy: here, the conviction is that animals can never give a consent, because animals cannot speak. All of a sudden, the differences between humans and animals do matter. As a result, the animal, who was just supposed to be included into the concept of responsibility in veterinary practice, becomes excluded again. Because of lacking the ability to speak and give consent, the responsibility of finding a consent is restricted to an inter-human relation, e.g. between veterinarians and clients. In regard to the principle of respect for the autonomy, the differences between humans and animals undermine the analogy of a physician and his patient, as a consequence, the picture of the vet as a garage mechanic shows up again. The human-animal divide can't be bridged in the aspect of autonomy. One could now oppose, that this argument uses the wrong measurements and that the animal patient has to be compared to an underage patient, not to a 'normal' adult person who is able to reason. Patients, who are unable to reason are not unique to veterinary medicine, but form a widely discussed issue in medical ethics. This argument has great potential and deserves further scrutiny, but the point of my argument is another one. Instead of seeking to establish another form of similarity between animal and marginal cases of human patients, I would suggest embracing the difference between both – in order to proceed with the inclusion of animals into medical ethics. In medical ethics, 'the patient – whether human or animal – is at heart of the decision-making process and we should be clear that ethical frameworks explicitly reflect that importance' (Johnston, 2013: 39). Respect for the patient's autonomy ranges high in this context, as it secures the inclusion of the individual patient's perspective into his medical treatment. But in veterinary ethics, the differences between human and animals are taken as an argument for its exclusion, stating that 'the informed opinion of the animal in question is never available', (Fentener van Vlissingen, 2001) or that 'animals cannot represent their own interests' (Yeates, 2010). As a consequence, the concept of 'patient advocacy' is promoted (Fettman and Rollin, 2002).

Respect for animal patients as subjects

This conclusion can be interpreted as a result from an ethical framework which operates by defining identities as basis for determining corresponding moral duties. The identity which is at stake, is the idea of a (human) patient that requires respect as an autonomous subject. The problem now is, that the determination of an animal as lacking the sufficient similarity to this concept, is taken as an answer to the question of a patient-centered ethics in Veterinary medicine. Because animals are different from humans, there is no way to shape and articulate a direct responsibility, that is realized in the relationship between a vet and its patient.

For a revision of these unsatisfactory conclusions, the critique against moral individualism and the logic of inclusion and exclusion that has been formulated by animal ethics is promising. Kelly Oliver puts her finger on the general problem: '[...] starting from the premise that it is our similarities and not our differences that matter, how can we even imagine any sort of ethics that encounters animals in terms of their own interests as they experience them?' (2009: 29) In this sense, I suggest to make otherness and difference the starting point and affirmed aim of veterinary ethics. The ethical problem of how to provide protection for animal patients thus implies the consideration of many different forms of non-human subjectivity. The post-humanist discourse which works on the decentering of subjectivity has already been made fruitful for this work. Instead of the 'existence of centered subjects being accepted', a focus is put 'on the processes which produce animal subjectivity' (Holloway, 2007: 1044). We might further attune our sensitivities to the fact that veterinary medicine is not something that happens to animals, but is instead something that takes place with them. In veterinary practice, many ways of relating, responding, asking and listening, of cross-species, inter-subjective relations are already proving that the analogy of the garage mechanic is misleading. Ethical reasoning thus could be linked to the lifeworld experience of vets working with their patients, most of them aware that their animal patients are, as Nick Taylor puts it, 'embodied individuals living their lives entangled with humans and their own wider environment' (2012: 40).

Conclusion

In conclusion, we have seen that the current moral demand on veterinary ethics to conceptualize a specific responsibility of veterinarians for their patients is producing an unsatisfying situation: the attempt to include animals in the status of patients by describing them in similarity to human subjects yields the opposite result and excludes animals from their potential patient-status. A source of this re-exclusion could be found in the binary logic of the human-animal-divide: subjecthood is traditionally designed as human subjecthood, characterized by language, thought, free will, etc., therefore animals cannot gain the status of subjects. Recently, this logic has been criticized, e.g. in philosophy, and various alternatives have been formulated in animal ethics, such as the concept of anthropo-zoo-genesis by Vinciane Despret (2004) or of Becoming-With by Donna Haraway (2008). To investigate these theoretical conceptualisations could contribute to further meet the ethical demand of including animal patients in veterinary ethics.

References

Ashall V, Millar K.M. and Hobson-West, P. (2017). Informed Consent in Veterinary Medicine: Ethical Implications for the Profession and the Animal 'Patient, in: *Food Ethics*, August 21, 2017, doi:10.1007/s41055-017-0016-2. Accessed 24 January 2018.

Batchelor, C.E.M. and McKeegan, D.E.F. (2012). Survey of the frequency and perceived stressfulness of ethical dilemmas encountered in UK veterinary practice, in: *Veterinary Record*, 170, 19. Available at: http://veterinaryrecord.bmj.com. Accessed 21 July 2015.

Beauchamp, T.L. and Childress, J.F. (2009). *Principles of Biomedical Ethics,* Oxford, Oxford University Press.

British Veterinary Association (BVA) (2016): *Vets speaking up for animal welfare: British Veterinary Association animal welfare strategy.* Available at: https://www.bva.co.uk/uploadedFiles/Content/News,_campaigns_and_policies/Policies/Ethic_s and_welfare/BVAanimal-welfare-strategy-feb-2016.pdf. Accessed 22 February 2018.

BTK (Bundestierärztekammer Deutschland) (2015). *Ethik-Kodex.* Available at: http://www.bundestieraerztekammer.de/index_btk_ethikkodex.php. Accessed 24 January 2018.

Despret, V. (2004). The Body We Care For, in: *Body&Society* 10, 111-134.

Fentener van Vlissingen, M. (2001). Professional ethics in veterinary science – considering the consequences as a tool for problem solving, in: *Veterinary Sciences Tomorrow,* 1, 1-8.

Fettman, M.J. and Rollin, B.E. (2002): Modern elements of informed consent for general veterinary practitioners, in: *Journal of the American Veterinary Medical Association*, 221, 10, 1386-1393.

Grimm, H. and Huth, M. (2017): One Health: Many Patients? A Short Theory on What Makes an Animal a Patient, in: Jensen-Jarolim, E. (ed.): *Comparative medicine: disorders linking humans with their animals.* Springer, Cham, 219-230.

Haraway, D (2008): *When Species Meet.* University of Minnesota Press, Minneapolis.

Holloway, L. (2007): Subjecting Cows to Robots: Framing Technologies and the Making of Animal Subjects, in: *Environment and Planning D: Society and Space,* 25 (n.d.), 1041-1060.

Johnston, C. (2013): Lessons from Medical Ethics, in: Wathes, C. M. *et al.* (ed.) (2013): *Veterinary and Animal Ethics. Proceedings of the First International Conference on Veterinary and Animal Ethics, September 2011*, Chichester: Wiley-Blackwell, 32-43.

Main, D.C. (2006): Offering the best to patients: ethical issues associated with the provision of veterinary services. *Veterinary Record* 158, 62-66.

Mullan, S. and Fawcett, A. (2016): *Veterinary Ethics. Navigating Tough Cases.* 5mpublishing, Sheffield.

Oliver, K. (2009). *Animal Lessons. How They Teach Us To Be Human*, Columbia University Press, New York.

Rollin, B.E. (2006). *An Introduction to Veterinary Medical Ethics: Theory and Cases*, 2nd edition, Blackwell Publishing.

Taylor, N. (2012). 'Animal, Mess and Method: Post-Humanism, Sociology, and Animal Studies.' In Birke, L. and Hockenhull, J. (eds) *Crossing Boundaries: Investigating Human-Animal Relations,* Brill, Leiden.

Tölle, M. (2015). Ethik-Kodex in der Kritik: Im Zweifel für die Wurst? in: *VETimpulse* 24, 7, 1-3.

Whiting, M. (2011). Justice of Animal Use in the Veterinary Profession, in: Wathes, Christopher M. *et al.* (ed.) (2013): *Veterinary & Animal Ethics. Proceedings of the First International Conference on Veterinary and Animal Ethics, September 2011*, Chichester: Wiley-Blackwell, 63-74.

Woods, A. (2013). The History of Veterinary Ethics in Britain, ca. 1870-2000, in: Wathes, C. M. et.al. (ed.): *Veterinary & Animal Ethics. Proceedings of the First International Conference on Veterinary and Animal Ethics, September 2011*, Chichester: Wiley-Blackwell, 3-16.

Yeates, J. (2010). Ethical Aspects of Euthanasia of Owned Animals. *In Practice* 32, 70-73.

39. Handle with care: an alternative view on livestock medicine

J. Karg and H. Grimm*
Unit of Ethics and Human-Animal-Studies, Messerli Research Institute, Vienna, Veterinaerplatz 1, 1210 Vienna; johanna.karg@vetmeduni.ac.at

Abstract

The basic idea of care ethics is still of little importance in veterinary practice. However, veterinary medicine is undergoing a radical gender change as the percentage of female graduates has risen from 15 to 80% in the last thirty years. Since the ethics of care originates from a largely feminist approach, the radical increase of female veterinarians may require a reevaluation. A broader perspective and an alternative view on care is urgently needed to drive the discussion about ethics into new territory. The question of this paper is: Is it possible to transfer this feminist criticism into the field of (traditional) veterinary ethics? We will discuss the response to a practical case in livestock practice, which originates from a classic work of veterinary ethics, written by a luminary in this field, Bernard Rollin. It deals with a pregnant cow with a squamous cell carcinoma that the owner doesn't want to be treated. We will identify criteria that differentiate feminist from traditional ethical theories, to challenge the response to this dilemma from an ethics of care position. We will highlight missing aspects in traditional positions, for instance the dimension of relationships and the role of feelings in moral decision making, which are of significance to care accounts in ethics.

Keywords: veterinary medicine, feminism, gender change, ethics of care

Introduction

Much progress has taken place in veterinary ethics during the past years, albeit from within traditional lines of thought. Alternative views on veterinary medicine got lost on the way.

The term 'Ethics of Care' was first used in 1981, when during a study of moral development psychologist Carol Gilligan made the observation that women seem to attach more importance to personal relationships, to attentiveness towards individuals and – most significant of all – to feelings in ethical conflicts (Gilligan, 1982). As most of the traditional ethical theories, utilitarian or deontological, were primarily written by men, they didn't attach high importance to women, or their experiences and thoughts, and hence their moral issues were largely neglected. This led feminist ethicists to a reformulation of these traditional, entrenched attitudes to morality.

In their recently published, long-awaited book 'Navigating tough cases', Siobhan Mullan and Anne Fawcett give the impression, that care ethics cannot be considered a valuable source for veterinary ethics. For instance, they state that care ethics has severe limitations and has been criticized for being 'confusing, vague and underdeveloped' (Mullan and Fawcett, 2017: 54). However, taking into account that they only spend two pages on it and refer to two publications only, we take this as an example that supports a very different claim, namely that care ethics is often sold under value. To state our claim more strongly, this example stands for a missing debate on the potential of 'Ethics of Care'-approaches in veterinary medicine.

It seems like care ethics must undergo the same development cycle as in traditional ethics and society in order to leave behind its reputation as a 'confusing, vague and underdeveloped' theory and to be established firmly in veterinary ethics.

EurSafe 2018 – DOI 10.3920/978-90-8686-869-8_39, © Wageningen Academic Publishers 2018

As opposed to a widespread belief we demonstrate some very practical consequences of the Ethics of Care in the following case study. For an in-depth discussion of moral dilemmas in veterinary practice, it is very common to work with such case studies. We will present one response of a distinguished philosopher, Bernard Rollin, to a typical dilemma in livestock practice, to demonstrate how answers to moral questions are given within the classic veterinary ethics frame. The case we refer to is part of his pioneering book on veterinary medical ethics. The first part of his work depicts the struggling profession of veterinary medicine and the attempt to establish philosophical theories in veterinary medicine. The second part consists of case studies of typical dilemmas occurring in veterinary practice. In this part the author tries to give guidance to veterinarians by using the theories mentioned in the first part.

We will subsequently examine one of these cases ('Cow with Cancer Eye') with criteria addressed in care ethics and contrast them on that basis with traditional accounts.

The case: cow with cancer eye

The practical case to be discussed here is presented by the author as follows:

> 'You examine a cow in late pregnancy that has keratoconjunctivitis, blepharospasm, and photophobia due to an ocular squamous cell carcinoma. You recommend enucleation [surgical removal of the tumor] or immediate slaughter. The owner wants to allow the cow to calve, wean the calf, and then ship the cow. He does not want to invest in surgery for a cow that will soon calve.' (Rollin, 2006: 106)

A so-called 'Cancer eye' is the most common growth disease in bovine practice, especially in Austria and Bavaria, where the most frequent race, *Simmental-Fleckvieh*, is predisposed due to unpigmented eyelids. The circumferencial growth leads to symptoms such as swelling of the eyelids, limited mobility of eyelid and bulbus, itching, contamination of the bulbus, ulcerated keratitis, infestation with flies, swelling of the regional lymph nodes, exhaustion and loss of appetite (Dirksen *et al.*, 2002).

In this case surgery and elimination of the tissue affected by the tumor would be the therapy of choice for the veterinarian. But surgery is an expensive procedure and sedation, stress or anaesthesia can cause a loss of pregnancy. This may be enough reason for the farmer not to invest in surgery even if his animal is suffering. The veterinarian, in contrast, will probably not reach his decision as quickly: Should he leave the cow untreated?

A feminist debate about the response to this case

We will debate the case by plotting the traditional response based on classical ethics and then present feminist criticisms and an alternative viewpoint. We will thereby proceed by considering how the accounts address four important criteria/corner-stones of moral thinking: the moral point of view, universality, rationality and feelings, and the quest for an external source of normativity.

The moral point of view

The case as described above presents a typical moral challenge in veterinary medicine, where the three involved parties – veterinarian, animal owner and animal – seem to have different interests. In a situation like this, traditional ethics plead for the so called 'moral point of view', which involves keeping a detached, objective and impartial position in a situation of conflict, regardless of one's own subjective preferences. This standpoint provides the only one for a valid ethical reflection and fair decision.

Feminist ethicists consider the idea of this so-called 'moral point of view' hypocritical, and in contrast believe in the moral significance of relationships. Relationships do have a moral value and the fact that we are all integrated in a social environment makes a significant difference when entering a moral situation (Wendel, 2003).

Rollin takes this case to be a 'classic example' of the 'Fundamental Question of Veterinary Medicine: Does the veterinarian have primary obligation to the animal or the owner?' (Rollin, 2006: 106). Traditional ethics focuses primarily on the balance of conflicting interests. The veterinary analyst in this case tries to understand the farmer's position right from the beginning while assuming, 'that it is not in the economic interest of the farmer to treat the cow, as, for example it would be if the untreated eye were to eventuate in an aborted calf' (Rollin, 2006: 106). But throughout the rest of the text, he seems to reduce the case to a problem of animal welfare, leaving all other interests aside in order to allow for a single best answer: Treatment of the cow is without any alternative. 'If the veterinarian can persuade the client that by doing good [i.e. surgery on the cow], he will also do well, the issue is resolved.' In fact, it's all about the conviction of the farmer and his 'lack' of a 'personal ethic' (Rollin 2006: 106).

Under the camouflage of a 'moral point of view' the author takes a partisan position in favor of the animal without even indicating an alternative solution. It seems clear that the viable solution can only be based on the norm of taking care for animal welfare.

If animal welfare should be the only parameter discussed in veterinary ethics, cases can be easily solved. However, veterinarians seem to struggle with their ambivalent and often conflicting responsibilities towards animals and owners. In practical terms, they rarely have the experience that only the animal can be given priority. Instead they are faced with the challenge of finding the right balance between animal welfare and other relevant aspects, such as economic profit.

Universality

For traditional ethicists, a significant goal is to provide universal guidelines that can be applied to moral conflicts. This idea pictures moral conflicts as clear structured problems. But not every conflict can be solved like an arithmetical problem. In practice, this context-independency often leads to a simplification of a conflict such that individual agents and motivations are put aside and the specific complexity of every single conflict becomes irrelevant.

An Ethics of Care aims to discuss and find flexible approaches for every single conflict by adapting and conforming to the respective situation. Therefore, it is thought of as a situational ethics (Gruen, 2015).

Although Rollin deals with particular cases, their analysis is carried out along abstract principles he argues for in the first part of the book. In case described above, Rollin talks about the 'The Fundamental Question of Veterinary Medicine' (Rollin 2006: 106) and does not allow for a description that takes the case's individual characteristics into account. For instance, why does the farmer not want to pay? Is he short on money or is he just not willing to pay for the treatment? How seriously is the cow deprived of welfare after the treatment because of the loss of one eye? Does the veterinarian have the medical abilities required for the surgery? Such information about the particular details of the case make a difference according to an Ethics of Care. These questions would complicate the case in an enormous way. However, these details also make the case more challenging. Care ethics highlights the risk of simplifying a case to the extent that the moral problem erodes due to the application of abstract principles.

Rationality and feelings

In contrast to traditional accounts in ethics, an Ethics of Care argues for respecting feelings such as sympathy, empathy and concern for the involved parties, as one element to evaluate the situation. These instruments can help to evaluate the non-verbally expressed interests, especially when the situation includes suffering and grievances on the part of every involved agent, and demands a prudent handling of all these components. A personal involvement with the being in question provides important moral insights. While this position is core to care ethics, traditional accounts describe feelings as a secondary or even problematic part of moral thinking (Weidemann-Zaft and Schochow, 2012). 'Emotions are simply bodily reactions, whereas reasoning involves complex intentionality' (Nussbaum, 1996: 79).

In our case, the author recommends that the veterinarian should remind the farmer of 'the profound nature of human ocular pain' to 'soften his position' (Rollin, 2006: 107). To put oneself in someone's shoes to feel what he or she feels is a classic example of empathy. In this case it is used as an instrument to convince the farmer of the right way. A feminist ethicist would appreciate this idea of using one's feelings, not as a rhetoric means to a given end, but as a valuable source of moral thinking in a given situation.

The quest for an external source of normativity

To 'arm' the veterinarian for the situation described and to look for 'guidance' for his actions, the author of our example refers to an external voice: 'The Veterinarian's Oath and The Federal Law in the United States' (Rollin, 2006: 106). This perspective is further fostered with the closing argument of the case: 'Thus veterinarians should embrace social and legal change mandating control of animal suffering, for only through this avenue can their authority be made commensurate with their responsibility.' (Rollin, 2006: 107). In other words, the answer to the moral question 'What should I as a vet do?' is given on the basis of external sources of normativity. The guiding idea to structure veterinary responsibility is: What does one expect from a veterinarian?

The case study leaves the various relationships between the involved agents as an untouched source of normativity. A relational ethics would highlight these dimensions and take a closer look at them. Involved agents and their background, their particular relations and corresponding duties serve as an internal source for reflecting on one's moral responsibility. In other words, relationships carry normative weight (Held, 2006). To care for someone who is in need of care, leaves me with different responsibilities if the person in need of care is my child in to contrast if it were someone else's child. To scrutinize such relational sources of normativity is core to an Ethics of Care, but not the traditional account. Applied to the case in question, this account certainly does not provide a clear-cut solution, which is often aimed at in ethics. Contrary to the traditional account, an Ethics of Care does not and cannot rely on external normative principles but develops ethical decisions, not only in light of the situation but out of the situation, as a source of normativity. This is much less straight forward than in traditional accounts.

Conclusion

In this contribution, we used a paradigmatic case in veterinary medicine to illustrate some differences between a traditional ethical account and an Ethics of Care. Veterinary ethics in the traditional mindset runs the risk of losing important features of the case due to the – laudable but problematic – idea of reducing complexity by applying to fixed principles. Our aim was not to provide an alternative that solves all problems, but develop another way of approaching and reflecting on challenging cases.

Care ethics still plays a minor role in veterinary ethics, as it appears in latest works, such as 'Navigating Tough Cases' (Siobhan Mullan and Anne Fawcett).

This view could broaden the debate by putting the focus in moral conflicts on different aspects, namely the thorough description of the veterinarian-owner-animal relationships and the various sources of moral thinking. What the appropriate response to the Cancer Eye case from the perspective of care ethics would look like is beyond the scope of this essay. But developing such a response is an important task. In light of the fact that more and more females are taking up the veterinary profession, a feminist Ethics of Care might have a promising future in veterinary ethics, and may find its way into textbooks to come.

References

Dirksen,G., Gründer, H. and Stöber, M. (2002). Innere Medizin und Chirurgie des Rindes. 4. Auflage. Blackwell Verlag GmbH, Berlin, Germany, pp. 1196-1197.

Gilligan, C. (1982). In a different Voice: Psychological Theory and Women's Development. Harvard University press, USA.

Gruen, L. (2015). Entangled Empathy: An alternative ethic for our relationships with animals. Lantern Books, NY, USA, pp.33.

Held, V. (2006). The Ethics of Care. Personal, Political, Global. Oxford University Press, NY, USA. Kohlen, H. and Kumbruck, C. (2008) Care-(Ethik) und das Ethos fürsorglicher Praxis, SSG Sozialwissenschaften, USB Köln.

Mol, A., Moser, I. and Pols, J. (2010). Care in Practice. On tinkering in clinics, homes and farmes. Transcript Verlag, Bielefeld, Germany.

Mullan, S. and Fawcett, A. (2017). Veterinary Ethics: Navigating tough cases. 5M Publishing, Sheffield, UK, pp. 52-54.

Nussbaum, M. C. (1994). The Therapy of Desire. Theory and Practice in Hellenistic Ethics. Princeton University Books, USA, 79 pp.

Pauer-Studer, H. (1996). Das Andere der Gerechtigkeit. Moraltheorie im Kontext der Geschlechterdifferenz; Akademie Verlag, Berlin, Germany.

Paul, E.S. and Podberscek, A.L. (2000). Veterinary education and students' attitudes towards animal welfare. Veterinary Record 146: 269-272.

Quinn, C., Kinnison, T. and May, S.A. (2012). Care and Justice orientations to moral decision making in veterinary students. Veterinary Record 171: 446.

Rollin, B. (2006). Veterinary Medical Ethics. Theory and Cases. 2. Edition. Blackwell Publishing, Iowa, USA, pp.106-107.

Stanford Encyclopedia of Philosophy. (2009). Feminist Ethics. Available at: https://plato.stanford.edu/entries/feminism-ethics.

Weidemann-Zaft, S. and Schochow, M. (2012). Strukturelemente von Ethikberatung, Ethik in der Medizin, pp. 335-338.

40. Being a veterinary patient and moral status: a disentanglement of two normative dimensions

E. Thurner[*], *M. Huth and H. Grimm*
Unit of Ethics and Human-Animal-Studies, Messerli Research Institute, Veterinaerplatz 1, 1210 Vienna, Austria; elena.thurner@vetmeduni.ac.at

Abstract

In this paper, we argue on the basis of the intuitive use of the term 'patient' in everyday language that the veterinary concept of the animal patient is not intrinsically linked with the concept of moral status. We confirm our thesis in three steps by means of the case of a hypothetical dairy cow who suffers from mastitis. In the first step, we point out that Grimm and Huth's (cf. 2017) normative theory of the animal patient contradicts the intuitive use of the term 'patient' with regard to the case of a dairy cow whose mastitis is treated by a vet. While Grimm and Huth deny that the cow can be understood as a patient, the cow qualifies as a patient according to the use of the term 'patient' in everyday language. In the second step, we sketch the main points of Grimm and Huth's theory in order to explain why the authors refuse to consider the dairy cow a patient. In the third step, we focus on the intuition that the cow should be seen as a patient and demonstrate by means of a heuristic method that linking the concept of the patient with the concept of moral status is mistaken. We argue that the end of a veterinary intervention to promote an animal's health-related interests is sufficient in order to qualify as an animal patient. Opposed to Grimm and Huth (cf. 2017), we claim that respecting the animal's moral status is not a necessary condition. In summary, the link between the concept of the animal patient and the concept of moral status will be questioned and the argument put forward that it cannot be plausibly sustained.

Keywords: animal patient, moral status, veterinary ethics

Introduction

The concept 'animal patient' has gained prominence in the recent debate in veterinary ethics (e.g. Gardiner, 2009; Grimm and Huth, 2017; Weich and Grimm, 2017; Yeates, 2013: 1-31). It has been suggested that an animal should only qualify as a patient if the treatment she receives complies with the end of medicine (cf. Grimm and Huth, 2017: 225). We believe that the concept of the patient can be used as a demarcation line to distinguish interventions a veterinarian performs as a medical professional from interventions a veterinarian is asked to perform for non-medical reasons. Herwig Grimm and Martin Huth have contributed to the debate on the concept with a normative account of the animal patient that emphasizes a link between the concept of the veterinary patient and the concept of moral status. The authors identify sentience as the crucial capacity for conferring moral status to an animal. Trying to explain the positive connotation of the 'patient' they stress that an animal can only be considered a patient if she has moral status and if her moral status is respected by the performance of a veterinary intervention (cf. Grimm and Huth, 2017: 222). We argue that this link is not compatible with the intuitive use of the term 'patient' in everyday language. Therefore, the aim of our paper is to demonstrate by means of the intuitive use of everyday language that the concept of the veterinary patient is not intrinsically linked with the concept of moral status. We will support this view using an illustrative example: a hypothetical dairy cow suffering from mastitis who is being treated by a vet.

We develop our argument in three steps. In the first step, we will sketch a few cases of veterinary practice to examine the use of the word 'patient' in everyday language. In this context, we point out that Grimm and Huth's theory contradicts the intuitive use of the term 'patient' with regard to the case of a dairy

EurSafe 2018 – DOI 10.3920/978-90-8686-869-8_40, © Wageningen Academic Publishers 2018

cow who suffers from mastitis. For this reason, this case serves as our paradigm to elaborate our thesis. In the second step, we will outline the central points of Grimm and Huth's theory of the veterinary patient to illustrate why the authors refuse to consider the cow and most animals in food chains patients. We argue that the conflict between Grimm and Huth's theory and the intuitive use of the term 'patient' is rooted in the misguided idea to link the concept of the patient with the concept of moral status. Hence, in the third step, we will explain why this link is mistaken.

Veterinary patient or not – five cases

In this section we picture different cases of veterinary practice to illustrate the use of the term 'patient' in everyday language. Case one: A cat who is taken to the vet and gets preventive treatment in form of vaccine against cat plague. Most people will agree that we are intuitively inclined to consider the cat a patient. Case two: Let us imagine a dog, who has been injured by a conspecific and is taken to the vet to treat her wounds. Also, in this case, most people will not have a problem referring to the dog as a patient. Case three: Consider a male piglet who is castrated by a vet to prevent his meat from becoming boar tainted, which consumers dislike (cf. Singer and Mason, 2006: 50). We believe that calling the piglet a patient in this case will intuitively leave many people with unease. Case four: Take a cow infected with foot-and-mouth disease (FMD) who is culled for being a vector by a vet to prevent the disease from spreading. Many people will feel uneasy in view of speaking about the cow as a patient.

How is the use of the concept 'patient' legitimized? Certain veterinary procedures provoke an intuitive reluctance to refer to the concerned animals as patients. We assume that this reluctance is owed to the fact that the concept of the patient in our pre-theoretical language use is positively connoted. Grimm and Huth aim at explaining the concept's positive connotation with reference to an animal's moral status. By linking the concept of the patient with the concept of moral status, they stress that speaking of an animal as a patient mirrors respecting her moral status in the course of performing a veterinary intervention (cf. Grimm and Huth, 2017: 222). This can be seen in cases one and two, but not in cases three and four and explains the use of the word 'patient'. However, let us focus on a fifth case: A dairy cow suffers from mastitis and is treated by a vet to cure her disease. Are we intuitively inclined or reluctant to consider her a patient? From the point of view of our common language use, we have a strong intuition that the cow is a patient. However, Grimm and Huth disagree. With respect to this case, Grimm and Huth argue that the cow should not be referred to as a patient although this position is incompatible with the intuitive use of the term 'patient'.

Grimm and Huth: respecting the veterinary patient's moral status

Grimm and Huth illustrate their account of the animal patient with several cases of veterinary practice. In this context, they introduce the aforementioned case of a hypothetical dairy cow who suffers from mastitis. Grimm and Huth argue that the promotion of the cow's health-related interests by curing the mastitis is not the end of the veterinary intervention, but solely a means to the end of restoring the cow's productivity. Clearly, the cow is not treated for her own but for others' sakes: 'If the health of the cow is cared for to sustain its productivity, the end of veterinary action is not the cow's presumed health-related interest but the farmer's (in productivity). The cow's health is only secondary and a means to economic ends.' (Grimm and Huth, 2017: 225) As the cow is not treated for her own sake, her moral status – understood as being respected by treating the cow for her own sake (cf. Grimm and Huth, 2017: 222) – is disrespected in the veterinary intervention. Therefore, Grimm and Huth argue that it is not justified to refer to the cow as a veterinary patient (cf. Grimm and Huth, 2017: 225). Summarizing their account, Grimm and Huth conclude that an animal is a veterinary patient if the end of a vet's intervention is the promotion of the animal's health-related interest for the animal's own sake: 'In a nutshell, the argument goes that we can only refer to animals as patients as long as they are treated with regard to medicine's end, which is to protect and promote health-related interests.' (Grimm and Huth, 2017: 225)

As already mentioned, Grimm and Huth's theory contradicts the intuitive use of the term 'patient' in everyday language with regard to the dairy cow. Grimm and Huth presuppose that the dairy cow is in possession of moral status, which is disrespected since the cow receives medical treatment for someone else's sake. However, the cow's health-related interests are in fact promoted by curing the mastitis. This seems to be the case because the vet is responsive to the health status of the cow and to her health-related interests even though the initial intention of the intervention might be supporting the owner's economic needs. Despite serving as a means to restore the cow's productivity, it is obvious that the veterinary intervention's primary end is the promotion of the cow's health-related interests. Hence, it is intuitively clear that the cow is a patient even though her moral status is disrespected because she is not cured for her own sake but for someone else's. Even if one takes the point further and argues, like Hsiao (cf. 2017), that farm animals – despite being sentient – do not possess moral status in virtue of their lack of rationality, it is intuitively still plausible to refer to the cow with mastitis as a patient if she receives treatment. Given the fact that it intuitively makes sense to speak about the dairy cow as a patient even if her moral status is disrespected or even denied, we conclude that the concept of the veterinary patient and the concept of moral status cannot intrinsically be linked in our language use.

The veterinary patient and the promotion of her health-related interests

In the following we reflect upon the commonly shared intuition that the dairy cow whose mastitis is treated by a vet should be considered a veterinary patient. We will analyse this intuition, which rests on the use of everyday language, by means of a heuristic method to demonstrate why linking the concept of the patient with the concept of moral status is mistaken.

We argue that at the core of the intuition to call the dairy cow a patient is the notion that the end of a veterinary treatment to promote an animal's health-related interests is decisive for declaring the animal a patient. Therefore, we suggest the following modification of Grimm and Huth's theory of the veterinary patient (cf. Grimm and Huth, 2017: 226): An animal is a patient if the end of a veterinary intervention is the promotion of an animal's presumed health-related interests for her own sake or for others' sakes. Hence, the animal is constituted as a patient if the veterinary intervention is – maybe among other aims and intentions – responsive to the animal's needs with regard to health. Although Grimm and Huth's theory and the modification we propose bear a strong resemblance, there is a significant difference between these two approaches. Both versions deem it a necessary condition for declaring an animal a patient that the end of the vet's intervention is to promote an animal's health-related interests. However, Grimm and Huth hold that the promotion of an animal's health-related interests presupposes that the animal's moral status is respected by performing the intervention for the animal's own sake. But, as indicated above, assuming that a veterinary treatment aims at the promotion of an animal's health-related interests it makes perfect sense to talk about the animal as a patient even if her moral status is disrespected or her possession of moral status is denied. Therefore, we reason that one can consider an animal a patient without taking her moral status into account. Nevertheless, we stick to the premise of Grimm and Huth's argument that sentience is a necessary precondition for having health-related interests (cf. Grimm and Huth, 2017: 222). For this reason, we argue that only sentient beings qualify as patients.

By means of a heuristic method, we confirm in three steps our thesis that the concept of the veterinary patient is intrinsically not linked with moral status but with the end of a veterinary intervention to promote an animal's health-related interests. In Table 1 we systematize the results gained. The use of the term 'patient' is depending on a specific end of a veterinary intervention, namely to promote the animal's health-related interests. If a veterinary intervention aims at promoting a healthy or a sick animal's health-

Table 1. Referring to an animal as a patient depends on a veterinary intervention's end to promote her health-related interests (HRI).

	animal: healthy	animal: sick
HRI	patient	patient
¬ HRI	¬ patient	¬ patient

related interests, then the animal is a patient. If a veterinary intervention does not aim at promoting an animal's health-related interests, then the animal does not qualify as a patient in everyday language.

In Table 2 it is pointed out that the concept of the veterinary patient is not intrinsically linked with the concept of moral status. If respecting an animal's moral status would be decisive for perceiving her as a patient, the animal would not be considered a patient if her moral status is disrespected by a veterinary intervention or denied. However, it can be read from Table 2 that animals should be declared patients in these cases under the condition that the veterinary treatment aims at promoting their health-related interests. Therefore, the concept of the veterinary patient is not intrinsically linked with the animal's moral status but with the end of a veterinary intervention to promote an animal's health-related interests.

In Table 3 light is shed on the fact that a healthy or a sick animal is a veterinary patient if a veterinary intervention aims at promoting her health-related interests irrespective of the animal's moral status. Even if a healthy or a sick animal's moral status is disrespected or denied, the animal can be perfectly termed a patient if the treatment aims at promoting her health-related interests. If a veterinary intervention does not aim at promoting an animal's health-related interests, then one cannot refer to the animal as a patient. Neither in the case that the healthy or the sick animal's moral status is respected or disrespected by a treatment nor in the case that the animal's moral status is denied.

Table 2. Linking the concept of the veterinary patient with the concept of moral status (ms) is mistaken.

	animal: ms	animal: ¬ ms
HRI	patient	patient
¬ HRI	¬ patient	¬ patient

Table 3. A healthy or a sick animal is a veterinary patient if a veterinary intervention aims at promoting her health-related interests (HRI) irrespective of her moral status (ms).

	animal: healthy, + HRI	animal: sick, + HRI
ms	patient	patient
¬ ms	patient	patient

Conclusion

In summary, it is not the fact that we respect an animal's moral status but aiming at the promotion of her health-related interests what makes an animal a patient: An animal is a patient if the end of a veterinary intervention is to promote the animal's health-related interests for her own or for someone else's sake. An animal cannot be considered a patient if promoting her health-related interests is not the end of a veterinary intervention. As the end of a veterinary intervention to promote an animal's health-related interests is independent of the animal's moral status, we conclude that the concept of the veterinary patient cannot be linked with the concept of moral status. Hence, it follows that there are not only veterinary patients whose moral status is respected but also veterinary patients whose moral status is disrespected or denied. Therefore, the concept of the veterinary patient's positive connotation expresses that the promotion of an animal's health-related interests is the end of a veterinary intervention. Contrary to the account of Grimm and Huth, the concept of the patient is not positively connoted because an animal's moral status is respected, but because an intervention aims at promoting an animal's health-related interests.

On this basis we can explain the intuitive use and positive connotation of the term 'patient' in everyday language. Take again the intuition that the dairy cow whose mastitis is treated medically should be termed a patient: As the veterinary intervention aims at promoting the cow's health-related interests, she is a sick patient, whose moral status is disrespected – in virtue of being treated for others' sakes – or denied. Besides, it becomes clear why we have had the intuitive impression in the second section that we should speak about a patient in the case of the cat and the dog but not in the case of the piglet and the cow infected with FMD. As the end of the preventive vaccination against cat plague is the promotion of the cat's health-related interests, the cat is a healthy patient whose moral status is respected. The same applies to the dog. As the end of treating the dog's wounds is promoting the dog's health-relating interests, the dog is a sick patient whose moral status is respected. Opposed to these cases, the end of castrating the piglet is not the promotion of the piglet's health-related interests but the economic interests of the farmer and the aesthetic interests of the consumers. As the intervention does not aim at promoting the piglet's health-related interests, he is a healthy non-patient whose moral status is disrespected by the treatment or denied. The same holds true for the cow who is culled by a vet due to being infected with FMD. Obviously, the cow does not have health-related interests in being culled but in being cured (cf. Mepham, 2001: 347). Therefore, the cow's health-related interests are not promoted by the cull but only by curing her disease. As the end of culling the cow is the prevention of the disease's spread and not the promotion of the cow's health-related interests, the culled cow is a sick non-patient whose moral status is disrespected by the treatment or denied. To conclude, the intuitive reluctance to consider the piglet and the cow infected with FMD as patients is linked to the fact that the respective intervention does not aim at promoting the animals' health-related interests.

Acknowledgements

This paper is reconsidering the argument presented in Grimm and Huth (cf. 2017) who act as co-authors of this contribution.

References

Gardiner, A. (2009). The Animal as Surgical Patient: a Historical Perspective in the 20th Century. History and Philosophy of the Life Sciences 31: 355-376.

Grimm, H. and Huth, M. (2017). One Health: Many Patients? A Short Theory on What Makes an Animal a Patient. In: Jensen-Jarolim, E. (ed.) Comparative Medicine. Disorders Linking Humans with Their Animals. Springer, Cham, Switzerland, pp. 219-230.

Hsiao, T. (2017). Industrial Farming is Not Cruel to Animals. Journal of Agricultural and Environmental Ethics 30: 37-54.
Mepham, B. (2001). Foot and mouth disease and british agriculture: ethics in a crisis. Journal of Agricultural and Environmental Ethics 14: 339-347.
Singer, P. and Mason, J. (2006). Eating. What we eat and why it matters. Arrow Books, London, UK.
Weich, K. and Grimm, H. (2017). Meeting the Patient's Interest in Veterinary Clinics. Ethical Dimensions of the 21st Century Animal Patient. Food ethics. DOI 10.1007/s41055-017-0018-0.
Yeates, J. (2013). Animal Welfare in Veterinary Practice. Wiley-Blackwell, Oxford, UK.

41. Manifold health: the need to specify One Health and the importance of cooperation in (bio)ethics

F.L.B. Meijboom[*] *and J. Nieuwland*
Centre for Sustainable Animal Stewardship, Faculty of Veterinary Medicine, Utrecht University, Yalelaan 2, 3584 CJ Utrecht, the Netherlands; f.l.b.meijboom@uu.nl

Abstract

It looks like One Health (OH) is here to stay, given its endorsement at the level of policy and the way it shapes curricula of relevant scientific disciplines. This has not gone unnoticed by philosophers who critically appraise the concept and its normative assumptions (cf. Capps and Lederman, 2015; Thompson and List, 2015; Verweij and Bovenkerk, 2016). Though applauded for bringing together a diversity of disciplines to deliver solution to multifarious problems, the concept involves ambiguity. On the one hand, its added value sometimes remains unclear when narrowly understood as cooperation between veterinary professionals and their human health counterparts, on topics such as zoonotic disease and antimicrobial resistance. On the other hand, more broadly interpreted, One Health encompasses everything related to health, which may lead some to question its relevance and applicability. To address these vulnerabilities of the One Health concept we suggest a four-way distinction of functions. This avoids a narrow understanding by pointing out the relevance of health promotion, while at the same time putting flesh on the bones of OH as a full-fledged perspective on interspecies health policy. We believe these functions provide a compelling specification of One Health, enriching practical application as well as ethical reflection. We complement this proposal with an outline for an ethical framework to support decision-making at different levels within a One Health perspective.

Keywords: health, humans, animals, nature

One Health: linking humans, animals and the environment

The idea of One Health (OH) explicitly puts human and non-human animal health (hereafter 'animal') against the background of their shared environment (Zinsstag *et al.*, 2011). In that sense, it represents an ecological perspective on public health (cf. Lang and Rayner, 2012; Coutts *et al.*, 2014). OH understands human beings, as well as animals, as inextricably part of ecosystems. This recognition of the interplay between individual health and the environment also has its historical precursors, going back for example to the writings of Hippocrates (Barrett and Osofsky, 2013). Nonetheless, it took until the beginning of the 21st century, facing an upsurge of emerging infectious diseases, for the idea of OH to gain a strong foothold. In 2004, the Wildlife Conservation Society made an effort to bring together relevant partners to discuss this threat of emerging infectious diseases arguing that:

> (r)ecent outbreaks of West Nile virus, Ebola hemorrhagic fever, SARS, monkeypox, mad cow disease, and avian influenza remind us that human and animal health are intimately connected. A broader understanding of health and disease demands a unity of approach achievable only through a consilience of human, domestic animal and wildlife health – One Health (Cook *et al.*, 2004).

'One World One Health' was officially established, along with the so-called Manhattan principles to inform health policy (Cook *et al.*, 2004). This was followed up some years later in 2008 by an endorsement of the idea of OH by the American Veterinary Medical Association (AMVA), describing it as the need for a 'collaborative effort of multiple disciplines – working locally, nationally, and globally –

Svenja Springer and Herwig Grimm (eds.) ***Professionals in food chains***
EurSafe 2018 – DOI 10.3920/978-90-8686-869-8_41,

to attain optimal health for people, animals and the environment' (2008). Eventually, the UN endorsed OH via a tripartite position paper (2010), involving the World Health Organization (WHO), the Food and Agricultural Organization (FAO) and the World Animal Health Organisation (OIE).

While OH as an idea spans a wide range of factors, much of how it is understood and operationalized relates to zoonotic diseases – pathogens with the ability to jump species – and other threats associated with animals and our interactions with them, such as antimicrobial resistance (Lapinsky *et al.*, 2014). Given that the increasing threat of emerging infectious diseases has played a remarkable role in the development and endorsement of OH, this focus does not come as a surprise. But it could obfuscate other relevant aspects and avenues of thought. Now that One Health is acknowledged at several levels – e.g. determining (inter)national health policy, shaping curricula of medical sciences and professions (Gibbs, 2014) – the question of what OH involves exactly becomes rather relevant. This in part involves value assumptions about whose health matters (cf. Nieuwland and Meijboom, 2015). We suggest a further specification of OH in terms of four functions with a two-fold aim: to get a better grip on the idea of interspecies health as well as to avoid blind spots in ensuing moral deliberations.

Interspecific threats

OH is often linked to health problems related to emerging infectious diseases (EID), especially in the case of novel threats to human health, for example Avian Influenza, AIDS, Ebola, and SARS. It is no coincidence that these examples are zoonotic with a wildlife origin, as the majority of EID's are zoonotic (Taylor *et al.*, 2001), with almost 75% of zoonotic emerging infectious disease events having a wildlife origin (Jones *et al.*, 2008). The development of antimicrobial resistance in animals such as pigs is another example of a substantial threat to health. Transmission of these resistant bacteria to human beings could lead to infections not susceptible anymore to antimicrobial treatment. As One Health connects the health between human beings and animals against the backdrop of their shared environment, this human-centric way of viewing animals as source of infectious threat is not the only way. We could also look at interspecies relations from the perspective of animals. Furthermore, the focus cannot be restricted to zoonosis. For OH reverse zoonotic diseases, or anthropozoonoses, are equally relevant (Hanrahan, 2014).

Interspecific benefits

Others have already pointed out the potential of One Health to extend beyond interspecific threats to health. Hodsgon and Darling put it as follows:

> One Health is not limited to the prevention of zoonoses; it also encompasses the human health benefits from animals. Benefits to humans include animals used in the production of food for human consumption, animals as models for research of human diseases, and pet-assisted therapy (2011: 189).

What the authors call *zooeyia* signifies the human health benefits that follow from interaction or use of animals. Establishing the benefits of interspecies interaction is not straightforward. The correlation between, for instance, better cardio-vascular functioning and having the companion of a dog may not reflect a causal relation. The American Heart Association carefully states, based on the available literature that '(p)et ownership, particularly dog ownership, is probably associated with decreased CVD [cardio-vascular disease] risk' (Levine *et al.*, 2013). Interspecies interaction may be one of the ways in which human health could be promoted. Animals are inextricably part of human societies, which makes interspecies interaction a possible social determinant of health. However, similar to the bi-directional nature of threats to health between species, benefits might also accrue for animals; OH involves, as some have argued, a 'two-way affair' (Sandøe *et al.*, 2014). Indeed, we could also consider anthropoeyia: the health benefits to animals that follow from their interaction with human beings. What sort of human behaviour promotes the health and longevity of companion animals? Recognition of the social determinants of health opens up new ways of looking at human-animal interaction. While OH first and foremost emphasized the role of animals with regard to emerging infectious diseases, it also provides a framework to look at the possible health benefits of human-animal interaction. This prompts research into hypothesized health benefits of interspecies interaction, the results of which could inform policy measures aimed at promoting health at the level of social determinants (e.g. Rock *et al.*, 2015). So interaction with animals could represent a social determinant of health for human beings, and vice versa. Furthermore, rather than representing a social determinant of health for each other, human beings and animals might also share certain social determinants of health. For example, Sandøe and colleagues (2014) discuss the extent to which obesity in humans and animals might have certain determinants in common; information that could prove very helpful to promote the health of both humans and their animal companions.

Epistemic challenge

Animals and their health represent a potential source of information and knowledge. The integrated approach to health that underlies OH provides an interesting perspective of generating interspecies health knowledge. As a result, we distinguish the epistemic function of One Health. To what extent can health knowledge be translated and used across species? Despite its relation to modern health care, the issue of animal research is less prominently part of the OH discourse compared to zoonotic diseases. But how could we possibly exclude the involvement of animals in research from the scope of OH? Granted, the links are comparative rather than causal, but this does not appear relevantly different. The extent to which human health policy relies extensively on animal models resonates with the way in which the idea of OH strives to bring out the relevant connections between human and animal health. Instead of viewing animal research as merely an element of OH, it should be considered a major part of interspecies health policy. If animal research indeed contributes to human health, its importance cannot easily be overstated. The involvement of animals is pervasive throughout human medicine, as it is regarded a necessary precursor to clinical trials in human beings. Virtually all human medication and medical technology that required a clinical trial has been tested on animals. A comparative approach to human and animal health is thus considered vitally important in human medicine looking at both institutional design as well as flow of resources. By providing an interspecies perspective on health, OH thinking should encourage investigating the epistemic value of animal models for the benefit of humans and animals. Again, this is not restricted to the potential health benefits of humans only. To avoid bias, like in the case of interspecies threats and benefits to health, OH as a descriptive approach should highlight relevant pathways of knowledge transferal across species. The subsequent task is then to make a moral judgment about how to deal with these possibilities.

Ecological challenge

One Health goes beyond the idea of One Medicine by explicitly putting the interconnections between human and animal health against the background of their shared (natural) environment. Changes to the natural environment may play an important role in the emergence of infectious disease (Patz *et al.*, 2004). If one does not engage with the ecological background of disease emergence, and primarily focuses on the transmission of disease between human beings and animals, any subsequent measures to protect health could very well prove inadequate, symptomatic, and incomplete.

Ecological processes are important in understanding and addressing disease emergence but there is more to the link between human health and ecology than that. Ecosystem functioning is vitally important to support human and animal health. Moreover, ecosystems not only benefit human and animal health in terms of the services they provide, they are fundamental in the sense of representing necessary conditions for health (Holland, 2008). Sketching out the relevant links, human health inevitably relies on clean water, breathable air, pollination, fertile soil, stable climate, and so forth. Some may argue that this takes OH thinking too far. It becomes an all-encompassing framework containing all things health-related. This objection, however, is itself question-begging. Why would it be a problem to include the ways in which human and animal health is supported by ecosystem services? Moreover, if one would exclude these considerations, we might indeed end up with symptomatic and ad hoc solutions that are profoundly incomplete. Taking OH to its logical consequence entails a socio-ecological perspective (Zinsstag *et al.*, 2011; Stephen and Karesh, 2014). Social and ecological factors together make up the fabric of our shared environment, affecting individual health in various ways. This opens up a wide array of new questions for health professionals. By linking different biological domains and bringing them to bear upon each other, OH confronts for example the veterinary profession with new issues and background considerations. For example, if veterinarians have a responsibility at the level of veterinary public health, should they not also engage (perhaps primarily on a collective rather than individual level) with the ecological impact of livestock production?

Ethical challenges

We believe that the four functions represent a plausible specification of OH. In doing so, we have also highlighted novel ethical challenges that such an interspecies perspective on health raises. Linking humans, animals, and the environment not by means of threats to health only but also taking into account the positive health effects broadens the range of ethical concerns. We do not believe that OH requires a new kind of or specialization in ethics. Instead, we better rely on already existing expertise, but doing so in a way that traverses disciplinary boundaries both within the natural sciences as in bioethics. Developing such integrated perspectives can be stimulated in various ways, for example by setting up specific journals, congresses and research collaborations. Our contribution in this paper, in addition to the aforementioned functions, is proposing an outline of an ethical framework to inform OH policy. We consider four aspects essential to any framework in this field (Meijboom and Nieuwland 2017).

First, it requires reflection on fundamental moral assumptions, for example about the moral status of human beings and non-human individuals or collectives, and the conception of health. Again, these are up for discussion rather than fixed values of OH. And they represent the gamut of moral philosophy, emphasizing the need for transcending boundaries between disciplines both within natural sciences and (bio)ethics.

Second, in addition to the moral assumptions and bio-ethical concerns, the framework should address questions of justice, the configuration of health-related institutions, and representation. This pushes the need for transcending the boundaries of disciplines even more, going from moral to political philosophy. Health policy involves decisions about the distribution of resources, which are matters of justice. To what extent should an interspecies and ecological perspective on health policy consider the boundaries of nation states or those between species morally relevant? In addition to the question of distribution is the issue of representation. Not only whose health matters – cashed out in terms of resource distribution – but also who decides on these matters politically?

Finally, more of a challenge than a characteristic, the framework should be able to support decision-making in often multi-disciplinary contexts with regard to concrete health issues. In doing so, it needs to pay sufficient attention to the various moral assumptions involved without ending up in endless philosophical debate. The same holds for the capacity to view health in all its complexity and interconnectedness, while still finding ways forward for distinct issues.

Conclusion

We have proposed a specification of OH to provide common ground amongst the various participants in health policy. These four functions already uncover new ethical considerations. To support ethical decision-making, we have highlighted three aspects of what it means to do ethics in the context of ecological and interspecies health policy. Some will question whether such a framework actually supports ethical decision-making, or that it demands too much of individual (health) professionals and policy makers; it only makes things more difficult. But perhaps making things more difficult is inevitable if we take OH seriously.

We should emphasize that an ethical framework based on these elements is not for ethicists only. Every OH professional is confronted by normative choices, which, from an ethical perspective, demand awareness, analysis and justification. The ethical framework can help to do so. This way, ethicists can work together with researchers, policymakers, and others to address the ethical conundrums involved in realizing an ecological and interspecies approach to health policy.

References

American Veterinary Medicine Association (2008). One Health: a New Professional Imperative. One Health Initiative Task Force final report. Available at: https://www.avma.org/KB/Resources/Reports/Pages/One-Health.aspx. Accessed 26/1/2018.

Barrett, M. A. and Osofsky, S. A. (2013). One Health: Interdependence of People, Other Species, and the Planet. In: Katz, D.L. Elmore, J.G. Wild, D.M.G. Lucan, S.C. (eds.). Jekel's Epidemiology, Biostatistics, Preventive Medicine, and Public Health (4th ed.). Elsevier/Saunders, Philadelphia, USA, 364-377.

Capps, B. and Lederman, Z. (2015). One Health and paradigms of public biobanking. Journal of Medical Ethics 4/3: 258-262.

Cook, R.A., Karesh, W.B. and Osofsky, S.A. (2004). The Manhattan Principles on 'One World, One Health'. At: http://www.oneworldonehealth.org/index.html. Accessed 26/1/2018

Coutts, C., Forkink, A. and Weiner, J. (2014). The Portrayal of Natural Environment in the Evolution of the Ecological Public Health Paradigm. Int. J. Environ. Res. Public Health 11: 1005-1019.

Gibbs, E.P.J. (2014). The evolution of One Health: a decade of progress and challenges for the future. The Veterinary Record 174/4: 85-91.

Hanrahan, C. (2014). Integrative Health Thinking and the One Health Concept: Is Social Work All for 'One' or 'One' for All? In: T. Ryan (ed.). Animals in Social Work. Palgrave MacMillan, Basinstoke, UK, pp: 32-47.

Herzog, H. (2011). The impact of pets on human health and psychological well-being: fact, fiction, or hypothesis? Current directions in psychological science 20/4, 236-9.

Hodgson, K. and Darling, M. (2011). Zooeyia: an essential component of 'One Health'. The Canadian Veterinarian Journal 52/2: 189-191.

Holland, B. (2008). Justice and the Environment in Nussbaum's 'Capabilities Approach' Why Sustainable Ecological Capacity Is a Meta-Capability. Political Research Quarterly 61/2: 319-332.

Jones, K.E., Patel, N.G., Levy, M.A., Storeygard, A., Balk, D., Gittleman, J.L. and Daszak, P. (2008). Global trends in emerging infections diseases. Nature 451/21: 990-994.

Lang, T. and Rayner, G. (2012). Ecological public health: the 21st century's big idea? BMJ 345 e5466 doi: 10.1136/bmj.e5466.

Lapinski, M.K., Funk, J.A. and Moccia, L.T. (2014). Recommendations for the role of social science research in One Health. Social Science and Medicine dx.doi.org/10.1016/j.socscimed.2014.09.048

Meijboom, F.L.B. and Nieuwland, J. (2017). Gezondheid in meervoud: over ethische aspecten bij One Health en de noodzaak tot samenwerking binnen de ethiek. Preadvies NVBe, Utrecht.

Nieuwland, J. and Meijboom, F.L.B. (2015). One Health as a normative concept: implications for food safety at the wildlife interface. In: Dumitras, E.D. Jitea, I.M. and Aerts, S. (eds.) Know your food: food ethics and innovation. Wageningen: Wageningen AP. 1-434.

Patz, J. A., Daszak, P., Tabor, G. M., Aguirre, *et al.* (2004). Unhealthy Landscapes: Policy Recommendations on Land Use Change and Infectious Disease Emergence. Environmental Health Perspectives 112/10: 1092-1098.

Rock, M.J., Adams, C.L., Degeling, C., Massolo, A. and McCormack, G.R. (2015). Policies on pets for healthy cities: a conceptual framework. Health Promotion International 30/4: 976-986.

Sandøe, P., Palmer, C., Corr, S., *et al.*, (2014). Canine and feline obesity: a One Health perspective. Veterinary Record, 175/24: 610-616.

Stephen, C and Karesh, W.B. (2014). Is One Health delivering results? Rev. sci. tech. Off. int. Epiz. 33/2: 375-379.

Taylor, L.H., Latham, S.M. and Woolhouse, M.E. (2001). Risk factors for human disease emergence. Phil. Trans. R. Soc. Lond. B 356: 983-989.

Thompson, P.B. and List, M. (2015). Ebola Needs One Bioethics. Ethics, Policy and Environment 18: 96-102.

Verweij, M.F. and B. Bovenkerk (2016). Ethical Promises and Pitfalls of OneHealth. Public Health Ethics 9/1: 1-4.

Zinsstag, J., Schelling, E., Waltner-Toews, D. and Tannera, M. (2011). From 'one medicine' to 'one health' and systemic approaches to health and well-being. Preventive Veterinary Medicine 101/3-4: 148-156.

42. Entangled health – reconsidering zoonosis and epidemics in veterinary ethics

M. Huth
Unit of Ethics and Human-Animal-Studies, Messerli Research Institute, Veterinaerplatz 1, 1210 Vienna, Austria; martin.huth@vetmeduni.ac.at

Abstract

This paper aims to demonstrate how the conception of zoonosis and epidemics determines the practices of combatting these diseases in an ethically significant way. In current debates, the paradigm of transmission is the most powerful and most visible one – albeit it is not entirely exclusive among veterinarians and other involved professionals. The logic behind this view is a kind of a management of 'natural processes' that focuses more or less exclusively on the avoidance of transmission of pathogens. Within this frame, human or animal bodies and pathogens are conceived of as separate objects. To prevent transmission (for the sake of animal health, public health or economic constraints as differentially accepted justifications) one can use medication, vaccination, keeping conditions with maximized biosecurity, and – regularly, though controversial – culling. However, a shift in perspectives is possible. The concept of 'entangled health' enables us to reconsider zoonosis and epidemics as phenomena emerging from particular configurations of coexistence as introduced e.g. by the actor-network-theory in Latour. Hence, different notions of health and alternative possibilities for prevention and combatting diseases become more visible. The paper focuses particularly on two central issues: First, health and disease are understood as inevitably conditioned by various contexts. The ordinary conceptualization of health in veterinary medicine widely ignores sociocultural contingency and the dependency on the particular constellation of people, animals, and pathogens. Second, this leads to the possible recognition of a particular disease as an outcome of intensive animal use with tight profit margins particularly in industrial farming contexts. Animals are immuno-compromised by accelerated growth rates, by stress due to housing conditions, and by the lack of ordinary confrontation with pathogens to develop a 'normal' immune system. Moreover, mass culling as supposedly necessary measure to fight e.g. food-and-mouth disease (not a zoonosis, thus done for economic reasons) becomes open for criticism because this view is also an outcome of specific conditions of human-animal relations.

Keywords: culling, disease diagram, transmission, pathogenicity

Introduction

This paper focuses on conceptions and practices concerned with epidemics and zoonosis. Basically, this topic is unambiguously regarded as an important field of research and practice in veterinary medicine (and beyond, since it is the underlying issue of the paradigm of One Health) since approximately 60% of the known human diseases are infectious diseases transmissible between different species. The same applies to 75% of newly discovered diseases (AVMA, 2008: 3). Microbial pathogens are a threat for humans – either in terms of contagion to humans or in terms of economic concerns (Cohen and Stassen, 2016).

At the first glance, relevant actions to combat epizootic and zoonotic diseases seem to root in undisputed facts and necessary parameters of action. Hence, the frequent criticism of current practices of (prophylactic) mass culling in public debates and in the academic ethical discourse (Cohen and Stassen, 2016; Mepham, 2016; Degeling *et al.*, 2016) is often countered with reference to the unavoidable need of these practices against the backdrop of scientific (and thus incontestable) evidence (Latour, 1999;

EurSafe 2018 – DOI 10.3920/978-90-8686-869-8_42, © Wageningen Academic Publishers 2018

Hinchliffe, 2016: 30). Veterinarians as executive body seem to have no choice than to contribute to the eradication of 'a host species, to prevent the pathogen entering and contaminating new individuals and populations' (Degeling *et al.*, 2016: 2). However, a thorough philosophical analysis of the tacit assumptions beneath the relevant conception of health and disease opens possibilities of an ethical critique and for a re-configuration of (ill) health. A raised awareness of the way in which health and disease are framed could serve as valuable source of ethical considerations in veterinary practice and could support prophylaxis against infectious diseases.

The aim of demonstrating keystones for this re-configuration will be pursued in four steps. First, drawing from Charles Rosenberg's famous considerations, health and disease will be described as frames with social significance. Second, this theoretical account will be applied to the current notion of infectious diseases in animals. Third, the pitfalls of such an understanding will be pointed out. Fourth, an alternative approach will be introduced that opens the possibilities of rethinking epidemics and zoonosis.

Framing epidemics and zoonosis

The topic of infectious diseases represents a major concern in the ethical debates about veterinary interventions. Since there are variegated concerns at stake like animal welfare, public health, food security, economic interests, a proper narrative as well as a decent reaction to outbreaks of infectious diseases could be seen as a hard problem (Hinchliffe, 2016: 29). This paper proceeds from the assumption that these debates occur against the background of specific conceptions of disease that pre-determine the scope of argumentation and of relevant practices: 'Disease is irrevocably a social actor, that is, a factor in a structured configuration of social interactions.' (Rosenberg, 1997: xx) An ethics of veterinary medicine should therefore not only focus on the different lines of argumentation and the presumed ethical positions regarding moral duties towards humans and animals. It should also dig deeper to reveal the nature and significance of the underlying health concept.

Generally speaking, a lot of controversies regarding the concepts of health and disease are part and parcel of the turf war between naturalistic and constructivist approaches (Nordenfelt, 2007). Naturalistic conceptions rely on the assumption that disease represents a nothing but a biological dysfunction (Boorse, 1997). Primarily, there is no ethical or political concern immanent in the concept of disease. The critique of this approach is directed towards the suppression of illness (the experience of feeling ill) and of normative significance of disease. No moral obligation follows from a diseased body itself. The counterpart, constructivism, is usually seen as a notion of disease that focuses on its social constitution. It is called into question by pointing out the supposed arbitrariness of 'social' diseases (Hinchliffe, 2016: 30).

However, another way to comprehend the concepts of health and disease has been introduced by Charles Rosenberg (Rosenberg, 1992; 1997): His conception of disease cuts across the alleged opposition between naturalism and constructivism. A social construction of disease represents a tautology (Rosenberg 1997: xiv), but it is not a mere (or even arbitrary) construction. Hence, Rosenberg coins the concept of the frame for a decent understanding of (ill) health. '[D]isease is at once a biological event, a generation-specific repertoire of verbal constructs reflecting medicine's intellectual and institutional history, an occasion of and potential legitimation for public policy, an aspect of social role and individual – intrapsychic – identity, a sanction for cultural values, and a structuring element in doctor [or veterinarian, MH] and patient interaction' (Rosenberg, 1997: xiii). Therefore, we need to 'focus on the connection between biological event, its perception by patient and practitioner, and the collective effort to make cognitive and policy sense out of this perception' (*ibid.*: xvi). Disease is thus a natural process as well as a social, political and ethical issue: 'Disease concepts imply, constrain, and legitimate individual behaviors and public policy.' (*ibid.*: xiv) In a similar vein, Hinchliffe *et al.* conceive of ill health in the context of veterinary medicine as framed by 'disease diagrams' understood as bundles of concepts and practices that sustain our notions of a particular disease (Hinchliffe *et al.*, 2017: 26).

It is important to note that a disease diagram is not a simple representation but a specific sort of bringing disease into being, of enacting a certain disease (*ibid.*: 27). (Veterinary) medicine imposes a speculative mechanism to an otherwise opaque body (Rosenberg, 1997: xvii). The frame constitutes disease *as* a disease; it is an epistemological foundation including practical and ethical implications. Moreover, it is of relevance to point out that a disease diagram is usually not a solitaire. Even though a particular frame is to be considered as a hegemonic frame that largely determines theories and practices with respect to a disease, there are virtually always alternative frames that play a subordinate role (Hinchliffe *et al.*, 2017). However, this does not mean that they do not play any role or that they are not (tacitly) efficacious in influencing notions and practices. In the upcoming section, we will focus on the dominant frame as regards infectious diseases and will then contrast it to the valuable but widely concealed diagram of entangled health.

The hegemonic frame: the transmission-diagram

One can understand the notion of a disease as outcome of a particular 'disease diagram' that might coexist or be in tension with other diagrams (Hinchliffe *et al.*, 2017). Disease management could then be conceived of as a response to a hegemonic diagramming. In the case of epizootics and zoonosis, disease is usually diagrammed as an event that is caused by the invasion of pathogens in a previously unadulterated, hence healthy body. Thus, the circulation of microbes represents the most significant and mainly targeted health concern. According to the relevant EU regulation, '[t]he free movement of safe and wholesome food is an essential aspect of the internal market and contributes significantly to the health and well-being of citizens, and to the social and economic interests.' (EC No 178/2002: L31/1) But (global trade) forms an overarching milieu 'in which microbes have much greater opportunities to create new niches, cross species boundaries, travel worldwide very quickly and establish new beachheads in the populations of people and animals' (AVMA, 2008: 3).

This understanding obviously traces back to the discovering of microbes by Louis Pasteur and others and the practical implications (Latour 1999: 124). Right after this discovery, techniques of controlling and, first and foremost, excluding microbes qua pathogens have been established. A network of hygienists, farmers, equipment, domestic animals, veterinarians and official institutions has been rising to respond to the basically always imminent danger of contamination. Pathogens are now framed as outsiders to healthy lives. Hence, the attention as well as the relevant practices shifted from subjects or bodies to virtually invisible discrete entities that exist outside the patient's body (Hinchliffe *et al.*, 2017: 31f). The common tacit assumption is that a pathogen-free body (country, world) is the goal of interventions. This is actually at odds with numerous biological and veterinarian insights about the entanglement of various living beings also in terms of health.[1] Biosecurity as a discourse and as material practice (Hinchliffe *et al.*, 2017: 5) has been emerging to pursue this goal. Notably the rhetoric of control, resistance, self-defense and immunity displays the individual as discrete entity, as prone to invasion from a dangerous outside, and as to be armored against incursion of pathogens (Esposito, 2011). This underlines that a disease diagram serves as a legitimation of treatments of individuals and populations in potentially drastic manners to armor particularly human populations against the invasion of the pathogens. Donna Haraway points out a particularly telling example related to outbreaks of avian influenza in Southeast Asia: 'Perhaps the Bangkok 'Post' on January 23, 2004, got the war of worlds, words and images right with a cartoon showing migratory birds from the north dropping bombs – bird shit full with avian flu

[1] Hinchliffe *et al.* (2017: 40) report a shift across sciences from pathogens as predictable, episodic and calculable entities to less determineable and contingent matters.

strain H5N1 – on the geobody of the Thai nation.' (Haraway, 2008: 269) Moreover, economic interests are related to the same rhetoric as the potential financial losses (e.g. in outbreaks of foot-and-mouth disease that is not a zoonosis) also seem to justify drastic measures in terms of stamping out (Cohen and Stassen, 2016: 139). Hence, it seems that the policies of prevention and treatment are sometimes directed against 'diseases of economy' rather than against diseases of bodily beings.

The pitfalls of the transmission diagram

The diagram of transmission is powerful but it is neither conceived of as an unequivocally unproblematic nor as the only detectable account. However, it seems to build the basis for a bulk of regulations and practices that are currently widespread.

The obviously most basic assumption of the hegemonic disease diagram of transmission is the germ theory tracing back to Pasteur. It presupposes bodies that are (or should be) initially pathogen-free. Only a microbial invasion leads to contamination and, hence, to disease or at least health risks for humans and animals. Consequently, measures of control like 'strict protocols for cleaning and disinfection of livestock houses as well as persons entering houses, minimalisation of visitors, hygienic disposal of animal waste, etc.' (Verweij and Meijboom, 2015: 145) have been established and progressively intensified. Biosecurity is pursued through the building of borders and constant action that should prevent and preempt danger (Hinchliffe *et al.*, 2017: 26; 36). The prevention should be as perfect as possible.

However, this seems to lead to a dialectic move (Esposito, 2011) that turns biosecurity into a possible biohazard that corrupts health (of animals and humans) in two different respects:

First, the narrow focus to form clear barriers between an alleged inside and outside potentially has disadvantages for health. The lack of exposure to ordinary pathogens presents a significant problem. Hinchliffe *et al.* report the example of chickens that are kept virtually isolated from any environmental influences by germs (Hinchliffe *et al.*, 2017: 98). Even though these animals are kept 'biosecure' the emergence of infectious diseases of whole flocks cannot be entirely prevented as indicated in the following quote of a poultry processor: 'Typically, if you've got broilers, say, at 40 days, it can be clear of *Campylobacter*, and then, within 24 hours, the whole flock can be [positive].' (cited in *ibid.*: 102)[2] Hinchliffe *et al.* interpret this as an outcome of the fact that the development of a normal immune system is impaired by the relative absence of pathogens. In turn, the 'pathogenicity' (Hinchliffe *et al.*, 2017: xiv) of pathogenic microbes is increased; disease is now understood as the outcome of a complex interplay between hosts, microbes and environments (*ibid.*: 5f.). Moreover, this pathogenicity could also be heightened by breeding and keeping conditions dictated by economic constraints (due to tight margins): 'While ostensible contained and biosecure, the birds are arguably immune-compromised as a result of accelerated growth rates and continuous throughput of feed.' (Hinchliffe, 2016: 32) In addition, the density of tightly packed bodies contributes to pathogenicity. 'Protective barriers to a diseased 'outside', on this view, are of little significance when thousands of confined animals in close proximity seem to provide a ready-made disease incubation chamber.' (Hinchliffe *et al.*, 2017: 98) Thus, human control might produce new microbial environments that might not be understood as source but as a significant support for disease outbreaks.

Second, the non-vaccination policy in the EU established in the 1990s displays the attempt to perfectly eradicate microbial threats (Cohen and Stassen, 2016; Mepham, 2016; Aerts and De Tavernier, 2016). Stamping out instead of vaccination is primarily an economic matter but often pictured as a matter of (alleged) security, too. Vaccinations might fail to immunize all or a majority of individuals. Moreover,

[2] *Campylobacter* infections in humans (a food-borne disease) cause diarrhoe.

there is a basic problem, here exemplified by the case of avian diseases: 'Mass poultry vaccination can lead to problems identifying birds that carry the virus but do not have disease, making it difficult to control the spread of the virus to other host, including people.' (Degeling *et al.*, 2016: 5) However, prophylactic mass culling finds little acceptance in the society (Cohen and Stassen, 2016) and is suspected to cause severe animal welfare problems 'by methods that, because of the panic situation, were often inhumane' (Mepham, 2016: 123). If millions of individuals are culled within short time (like in outbreaks of avian influenza as well as BSE or FMD; Aerts and De Tavernier, 2016: 177), catching, stunning or killing could be performed in inappropriate ways.

Entangled health – an alternative frame

With the concept of 'entangled health', one could put forward a shift in perspectives that promises to mitigate these ethical concerns. In what follows, a conception of disease will be introduced that focuses on 'disease situations' (Hinchliffe *et al.*, 2017: 52) rather than on the mere transmission of microbes. The point of departure is no longer contamination but configuration (Rosenberg, 1992: 293). Hence, disease events are dependent on relations and interactions between bodies.[3] 'Health is more than locating disease threats or contaminants, it is understanding the ecologies and configurations of viruses, hosts, responsive bodies, reagents and the stresses of production systems.' (Hinchliffe, 2016: 33) Instead of pathogens, the concept of pathogenicity (as explained above) can be emphasized. 'A healthy host within a healthy population and environment is likely, for example, to reduce the pathogenicity of a microbe.' (Hinchliffe *et al.*, 2017: xiv) Consequently, disease emerges from a particular constellation in which pathogens play an important role but there is not necessarily a mono-causal determination. This also leads to an understanding of health and disease as a continual matter. Drawing from Canguilhem's influential theory of pathology (Canguilhem, 1991), one can understand health in a less dichotomous way than it is often conceived of in the context of epidemics and zoonosis: Disease is not the sheer opposite of health but a shift in the interplay between various actants (humans, animals and microbes). Health is thus not the absence of pathogens but the tolerance of a body as regards infractions of its normality. This tolerance relies on a homeostasis that extends over time and space; it has its history (exposure to pathogens to develop of a decent immune system) and exceeds the borders of the individual body as it is dependent on the mentioned relations of bodies, pathogens and environments.

A relational approach to infectious diseases must also mind the role of governance, infrastructures, scientific paradigms and economic structures. Consequently, one has to take into account a fundamental temporality and contingency and has to give up essentialist disease diagrams that take pathogens as ahistorical entities independently of relations and historical situations (Latour 2004; Hinchliffe *et al.*, 2017: 33). With regard to animals' and human health, it is crucial how we conceive of disease and health in a certain historical situation and a particular socio-cultural context. Policies and connected treatments of animals are deeply determined by the historically changeable notion of disease as an economic matter, as a matter of individual unadulterated (as in the Pasteur paradigm) or related bodies, as a matter of different sorts and degrees of obligations in a normative infrastructure, etc.

Conclusion

This paper has argued for a shift in perspectives regarding infectious diseases in animals. Ethical considerations are often limited to treatments (from vaccination to cure or to culling) but do not take into account disease situations and the relations and interactions between organisms as actants. Since the

[3] In this view, bodies are conceived of as acting in the sense of actants (Latour, 2004: 75). Inter-actions are not only possible for subjects in the traditional sense of autonomous beings but also for beings that are influential for the assemblage of beings they are in.

transmission paradigm tends to reduce health to pathogen-freedom and potentially provokes immune-compromised in sanitized, closed environments, a deep analysis of tacit assumptions and structures seems of ethical significance. A quote by Steve Hinchliffe seems to provide a decent concluding sentence:

> Health is more than locating disease threats or contaminants, it is understanding the ecologies and configurations of viruses, hosts, responsive bodies, reagents and the stresses of production systems. (Hinchliffe, 2016: 33)

References

Aerts, S. and De Tavernier, J. (2016). Killing animals as a matter of collateral damage. In: Meijboom, F.L.B. and Stassen E. (eds.) The end of animal life: a start for an ethical debate. Wageningen Academic Publishers, Wageningen, the Netherlands, pp. 167-186.

AVMA (2008): One Health: a new professional imperative. Available at: https://www.avma.org/KB/Resources/Reports/Documents/onehealth_final.pdf. Accessed 12 January 2018.

Boorse, C. 1997. A Rebuttal on Health. In: Humber, J.M and Almeder, R.F. (eds.) What is Disease?, 1-134. New York: Humana Press, USA, pp. 1-134.

Canguilhem, G. (1991). The normal and the pathological. Zone Books, New York, USA.

Cohen, N.E. and Stassen, E. (2016). Public moral convictions about animals in the Netherlands: culling healthy animals as a moral problem. In: Meijboom, F.L.B. and Stassen E. (eds.) The end of animal life: a start for an ethical debate. Wageningen Academic Publishers, Wageningen, the Netherlands, pp. 137-148.

Degeling, C., Lederman, Z., and Rock, M. (2016). Culling and the common good: re-evaluating harms and benefits under the one health paradigm. In: Public Health Ethics, 2016, pp. 1-11.

Esposito, R. (2011). Immunitas. The protection and negation of life. Polity Press, Cambridge.

Haraway, D. (2008). When Species Meet. University of Minnesota Press, Minneapolis, USA.

Hinchliffe, S. (2016). More than one world, more than one health: Re-configuring interspecies health. Social Science and Medicine, 129, pp. 28-35.

Hinchliffe, S., Bingham, N., Allen, J. and Carter, S. (2017). Pathological lives. Disease, space and biopolitics. Wiley Blackwell, Oxford, UK.

Latour, B. (2004). Politics of nature. How to bring the sciences into democracy. Harvard University Press, London, UK.

Mepham, B. (2016). Morality, morbidity and mortality: an ethical analysis of culling nonhuman animals. In: Meijboom, F.L.B. and Stassen E. (eds.) The end of animal life: a start for an ethical debate. Wageningen Academic Publishers, Wageningen, the Netherlands, pp. 115-136.

Nordenfelt, L. (2007). The Concepts of Health and Disease Revisited. In: Medicine, Health Care and Philosophy 10, pp. 5-10.

Rosenberg, C. (1992). Explaining epidemics and other studies in the history of medicine. Cambridge University Press, New York, USA.

Rosenberg, C. (1997). Introduction: Framing Disease. Illness, society, and history. In: Rosenberg C. and Golden, J. (eds.) Framing Disease. Rutgers University Press, New Brunswick, USA.

Verweij, M. and Meijboom, F.L.B. (2015). One Health as Collective Responsibility. In: Dumitras, D.E., Jitea, I.M. and Aerts, S.: Know your Food. Food Ethics and Innovation, Wageningen Academic Publichers, Wageningen, Netherlands: pp. 144-149.

Section 7.
Veterinary ethics: in practice

43. Veterinary responsibilities within the One Health framework

J. van Herten[1,2*] *and F.L.B. Meijboom*[3]
[1]*Wageningen University and Research, Hollandseweg 1, 6706 KN Wageningen, the Netherlands;*
[2]*Royal Veterinary Association of the Netherlands, De Molen 77, 3995 AW Houten, the Netherlands;*
[3]*Utrecht University, Faculty of Veterinary Medicine, Yalelaan 2, 3584 CJ Utrecht, the Netherlands;*
joost.vanherten@wur.nl

Abstract

In many cases a One Health strategy in zoonotic disease control results in the promotion of health of humans, animals and the environment. However, there are also situations where this mutual benefit is not that evident. For instance, when restrictions of veterinary use of antimicrobials to protect public health negatively affects animal health. This confronts veterinarians with a conflict of professional responsibilities. Society expects veterinarians to safeguard animal health and welfare as well as veterinary public health. But if a One Health strategy to combat zoonotic disease threats leads to a conflict of these interests, whose health should veterinarians protect? Working in the context of the food industry, veterinarians are sometimes forced to favour the economic interests of farmers over those of animals. With the One Health framework emphasizing the threat of zoonotic diseases for public health, veterinarians seem to have yet another reason to let human interests prevail. Our claim is that a more holistic perspective on One Health gives veterinarians an opportunity to strengthen their position as animal advocates. We will argue that the best way to safeguard human health is to promote the health of animals and the environment because prevention is better than cure. This implies the veterinary profession should promote animal husbandry systems in which the health and welfare of animals are integrated with attention to public health and the environment. We will substantiate this empirical claim by elaborating the role of veterinarians in antimicrobial resistance policies in the Netherlands and conclude that to serve public health, the central responsibility of veterinarians should be to champion animal health and welfare.

Keywords: veterinarian, public health, animal health, antimicrobial resistance

Introduction

The motto of the Royal Veterinary Association of the Netherlands (RVAN) is: 'hominem animalumque saluti', which stands for: 'to the health of humans and animals'. When taken literally one could argue that human health is the primary concern of veterinarians, because this is mentioned first. However in our opinion, the One Health paradigm gives us reasons to question this.

One Health can be defined as the collaborative effort of multiple disciplines – working locally, nationally, and globally – to attain optimal health for people, animals and our environment (American Veterinary Medical Association, 2008). In this view, especially veterinary and human medicine should cooperate to address health challenges at the human-animal-environmental interface. One Health's starting point is the awareness that the health of humans, animals and the environment are interconnected. Up till now, ideas about One Health have been developed beyond zoonotic diseases (Lerner and Berg, 2015; Meijboom and Nieuwland, 2017). Comparative or translational medicine and the positive effects of interaction with animals on the health of humans in hospitals, prisons or elderly homes, can all be covered under the One Health umbrella.

However, in this paper the focus is on the added value of One Health in the discussion on veterinary responsibilities in the context of zoonotic disease threats from intensive livestock farming. Our position

EurSafe 2018 – DOI 10.3920/978-90-8686-869-8_43, © Wageningen Academic Publishers 2018

is that One Health calls upon veterinarians to publicly address the underlying causes of the negative externalities of animal husbandry. As a consequence, veterinarians should promote animal husbandry systems based on animal health and welfare. The claim is that keeping our animals healthy and caring for their welfare will automatically reduce the risk of zoonotic disease transfer and thereby indirectly promote human health. To substantiate this, we will elaborate the problem of antimicrobial resistance.

One Health stresses that both animal and human health should be addressed in an integrated way, yet unfortunately the concept does not automatically provide moral guidance in situations of conflict. Nonetheless, the One Health concept can still be of relevance here. Although it is not an ethical framework, it helps to search for innovations to deal with these conflicts, such as the switch to a more preventive approach towards animal and human health.

Public health risks

In a small country like the Netherlands, the number of farm animals (125 million) exceeds the number of humans (17 million) by a sevenfold (Statistics Netherlands, 2016). In this context, animals and humans have a direct and mutual impact on each other's health. Especially food producing animals are associated with several public health risks. Some of these public health threats start from animal health problems that are related to intensive production systems. A one-sided focus on productivity and efficiency increases the incidence of diseases like mastitis in dairy cows and post-weaning diarrhoea in piglets (Fleischer *et al.*, 2001; Rhouma *et al.*, 2017). To correct these negative side-effects of intensive animal production systems, veterinarians started to prescribe considerable amounts of antimicrobials. Thus, highly efficient animal production systems became possible thanks to (prophylactic and curative) veterinary interventions.

It is clear that the use of antimicrobials in animals causes antimicrobial resistance in pathogenic and commensal bacteria in animals. Furthermore, there is scientific and public consensus on the importance of restrictive use of antimicrobials in animal husbandry due to its potential impact on public health (World Health Organization, 2017). This has also been acknowledged by the Dutch government. In the past years the Netherlands implemented strict reduction policies on the basis of the precautionary principle. This resulted in a 64,4% decrease of antimicrobial use in the period 2009-2016 (Veterinary Medicines Institute, 2017). Although farmers and veterinarians were worried about the negative effects of these restrictions, it seems that, to date, there is little evidence that animal health is seriously affected by the reduction of antimicrobials in food producing animals (Council on Animal Affairs, 2016). Apparently, veterinarians and farmers were able to safeguard animal health and hence food quality in other ways, such as vaccination and improving biosecurity. This seems to fit perfectly in a One Health perspective. However, if we consider that this case should be the rule rather than the exception in dealing with public health risks, this requires reflection on the responsibility of veterinary professionals.

Veterinary responsibilities

Societal expectations with regard to the professional responsibilities of veterinarians are changing. The work of veterinarians is no longer restricted to curative medicine in the interest of individual animals and their owners. A modern veterinarian must also have the competence to take collective and global perspectives into account and has responsibilities to care for animal welfare and public health as well (Meijboom, 2017). In reality, many veterinarians are struggling with this plethora of responsibilities. In daily practice veterinarians often have to deal with situations where human and animal interests are in conflict and no easy solutions are available.

Within the profession this has opened the discussion on veterinary responsibilities and how to deal with expectations from society. In the Netherlands, for instance, an open letter of a group of veterinarians organised in The Caring Vets led to an intense internal debate about the role of the profession in the transition towards a more sustainable type of animal husbandry (Burgers, 2017). The Caring Vets claim that a veterinarian's main interest should be animal welfare. In contrast, many Dutch farm animal veterinarians felt a responsibility for food safety, public health and increasing the economic profit of farmers, too. The tendency towards a greater veterinary concern for animal welfare is also reflected by a position paper of the British Veterinary Association (BVA), in which the authors argue that neither emotions nor economic factors may trump animal welfare considerations (British Veterinary Association, 2016). The BVA argues that improving animal welfare should be the profession's primary aim and motivation. The profession's increased concern for animal welfare seems one of the answers to the societal expectations regarding veterinary responsibilities. Although the recent attention focuses mainly on animal welfare – of which health is an essential part –, veterinarians are still expected to promote public health and health of the environment as well. This is not only a responsibility that is placed on the veterinarian, but is acknowledged by the profession too. It is broadly recognized that with their cross-species pathobiological expertise, veterinarians can make an essential contribution to public health.

However, to make these responsibilities operational is not easy. It is within the context of this internal and public debate, veterinarians have to deal with possible conflicts between these assumed professional responsibilities. For instance, to safeguard the health and welfare of piglets veterinarians treat post-weaning diarrhoea with antimicrobials like colistin, that is also considered a critical important antimicrobial for human health. Many veterinarians realize this, but feel they have no other options on the short term.

Although captured in the One Health concept, this framework does not yet prescribe how exactly the relation between human, animal and ecosystem health should be shaped. On one aspect, there seems consensus: it requires a closer cooperation between veterinary and human medicine. This resulted in successfully sharing and assessing signals of zoonotic diseases. However, this only covers part of the responsibility, because (1) the cooperation as such is not novel and (2) it represents a reactive strategy to address zoonotic health risks.

With regard to the first point, collaboration between veterinarians and human medicine is not as new as presented in One Health advocacy. In an historic analysis of zoonotic disease control in the Netherlands (1898-2001) Haalboom concludes that this cooperation was actually not the biggest concern. She claims that the real problem of ineffective zoonotic disease control was an underlying conflict of interest between public health and economic interests of the food industry (Haalboom, 2017).

This leads to the second restriction: the reactive strategy to public health risks. Some first interpretations of the One Health concept seem to suggest that veterinarians should mainly identify and combat zoonotic disease threats for the sake of human health (National Institute for Public Health and Environment, 2018). In that case One Health would be just another form of public health. In our opinion we should start from a broader, more holistic perspective of One Health that does not consider animals as mere sources of infectious diseases or other public health risks, but that recognizes the independent value of animal health and environmental health. Such a frame offers opportunities for veterinarians to address underlying reasons for public health threats of animal husbandry. This is a more effective strategy to deal with public health issues than reactive strategies. To structurally improve human and animal health, it is necessary to publicly discuss the negative externalities of animal husbandry. Based on their expertise veterinarians should play an important role in this debate, e.g. by stressing the animal and human health costs of animal production systems that are only focussed on cost efficiency. Moreover, in the long run this strategy could help veterinarians to prevent some value conflicts between animal health and public health. Rather than solving all problems, this implies that veterinarians should share their concerns with society via advices to public policy, formulating and communicating position papers and hands-on advice to animal owners. The Dutch approach to tackle antimicrobial resistance is an example of how this could work.

One Health in practice

The risk of transmission of zoonotic diseases from farm animals to humans is greatly influenced by the current system of intensive animal husbandry (Kimman *et al.*, 2013). Over the past decades, intensification of animal husbandry in the Netherlands is predominantly driven by reducing costs. Because of the pressure of global markets on food prices, retail and food producers force farmers to scale up and focus on maximizing production against costs that are as low as possible. Within this system, veterinarians and farmers still aim to keep animals healthy. However, due to these suboptimal economic circumstances addressing health problems with high use of antimicrobials seemed to be most effective. It was a cheaper option to (preventively) treat bacterial infectious diseases than to invest in biosecurity, feed or housing. As a result, until 2011 the Netherlands led the European charts of antimicrobial use in farm animals. Since then, the Dutch government issued strict policies for antimicrobial use in animals because of scientific and societal worries about public health risks. This resulted in the so-called 'Dutch Model' to reduce antimicrobial use. This approach appeared to be very effective and led to a significant reduction of antimicrobial us. as the policy has even been presented as an example for reduction policies in animal husbandry (Sheldon, 2016). This success of the strategy can be partly attributed to a stronger position of veterinarians on farms (one-on-one contract, herd health plan and regular farm visits), registration and benchmarking of antimicrobial use, vaccination strategies and better biosecurity.

In spite of the success, the current question in the Netherlands is whether it is possible to reduce antimicrobial use any further. The Dutch government and the Veterinary Medicines Authority both indicate that additional reduction is possible, certainly on farms that are registered as structural high users (Ministry of Economic Affairs and Ministry of Health, 2016; Veterinary Medicines Institute, 2017). This will probably be true. However, under current circumstances veterinarians often have no choice but to treat sick animals with antimicrobials. Instead of attempting to maintain animal health with the use of antimicrobials under difficult circumstances, we argue that veterinarians could achieve more for animal health – and therefore human health – if they publicly addressed the underlying system errors. Veterinarians could use their Aesculapian authority (cf. Rollin, 2002) to inform the public about the downside of intensive food production, like antimicrobial resistance.

There are many definitions of human, animal and environmental health. In veterinary medicine health is sometimes defined in terms of production or reproduction (Gunnarsson, 2006). From this perspective, one could maintain that a focus on production automatically implies that animals have to be healthy. However, it is clear that pushing animals towards high productivity increases the risk of certain diseases (Rauw *et al.*, 1998). Within the context of zoonotic diseases and antimicrobial resistance it is perhaps more appropriate to understand One Health as the absence of disease (cf. Boorse, 1977). This definition is understandable for veterinary practioners and gives them a reason to promote disease prevention. However, within the One Health framework we would suggest to understand health as resilience: the capacity or ability of an individual or a system to react to an external force and to maintain or return to a state of equilibrium (Bhamra *et al.*, 2011). This concept can be applied to all components of One Health: humans, animals and ecosystems. This approach provides grounds to transform animal husbandry systems in a way that resilience of animals is promoted. Consequently human and environmental health will benefit from this.

With their specific expertise, the veterinary profession is qualified to guide this transition towards a sustainable animal husbandry in which antimicrobial use can be minimized further.

For individual veterinarians starting this process can be difficult. The problem is not rooted in moral indifference. For most veterinarians their main concern is the health and welfare of the animals under their care. However, the influence of farm animal veterinarians is limited, because they provide veterinary care on request of farmers. These interventions are often directed at solving disease problems on the short-term.

Of course, veterinarians will advise farmers how to prevent diseases and have healthier animals on the long-term as well. This can be problematic because veterinarians are dependent on the will and the (financial) possibilities of farmers to change circumstances for the better of their animals. Therefore, we propose to start this change at the level of professional organisations of veterinarians. They have a different position. They can translate their concerns into position papers that transcend the problems on individual farms. Veterinary organisations should publicly address structural underlying causes in the food production chain that precede diseases in animals, like the transport of veal calves or growth rate in broilers. They should call on all responsible parties in the food production chain (farmers organisations, food producers, banks and retail) as well as the government and consumers to take action. By doing so veterinary associations will support individual veterinarians in their daily work to improve animal health and welfare on farm level. Veterinarians truly contribute to the idea of One Health if they collectively promote animal husbandry systems that are not primarily focussed on production but on animal health and welfare.

Conclusion

Current intensive livestock farming in the Netherlands is at odds with animal health as well as public health. In our opinion, to optimally promote One Health the primary interest of veterinarians should be to promote animal health. Instead of the current curative and control focussed perspective of many One Health strategies to address zoonotic diseases, veterinarians should advocate a more preventive strategy with healthy animals as a precondition of animal husbandry. This starts with veterinarians who give animal health and welfare priority over economic interests of the food production chain if human health is at stake. On an individual level this can be a difficult task for veterinarians since they are financially dependent on farmers. Professional organisations of veterinarians, however, can play an important role: they can publicly address these issues, call for the necessary structural changes in animal husbandry systems and support individual farmers in making a step from a curative control based approach to a strategy that focus on prevention. In the long run, such a preventive approach can mitigate possible conflicts between animal health and public health veterinarians are confronted with.

References

American Veterinary Medical Association (2008). 'One Health: A New Professional Imperative.' Www.Avma.Org. 2008. Available at: https://www.avma.org/KB/Resources/Reference/Pages/One-Health94.aspx.

Bhamra, R., Dani S. and Burnard K. (2011). 'Resilience: The Concept, a Literature Review and Future Directions.' *International Journal of Production Research* 49 (18): 5375-93. https://doi.org/10.1080/00207543.2011.563826.

Boorse, C. (1977). 'Health as a Theoretical Concept.' *Philosophy of Science* 44 (4): 542-573.

Burgers, A. (2017). 'Kom in Verzet, Dierenarts. Dit Is Geen Dierenwelzijn.' NRC. June 2017. Available at: https://www.nrc.nl/nieuws/2017/06/26/kom-in-verzet-dierenarts-dit-is-geen-dierenwelzijn-11289973-a1564575.

British Veterinary Association (2016). 'Animal Welfare Policy Position.' Available at: https://www.bva.co.uk/News-campaigns-and-policy/Policy/Ethics-and-welfare/Animal-welfare.

Council on Animal Affairs (2016). 'Antibiotic Policy in Animal Husbandry: Effects and Perspectives.' Available at: https://english.rda.nl/publications/publications/2016/03/01/antibiotic-policy-in-animal-husbandry-effects-and-perspectives.

Fleischer, P., Metzner M., Beyerbach M., Hoedemaker, M. and Klee, W. (2001). 'The Relationship between Milk Yield and the Incidence of Some Diseases in Dairy Cows.' *Journal of Dairy Science* 84 (9): 2025-2035.

Gunnarsson, S. (2006). 'The Conceptualisation of Health and Disease in Veterinary Medicine.' *Acta Veterinaria Scandinavica* 48 (1): 20. https://doi.org/10.1186/1751-0147-48-20.

Haalboom, A.F. (2017). *Negotiating Zoonoses: Dealings with Infectious Diseases Shared by Humans and Livestock in the Netherlands (1898-2001)*. Utrecht University.

Kimman, T., Hoek M. and De Jong M.C.M. (2013). 'Assessing and Controlling Health Risks from Animal Husbandry.' *NJAS – Wageningen Journal of Life Sciences* 66 (November): 7-14. https://doi.org/10.1016/j.njas.2013.05.003.

Lerner, H. and Berg, C. (2015). 'The Concept of Health in One Health and Some Practical Implications for Research and Education: What Is One Health?' *Infection Ecology & Epidemiology* 5 (1): 25300. https://doi.org/10.3402/iee.v5.25300.

Meijboom, F.L.B. and J. Nieuwland (2017). 'Gezondheid in Meervoud.' Utrecht: NVBE.

Meijboom, F.L.B. 2017. 'More Than Just a Vet? Professional Integrity as an Answer to the Ethical Challenges Facing Veterinarians in Animal Food Production.' *Food Ethics*, August. https://doi.org/10.1007/s41055-017-0019-z.

Ministry of Economic Affairs, and Ministry of Health (2016). 'Letter to the Parliament on Forthcoming Policy on Antimicobial Use in Livestock.' Available at: https://www.rijksoverheid.nl/documenten/kamerstukken/2016/07/08/kamerbrief-over-vervolgbeleid-antibiotica-in-de-veehouderij.

National Institute for Public Health and Environment (2018). 'One Health Portal.' Accessed March 23, 2018. https://onehealth.nl/Over_One_Health.

Rauw, W.M., Kanis, E., Noordhuizen-Stassen, E.N. and Grommers, F.J. (1998). 'Undesirable Side Effects of Selection for High Production Efficiency in Farm Animals: A Review.' *Livestock Production Science* 56 (1): 15-33.

Rhouma, M., Fairbrother, J.M., Beaudry, F. and Letellier, A. (2017). 'Post Weaning Diarrhea in Pigs: Risk Factors and Non-Colistin-Based Control Strategies.' *Acta Veterinaria Scandinavica* 59 (1). https://doi.org/10.1186/s13028-017-0299-7.

Rollin, B.E. (2002). 'The Use and Abuse of Aesculapian Authority in Veterinary Medicine.' *Journal of the American Veterinary Medical Association* 220 (8): 1144-1149.

Sheldon, T. (2016). 'Saving Antibiotics for When They Are Really Needed: The Dutch Example.' *BMJ*, August, i4192. https://doi.org/10.1136/bmj.i4192.

Statistics Netherlands (2016). 'Compendium Voor de Leefomgeving – Ontwikkeling Veestapel Op Landbouwbedrijven, 1980-2015.Pdf.' Centraal Bureau Statistiek. http://www.clo.nl/indicatoren/nl2124-ontwikkeling-veestapel-op-landbouwbedrijven-.

Veterinary Medicine Institute (2017). Usage of Antibiotics in Agricultural Livestock in the Netherlands in 2016. Available at: http://www.autoriteitdiergeneesmiddelen.nl/en/home.

World Health Organization (2017). 'WHO Guidelines on Use of Medically Important Antimicrobials in Food-Producing Animals.' Available at: http://www.who.int/foodsafety/areas_work/antimicrobial-resistance/cia_guidelines/en.

44. The role of Canadian veterinarians in improving calf welfare

*C.L. Sumner and M.A.G. von Keyserlingk**
Animal Welfare Program, Faculty of Land and Food Systems, University of British Columbia, 2357, Main Mall, Vancouver, BC, V6T 1Z6, Canada; marina.vonkeyserlingk@ubc.ca

Abstract

As advisors to dairy farmers, veterinarians are ideally positioned to influence the health of the dairy herd. Recent studies have demonstrated that dairy cattle veterinarians are also concerned about animal welfare, specifically on issues related to the housing environment, painful conditions and procedures, and managing disease of the adult animals. However, less is known regarding their perspectives on calf welfare. The goal of this study was to engage cattle veterinarians in an in-depth discussion to gain a better understanding of what they think about calf welfare, as well as provide clarity on what they feel is their responsibility in improving the welfare of dairy calves. Focus groups (n=5), that collectively had 33 participants representing five Canadian provinces and different geographical regions, were conducted as part of a continuing education workshop for Canadian cattle veterinarians. Two trained individuals undertook exploratory data analysis using applied thematic analysis where initial themes were identified and used to develop a detailed codebook to guide the coding process. All transcripts were coded twice to further test the validity of the initial codes and themes. Four major themes were identified: (1) Veterinarians prioritized calf health; frequently trading off this issue for other issues such as the calf's social needs. (2) Veterinarians see their role in improving calf welfare within the context of shifting norms of calf management, believed to be consequence of pressure from within their profession, but also arising from pressure from their clients and the public. (3) Veterinarians see their role as one of exerting social influence, primarily as an educator of their clients. Finally, (4) veterinarians see their responsibility in improving calf welfare as shaped by their personal values and professional ethics. Our results indicate that the veterinarians participating in this study are concerned about a range of calf welfare issues, believe they should be more involved in calf management on farms, and see their role in improving calf welfare as shaped by their own values, the needs of their clients, and the concerns of the public.

Keywords: focus groups, dilemmas, professional ethics, social influence

Introduction

Improvements in animal welfare at the farm level need to consider all relevant social actors that influence the farmer (Shortall *et al.*, 2016). Since dairy farmers consider the veterinarian to be an important advisor for animal welfare (Broughan *et al.*, 2016), it is worthwhile to consider the veterinarian's perspective on calf welfare. To our knowledge, no in-depth study has been conducted on Canadian veterinarian perceptions about improving calf welfare on farms, thus, the goal of this study was to learn (1) what Canadian dairy cattle veterinarians think are calf welfare issues, and (2) what they think is their responsibility in improving calf welfare. In this extended abstract we present our preliminary findings.

Methods

This study was approved by the University of British Columbia Behavioural Research Ethics Board under: *#H16-00421* and all participants provided written consent prior to participation. A one-hour focus group session was conducted during a continuing education workshop for Canadian cattle veterinarians on the topic of dairy cattle welfare. We created five focus groups with 33 participants (5 women, 28 men) from five Canadian provinces (British Columbia, Alberta, Ontario, Quebec, and Maritimes). All focus groups were audiotaped and transcribed verbatim. Using Guest *et al.* (2012), two

EurSafe 2018 – DOI 10.3920/978-90-8686-869-8_44, © Wageningen Academic Publishers 2018

trained individuals undertook exploratory data analysis using applied thematic analysis to answer our research questions. We first identified emergent themes and codes in the transcripts, then developed a codebook to code all transcripts, and finally recoded all transcripts to confirm the validity of the finalized codebook.

Results and discussion

Four major themes emerged during data analysis that given the changing norms of calf management within the dairy industry, our participants see themselves as having the ability and duty to play a proactive role in improving calf welfare. Theme 1 indicates that participant concerns are in alignment with other veterinarian concerns about calf health (Bauman *et al.*, 2016) and treating pain (Fajt *et al.*, 2011). Participants also expressed concerns about bull calf management, hunger, and inadequate nutrition; all having received little attention in the literature. Not surprising, participants prioritized calf health; frequently trading off this issue for other issues such as the calf's social needs. This view is consistent with other studies showing veterinarians are primarily concerned with production-related welfare concerns (Verbeke, 2009). Veterinarians should, however, not view welfare concerns such as social housing and health as binary issues and our study provides evidence that concerns about hunger, nutrition, and social needs are worthy of further exploration from the veterinary perspective.

Theme 2 indicates that participants see pressures external and internal to the dairy industry as primary drivers of change in calf management on farms. External pressures included negative public perceptions about cow-calf separation. Participants also felt that increased public awareness would alleviate these concerns, however, focusing solely on increasing awareness of dairy management practices that may not align with public values has been shown ineffective (Ventura *et al.*, 2016a). Internal pressures from within the dairy industry were primarily viewed as coming from farmers and veterinarians both placing pressure on farmers to improve. The social influence of farmers and veterinarians on farmer adoption of practices has been noted in other studies. Farmers indicate that veterinarians do have an important social role in advising clients on welfare (Kauppinen *et al.*, 2010); farmers exert social pressure on each other to adopt improved disease management practices on farms (Swinkels *et al.*, 2015). However, the social pressure from veterinarians exerted on farmers to improve calf welfare remains underexplored both in Canada and elsewhere.

Theme 3 indicates that participants embody different roles and use a variety of strategies to promote calf welfare improvements, most frequently as an educator responsible for teaching their clients. Cattle veterinary practitioners and researchers have reported inadequacy in improving welfare because they lacked knowledge on how to do so (Ventura *et al.*, 2016b). Our study participants offered a variety of strategies on how to improve calf welfare on farms, suggesting that they may be more comfortable or experienced in employing successful strategies. Participants discussed the need for face-to-face interactions with their clients to discuss their overall farm goals. This shift towards tailored advisement approaches that consider overall farm goals with farmer decision-making are needed for future success with veterinarian-farmer communication (Kristensen and Jakobsen, 2011). Further work on identifying ways that veterinarians address the diverse aspects of calf welfare would provide needed description of practices that could be promoted in the dairy industry.

Theme 4 indicates that participants have a range of judgements about calf welfare and their professional obligations towards calves, indicating that although veterinarians are primarily responsible for the health of the lactating herd (LeBlanc *et al.*, 2006), they extend their professional obligation and moral concern to the dairy calf. Ventura *et al.* (2016b) identified that a poorly understood conception of welfare was a primary barrier likely preventing improvements in cattle welfare. We suggest that understanding veterinarian judgements about calf welfare can help identify issues most amenable to change, for example, focusing on increasing veterinary advocacy for the use of pain relief during dehorning.

Participants also described a range of normative claims about dairy farmer values related to calf welfare including farmer motivation to improve welfare as related to both economics and concern for the calf. Understanding veterinarian claims about farmers is needed to identify how these two stakeholders can cooperate to improve calf welfare. Farmers feel veterinarians are unaware or dismissive of what they value for their farm goals (Derks *et al.*, 2012); in turn, veterinarian perceptions of what is important to the farmer can inhibit approaching a topic (Shortall *et al.*, 2016). What these examples, in addition to our study, imply are missed opportunities to improve welfare based on veterinarian assumptions about clients' perceptions of calf welfare. Participants described balancing professional obligations and moral concerns to improve calf welfare, and logistics such as time management. Dilemmas in balancing conflicting responsibilities have been described elsewhere with cattle veterinarian practitioners (Ventura *et al.*, 2016b). Our study participants' descriptions of navigating these dilemmas indicate that practices that improve calf welfare shift as the veterinarians' values shift, notably that treating pain is a moral duty for veterinarians and recognizing this is important to their clients.

Implications

This study provides an in-depth description of a diverse group of Canadian dairy cattle veterinarians think about calf welfare and what they should do to improve it on both professional and moral grounds. The views described in this study do not represent all Canadian veterinarians. Further explorations of the demonstrated concerns among a larger sample of veterinarians or veterinarians in different regions or countries would provide valuable insight about the extent that these are shared, and the potential influence of cultural contexts in shaping how veterinarians view improving calf welfare.

References

Bauman, C.A., Barkema, H.W., Dubuc, J., Keefe, G.P. and Kelton, D.F. (2016). Identifying management and disease priorities of Canadian dairy industry stakeholders. Journal of Dairy Science 99: 1-10.

Broughan, J.M., Maye, D., Carmody, P., Brunton, L.A., Ashton, A., Wint, W., Alexander, N., Naylor, R., Ward, K., Goodchild, A.V., Hinchliffe, S., Eglin, R.D., Upton, P., Nicholson, R. and Enticott, G. (2016). Farm characteristics and farmer perceptions associated with bovine tuberculosis incidents in areas of emerging endemic spread. Preventive Veterinary Medicine 129: 88-98.

Derks, M., van de Ven, L.M.A., van Werven, T., Kremer, W.D.J. and Hogeveen, H. (2012). The perception of veterinary herd health management by Dutch dairy farmers and its current status in the Netherlands: A survey. Preventive Veterinary Medicine 104: 207-215.

Fajt, V.R., Wagner, S.A. and Norby, B. (2011). Analgesic drug administration and attitudes about analgesia in cattle among bovine practitioners in the United States. Journal of the American Veterinary Association 238: 755-767.

Guest, G., MacQueen, K.M. and Namey, E.E. (2014). Applied thematic analysis. SAGE, Thousand Oaks, CA.

Kauppinen, T., Vainio, A., Valros, A., Rita, H. and Vesala, K.M. (2010). Improving animal welfare: Qualitative and quantitative methodology in the study of farmers' attitudes. Animal Welfare 19: 523-536.

Kristensen, E. and Jakobsen, E.B. (2011). Challenging the myth of the irrational dairy farmer; understanding decision-making related to herd health. New Zealand Veterinary Journal 59:1-7.

LeBlanc, S. J., Lissemore, K.D., Kelton, D.F., Duffield, T.F. and Leslie, K.E. (2006). Major advances in disease prevention in dairy cattle. Journal of Dairy Science 89: 1267-1279.

Shortall, O., Ruston, A., Green, M., Brennan, M., Wapenaar, W. and Kaler, J. (2016). Broken biosecurity? Veterinarians' framing of biosecurity on dairy farms in England. Preventive Veterinary Medicine 132: 20-31.

Swinkels, J.M., Hilkens, A., Zoche-Golob, V., Krömker, V., Buddiger, M., Jansen, J., and Lam, T.J.G.M. (2015). Social influences on the duration of antibiotic treatment of clinical mastitis in dairy cows. *J. Dairy Sci.,* 98: 2369-2380.

Ventura, B.A., von Keyserlingk, M.A.G., Wittman, H. and Weary, D.M. (2016a). What difference does a visit make? Changes in animal welfare perceptions after interested citizens tour a dairy farm. PLoS ONE 11(5): e0154733.
Ventura, B.A., Weary, D.M., Giovanetti, A.S. and von Keyserlingk, M.A.G. (2016b). Veterinary perspectives on cattle welfare challenges and solutions. Livestock Science 193: 95-102.
Verbeke, W. (2009). Stakeholder, citizen and consumer interests in farm animal welfare. Animal Welfare 18: 325-333.

45. The vet in the lab: exploring the position of animal professionals in non-therapeutic roles

V. Ashall[*] and P. Hobson-West
Centre for Applied Bioethics, School of Veterinary Medicine and Science, University of Nottingham, Sutton Bonington Campus, LE12 5RD, United Kingdom; vanessa.ashall@nottingham.ac.uk

Abstract

The role of the veterinary professional is complex but, until recently, has not been the subject of much academic research. Vets play an important role in a wide variety of social contexts, including in 'non-therapeutic' roles, for example in facilitating the use of animals in sport or for food production. This paper focuses on a further non-therapeutic example, namely the role of the vet in laboratory animal research. The research arena itself is characterised by ethical tensions and polarised opinion regarding the significance of, and justification for, using animals in harmful research for the primary benefit of humans. The aim of this paper is to explore some of the ethical questions raised by the position of the veterinary surgeon in the lab, focusing on the UK example. First, we outline the legislative responsibilities for the Named Veterinary Surgeon (NVS) under the Animals Scientific Procedures Act (ASPA) 1986, and compare this with the professional guidance for UK vets in all settings (RCVS Code of Conduct), making the argument that the NVS responsibilities to a scientific establishment under ASPA should be viewed as additional to the multiple responsibilities which are faced by all veterinary surgeons. A critical review of published and grey literature is then used to highlight how poorly the nature of this role is understood. In line with recent calls for improved sociological understanding of the veterinary profession, particularly in the case of potentially harmful veterinary interventions, we raise the need for careful empirical work which is focused on this complex professional role, and identify three preliminary research themes: The embodiment of professional expertise; the relocation of veterinary procedures; and reframing the veterinary 'patient'. Finally, we conclude more broadly that the current articulation of a veterinarian's role may not fit well with some aspects of the varied non-therapeutic work which is undertaken by the profession.

Keywords: responsibility, conflict of interest, veterinary patient, animal research, Named Veterinary Surgeon

Introduction

Whilst the image of veterinary surgeons as specialised physician for pets is reportedly gaining traction (Hueston, 2016), contemporary veterinary professionals, in fact, play a role in a wide variety of societal interactions with animals. However, the complex societal and ethical role played by the veterinary professional is only just beginning to emerge as a focus for sustained academic thought (Ashall and Hobson-West, 2017; Ashall *et al.*, 2017; Hobson-West and Timmons, 2015). Animal research is a field characterised by ethical tensions and polarised opinion regarding the significance of and justification for using animals in research for the primary benefit of humans (Davies *et al.*, 2016). Through a documentary analysis of UK legislation and peer reviewed literature this paper explores some of the ethical questions which are raised by the required involvement of veterinary professionals in UK laboratory animal research. The paper identifies three themes for future empirical research and draws preliminary conclusions which also have broader relevance for other 'non-therapeutic' veterinary roles, such as veterinary professional involvement in the production of food and in animal sports.

EurSafe 2018 – DOI 10.3920/978-90-8686-869-8_45, © Wageningen Academic Publishers 2018

The Named Veterinary Surgeon

All vertebrate animals used in research procedures conducted in Europe are protected under European Directive 2010/63/EU. This Directive is implemented in the UK through the Animals (Scientific Procedures) Act (ASPA, 1986), requiring triple level licensing of research establishments, research projects and individuals who are performing research procedures. Under Section 2C(5) of the Act, establishment licenses must specify five named individuals who hold specific legal responsibilities. It is this Section of the Act which requires all licensed research establishments to employ a Named Veterinary Surgeon (or NVS). With regard to the legal responsibilities of the NVS, section 8.6.1 of the Guidance on the Operation of the Animals (Scientific Procedures) Act states that:

> You are accountable to the establishment license holder for fulfilling your duties and responsibilities. In addition NVSs should also observe their professional responsibilities to the animals under their care, to other veterinary surgeons, to the public and to the Royal College of Veterinary Surgeons, as set out in the RCVS Code of Professional Conduct for veterinary Surgeons.

This role is therefore particularly complex in terms of accountability and professional responsibility, since the NVS is accountable to both the establishment license holder (under A(SP)A), whilst *also* having professional responsibilities to the animals under their care, the public, other veterinary surgeons and the Royal College of Veterinary Surgeons (under the Veterinary Surgeons Act). By comparing the guidance provided for vets working under A(SP)A or the Veterinary Surgeons Act, we therefore argue that the NVS responsibilities to a scientific establishment under A(SP)A should be viewed as additional to the multiple responsibilities which are faced by all veterinary surgeons, and which have previously been identified as a potential source of ethical conflict (Morgan and McDonald, 2007).

Documentary analysis and preliminary research themes

Both the RCVS Code of Conduct and the Guidance on the Operation of the Animals (Scientific Procedures) Act provide some detail on the legal role of the NVS. However, literature is sparse: In October 2017 a database search using CAB Abstracts, Medline, Pubmed and Scopus and the terms 'animal research' or 'laboratory' and 'veterinary profession' resulted in very few examples of literature focused on the role of the NVS. Thematic analysis of these documents has enabled the identification of three preliminary research themes which, we argue, raise important ethical questions and develop a focus for more careful and detailed empirical investigation.

In line with recent calls for improved sociological understanding of the veterinary profession (Hobson-West and Timmons, 2015), particularly in the case of potentially harmful veterinary interventions (Ashall and Hobson-west, 2017), this paper makes use of examples of qualitative social science (Hobson-West and Davies, 2017), working party reports (Poirier *et al.*, 2015), opinion published in veterinary journals (Anonymous, 2004) and published conference proceedings (Jennings and Hawkins, 2015; Gilbert and Wolfensohn, 2012) to illustrate our proposed themes.

The embodiment of professional expertise

Section 8.6.2 of the ASPA Guidance specifies that an NVS has a responsibility to advise the scientific establishment on implementing the 3Rs (Reduction, Replacement, Refinement) of humane scientific technique (Russell and Burch, 1959). This includes, for example, advising researchers who may have little surgical experience on both general and experimental surgical techniques. This arrangement appears designed to promote the welfare of research animals through bringing veterinary professional knowledge and experience to the design and conduct of research procedures. Whilst we were unable to find published work reporting public views on this arrangement, a recent conference paper based on informal surveys and workshops with members of the public argues that greater clarity is needed about vets' advisory role and responsibility for animal welfare (Jennings and Hawkins, 2015).

Furthermore, it is made very clear in both the ASPA Guidance and the RCVS code of conduct that the NVS may not physically undertake a research procedure, unless they are additionally licensed and therefore 'acting' as a researcher themselves (RCVS Code of professional Conduct for Veterinary Surgeons. Supporting Guidance 24.51).

A clear legislative boundary therefore exists between the permissibility of a veterinary professional advising a researcher as opposed to physically performing the research procedure, even though the latter might be argued to more reliably improve research animal welfare. Indeed, members of the public have also apparently expressed the belief that veterinary professionals do conduct all invasive animal research procedures in the UK (Jennings and Hawkins, 2015).This first research theme raises interesting ethical questions concerning why embodied, but not advisory involvement in harmful animal research, is currently considered outside the boundaries of veterinary professionalism in law. Further, it contributes to a broader ethical discussion around the definition of, and justifications for, 'harming' and 'benefitting' animals in the veterinary professional context (Ashall, 2017).

The relocation of veterinary procedures

An alternative understanding of the advisory role of the NVS is that it enables the relocation of specific veterinary procedures out of the clinic and into the research environment. In a working party report which focuses on the roles and responsibilities of the NVS it is argued that:

> Veterinarians have a solid base of knowledge and expertise in comparative pathology, diagnosis, prognosis, disease prevention and treatment, anaesthesia and surgery, pain recognition and control, breeding control, and euthanasia that is relevant to laboratory animals [...] They are therefore uniquely qualified to provide training, assessment and supervision on what is considered to be veterinary interventions for scientific procedures. Consequently it is strongly recommended that they are included in the provision of training and supervision to others. (Poirier *et al.*, 2015: 93 emphasis added)

The concept of 'veterinary interventions for scientific procedures' in the above quote is a really interesting example of how a clinical procedure may be reimagined and repurposed following its physical transfer to another setting (and see Ashall, 2017). The concept of relocation raises specific questions concerning the limitations of such 'transfers', including what is known of the social and ethical transferability of specific clinical practices (and see Ashall *et al.*, 2017). For example, in response to Poirier and colleagues (2015) we might question whether and how the veterinary act of euthanasia is relevant when performed by non-veterinarians in the research setting.

Reframing the veterinary 'patient'

Existing literature suggests that NVSs consider their treatment of research animals to be in many ways the same as that of other veterinary 'patients':

> Like a large animal vet, I have a duty to keep animals healthy until their usefulness has come to an end – in this case, the end being euthanasia after an experimental procedure, as opposed to slaughter and consumption. (Anonymous, 2004: 280)

In this example, the NVS draws little distinction between the vet's care of research animals and that of other societally 'useful' animals, an approach which is also taken by other authors:

> Veterinarians should endeavour to ensure that each animal under their care has a life that is worth living, whether in producing food, offering companionship, enhancing the diversity of our ecosystems or advancing science. (Gilbert and Wolfensohn, 2012: 169)

Furthermore, arguments have been made that the veterinary profession are also benefitting from animal research through applying the results of scientific progress to the treatment of their animal 'patients':

> All veterinary clinicians need medicines for their patients and must rationalise the use of laboratory animals to acquire knowledge leading to better treatments for other animals. In effect, along with human clinicians and all those who accept recommended treatments, they are end-users of the system. (Gilbert and Wolfensohn, 2012: 168)

The veterinary profession, and individual vets, are therefore framing both animals who are harmed to produce science and animals who benefit from consuming science as 'patients'. Future work should explore the cyclical nature of animals as veterinary patients, both inside and outside the laboratory, and the potential conflicts of interest and ethical tensions this may present. Such work will help address broader definitional questions of what it means to be a veterinary 'patient' or a veterinary professional (Ashall *et al.*, 2017; BVA, 2016).

Conclusion

This paper should appeal to veterinary professionals working in traditional and non-therapeutic roles, as well as social scientific researchers and ethicists, and encourage an improved understanding of the role of the NVS and other complex veterinary roles. In particular, the research themes identified demand a specific focus on the ethics of advisory as opposed to embodied veterinary labour, the relocation of veterinary procedures into non-clinical settings and the potentially multiple meanings of the term 'veterinary patient'. We propose that qualitative research on these themes offers a valuable opportunity to more fully understand a crucial area of science and develop social and ethical understanding of the veterinary profession.

The paper has identified the role of the NVS as ethically complex due to additional responsibilities which extend beyond, complicate, and potentially conflict with those held by all veterinary professionals. The RCVS Code acknowledges the potential for conflicting professional responsibilities and maintains that a vet's first priority is animal welfare:

> On occasions, the professional responsibilities in the Code may conflict with each other and veterinary surgeons may be presented with a dilemma. In such situations, veterinary surgeons should balance the professional responsibilities, having regard first to animal welfare. (RCVS, 2014: 5).

We argue that the existing challenge of prioritising animal welfare (Ashall *et al.*, 2017) is further complicated in non-therapeutic veterinary roles which involve additional responsibilities to specific publics or institutions.

Finally, recent claims that the UK veterinary profession should view their responsibilities to animal patients as equal to that of a paediatrician (BVA, 2016) appear complicated by the profession's involvement in fields where harm is intentionally inflicted. We therefore conclude that roles which involve veterinary professionals working in non-therapeutic contexts may challenge the wider professional norms of what it means to be a veterinary surgeon in the 21st century, and agree with calls (Rollin 2002) for a more complex analysis of the veterinary professional's moral role.

References

Anonymous (2004). Working as a Named Veterinary Surgeon In Practice; 26:5 279-281.

Ashall, V. (2017). 'Veterinary donation: To what extent can the ethical justifications for living human donation be applied to living animal donation?' Unpublished PhD Thesis, University of Nottingham.

Ashall, V. and Hobson-West, P. (2017). 'Doing good by proxy': human-animal kinship and the 'donation' of canine blood. Sociology of Health and Illness, 39, 908-922.

Ashall, V., Millar, K.M. and Hobson-West, P. (2017). Informed Consent in Veterinary Medicine: Ethical Implications for the Profession and the Animal 'Patient'. Food Ethics, https://doi.org/10.1007/s41055-017-0016-2.

ASPA. (1986). Animals Scientific Procedures Act [Online]. Available at: http://www.legislation.gov.uk/ukpga/1986/14/introduction. Accessed 1 October 2017.

BVA (2016). Vets speaking up for animal welfare: BVA animal welfare strategy. Available at:https://www.bva.co.uk/uploadedFiles/Content/News,_campaigns_and_policies/Policies/Ethics_and_welfare/BVA-animal-welfare-strategy-feb-2016.pdf. Accessed 20 March 2018.

Davies, G. F., Greenhough, B. J., Hobson-West, P., *et al.* (2016). Developing a Collaborative Agenda for Humanities and Social Scientific Research on Laboratory Animal Science and Welfare. PLoS ONE, 11:7.

Gilbert, C. and Wolfensohn, S. (2012). Veterinary ethics and the use of animals in research: are they compatible? In: Wathes, C. M., Corr, S. A., May, S. A., McCulloch, S. P. and Whiting, M. C. (eds.). Veterinary & Animal Ethics: Proceedings of the First International Conference on Veterinary and Animal Ethics Wheathampstead: Universities Federation for Animal Welfare (UFAW).

Hobson-West, P. and Timmons, S. (2015). Animals and anomalies: an analysis of the UK veterinary profession and the relative lack of state reform. The Sociological Review, 61, 1, 47-63.

Hobson-West, P. and Davies, A. (2017). Societal Sentience: Constructions of the Public in Animal Research Policy and Practice Science, Technology, & Human Values https://doi.org/10.1177%2F0162243917736138.

Hueston, W. (2016). Veterinary medicine: public good, private good or both? Veterinary Record 178, 98-99.

Jennings, M. and Hawkins, P. (2015). 'Public expectations of the NVS'. Paper presented to Laboratory Animals Veterinary Association Annual Conference, October 2015, Availabe at:https://www.researchgate.net/publication/303333752. Accessed on 15 March 2018.

Morgan, C. A. and McDonald, M. (2007). Ethical dilemmas in veterinary medicine. Veterinary Clinics of North America: Small Animal Practice, 37, 165-179.

Poirier, G. M., Bergmann, C., Denais-Lalieve *et al.* (2015). ESLAV/ECLAM/LAVA/EVERI recommendations for the roles, responsibilities and training of the laboratory animal veterinarian and the designated veterinarian under Directive 2010/63/EU. Laboratory Animals, 49, 89-99.

RCVS (2014). RCVS Code of Professional Conduct for Veterinary Surgeons [Online]. Available at: http://www.rcvs.org.uk/advice-and-guidance/code-of-professional-conduct-for-veterinary-surgeons/supporting-guidance/miscellaneous. Accessed 1 October 2014.

Rollin, B.E. (2002). The use and abuse of Aesculapian authority in veterinary medicine. JAVMA 220, 8, 1144.

Russel, W.M.S. and Burch, R.L. (1959). The Principles of Humane Experimental Technique. London: Methuen.

46. Antimicrobial resistance and companion animal medicine: examining constructions of responsibility

C. Cartelet[1], P. Hobson-West[1], S. Raman[2] and K. Millar[1]*
[1]Centre for Applied Bioethics, School of Veterinary Medicine and Science and School of Biosciences, University of Nottingham, Sutton Bonington Campus, Sutton Bonington, LE12 5RD, United Kingdom; [2]Institute for Science & Society (ISS), School of Sociology & Social Policy, University of Nottingham, Nottingham NG7 2RD, United Kingdom; clio.cartelet@nottingham.ac.uk

Abstract

Antimicrobial resistance (AMR) is increasingly recognised as a critical problem in both human and veterinary medicine. One of the key strategies to slow down and limit its development and spread is to promote good antimicrobial stewardship by medical professionals, i.e. doctors and veterinarians. Multiple studies have endeavoured to understand the nature and construction of good stewardship in human medicine – and to some extent in farm animal practice – and it is increasingly realised that different constructions of ethical responsibility are affecting prescribing practice. Equivalent work has seldom been carried out with companion animal veterinarians. This paper will describe some of the ethical issues raised by antimicrobial prescribing in order to explore the interplay between constructions of responsibility. Data were collected through face-to-face interviews with veterinary practitioners, then coded and analysed thematically. This paper will focus on a sub-set of the empirical results illustrating the challenges companion animal veterinarians face while handling antimicrobials, and how they construct the notion of ethical responsibility in antimicrobial prescribing. This initial work presents the perspectives of veterinarians on antimicrobial prescription, and promotes a wider appreciation of professional constructions of responsibility. Results are intended to inform the development of new clinical guidelines and strategies to promote antimicrobial stewardship.

Keywords: veterinary ethics, companion animals, pets, veterinary medicine, antimicrobial stewardship

Introduction

Antimicrobial Resistance (AMR) is one of the most important current public health issues, and has been described as the 'quintessential one health issue' (Robinson *et al.*, 2016: 377), as it involves complex links between use of antimicrobials in humans and in animals, as well as in the environment. The main strategy to try and counter the development of resistance in clinical practice is to adopt good antimicrobial stewardship that promotes principles of prudent use (e.g. only use antimicrobials if absolutely necessary, use fewer critical drug classes first, always finish course, etc.) (Teale and Moulin, 2012).

Clinical practice, however, both in human and veterinary medicine, has struggled to adhere to these principles. In companion animal practice, for example, antimicrobials have been shown to be overused and inappropriately used (Knights *et al.*, 2012). Using a qualitative inductive approach, this study sheds some light on why this might be the case by unpacking how companion animal veterinarians construct and manage competing responsibilities of their professional role and how these in turn may affect antimicrobial prescribing.

The role of the veterinarian through an ethical lens

Before focusing on the case of AMR it is useful to initially present the professional responsibilities of veterinarians in terms of ethics and examine why an understanding of the ethical dimensions of

Svenja Springer and Herwig Grimm (eds.) ***Professionals in food chains***
EurSafe 2018 – DOI 10.3920/978-90-8686-869-8_46, © Wageningen Academic Publishers 2018

clinician practice can support the development of professional codes of conduct. The core issue studied by veterinary ethics has been summarised by Rollin (2013: 15) as follows:

> Veterinarians find themselves enmeshed in a web of moral duties and obligations that can and often do conflict. In the first place, veterinarians obviously have an obligation to their clients. Second, veterinarians have an obligation to their peers in the profession. Third, veterinarians have, in virtue of their special social role, an obligation to society in general. Fourth, as is often forgotten, veterinarians, like all human beings, have an obligation to themselves. Fifth, and most obscurely, veterinarians have an obligation to animals.

This 'obscurely' qualification can be explained, given the extent and nature of the veterinarian's 'obligation to animals' will be influenced by wider societal views on animals. Consequently, veterinary ethics is linked to animal ethics, a field that has evolved immensely over the past few decades with discussions of concepts of 'animal rights' and 'animal welfare' (Garner, 2008), increased obligations towards animals both at individual and societal levels (Singer, 2006), and an emphasis on animals' sentience (Broom, 2010) and their ability to feel pain (Rollin, 1998).

This complexity of the veterinarian's professional role has also been recognised by other ethicists (Yeates, 2009), and has been highlighted in literature by various works, focusing on animal welfare (Morgan, 2009), pet euthanasia (Morris, 2012), and informed consent (Ashall *et al.*, 2017). Other work has also emphasised the individual and varied nature of veterinarians' beliefs and attitudes and how this informs how they tackle ethical challenges in their professional life (De Graaf, 2005). In this paper, we draw from and extend the scope of these debates to consider how antimicrobial stewardship poses ethical challenges for veterinarians in companion animal practice, and how these present similar challenges to other areas of veterinary work but may also give rise to some novel dilemmas as veterinarians manage their diverse roles and responsibilities

Methodology

This presentation is based on an ongoing study which aims at investigating and describing the ethical dimensions involve in clinical decision-making by small animal veterinarians in the UK, particularly as it pertains to antimicrobial prescription as inscribed in the wider context of AMR.

Twenty-five semi-structured interviews of companion animal vets working in practice in the UK were carried out using a qualitative inductive approach. Two of the veterinarians were locums working at a variety of practices, the others worked in ten different companion animal practices of varying size and structure.

Participants included graduates from 7 UK vet schools opened before 2007, as well as a couple of non-UK graduates. Years of graduation ranged from 1983 to 2016. Seven participants were male and 18 were female reflecting the feminisation of the veterinary profession (Irvine and Vermilya, 2010), particularly in the companion animal sector. Interviews lasted between thirty and seventy-five minutes and were recorded, then transcribed. They focused on veterinarians' experience while prescribing antimicrobials as well as their views and attitudes about antimicrobials and the issue of AMR generally and as it pertains to public health. For reason of space, quotes are not included in this extended abstract.

Results

The interviews provided in-depth and detailed accounts of the perspectives of veterinarians which were analysed under several thematic headings. This paper focuses on one theme: the question of constructs of responsibility in practice. During coding, this was divided into several, specifically: (1) uncertainty and decision making; (2) professionalism; (3) the role of the client; and (4) stress vs stewardship.

Clinical decision-making: a risk management exercise?

During interviews, veterinarians were keen to point out that medicine is not an exact science, but rather a combination of art, science and experience, that always involves some degree of risk. They also recognise that, in some cases, antimicrobials are not ideally handled but that circumstances and limitations they frequently encounter make it difficult to practice good stewardship. One common example is the case of a patient with a high fever of unknown origin; with little else to go on, most vets agree that antimicrobial administration seems appropriate even if they are not sure what they are treating.

Financial considerations were another recurring theme brought up in interviews with vets emphasising that while they can explain why gold standard diagnostics are recommended to clients, treatment is often decided by what option the owners can afford or agrees to pay for. Cost is presented as one of the main barriers to carrying out more culture and sensitivity testing, for example.

While some of the above might seem obvious, it is important to remember that the reality of practice as experienced by veterinary surgeons is complex and varies with each case. This complexity is in turn reflected by the difficulties clinicians reportedly encounter while trying to apply clinical guidelines to their work or practice evidence-based medicine. Veterinarians will often err on the side of caution and make the treatment decision that they perceive as being the least risky for the patient and best associated with their construction of their ethical duties, while likely to lead to a positive outcome. This is supported by existing literature, such as Knights *et al.* (2012) study about perioperative usage of antimicrobials by UK veterinarians, 79.9% of veterinarians agreed that if they were not sure if antimicrobial prophylaxis was needed before a surgery they would administer it.

In summary, these results suggest that clinical decision-making is fraught with uncertainty and that veterinarians' focus is usually on reducing the perceived risk to the patient, within a practical frame dictated by circumstances, such as clients' financial limitations. Indeed, Schnobel (2014) has argued that the current legal framework surrounding veterinary medicine is too focused on the duty of the veterinarian to fulfil clients' wishes – treating pets strictly as property – and fails to recognise and consider the importance of the veterinarian's duty of care to the animal patient and to the emotional dimension of the animal-human bond. Likewise, arguments have been made in human medicine for a patient-centred rather than evidence-based approach to clinical work (Parker, 2001). Similarly, considering veterinarians' own narratives on clinical decision-making, there is room to discuss the role of clinical guidelines and evidence in veterinary medicine, in particular how prescriptively they should be used, while taking into account specific circumstances pertaining to both the patient and the animal-owner bond.

Veterinarians as professionals – authority vs stewardship?

In these interviews, veterinarians argued that while clients have a right to choose between treatment options, antimicrobial prescription should not be dictated by client preference. Interestingly though, this is framed as defending the knowledge and authority that is seen as belonging to the profession, rather than as an effort to protect antimicrobials from being overused.

Although the need to follow stewardship guidelines while prescribing antimicrobials to avoid overuse and inappropriate use was mentioned by the vets interviewed, it did not feature as prominently in their discourse, particularly in situations that could be constructed as challenging to the veterinarian's role. Affirming the veterinarian's authority – and consequently social standing – and their unique prescription privileges was a much more common reaction. This could be due to the lack of interest and involvement of companion animal vets in public health (Wohl and Nusbaum), and / or a perceived loss of occupational prestige (Hobson-West and Timmons, 2016), possibly due in part to the feminisation of the sector (Irvine and Vermilya, 2010). Interestingly, this loss of occupational prestige is echoed by some interviewees who call for the profession as a whole to be promoted in the public eye.

In summary, veterinarians are protective of their prescribing privileges – particularly when it comes to drugs often perceived as desirable – such as antimicrobials. This protectiveness, however, seems driven mostly by a desire to preserve the authority linked to the expert knowledge of the profession rather than by a motivation to reduce antimicrobial misuse and overuse through careful stewardship. Furthermore, it is interesting to note that while the loss of occupational prestige of many professions has been recognised, how and why it has happened, as well as how it impacts the expectations and lived experience of professionals is still unclear.

Client as both adversary and collaborator

Overall, veterinarians reported during interviews that most clients did not explicitly request antimicrobials for their pets, or if they did, would often accept following discussion that they might not be needed. Some of the language used by veterinarians to describe challenging consultations is, however, quite enlightening as they often spoke of 'strength' and 'courage' as being necessary to deny a client's request. Refusing a client or admitting that an examination is unfruitful are seen as difficult endeavours that require fortitude on the part of the veterinarian. It is likely that communication challenges vary between professionals as well as being based on age, experience and gender.

Another dimension of antimicrobial prescribing that has been demonstrated in human medicine, but is difficult to evaluate through interviews, is the fact that clinicians might perceive a patient as desiring antimicrobials even if they do not explicitly request them (Altiner, 2004). While it is likely that similar situations arise in veterinary medicine, if and how they influence veterinarians, however, remains unknown. Although beyond the remit of this paper, it should be noted that veterinarians were overall very supportive of clients who had difficulties administering medication to their pets and would freely change which antimicrobial or antimicrobial formulation was used to help ensure compliance. For example, they would prioritise making sure the animal was getting a full course of antimicrobials over other considerations.

Stewardship and stress: a difficult balance to strike?

As noted above, veterinarians' relationships with their clients are sometimes described as adversarial, especially when clinicians feel challenged in their knowledge and authority. Some veterinarians recognised that they could be convinced to prescribe antibiotics in specific circumstances if they believe it to be harmless to the animal and the client is being 'difficult.' This tension appears to trump constructions of public health responsibility and these types of decisions are made mostly to try and avoid negative consequences for the veterinarians themselves.

The gender, age and experience of the veterinarians were also reported to influence how much pressure they might perceive from clients while making prescription and treatment decisions. For example, male veterinarians expressed that they were less often challenged by clients compared to their female colleagues; or older, more experienced, veterinarians believed that clients would more readily accept advice from them compared to from a younger colleague.

In summary, different vets might have widely different experiences depending on personal circumstances and the stage of their career. This might help inform how to best support clinical veterinary practice and highlights the complexity and changing nature of individual veterinarians' experience while practicing. This is particularly important considering the high levels of stress reported by veterinarians (Batchelor and McKeegan, 2012) and the historically high rate of suicide in the profession (Bartram and Baldwin, 2010).

Conclusion

Good antimicrobial stewardship is regarded as key to slowing down the spread and worsening of antimicrobial resistance and this also applies in the companion animal medicine field. This empirical analysis highlights the importance of grounding understanding of stewardship within the complex and intersecting responsibilities faced by veterinarians as well as the varied nature of their experiences and beliefs. Veterinarians are fond of highlighting the specificity of their work, for example using the saying 'cats are not small dogs'; similarly, policy-makers must recognise that veterinary work comes with its own challenges that cannot be addressed using generic guidelines. Guidelines must be tailored to the cases presented, recognise the role of veterinary professional judgment and allow them to address the specific ethical dilemmas they encounter. While this complicates the implementation of stewardship guidelines, policy development can be strengthened by working with rather than against the complexity of professional practice, specifically recognising the need to understand the ethical complexity of clinical decision-making. Future work might explore ways of building on ethically informed veterinary expertise to develop better approaches to antimicrobial stewardship.

Acknowledgements

We would like to thank all the participants for their time and involvement in the study and the Leverhulme Trust for funding this PhD as part of the 'Making Science Public: Challenges and Opportunities' research programme [RP2011-SP-013].

References

Altiner, A. (2004). Acute cough: a qualitative analysis of how GPs manage the consultation when patients explicitly or implicitly expect antibiotic prescriptions. Family Practice, 21: 500-506.

Ashall, V., Millar, K.M. and Hobson-West, P. (2017). Informed Consent in Veterinary Medicine: Ethical Implications for the Profession and the Animal 'Patient'. Food Ethics. [Online]. Available at: https://link.springer.com/article/10.1007/s41055-017-0016-2/fulltext.html. Accessed 31 January 2018.

Bartram, D.J. and Baldwin, D.S. (2010). Veterinary surgeons and suicide: a structured review of possible influences on increased risk. Veterinary Record, 166: 388-397.

Batchelor, C.E. and McKeegan, D.E. (2012). Survey of the frequency and perceived stressfulness of ethical dilemmas encountered in UK veterinary practice. Veterinary Record, 170: 19.

Broom, D. M. (2010). Cognitive ability and awareness in domestic animals and decisions about obligations to animals. Applied Animal Behaviour Science, 126: 1-11.

De Graaf, G. (2005). Veterinarians' discourses on animals and clients. Journal of Agricultural & Environmental Ethics, 18: 557-578.

Garner, R. (2008). The Politics of Animal Rights. British Politics, 3: 110-119.

Hobson-West, P. and Timmons, S. (2016). Animals and anomalies: an analysis of the UK veterinary profession and the relative lack of state reform. The Sociological Review, 64, 47-63.

Irvine, L. and Vermilya, J.R. (2010). Gender Work in a Feminized Profession: The Case of Veterinary Medicine. Gender & Society, 24: 56-82.

Knights, C.B., Mateus, A. and Baines, S.J. (2012). Current British veterinary attitudes to the use of perioperative antimicrobials in small animal surgery. Veterinary Record, 170: 646-646.

Morgan, C.A. (2009). Stepping up to the plate: Animal Welfare, Veterinarians and Ethical Conflicts. Philosophy Thesis, The University of British Columbia.

Morris, P. (2012). Blue juice: Euthanasia in veterinary medicine, Temple University Press.

Parker, M. (2001). The ethics of evidence-based patient choice. Health Expectations, 4: 87-91.

Robinson, T.P., Bu, D.P., Carrique-Mas, J., Fevre, E.M., Gilbert, M., Grace, D., Hay, S.I., Jiwakanon, J., Kakkar, M., Kariuki, S., Laxminarayan, R., Lubroth, J., Magnusson, U., Ngoc, P.T., Van Boeckel, T.P. and Woolhouse, M.E.J. (2016). Antibiotic resistance is the quintessential One Health issue. Transactions of the Royal Society of Tropical Medicine and Hygiene, 110: 377-380.

Rollin, B.E. (1998). The unheeded cry: animal consciousness, animal pain, and science, Ames, Iowa State University Press.

Rollin, B.E. (2013). An Introduction to Veterinary Medical Ethics Theory and Cases, Somerset, Wiley.

Schnobel, S. (2014). Veterinary Negligence: Ethical and Legal Perspectives on Formulating a Duty of Care. ReValuing Care Research Network [Online]. Available at: http://revaluingcare.net/veterinary-negligence-ethical-and-legal-perspectives-on-formulating-a-duty-of-care. Accessed 19 January 2018.

Singer, P. (2006). In defense of animals: the second wave, Malden, MA, Blackwell Pub, 264 pp.

Teale, C.J. and Moulin, G. (2012). Prudent use guidelines: a review of existing veterinary guidelines. Revue scientifique et technique (International Office of Epizootics), 31: 343-54.

Wohl, J.S. and Nusbaum, K.E. (2007). Public health roles for small animal practitioners. J Am Vet Med Assoc, 230: 494-500.

Yeates, J.W. (2009). Response and responsibility: An analysis of veterinary ethical conflicts. Veterinary Journal, 182: 3-6.

47. What challenges is the veterinary profession facing – an analysis of complaints against veterinarians in Portugal

M. Magalhães-Sant'Ana[1,2*], *M. Whiting*[3], *G. Stilwell*[1,2] *and M.C. Peleteiro*[1,2]
[1]*Ordem dos Médicos Veterinários, Rua Filipe Folque, no.10J – 4°Dt°, 1050-113 Lisboa, Portugal;* [2]*CIISA, Faculdade de Medicina Veterinária, Universidade de Lisboa, Avenida da Universidade Técnica, 1300-477 Lisboa, Portugal;* [3]*Royal Veterinary College, Hawkshead Lane, North Mymms, Herts AL9 7TA, United Kingdom; mdsantana@gmail.com*

Abstract

Ethics is vital to the future of the veterinary profession. Despite its importance, there has been a lack of applied research on the ethical challenges faced by veterinarians and to what extent they might affect their professional roles. In particular, little is known about reasons why veterinary professionals are subject to disciplinary complaints. This study aims to shed light on the nature of complaints against veterinarians in Portugal and on the disciplinary measures that have been taken. A retrospective investigation of the disciplinary processes within the archives of the Portuguese Veterinary Order (*Ordem dos Médicos Veterinários* – OMV) was performed, using both quantitative and qualitative research methods. A total of 60 (out of 219) disciplinary processes, between 2012 and 2015, where analysed. Results show that complaints fall within nine categories of offence: (1) professional negligence; (2) failing to protect animal welfare; (3) certification misconduct; (4) advertising and promotions; (5) conflicting interests; (6) costs and payments; (7) lack of informed consent; (8) conflicting working relationships; (9) itinerant clinical practice. These areas highlight the plethora of ethical challenges faced by veterinarians in their daily work. Addressing these challenges and improving the standard of practice requires a combination of approaches, including life-long learning opportunities in both scientific and ethical decision-making. Gaps in governance have been identified, including the promotion of animal welfare, the responsible use of social media, advertising, mobile and distance medical services and effective communication. These findings seem to suggest that a revision of current veterinary regulations and ethical guidance in Portugal is warranted.

Keywords: animal welfare, disciplinary processes, professionalism, veterinary ethics, veterinary law

Introduction

Ethics is vital to the future of the veterinary profession and veterinarians are required to perform their duties in accordance with the highest standards. However, there has been lack of applied research on the ethical challenges faced by veterinarians and to what extent these might affect their professional roles. In particular, little is known about the type of complaints made against veterinarians, in Europe or elsewhere.

In Portugal, there are almost 6,000 registered veterinarians. Similarly to other European countries, the veterinary profession is regulated by a professional body, the Portuguese Veterinary Order (*Ordem dos Médicos Veterinários*, OMV), established in 1994. In order to practise, all veterinarians must be registered with the OMV, abide to its statutes (Law 125/2015, of Sept 3rd) and to the Code of Professional Conduct (*Código Deontológico*, CD-OMV).

The OMV Disciplinary Committee (*Conselho Profissional e Deontológico*, CPD) is responsible for overseeing practice standards of veterinarians working in Portugal. The CPD is composed by seven members, elected every four years (up until 2015, elections took place every three years), assisted by one

EurSafe 2018 – DOI 10.3920/978-90-8686-869-8_47, © Wageningen Academic Publishers 2018

barrister of law. CPD members work voluntarily, *pro buono publico*. The CPD president is responsible for screening the complaints for validity and relevance, and only those considered sound are investigated. Each process is allocated to one member, who then is responsible for its diligences and for suggesting an outcome (filing or sanction). Decisions are taken collectively, by simple majority.

Improving standards of practice within the veterinary profession can only be achieved if its problems are recognized and the corresponding challenges explored. Drawing from a retrospective investigation of a sample of disciplinary cases within the archives of the OMV, the aim of this preliminary study is twofold: (1) to characterize the disciplinary complaints made against veterinarians in Portugal, and (2) to typify the reasons why veterinary professionals working in Portugal are subject to disciplinary complaints.

Methods

Sampling

A retrospective investigation of the disciplinary cases within the archives of the OMV was performed using a mixed methods approach. From a total of 219 processes identified between 2012 and 2015, a randomized sample of 60 processes (15 per year) was selected (using Ultimate Randomizer© App). The following parameters were investigated: duration, identity of claimant (i.e. client, colleague, authorities), identity of defendant (position, gender, age and number of years in register), animals involved, reasons for complaining, testimonials and additional proofs of evidence, and the final decision (filing or conviction).

Data handling and analysis

For the quantitative material, Microsoft Excel (Microsoft Corporation, 2010) was used for data handling. Descriptive and inferential statistical analyses were conducted using SPSS Statistics V.24 (IBM Corporation, 2016). Fisher's exact test, which computes the Chi-square (χ^2) distribution of small samples, was used to measure the association between type of complaints and the demographic variables, and $P<0.01$ was considered statistically significant.

In terms of qualitative material, excerpts of complaints were transcribed verbatim and inserted into NVIVO 11 (QSR International, 2016). Thematic analysis was conducted using the data immersion/reduction technique proposed by Forman and Damschroder (2008). Guided by the research questions, a preliminary list of themes was generated after the initial coding run by the first author (MM-S) and discussed with co-authors. The list of themes was refined in subsequent coding runs and the process was repeated iteratively until theoretical saturation was reached.

Results and discussion

Demography of complaints

Processes took between 123 and 1,352 days to completion (median ($\tilde{x}$)=554 days, mean (μ)=589 days, standard deviation (σ)=317 days), measured from the day of receipt until the date the dispatch with the final decision was signed. The discrepancy in length between processes is notorious. Although the duration of cases might change according to their complexity, civil and criminal cases in Portugal are known for their overall slowness. Official figures say that in 2016 civil cases (which account for 89% of legal cases in Portugal) took an average 33 months to complete (DGPJ, 2017). By taking these figures as point of reference, the average time of almost 20 months found in this study is well within reasonable limits.

Complaints against veterinarians were submitted by clients (43% of claims), veterinary colleagues (30%), the Governmental Agriculture and Veterinary Department (*Direção Geral de Alimentação e Veterinária*, DGAV) (21.7%). Other claimants included police authorities and the regulatory body from another jurisdiction (5%). Regarding the defendant, 50% were female, age ranging between 26 and 66 years' old ($\tilde{x}$=38 years old, μ=40.7 years old, σ=10.7 years) and having been registered in the OMV for 1 to 23 years ($\tilde{x}$=12 years; μ=11.9 years, σ=6.2 years). No statistically significant association between the type of complaint and the gender of the defendant was found.

Almost half of complaints (48.4%) were made against practice directors from registered veterinary premises (Practice, Clinic or Hospital), 45% against clinicians other than practice directors, and 6.5% against official veterinary officers. In 28% of cases, no animal was involved. When animals were involved, dogs accounted for 64% of cases. Other animals included cows (21%), cats (11%), one rabbit and one horse. The predominance of dogs may be explained by their popularity; according to a recent poll (GFK *Track.2Pets* 2016), dogs can be found in one third of Portuguese homes, more than any other pet. Moreover, these results can also reflect a greater moral consideration towards 'man's best friend', especially when compared to other companion animals such cats and horses.

Typification of complaints

According to the thematic qualitative analysis, complaints against veterinarians in Portugal fell within nine main categories of offence:
1. professional negligence (30% of total of complaints);
2. failing to protect animal welfare (18%);
3. certification misconduct (11%);
4. advertising and promotions (11%);
5. conflicting interests/unfair competition/usurpation of power (10%);
6. costs and payments (10%);
7. lack of informed consent (3.5%);
8. conflicting working relationships (3.5%);
9. itinerant clinical practice (3%).

In 30% of processes complaints fell within two or more of these categories. A brief explanation follows and some examples, taken from the research sample, will be used to illustrate each of the aforementioned categories.

1. Professional negligence emerged as the main complaint against veterinarians in Portugal. Clients are more likely to complain against professional negligence than to anything else (χ^2=38.0, P=0.000). Typical examples involve the death of a healthy animal following elective caesarean section or poor clinical outcome following orthopaedic surgery. Recognising flaws in our own clinical and ethical decision-making is by no means straightforward, especially if harm is done by omission (i.e. when veterinarians fail to do what should have been done). In effect, the CD-OMV has been shown to focus on preventing harm and wrongdoing rather than in promoting responsible and autonomous attitudes (Magalhães-Sant'Ana *et al.*, 2015). Despite their relevance to the reputation of the veterinary profession, no convictions were reached in cases involving professional negligence. Several reasons may help explain this result including unfounded accusations, difficulty in gathering relevant evidence or lack of supporting documents, a culture of defending peers against a third party or an insufficient regulatory framework. These hypotheses require further exploration using in-depth content analyses. Another explanation may involve the reduced sample size, for which a broader study is in development.
2. Veterinarians were considered to fail to protect animal welfare mostly when issuing certificates of transport for emergency and casualty slaughter (χ^2=34.8, P=0.000). This includes all the nine

complaints involving production animals (i.e. bovines), which were remitted by the DGAV. European regulations state that no animal shall be transported unless it is fit for the intended journey (European Council, 2004). Three cases resulted in convictions, although the assisting veterinarians had claimed that transport followed by immediate slaughter was in the best interest of the animal, despite of it being injured. Ethical conflicts such as these are thought to be common place in farm animal practice (Hernandez *et al.*, 2018) and arise when personal beliefs collide with the regulatory framework and with competing economic interests; in effect, a recent investigation from Ireland explored the complexity of issues involved when veterinarians decide to transport and slaughter injured livestock, including the role of client pressure to have the animal slaughtered at the abattoir, the lack of availability of on-farm emergency slaughter and defective regulations (Magalhães-Sant'Ana *et al.*, 2017). Current ethical guidance provides little support to dealing with such complex scenarios, since it is known that the CD-OMV has not adapted to societal changes regarding the moral and legal status of animals (Magalhães-Sant'Ana *et al.*, 2015).

3. Veterinary certificates include all formal declaration issued by veterinarians (FVE, 2014). Certification is one of the most important duties held by veterinarians, which is enshrined into European veterinary codes of conduct (Magalhães-Sant'Ana *et al.*, 2015). Cases within this category include accusations of failing to register animals after electronic identification, false claims on insurance policies and incomplete or incorrect vaccination registers. The only convicted case involved a veterinarian whom had been convicted at another jurisdiction for similar facts.
4. Over one third (34.8%) of claims made by veterinarians against their colleagues concerned advertising and promotions. In Portugal, there are policies regarding what is or is not permissible in terms of publicity of veterinary services. Messages that are considered persuasive or aggrandized are not allowed by the CPD. In the same vein, advertising prices or discounts in veterinary services is considered inadmissible because it is believed that competence (or excellence) should be the only quality to attract clients towards veterinary services. These policies are probably grounded in a virtue ethics framework, in which a virtuous veterinarian will voluntarily refrain from gaining an advantage over others, and instead will focus solely on the benefit of animals and society. However, this normative approach is not without dispute and has resulted in a wide debate within the Portuguese veterinary community regarding the acceptable limits of advertising, especially using social media. Two of these cases have resulted in convictions.
5. Veterinarians were also found to claim against colleagues for using their position unfairly (30% of claims from veterinarians). This can include low cost or gratuitous neutering of companion animals (unfair competition), employees who allegedly used the resources from the employer for personal gain (usurpation of power) or official veterinary officers who accumulate the exercise of private veterinary practice with civil service (conflicting interests). In the last few decades, the Portuguese veterinary community has grown significantly (Chaher and Pereira, 2014), followed by a dramatic change in professional roles. Nowadays, the vast majority of veterinarians in Portugal work in urban companion animal practices. The increasing number of veterinary professionals working mostly in the same area of activity may have created additional situations of conflict with peers, which often result in disciplinary complaints. However, none of the nine cases resulted in convictions.
6. Almost 20% of claims made by clients involved costs of services or how payment was processed. All of these cases were secondary to a claim for negligence or lack of informed consent and no convictions were made. Examples include clients who demand a refund after losing their pets following elective surgery (following a negligence claim) and those who complain against the prices of services which they claim had not been presented or discussed beforehand (following 'lack of informed consent'). Effective communication has traditionally been considered a neglected skill in veterinary care, including communicating costs (Gray and Moffett, 2010). Being able to openly discuss costs of services, especially when procedures do not work according to plan, requires transparency and accountability, which seems to collide with the paternalistic view of doctors-as-authorities.

7. Veterinarians were accused of failing to effectively inform clients in three occasions: in one occasion clients felt that they were forced to allow invasive surgery in their queen and that they were not properly informed of its risks; in another occasion, the client accused the veterinarian of failing to perform the required analytical tests to reach a diagnostic, and for withholding relevant information from him. The third case has been described above (cf. point 6). The three cases were filed because no evidence was found to substantiate the claims. Again, these cases seem to emphasise the challenge faced veterinarians in communicating effectively and in having their authority called into question by the client.
8. Two cases involved conflicting working relationships, namely between employer and employee. These cases were eventually filed. Another case, involving the use of offensive comments in social media regarding a veterinary colleague, resulted in the conviction of the offender. The responsible use of social media is yet to be included as a topic in current governance.
9. Finally, two cases involved accusations, eventually filed, of itinerant clinical practice. These claims were made by veterinary colleagues who felt that the service being provided by the defendant was in breach of current rules of practice that prevent veterinary practitioners from providing home services without being called. Complaints such as these may become increasingly common since younger generations of veterinarians look at mobile and distance medical services (including telemedicine, which current veterinary regulations fail to address) as a business opportunity.

Disciplinary decisions

From the 60 cases, seven (12%) resulted in convictions while twenty-two (37%) prescribed, and were consequently filed, before a decision was made. Testimonials have been used in six cases (10%) and further evidence such as clinical records or additional clarifications have only been requested in nine cases (15%). Despite the fact that the relevance of the submitted complaints is outside the scope of this paper, it seems that faced with two arguably contradictory versions of the same facts, the version from the defendant has prevailed. This is in line with the defendant favourable doctrine *in dubio pro reo* in which wrongful convictions are considered to have larger detrimental impacts than wrongful acquittals (Nicita and Rizzolli, 2014).

Furthermore, the preliminary thematic analysis also shows that often a decision was reached without exhausting possible sources of evidence, such as witnesses' testimonials or by further inquiring the defendant. In particular, the final decision often lacks the sufficient detail to demonstrate that the conclusions that were reached, and the processual measures involved, were always in the best interest of the claimant. These findings cannot be explored in any detail in this paper and warrant further investigation.

Final remarks

This preliminary investigation characterized the disciplinary complaints made against veterinarians in Portugal between 2012 and 2015. The nine identified reasons why veterinarians are subject to disciplinary complaints highlight the plethora of ethical challenges faced by veterinarians in their daily work. Addressing these challenges and improving the standard of practice requires a combination of approaches, including life-long learning opportunities in both scientific and ethical decision-making. Moreover, the current findings seem to suggest that a revision of current veterinary regulations and ethical guidance in Portugal is warranted. Identified gaps in governance included the promotion of animal welfare, the responsible use of social media, advertising, mobile and distance medical services and effective communication. A more detailed investigation is under way in order to explore these issues in further detail.

Conflicting of interests and ethical statement

The first (MM-S) and last (MCP) authors are current members of the OMV-CPD (2016-2019). The third author (GS), who was member of the CPD between 2010 and 2012, was not directly involved in analysing the processes and did not influence the analysis. Permission was sought to analyse the disciplinary processes, conforming to an Ethical Review Form (Reference number: 673/CPD/2017). The dataset were anonymised and no potentially identifiable information was stored.

Acknowledgements

The authors are thankful to Ana Sousa for all her help in analysing the disciplinary processes.

This paper is part of the research project 'VETHICS 2022: A structured approach to describing and addressing the ethical challenges of the veterinary profession in Portugal', funded through grant SFRH/BPD/117693/2016 from the Fundação para a Ciência e Tecnologia (FCT), Portugal. MM-S thanks OMV and FCT for financial support.

References

Chaher, E.M. and Pereira, P.A. (2014). Desafios da evolução demográfica veterinária em Portugal. *Veterinaria Atual*, 73:26-28.

DGPJ. (2017). Estatísticas Oficiais da Justiça, Direção-Geral da Política de Justiça. Availabe at: http://www.siej.dgpj.mj.pt. Accessed 3 January 2018.

European Council. Council Regulation (EC) No1/2005 of December 2004 on the protection of animals during transport and related operations and amending Directives 64/432/EEC and 93/119/EC and Regulation (EC) No 1255/97 (2004).

Forman, J. and Damschroder, L. (2008). Qualitative Content Analysis. In *Empirical Methods for Bioethics: a primer* (Vol. 11, pp. 39-62). Oxford, UK: Elsevier.

FVE. (2014). FVE 10 principles of veterinary certification. Federation of Veterinarians of Europe. Available at: http://www.fve.org/news/index.php?id=163.

Gray, C. and Moffett, J. (2010). *Handbook of Veterinary Communication Skills*. Oxford, UK: Wiley-Blackwell.

Hernandez, E., Fawcett, A., Brouwer, E., Rau, J. and Turner, P.V. (2018). Speaking Up: Veterinary Ethical Responsibilities and Animal Welfare Issues in Everyday Practice. *Animals*, *8*(1), 15.

Magalhães-Sant'Ana, M., More, S.J., Morton, D.B. and Hanlon, A.J. (2017). Challenges facing the veterinary profession in Ireland: 3. emergency and casualty slaughter certification. *Irish Veterinary Journal*, *70*, 24.

Magalhães-Sant'Ana, M., More, S.J., Morton, D.B., Osborne, M. and Hanlon, A. (2015). What do European veterinary codes of conduct actually say and mean? A case study approach. *Veterinary Record*, *176*, 654.

Nicita, A. and Rizzolli, M. (2014). In Dubio Pro Reo. Behavioral Explanations of Pro-defendant Bias in Procedures. *CESifo Economic Studies*, *60*(3), 554-580.

48. Clinical ethics support services in veterinary practice

S. Springer[1], U. Auer[2], F. Jenner[3] and H. Grimm[1]*
[1]Unit of Ethics and Human-Animal-Studies, Messerli Research Institute, Veterinaerplatz 1, 1210 Vienna, Austria; [2]Clinical Unit of Anaesthesiology and perioperative Intensive-Care Medicine, Vetmeduni, Vienna, Veterinaerplatz 1, 1210 Vienna, Austria; [3]University Equine Hospital, Vetmeduni, Vienna, Veterinaerplatz 1, 1210 Vienna, Austria; svenja.springer@vetmeduni.ac.at

Abstract

Decision-making processes in veterinary medicine are determined not only by medical aspects, but also legal, ethical, emotional, financial and social factors have a lasting effect on treatment decisions for animal patients and lead to a growing complexity of veterinary patient care. The combination of all these factors and the personal values of the parties involved can lead to conflicting priorities and insecurity regarding the optimal treatment choice for each individual patient. Against this background, the demand for services supporting clinicians in their multifactorially influenced decisions is increasing. In human medicine, clinical ethics support services have long been established with the purpose of assisting health professionals when they are confronted with ethical questions and concerns. These services help identify and analyse the value conflicts and uncertainties of the parties involved, which often originate from insecurities in regard to patient care, problems within the clinic team or financial issues. Two main methods for clinical ethics support services – 'Moral Case Deliberation' and 'Clinical Ethics Consultations' – can be distinguished. They differ in terms of the function and the aspired goal of the consultation. An Equine Hospital Ethics Working Group was established at the University of Veterinary Medicine, Vienna in 2015. This group mainly follows the clinical ethics consultation in a small group, whereby an ethicist, veterinarians and animal caretakers participate in the meetings. The group meets when potentially problematic patients are identified, for whom questions regarding life-death, suffering and quality of life questions arise. Involved veterinarians, animal caretakers and the head of the equine clinic meet with an expert from the field of philosophy in order to explore and discuss a complex case from different points of view, identify and structure responsibilities and try to find a consensus or compromise regarding further diagnostics and therapies among all participants in the discussed clinical cases. Based on the experiences with this established working group, the paper demonstrates possibilities and used methods of clinical ethics support services at the equine hospital of the Vetmeduni, Vienna.

Keywords: veterinary medicine, clinical ethics consultation, support, benevolence towards animal patients

Introduction

In the early 1970s, clinical ethics support services in the field of human medicine were established in the USA to assist health professionals when faced with ethical questions and concerns (cf. Rosner, 1985; Fournier, 2015). Fournier (2015) consider the technological possibilities, changing expectations of patients regarding the clinician-patient relationship and respect for their values to be the reason for increasing ethical dilemmas health professionals are confronted with in their daily working life. Against this background, a lot of hospitals launched so-called clinical ethics committees, which were characterized by a multidisciplinary composition of physicians, philosophers, lawyers, theologians, psychologists and sociologists (Fournier, 2015). At the same time, the approach of clinical ethics consultation was established, whereby only one person – usually from the field of biomedical ethics – tried to apply their expertise with ethically challenging situations (Fournier, 2015). Over time, various methods and models of clinical ethical support services have emerged and proved more or less successful

 Svenja Springer and Herwig Grimm (eds.) ***Professionals in food chains***
EurSafe 2018 – DOI 10.3920/978-90-8686-869-8_48,

in the field of human medicine. However, these services share the common ground that they should help identify and analyze the value conflicts and uncertainties of all parties involved.

Meanwhile, the discussion on ethical issues is a core part of veterinary practice, which is embedded in every clinical encounter between veterinarians, animal owners, animals, as well as surrounding colleagues (cf. Mullan and Fawcett, 2017). Especially the change in the human-animal relationship, rapidly increasing technical possibilities in diagnosis and therapy, as well as the increasing specialization lead to fundamental quandaries in veterinary practice: Is it right to continue the treatment of a patient despite prolonged, untreatable or unavoidable suffering, if the owner does not want to part with the animal because of their emotional attachment? Is there a moral obligation to use advanced veterinary methods, even when the owner's financial possibilities are exhausted? Is it legitimate to carry out a highly innovative and new treatment, even though there are alternative options and it is not clear how the patient's well-being will be affected by the new method? The inseparability of medical, social, legal, financial, emotional and moral considerations leads to increasingly complex veterinary decision-making processes and seemingly irreconcilable dichotomous dilemmas. The combination of all these factors and the personal values of the parties involved can lead to conflicting priorities and insecurity regarding the optimal treatment choice for each individual patient. Against this background, the demand for services supporting veterinary clinicians in their multifactorially influenced decisions is increasing.

This paper aims to explore the possibilities of ethical consultation in the field of veterinary practice. Against the background of the already well-established clinical ethics support services in the field of human medicine, models and methods which seem to be of relevance for the veterinary practice will be outlined in a first step. Further, based on an established working group at the University of Veterinary Medicine, Vienna, methods and possible approaches of ethical counselling will be presented in a practical context. Based on the experiences with this working group, the paper will demonstrate the possibilities of clinical ethics support services in veterinary medicine in a last step and show their limits and challenges at the same time.

Methods and models of clinical ethics supporting services

In recent years, different methods have been established in order to provide support and help with ethically challenging situations for clinicians in their daily working life. Consultations depend on the location and context and can differ, depending on whether they are case-related or not and on the parties present. In a paper concerning methods in clinical ethics, Fournier (2015) distinguished between two main methods of clinical ethics which have been developed based on the historical background: firstly, 'Moral Case Deliberation' (MCD), and secondly, 'Clinical Ethics Consultation Services' (CEC).

MCD focuses on a reflection on non-case-related ethical issues that clinicians face in their daily routine. This model offers a discussion of principles, norms and values within the profession, which are separately reflected starting from specific cases. Molewijk *et al.* (2008) described MCD as a method which should focus on answering the question 'What should we consider as the morally right thing to do when facing this particular problem?' Against this background, the method involves discussing moral cornerstones and investigating the question what we should give normative validity to. It explores the meaning of the content of moral intuitions that occur in an ethically challenging context and aims to identify and formulate moral questions which focus on the moral discomfort of participants and help them understand it better.

In comparison, CEC provides prompt assistance in clinical cases which require a specific medical decision for a patient (Fournier, 2015; Molewijk *et al.*, 2015). This approach applies norms, values and principles which seem to be of relevance to the clinical case and have to be weighted in order to make a decision in a certain context. Consequently, depending on the format, different levels and contents of methods for ethical services in the clinical context can be described (Figure1).

Even though Fournier (2015) summarized in her article that in CEC 'the central focus is on the patient and on the medical decision to be made' and in MCD 'the central focus is on the health care team and on its moral competency' (Fournier, 2015: 556), it can be stated that the overall intended objective lies in the principle of benevolence towards patient care, as well as developing awareness and support for practicing veterinarians.

Clinical ethics support services in veterinary practice: why and how?

In everyday clinical routine, practicing veterinarians are confronted with decisions which they are to be held responsible for, on the one hand, and which need to be justified based on several reasons, on the other hand (cf. Salomon, 2011). During their studies and daily practice, veterinarians learn to handle medical information regarding their patient in order to decide about and justify further interventions during patient care. Jonsen *et al.* (2015) wrote about a fixed structure which helps organize and identify important aspects of the clinical case 'in order to carry out the reasoning process necessary for diagnosis and therapy' (Jonsen *et al.*, 2015, 3). In veterinary practice, the process flow – from anamnesis, physical examination, laboratory studies, etc. to treatment – represents an integral part which is strictly followed in decision-making processes. At this point, however, questions about the factors that influence and determine decision-making processes in veterinary practice arise.

In comparison to the primary dialogic structure between the physician and the patient in human medicine, veterinary practice is characterized by a threefold relationship between the veterinarian, the animal and the animal owner (May, 2013; Yeates, 2013). The three parties relate to each another, neither in a vacuum nor without conflicts. All actions occur in a social context and under consideration of the existing legal framework. Beyond that, health care professionals always act within a certain moral framework and are ruled by specific norms and principles, e.g. the avoidance of suffering and the promotion of the patients' well-being. In veterinary practice, however, external factors such as the owner's financial situation and emotional bond with the animal strongly influence decision-making processes beside these given frameworks. As a result, decisions for or against different veterinary actions depend on several factors and not only on the animal's interests and needs and medical indications, which often leads to ethical dilemmas in veterinary practice. Against this background, decision-making processes should not only be structured by patterns including clinical aspects, but also arising ethical issues should be identified and managed on the basis of a certain structure (Jonsen *et al.*, 2015). In this light, Jonsen *et al.* (2015) note: 'Just as clinical cases require a method for sorting data, so too ethical cases must have

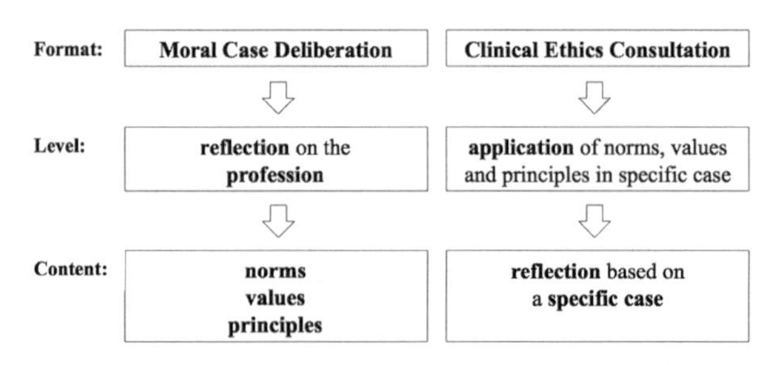

Figure 1. Comparison of the level and content of moral case deliberation and clinical ethics consultations as methods of clinical ethics support services in veterinary practice.

some method to collect, sort, and order the facts and opinions raised by the case. [...] Clinical reasoning begins with the facts of the case and moves toward presumptive diagnosis by sorting those facts into reasonable patterns of causality. Similarly, clinical ethical reasoning starts with the facts. A statement of the ethical problem in case follows a clear and complete collection of the facts of the case.' (Jonsen *et al.*, 2015: 3). In this context, clinical ethics support services can offer a method for structuring cases from an ethical point of view in order to identify, sort and resolve ethical aspects emerging during patient care.

There is a broad range of methods which can offer support for people confronted with ethical issues. Thereby, the choice of a method depends on the aspired aim, as well as several other conditions (e.g. cultural, social, institutional) which define and limit the scope of action. In the following, the Equine Hospital Ethics Working Group will be presented as a clinical ethics supporting service at the University of Veterinary Medicine, Vienna. On the basis of the theoretical considerations mentioned above, several aspects of the group will be pointed out (e.g. method, aim, function). Chances and limits of this method will be discussed by means of this practical example.

Equine Hospital Ethics Working Group

Establishment and aim of the working group

An Equine Hospital Ethics Working Group was established at the University of Veterinary Medicine, Vienna in 2015. The group comprises veterinarians from different disciplines, animal caretakers, the head of the equine clinic, as well as an ethicist. The first meeting of the group was convened to evaluate a questionnaire on 'Euthanasia in Equine Practice'. This study is based on an already existing questionnaire in the field of small animal medicine (Hartnack *et al.*, 2016). It was developed and successfully carried out in a graduation project (Weber-Schallauer, 2017). During the elaboration and evaluation process of the study, it became clear to what extent veterinarians see a need for discussions regarding euthanasia and questions on end-of-life treatment in equine practice. Against this background, the working group was set up with the aim of reflecting on ethical issues arising in the daily routine at the Equine Hospital and dealing with them in an interdisciplinary manner.

Methods and tools of the working group

In general, the group is convened in situations in which veterinarians are faced with ethically difficult decisions due to various factors related to a specific case. Consequently, this group works with the CEC method, in particular with the model of ethics consultants of a small group (cf. Molewijk *et al.*, 2015), in which veterinarians are supported in specific decision-making processes. According to Fournier's (2015) CEC classification, the working group can be assigned to the 'facilitating' model, which is suggested by the American Society for Bioethics and Humanities (ASBH, 2011). The meetings are held to provide assistance with identifying and analysing uncertainties in further patient care, whereby the consultant takes a neutral stance. In the discussion, various admissible and legal options are clarified. The consultant supports the involved stakeholders in developing ethical solutions (Fournier, 2015).

A need for discussion usually arises for patients who require difficult or prolonged treatment with an uncertain or poor prognosis for quality of life and/or survival and when compelling arguments can be found for both initiating or continuing treatment and euthanasia. In can be considerably stressful for the veterinarian to balance between state-of-the-art therapy and overtreatment and the effect on the patient's interests and needs, especially regarding short- and long-term quality of life. This is further exacerbated by emotional and financial considerations of and for the owner. In most cases, the working group is convened after treatment of a critical patient has commenced. The aims are to define the milestones of treatment (including pain management) success or failure and quality of life, as well as to find a consensus regarding cut-off points and inform future decisions regarding the continuation of treatment or euthanasia. By means of different tools and methods of applied ethics the discussion is guided to crucial core issues and factors, which veterinarians urge in uncertain situations of their veterinary practice. For instance, with the help of a responsibility check (cf. Beck, 2015) involved parties identify and structure their responsibility along five analytical questions: Who is responsible? For what is someone responsible (e.g. actions, consequences)? Because of what is someone responsible (e.g. values, norms, law)? When is someone responsible (prospective, present, retrospective) And, to whom is someone responsible (e.g. employer, profession)? These questions led to separate meetings in addition to ad-hoc case-based discussion, which focused on the reflection of the profession: Which responsibilities must veterinarians bear? Which principles must be followed? To what extent must external factors be taken into account in decision-making processes? And, to what extent do veterinarians bear responsibility for these external factors? Molewijk *et al.* (2015) describe that '[t]he (dis)connection between the facts of the case and the normative reasoning of the participants is an important analytical aspect of MCD.' (Molewijk *et al.*, 2015, 566) Against this background, the method of the case-related CEC was transferred to the general dialogue about norms, values and principles, which led to a transition to the MCD method, where the discussion has moved away from a specific case and, thus, ended with a reflection on general ethical issues that arise in the daily working life of veterinarians.

Conclusions

Based on the experiences of the working group, clinical ethics support services in veterinary practice can lead to a broad range of opportunities in regard to patient care, development of awareness for ethical issues, as well as the improvement of institutional culture and multidisciplinary cooperation (cf. Molewijk *et al.*, 2015). In line with Molewijk *et al.* (2015), the method of clinical ethics consultation in small groups enables a spontaneous gathering and dealing with critical cases. The establishment of such groups allows for time and space to discuss challenging aspects with all parties involved and monitor the case over several days, weeks or even months. Developing scenarios with critical control points along with a tight monitoring scheme turned out to be a positive way forward. Furthermore, it became clear that on the basis of already discussed cases so-called paradigm cases were established, to which veterinarians refer when confronted with similar new cases. Jonsen *et al.* (2015) specified paradigm cases as 'examples of serious assessments in prior, similar cases, to which the current case can be compared, in order to guide the clinician in this case.' (Jonsen *et al.*, 2015: 6) Further, the authors wrote that consultants should keep such paradigm cases in mind in order to recognize differences compared to a current case (Jonsen *et al.*, 2015). We observed that not only the clinical ethicist in the group benefited from paradigm cases, but also veterinarians utilized the knowledge previously gained, which instantaneously can be applied for a similar current situation. A main positive aspect of the Equine Hospital Ethics Working Group is its flexibility. Our experience shows that a pool of cases, thoroughly dealt with, allows for a time-efficient dealing with new cases. In other words, whereas the first cases were rather time-consuming, regular meetings on particular cases help speed up the process. This leads, to some extent, to a form of conservatism, whereas participants of the working group can relate the actual discussion to previous casuistry. Further, regular meetings with fixed intervals detached from a specific case would probably even deepen moral competencies, as well as promote ethical reflection even more. Taking this into account, it is only fair to mention that the hospital is situated in an institution designed for research and teaching and, thus, in a privileged situation on the one hand. On the other, it is also fair to assume that a considerable number of complicated cases involving new technologies, etc., will not be present itself in the average practice. Nevertheless, there are certainly enough ethical challenges left in general practice and it is a consequent challenge to design formats and schemes that support practicing and self-employed veterinarians, who do not have the privilege of an institutional background allowing for clinical ethics support services.

Against the background of the theoretical considerations and gained experiences of the working group, clinical ethics support services in veterinary practice offer the possibility to bridge the gaps between medical knowledge and a medically safe decision in uncertain situations in order to support a justified assumption of responsibility.

References

ASBH (2011). Core competencies in health care ethics consultations. Glenview, American Society for Bioethics and the Humanities.

Beck, V. (2015). Verantwortung oder Pflicht? Zeitschrift für Praktische Philosophie 2: 165-202.

Fournier, V. (2015). Clinical Ethics: Methods. In: ten Have, H.: Encyclopedia of Global Bioethics. Springer publishers, pp. 553-562.

Hartnack, S., Springer, S., Pittavino, M. and Grimm, H. (2016). Attitudes of Austrian veterinarians towards euthanasia in small animal practice: impacts of age and gender on views on euthanasia. BMC Veterinary Research 12:26.

Jonsen, A.R., Siegler, M. and Winslade, W.J. (2015). Clinical ethics: A practical approach to ethical decisions in clinical medicine. Mcgraw Hill Medical publishers, 244 pp.

Molewijk, B., Abma, T., Stolper, M. and Widdershoven, G. (2008). Teaching ethics in the clinic: The theory and practice of moral case deliberation. Journal of Medical Ethics, *34*, 120-124.

Molewijk, B., Slowther, A. and Aulisio, M. (2015). Clinical Ethics: Support. In: ten Have, H.: Encyclopedia of Global Bioethics. Springer publishers, pp. 562-570.

May, S.A. (2013). Veterinary Ethics, Professionalism and Society. In Wathes, C.M., Corr, S.A., May, S.A., McCulloch, S.P., Whiting, M.C. (eds.) Veterinary and Animal Ethics: Proceedings of the First International Conference on Veterinary and Animal Ethics. September 2011, Wiley-Blackwell publishers, pp. 44-58.

Mullan, S. and Fawcett, A. (2017). Veterinary Ethics: Navigating Tough Cases. 5M Publishing publishers, 526 pp.

Rosner, F. (1985). Hospital medical ethics committees: a review of their development. Journal of the American Medical Association 253: 2693-2697.

Salomon, F. (2011). Wie sinnvoll ist Ethikberatung im Krankenhaus? Ethikforum der ÄKWL, Münster. Available at: http://slideplayer.org/slide/2642435. Accessed 19 January 2018.

Weber-Schallauer (2017). Euthanasie in der Pferdepraxis. Diploma thesis, Vetmeduni, Vienna.

Yeates, J. (2013). Animal Welfare in Veterinary Practice. Wiley-Blackwell publishers, 200 pp.

Section 8.
Veterinary ethics: in teaching

49. Log-in for VEthics – applying e-learning in veterinary ethics

C. Dürnberger[1], K. Weich[1], S. Springer[1] and M. Wipperfürth[2]*
[1]Unit of Ethics and Human Animal Studies, Messerli Research Institute, Veterinaerplatz 1, 1210 Vienna, Austria; [2]Paris-Lodron University Salzburg, Department of English and American Studies, Didactics, Paris-Lodron University Salzburg, Erzabt-Klotzstraße 1, 5020 Salzburg, Austria; christian.duernberger@vetmeduni.ac.at

Abstract

The project 'VEthics E-Portfolio', funded by the Austrian Ministry of Health and carried out by the Messerli Research Institute, Unit of Ethics and Human Animal Studies, currently develops an e-learning course on ethics in the field of veterinary medicine. The paper gives an insight into the project in four steps: (1) It briefly introduces the didactic concepts of e-portfolio and e-learning within veterinary education/further training; (2) it discusses the role of ethics for veterinarians as it is understood in the project; (3) it delineates the didactic and technical implementation of the e-learning course and (4) it presents selected results of the first evaluation.

Keywords: veterinary ethics, teaching, e-learning

Log-in: e-learning with e-portfolios within veterinary education/further training

In the wake of the current 'animal turn', professionals in the field of veterinary medicine are confronted with ethically challenging situations. Against this background, veterinary faculties increasingly integrate ethics courses into their degree programs. In addition to lectures and seminars, e-learning tools offer supplementary didactic possibilities to make ethical deliberation more sustainable and practice-oriented.

Generally, e-learning (short for 'electronic learning') combines learning as a 'cognitive process for achieving knowledge' (Aparicio *et al.*, 2016: 292) and a technology (currently mostly online-tools) as 'an enabler of the learning process'. (*ibid.*) today, e-learning is often synonymous with online-learning and understood as a process that takes place partially or entirely over the internet (cf. Sun *et al.*, 2008). The potentials of e-learning are manifold (cf. Bora and Ahmed, 2013: 10): Most prominent is the promise that it makes information available to users disregarding time restrictions and geographic proximity. Furthermore, e-learning-tools take advantage of a variety of attractive didactic tools such as short films, graphical illustrations or interactive rating tasks, in order to present the content in an appealing, 'easy-to-learn' way, thereby inviting the user to an active participation, for example via quizzes, short-tests or votes.

The format of e-portfolios takes interactive and individualised learning a step further and is therefore highly attractive for reflective learning processes of practitioners. E-portfolios facilitate to reach out into the practical experiences of professionals and incorporate those into the learning process, thus allowing for a deeper reflection of practice and for the professionals to incorporate the new input into their practice. As a basic definition e-portfolios serve as personal multi-media documentation of a long-term learning process in order to enhance and illustrate learning outcomes (cf. Hornung-Prähauser, 2007 and Chapter 3). Improving practice and strengthening practitioners in their decision-making processes and their professional identity is one fundamental goal of any further training in veterinary ethics.

The concept of the reflective practitioner, which we also follow here, was originally coined by Schön (1984) and has since been applied to and further developed for a wide range of professions, including veterinary medicine (cf. e.g. Figgis *et al.*, 2009). As further explained below, successful ethical decision-

EurSafe 2018 – DOI 10.3920/978-90-8686-869-8_49, © Wageningen Academic Publishers 2018

making requires more than the acquisition of knowledge. 'VEthics E-Portfolio' takes into account findings on inservice-learning, and therefore includes formats that enhance situated learning, active participation and reflective processes by interlocking reflective phases with practice phases and professional growth in general.

Loading: veterinary ethics in the field of e-learning

One of the key issues in planning any learning process is the specification of the knowledge, skills or attitudes the students should acquire, as understanding the structure of the learning goals gives crucial hints at the structure of the learning process and adequate tasks (cf. Gruber and Stöger, 2011), as will also be shown further in the text.

For example, various e-learning courses in the private sector aim at a higher safety standard or a higher efficiency of certain work steps; other online-based courses, such as in the context of veterinary medicine, can impart the necessary knowledge about a specific field of the later job (such as making the correct diagnosis of fractures based on x-ray images). But what kind of knowledge should an ethics course impart to veterinarians?

Since veterinary ethics focuses on problems arising in practice, it can be subsumed under the term 'applied ethics'. (cf. Bayertz 1991: 23) However, the specific tasks of applied ethics are understood rather differently, from reconstructing 'minimal morals' on which a vast majority is able to agree (cf. Ropohl 1998: 281), over providing answers to problems of practitioners (Chadwick and Schroeder, 2002: 1) to describing moral systems, conventions and norms, making clear the relevant arguments in a debate and their tradition (cf. Fischer, 2001). Roughly speaking, the e-learning course presented here follows the last-mentioned (more descriptive) conception. The ethicist is less understood as an expert giving answers to moral questions, but rather as a 'travel guide' through the world of moral problems (cf. Reiss, 2005). Ethics, as sketched in the project, should equip veterinarians with 'the necessary knowledge and skills to analyse a difficult situation from different ethical perspectives (...) and to take into account different values involved' (Magalhaes-Sant'Ana *et al.*, 2009: 201); it should help them to 'understand their different roles and responsibilities as well as appreciating the expectations of different stakeholders in modern society.' (Magalhaes-Sant'Ana *et al.*, 2009: 197) It is a desideratum of research to what extent the described goals of the course are aligned to what users such as students at the VetMedUni Vienna or Austrian veterinary officers seek form such a course. However, discussions within the project 'VEthics' (cf. Dürnberger and Weich, 2016) gave the hint that such expectations vary widely (from 'learning a given code of conduct' to 'exchange moral relevant experiences'). Furthermore, the (positive) evaluation (see below) could be interpreted in a way that the course does meet the expectations of the users.

Within such a conception, veterinarians should neither internalise simple codes of conduct nor simply memorise theoretical facts. This entail the didactic challenge that rating tasks, self-tests or quizzes – as usually used in e-learning courses – are not suitable, neither as learning tools nor for self-monitoring the learning success. Rather than finding a 'clear' correct answer to a specific question, the learners in the present context need to be introduced to and guided through reflective processes. An e-learning course about ethics in the explained sense has to take on this challenge. The main choices are outlined in the following.

Ready: the Messerli e-learning course 'Veterinary Ethics'

The currently developed e-learning course aims at three target audiences: Veterinary medicine students, veterinarians and veterinary officers. Although veterinary officers have partially to deal with specific ethical questions (cf. Dürnberger and Weich, 2016), other topics address both, veterinary officers and (later) veterinarians and can be prepared in such a way that they attract all target groups. The above mentioned benefit of e-learning settings that users can learn independently of seminar times and locations make this learning environment particularly interesting for veterinarians and veterinary officers: These are two groups, who are faced with ethically challenging situations in their professional practice and who, however, normally do not have the chance or the time to attend seminars at the university.

The final e-learning course will consist of at least ten sessions. Each session will require about 25 to 30 minutes and tackle one topical issue each, as the following overview shows: (1) Introduction; (2) Variety of roles. What does professionalism mean?; (3) Ethics and morals; (4) Schools of ethical theories; (5) Livestock farming and society. How to better understand conflicts? (6) The Ethical Matrix. A tool for structuring ethical controversies; (7) Euthanasia; (8) Animal diseases. What does professional responsibility mean?; (9) Ethical perspectives on slaughter; (10) Animal Hoarding.

The topics were chosen on the basis of two sources: (1) On the curriculum of the VetmedUni Vienna in regard to veterinary ethics and (2) on the most important results of the project 'VEthics', which focused on ethically challenging situations of veterinary officers (cf. Dürnberger and Weich 2016). Sessions 2-6 are (as of January 2018) finished. All contents are in German. In the long term, a translation into English is planned. Technically, the course is integrated in 'Vetucation®', the e-learning platform of the VetMedUni Vienna, available only for students and (in future) for invited guests, and is implemented with the program 'Adobe Captivate®', an authoring tool for creating e-learning content.

The didactics were worked out with Manuela Wipperfürth, an expert for in-service learning, reflective practice and cooperative learning processes of professionals. The following aspects were considered to be of central importance:

1. Every session starts with an overview of the (a) specific content (a summary in two or three sentences), (b) the methodical procedure (the estimated duration and technical notes) and (c) the targeted output, in order to function as advance organizers that help increase motivation and learning output (cf. Roohani *et al.*, 2015). Advance organizers were originally introduced by David Ausubel and – as concise summaries or visualisations – should link prior knowledge of learners to upcoming new concepts, highlighting and structuring the most relevant aspects of what is to be learnt. The mentioned point (c) is vital with in-service learning settings since firstly practitioners need to accept a learning situation as profitable and effective first. Secondly, learners will activate prior knowledge and can thus more easily incorporate and adjust to new concepts. Thirdly, professionals are very respectfully positioned as autonomous learners and can evaluate the learning outcome also for themselves.
2. If possible, every session gives examples from everyday working life of veterinarians and/or veterinary officers in order to illustrate arguments or stimulate questions, thereby allowing for case-based, situated learning. In complex professional settings, such as ethical decision-making in veterinary medicine, the knowledge needed for expert performance can be described as 'encapulsated knowledge' (Boshuizen and Schmidt, 1992), meaning that over a large number of practical situations, in which professionals applied their knowledge from various sub-disciplines, they could develop a deep understanding of the professional problem. This complex professional knowledge can best be activated through cases and situations that come as close as possible to real-life situations (cf. Kopp *et al.*, 2007).
3. Didactically, every session comprises (a) different ways of input (mainly slides with voice-over, but for example also brief videos) and (b) tasks for the users to become active such as short assessments of case scenarios on the basis of a given Likert-scale or ranking (for example 'What are the most important tasks of agriculture? Please rank the given answers.'). However, the most important tool in regard to the user's active participation is the so called 'E-Portfolio'. This recordable file not only

offers a script of the whole session in which the user can look up every segment he is interested in, it also provides space for own reflections (see below).

A third learning concept – after reflective practice and situated, case-based learning – needs to be introduced and has been used to create the e-portfolio, namely that of deliberate practice: In order to improve skills and knowledge in the professions – or any other complex skill – learners need deliberate practise, meaning the willingness, motivation and stamina to systematically work on their skills over a longer period of time (cf. Ericsson *et al.*, 1993). The concept of portfolio work has been established and well researched in the field of teacher education on the basis of a complex learning model, including the concepts of situated case-based learning and a reflective approach (cf. Häcker, 2012; Wipperfürth 2015). The medical and the teaching profession both require a high degree of reflective practice, reached through increasingly merging technical knowledge and value-based, complex decision-making in practice (cf. Gruber *et al.*, 2006), so that methods can ideally be adjusted from one to the other.

This is not the place to retell an entire session, but in order to fill the rather abstractly described didactics with life and in particular to clarify the role of the 'e-portfolio', a few impressions follow:

Session 2 'Variety of Roles. What does professionalism mean?' starts with the above-mentioned overview of its content, its methodical procedure and the targeted output. Since the last point was emphasized as crucial, it is to be fully quoted: 'As a result of the session, you will be able to differentiate 'work', 'job' and 'profession'. You gain a deeper understanding of the term 'professionalism' and learn strategies for dealing with (role) conflicts.'

After that overview, the session starts with a short video, showing a veterinary officer who retells a situation in which he was insulted by a farmer in such a bad way that he started to insult him in turn (it is made clear that the veterinary officer is played by an actor, but that the described situation is based on real stories). After the video, the user is invited to assess the behaviour of the veterinary officer on a predefined Likert-scale by agreeing or disagreeing with two statements: 'The action of the veterinary officer is comprehensible'; 'The veterinary officer acted professionally.'

Having introduced this last term, the session prompts the user to think about his own understanding of 'professionalism' and requests him to describe it in two or three sentences in his 'e-portfolio' at the provided spot. So, at this point, the user is asked to open his 'portfolio'-file (which is also part of the online course and easily downloadable). In other words: Working through a session, the user has to open two windows, the 'normal' session and the 'e-portfolio'. the 'e-portfolio' is like a script, summarizing the content of the session, but it also integrates specific 'blanks' to be filled individually. At the end, this saved file documents both, the given content of the course and the individual answers and reflections of the learner.

After entering his answers into the 'e-portfolio', the user can continue with the session. In the following, he learns to differentiate between the terms 'work', 'job' and 'profession' and between different definitions of professionalism (which he then is asked to compare with his own thoughts earlier); the user is then and step-by-step guided to a more differentiated perspective of the given case – through two more videos and reflective tasks: at two further positions in the session, the user watches videos in which the veterinary officer continues to tell his story and – together with the user – comes to the conclusion that the farmer insulted him in his role as a representative of state while he felt personally offended and referred it to himself as private person – which is maybe comprehensible but unprofessional.

However, in the context of this paper the specific content is not so important, more important is that these impressions should have made a bit clearer the interlocked composition of input and the activity of the user.

Log-out: evaluation and outlook

The first already completed sessions were evaluated by veterinary students of the 3rd and 5th semester of the Vetmeduni Vienna. Students of both semesters have already attended several ethics courses in the classical seminar or lecture format, which must be completed as compulsory courses during their veterinary study. Against this background, they were already fundamentally familiar with ethical topics concerning their professional life.

The evaluation questionnaire included 23 questions and can be divided into three sections: Section (1) focused on the session's technique and design, section (2) comprised didactic questions (such as on the comprehensibility of the tasks) and section (3) addressed general items for the assessment of veterinary ethics education, in particular in the field of e-learning. In total, 355 students were invited via an email to take part in the evaluation process, out of which 45 worked through one self-chosen session and fully completed the corresponding evaluation questionnaire. Among participants, ca. 80% were female and ca. 20% male; ca. 42% were in the 3rd semester and ca. 48% in the 5th semester (rest 'unanswered').

Like every evaluation, this one also has its limitations. Above all, it can be assumed that students who take a total of 45 to 60 minutes (for working on the session and doing the evaluation) in their free time are students who are fundamentally interested in the topic of ethics in their professional life. To mitigate this aspect and also reach students who are not that much interested in ethics at the university, an incentive was set: Three restaurant vouchers were raffled among all participants.

Selected results of the first evaluation round can be presented: Concerning the question 'How would you rate the following aspect of the edited session: Overall ease of use?', 88,86% rated the ease of use as 'very good' or 'good'. Especially given examples from the field of veterinary medicine were rated as an important part for understanding the taught content. For this aspect, 86,5% 'agreed completely' or 'agreed' that the given examples were helpful.

The aspect whether the 'e-portfolio' is easy to work with and easy to handle technically was designed as open-ended question. In general, students stated that it helps to 'deal with the topic self-reliantly' and that tasks, which have to be completed during the session, 'take you out of passivity'.

Despite the very positive results in crucial aspects, limitations of e-learning tools for ethical learning are looming. Especially at open questions students stated that they appreciate the online course as 'addition' to classical formats. In their opinion, ethical reflection benefits from face-to-face-communication between students and teacher and among the students. Such results indicate the usefulness of so called 'blended learning' concepts as a combination of classroom-based courses and e-learning tools, and take a critical view of any effort to replace classroom events with e-learning courses, in particular in the field of ethics for practitioners.

The project 'VEthics E-Portfolio' will develop the full online course until the end of 2018. A second evaluation round will include veterinary officers. It is planned that the final course will be part of a 'blended learning' concept.

References

Aparicio, M., Bacao, F. and Oliveira, T. (2016). An e-Learning Theoretical Framework. Educational Technology & Society 19 (1): 292-307.

Bayertz, K. (1991). Praktische Philosophie als angewandte Ethik. In: Bayertz, K. (ed.) Praktische Philosophie. Grundorientierungen angewandter Ethik. Rowohlt Taschenbuch, Reinbek bei Hamburg, Germany, pp. 7-47.

Bora, U.J. and Ahmed, M. (2013). E-Learning using Cloud Computing. International Journal of Science and Modern Engineering (IJISME), Volume-1, Issue-2, January 2013: 9-13.

Boshuizen, H. and Schmidt, H. (1992). On the Role of Biomedical Knowledge in Clinical Reasoning by Experts, Intermediates and Novices. Cognitive Science. A multidisciplinary Journal, Volume 16, Issue 2, April 1992: 153-184.

Chadwick, R. and Schroeder, D. (2002). General introduction. In: Chadwick, R., Schroeder, D. (ed.) Applied ethics. Critical concepts in philosophy. Volume I: Nature and cope. Routledge, New York and London, USA, pp. 1-20.

Dürnberger, C. and Weich, K. (2016): Conflicting norms as the rule and not the exception – ethics for veterinary officers. In: Olsson, A.S., Araújo, S.M., Vieira, M.F. (ed.) Food futures: ethics, science and culture. Wageninen Academic, Wageningen, pp. 285-290.

Ericsson, K.A., Krampe, R.T., Tesch-Romer, C. (1993). The Role of Deliberate Practice in the Acquisition of Expert Performance. Psychological Review 100(3): 363-406.

Figgis, J. (2009). Regenerating the Australian Landscape of Professional VET Practice: Practitioner-Driven Changes to Teaching and Learning. National Centre for Vocational Education Research, Adelaide, Australia, 35 pp.

Fischer, J. (2001). Die Begründungsfalle. Plädoyer für eine hermeneutisch ausgerichtete theologische Ethik. Zeitschrift für Evangelische Ethik 46: 163-167.

Gruber, H., Harteis, C. and Rehrl, M. (2006). Professional Learning: Erfahrung als Grundlage von Handlungskompetenz. Bildung und Erziehung 59 (2): 193-203.

Gruber, H. and Stöger, H. (2011). Experten-Novizen-Paradigma. In: Kiel, E. and Zierer, K. (ed.) Basiswissen Unterrichtsgestaltung. Bd. 2. Unterrichtsgestaltung als Gegenstand der Wissenschaft. Schneider-Verlag, Hohengehren, Germany, pp 247-264.

Häcker, T. (2012). Portfolioarbeit im Kontext einer reflektierenden Lehrer/innenbildung. In: Egger, R and Merkt, M. (ed.) Lernwelt Universität. VS Verlag, Wiesbaden, German: pp. 263-289.

Hornung-Prähauser, V., Geser, G., Hilzensauer, W. and Schaffert, S. (2007). Didaktische, organisatorische und technologische Grundlagen von E-Portfolios und Analyse internationaler Beispiele und Erfahrungen mit E-Portfolio-Implementierungen an Hochschulen. Salzburg Research Forschungsgesellschaft, Salzburg, Austria, 180 pp.

Kopp, V., Stark, R. and Fischer, M.R. (2007). Förderung von Diagnosekompetenz in der medizinischen Ausbildung durch Implementation eines Ansatzes zum fallbasierten Lernen aus Lösungsbeispielen. GMS Z Med Ausbild 2007;24 (2) pp Doc107.

Magalhães Sant'Ana, M., Baptista, C.S., Olsson, I.A.S., Millar, K. and Sandøe, P. (2009). Teaching animal ethics to veterinary students in Europe: examining aims and methods. In: Millar, K., West, P.H. and Nerlich, B. (ed.) Ethical futures: bioscience and food horizons. Wageningen Academic Publishers, Wageningen, the Netherlands, pp. 197-202.

Reiss, M. (2005). Teaching animal bioethics: pedagogic objectives. In: Marie, M., Edwards, S., Gandini, G., Reiss, M. and Von Borell, E. (ed.) Animal Bioethics: Principles and teaching methods. Wageningen Academic Publishers, Wageningen, the Netherlands, pp. 189-202.

Roohani, A., Jafarpour, A. and Zarei, S. (2015). Effects of Visualisation and Advance Organisers in Reading Multimedia-Based Texts. 3L: Language, Linguistics, Literature® 21.2: pp. 47-62.

Ropohl, G. (1998). Technikethik. In: Piper, A., Turnherr, U. (ed.) Angewandte Ethik. Eine Einführung. Beck, München, Germany, pp. 264-287.

Schön, D.A. (1984). The Reflective Practitioner: How Professionals Think In Action. Basic Books, New York, USA, 384 pp.

Sun, P. C., Tsai, R. J., Finger, G., Chen, Y. Y. and Yeh, D. (2008). What drives a successful e-Learning? An empirical investigation of the critical factors influencing learner satisfaction. Computers & Education, 50(4): 1183-1202.

Wipperfürth, M. (2016). Sprachlosigkeit in der LehrerInnenbildung? Reflective best practice in dialogue. Teaching Languages – Sprachen lehren (2016): 123-145.

50. Filling the gap: teaching human-animal studies in European vet departments

P. Fossati[1*] *and A. Massaro*[2]
[1]*Department of Health, Animal Science and Food Safety, University of the Studies of Milan, Via G. Celoria, 10, 20133 Milan, Italy;* [2]*Independent Researcher, Via Assarotti 23/8, 16122 Genova, Italy; paola.fossati@unimi.it*

Abstract

Human-animal studies (HAS) represents an important research tool for understanding human culture by revealing the complexity of animal-human interfaces. It offers a fresh critical perspective on the history of the interaction between humans and animals. Recognizing the deep influence animality has on the human realm, HAS can help in broadening our knowledge about human beings, seen not as an alien species with respect to nature but as an interacting part of it. Overall, by discussing the intersections of human and nonhuman animals in society, HAS offers the opportunity to expand humans' views regarding other living beings, while helping humanity to understand its relationships with nonhuman animals and to work out the implications of those relationships. Hence the importance of bringing the discipline of human-animal studies into the realm of scholarly inquiry and the educational system. However, there is some evidence of a lack of HAS instruction in European academic curricula, particularly in the provision of programmes and curricula by faculties of veterinary medicine. Based on this premise, the Authors have investigated the state of HAS in the curricula and teaching programmes offered by European faculties of veterinary medicine, in order to assess the trend in the field. Furthermore, for the purpose of completing the framework, the project *Cibo: la vita condivisa* ('Food: shared life'), an innovative cycle of Summer Schools designed to enhance teaching and debating activities in the field of HAS within academia, is presented. In their conclusion, the Authors argue that the format of this project is a pioneering instrument to incorporate this subject in veterinary curricula, from which new specific courses and other forms of educational offerings on the issue might be developed.

Keywords: human-animal relations, educational offering, veterinary studies, European universities, Italian universities

Introduction

Human-animal studies (HAS) is a rapidly growing interdisciplinary field that focuses on the place of animals in human cultures and societies (DeMello, 2012). The importance of HAS arises from its exploration and critical evaluation of the complex field of human-animal relations (Dupré, 2006). Despite this, there are still problems facing the field, perhaps the biggest of which is the lack of attention given to this newly developing discipline in academic institutions (Taylor, 2013). As argued by Ken Shapiro and Margo DeMello, founders of the Animals and Society Institute (ASI), which is an American research and educational organization that advances the status of animals in public policy and promotes the study of human-animal relationships, this may be due to the conservative culture of universities combined with the interdisciplinary nature of the field. These two factors make it difficult to locate any theoretical paradigms or methodologies inherent to specific disciplines amenable to the study of human-animal relations (Shapiro and DeMello, 2010). In any event, it's a fact that over the last ten years human-animal studies have gained in importance in the research landscape (McHugh and Marvin, 2014; McHugh, 2016). But the institutionalization of HAS as a discipline has developed unevenly in Europe, and, most importantly, it has not yet become widespread.

EurSafe 2018 – DOI 10.3920/978-90-8686-869-8_50, © Wageningen Academic Publishers 2018

In Italy, a pioneering instrument to combat this issues in veterinary departments has been developed. From 2015, the Faculty of Veterinary Medicine of the University of Milan has hosted a cycle of Summer Schools, designed by the Authors of this paper themselves, devoted to human-animal studies (Fossati, 2017; Massaro and Fossati, 2015). The Summer Schools are not organised by a department, but instead result from collaboration between a single university researcher and an independent organization that organizes and finances itself. The next step would be to institutionalize this field of research, in order to improve the availability of course offering and further foster interest for the issue.

About HAS

HAS is an interdisciplinary field of study that focuses on both nonhuman and human animals (DeMello, 2012; Freeman, 2010; Ingold (Ed.), 1994), exploring ways they share spaces in human social and cultural worlds. The terms nonhuman and human animals identify humans as animals and emphasize commonality between humans and animals who aren't human. This wording helps to indicate that humans and animals are not two unrelated categories. HAS aims to foreground the many different dimensions of humans' interactions and relationships with non-human animals, developing perspectives on living beings and, in particular, expanding humans' views regarding other living beings (Andersson *et al.*, 2014).

This is challenging because of the existence of a wide variety of animals with which different kinds of people have engaged in many contexts, so that human-animal relations have taken many forms throughout history and continue to evolve. Account should be taken of the sociocultural, economic, and political circumstances in which human-animal relations occur. In addition, the diverse types of animals and their particular status within human societies must be carefully charted (Brantz, 2010).

Furthermore, it has been argued that access to the (scientific and non-scientific) knowledge of animals is through human-constructed sources, and 'then we never look at the animals, only ever at the representation of the animals by humans' (Fudge, 2002). This remark makes it necessary to point out that humans have carried out their attempts to understand the other living beings according to many different approaches, with the result that humans' transmission and sharing of views about nonhumans is incredibly layered, in relation to the various humans' resources and abilities (Cherry, 2017).

Such a scenario is well summarized in the following, recent description of human-animal studies as 'an international phenomenon, with animal-related puzzles engaging the minds of inquiring scholars the world over' (O'Sullivan and Bennison, 2011).

Aim

Although human-animal studies have valuable potential for analysing the many ways that cultures are now interacting with nonhuman animals, there is a lack of attention given to programmes of studies and research in this field in the context of European academia.

The aim of this paper is to point out the situation regarding the state of human-animal studies in Europe, with a focus on faculties of veterinary medicine, and then to highlight the Italian project *Cibo: la vita condivisa* ('Food: shared life'). This project was designed to enhance teaching and debating activities in the field of HAS, and to develop a teaching format that should be included in the curricula in order to help to introduce HAS into Vet academic programs. Therefore, the Authors propose an analysis of the state of HAS in the curricula and teaching programmes offered by the European faculties of Veterinary Medicine, in order, firstly, to understand how widespread this topic is in Europe, and, secondly, to provide a better understanding of how these studies can be promoted into the traditional veterinary curriculum – which would be an improvement and a development of the format of the Summer Schools held so far, thanks to the aforementioned project.

Methodology

As part of the aim of investigating and understanding the current availability of courses on HAS into European veterinary curricula, the Authors have identified the courses offered by faculties and schools of veterinary medicine in Europe. The inclusion of topics such as 'anthrozoology', 'animal assisted therapy or interventions (AAT – AAI)', 'human animal studies (HAS)', or 'animals and societies' as subjects on the curricula was treated as a reasonable indicator of the existence of the provision of education and training in the field of human-animal studies. Indeed, these topics suggest that an approach to the study of HAS is offered to students.

The lack of a study and monitoring center related to the status of HAS in Europe has led the Authors to consider the Animal and Society Institute (ASI) as a primary source from which to gather information regarding the teaching of HAS at European Veterinary universities. However, in order to understand the validy of this criterion, they have conducted a parallel analysis of the data included in the ASI's list of Italian veterinary universities by performing an autonomous search inside all Italian veterinary universities' websites; they have done the same for UK veterinary universities. The choice of these countries was made on the basis of the broadest offer of veterinary curricula in the European landscape (9 universities in the UK; 13 universities in Italy), and yet it is not fully included in the ASI's list. To this end the Authors have searched inside the universities' websites for the aforementioned key terms that they treat as a sign that a real human animal perspective is being offered to students. These comparisons demonstrate that the ASI website is not inclusive of all HAS courses offered in Italy (it includes only one of the five active in Italy), but it is inclusive of all those offered in UK. Therefore, the Authors conclude that the ASI list is not completely representative of what is offered in European veterinary universities, but it can be considered a first tool, useful for giving a glance of what is happening, while waiting for a more accurate monitoring center of HAS activities conducted in Europe. While conscious that it is not inclusive of all courses offered in the EU veterinary curricula, the Authors have used the list of the Universities that offer HAS courses in Europe that is present on the ASI website as initial tool for research,.

Thus, the Authors analysed the websites of the seven European faculties and schools of veterinary medicine (Edinburgh, Liverpool, Wien, Berlin, Pisa, Leon and Murcia), included in the ASI's list (research updated at January 2018), using a double standardized methodology. First, the websites of the courses or faculties were checked, in order to verify that the existence of the courses on HAS was immediately clear to users. Second, when the information on the website was not clear enough, the curricula were analysed. Then the autonomous search inside all veterinary universities' websites of Italy and UK was performed.

Finally, the project *Cibo la vita condivisa* ('Food: shared life') was described with respect to both its background and rationale, and its first concrete results.

Results

According to the research started from the ASI's website it was found that five of the seven European universities listed on the ASI website offer this kind of education to their students. Three universities, in Edinburgh, Pisa, and Wien (Messerli Institute), offer master courses, and include a HAS approach among traditional perspectives, while two universities, in Liverpool and Berlin (Department of Veterinary Medicine of the Freie university), offer at least a teaching module focused on HAS.

The situation seems different in the Spanish veterinary departments of Leon and Murcia listed on the ASI website. From a check of their websites and curricula, it was impossible to determine what kind of contribution they make to the HAS approach. The Authors contacted these departments in order to better understand their programmes of study. The reply from the Veterinary Department of Murcia University made it impossible to understand whether HAS was included or not since the person who replied to the email did not know what HAS is. After this was explained, they did not return the message. There has been no reply from Leon University. It is therefore impossible for users to assess whether these departments really offer a HAS approach.

Regarding the state of the art of HAS in the UK, the University of Edinburgh offers a Master in 'Animal Welfare, Ethics, and Law' which started in 2017. Students can choose an optional course devoted to Anthrozoology. It is interesting to notice that students intending to go on to the Diploma or MSc in the School of Veterinary Studies are strongly recommended to take Anthrozoology in their certificate year. In its Postgraduate Diploma/MSc in Veterinary Professional Studies, the University of Liverpool offers the module 'Animals & Human Society'. Unfortunately, the research provided no further evidence of other offers of HAS courses in the other universities that provide veterinary curricula.

In Italy the research for the chosen keywords on websites and in the programmes of the thirteen universities offering veterinary curricula has revealed a similar limited availability of HAS courses. Indeed, only five universities endorse HAS. The Department of Veterinary Science of Pisa houses the 'Laboratorio di Etologia e Fisiologia Veterinaria' (Laboratory of Veterinary Ethology and Physiology), which organizes a Master in 'Etologia clinica veterinaria' (Veterinary Clinical Ethology), and includes a module on 'Antrozoologia didattica e consulenziale' (Anthrozoology for education and consulting) and another one on 'Terapie Assistite con gli Animali' (Animal Assisted Therapy). The University of Naples offers a Master in 'Zooantropologia sanitaria per gli interventi assistiti dagli animali (Anthrozoology and AAI)'. The University of Sassari offers a degree course in Veterinary Medicine that includes a teaching module that includes the topic 'anthrozoology'. The University of Parma offers a Master in canine behaviour and training, and the programme includes issues in the field of anthrozoology. The University of Perugia offers a Master in 'Veterinary pain management and behavioural medicine' which includes elements of anthrozoology.

Project *Cibo la vita condivisa* ('Food: shared life')

The project *Cibo: la vita condivisa* ('Food: shared life') consists of an innovative, interdisciplinary cycle of Summer Schools, devoted to human-animal studies and targeted to young graduates and scholars, but also practitioners, lawyers, journalists, teachers interested in the issue. The project takes a holistic approach to updating, widening and deepening education and training initiatives in animal law and animal ethics themes, in a European landscape where these topics are covered by isolated training options.

The key characteristic of the project is, at present, a unique way of fostering interest in human-animal studies, promoting research in this field, and extending the visibility of this issue in Italy (and possibly even in Europe, opening it up to European Universities and the EU society at large). The Schools aim to offer an integrated, multidisciplinary approach that operates on the dimension of 'giving knowledge' while fostering critical thinking about complexity of the disciplines of interest. As a multidisciplinary course, the project faces the challenge of offering a multifaceted approach to animal studies, in contrast to current practice, in which a 'single-theme' approach has been prevalent (i.e. anthrozoology, animal assisted therapy, animals and society). Interaction between excelling scholars, institutional representatives, and NGOs' representatives from Italy and from several countries in Europe, is proposed through lectures, seminars and workshops. Furthermore, thematic Round Tables open to the public are organized to discuss selected themes of interest, reaching a wider audience. The cycle of Summer Schools is designed to enhance teaching and debating activities in the field of human-animal studies and it is expected to be the first step towards developing a line of comprehensive education in the Italian (and European) landscape of human-animal studies and institutionalizing course offerings in the area.

Each Summer School is comprised of five intensive courses (morning – daily) and a series of five specialist practical seminars, held with the intervention of distinguished specialist guest lecturers, which constitute the 'core' of the teaching programmes.

Three Summer Schools have already been carried out. The courses were held at the Faculty of Veterinary Medicine of Milan, but a learning space for participants from diverse professional and cultural backgrounds – such as philosophy, sociology, theology and law – was offered. In the first edition (Food: shared life, 2015), the foremost socio-economic, juridical and ethical issues that flow from human food choices were investigated; in the second year (Food: shared life. Beyond the dish, 2016) the Summer School was focused on both animal welfare and the sustainability of animal production; the third edition (Food: shared life. Communicating animal welfare, 2017) was devoted to investigating how animal welfare is communicated, analysing the strategies of communication regarding food of animal origin and its production. Special attention was given to the role of communication in influencing consumer behaviour and food choices. Participation was very successful: active and ensuring satisfactory learning outcomes, as emerged from final assessments and evaluation of the satisfaction of participants.

Discussion

The previous investigations have confirmed the initial supposition about the lack of teaching of HAS in European academic curricula, and in particular the existing gap in the provision of programmes and curricula by faculties of veterinary medicine. As previously noticed, there are probably other European veterinary universities offering a HAS approach and it is probably true that other faculties of veterinary medicine offer some other courses on this subject, but from a simple google search it is really tricky to discover.

An example of this 'offer out of sight' is the Italian project *Cibo: la vita condivisa* ('Food: shared life'), an innovative cycle of Summer Schools designed to enhance teaching and debating activities in the field of HAS within academia.

Furthermore the authors suggest that the creation of a European center for the study and monitoring of this subject would be required in order to help understanding the statusof HAS in the EU and with the broad aim of ensuring a consistent framework for all concerned. A similar monitoring center would also help young students willing to explore the HAS approach to find the right course of study.

Conclusion

Human-animal studies are emerging as a fast-growing field of research, capable of bringing forward studies on non-human animals while exploring the spaces that they occupy in human societies and cultures. HAS encourages critical impulses and fosters debate throughout society. Greater focus is required on the issue, in order to provide further opportunities and means to deepen, develop and broaden study/training, as well as inquiry, in the European landscape, filling the gap in teaching courses and curricula on the subject. From this perspective, appropriate proposals should be developed in order to create specific pedagogical methodologies, training tools and evaluation, as instruments to orientate the stakeholders to better understand and tackle HAS.

The goal could be the promotion, institutionalization and establishment of a new range of courses in the field of HAS, within the academia, which might have the added value of enhancing the establishment of a new line of research in such field, developing new specific curricula, and creating new academic positions.

One model may be provided by the project *Cibo: la vita condivisa* via its proposal for thematic Summer courses, which have been demonstrated to be an effective platform for truly interdisciplinary work, suitable to draw the multi-faceted development of a subject that is increasingly relevant in contemporary societies.

This format could be improved, and the exchange of knowledge and best practice with those veterinary departments in Europe today that have already developed curricula and established HAS at teaching and research facilities will be valuable.

Furthermore, paying greater attention to inquiring and teaching the discipline of HAS in veterinary departments will not only allow the development of the related field of research, but also will make it possible to provide future veterinarians with a new competence, by raising their awareness of the importance of studying human-animal interactions.

With regard to scientific literature, we hope for greater attention on HAS teaching inside European veterinary departments and the development of further investigations to better understand the status of HAS inside academic curricula.

References

Andersson Cederholm, E., Björck, A., Jennbert, K. and Lönngren, A-S. (Eds.) (2014). Exploring the Animal Turn: Human-animal relations in Science, Society and Culture. (The Pufendorf Institute for Advanced Studies). The Pufendorf Institute of Advanced Studies, Lund University.

Brantz, D. ed. (2010). Beastly Natures. Animals, Humans, and the Study of History. Charlottesville, VA: University of Virginia Press.

Cherry, E. (2017). The Sociology of Non-human Animals and Society. In: The Cambridge Handbook of Sociology, Chapter 10: 95-104.

DeMello, M. (2012). Animals and Society. An Introduction to Human-Animal Studies; Columbia University Press: New York, NY, USA.

Dupré, J. (2006). Humans and Other Animals. Oxford: Clarendon Press.

Fossati, P. (2017). Food: shared life. Editorial. Relations. Beyond Anthropocentrism, 5.1:7-10.

Freeman, C. P. (2010). Embracing humanimality: Deconstructing the human/animal dichotomy. In G. Goodale and J. E. Black (Eds.). Arguments about Animal Ethics (pp. 11-30). Lanham, MD: Lexington Books.

Fudge, E. (2002). A left-handed blow: Writing the history of animals. In Representing Animals, 3-18, ed. N. Rothfels. Bloomington, IN: Indiana University Press.

Ingold, T. (1994). What is an animal?. London, Routledge.

Marvin G. and McHugh, S. (Eds.) (2014). Routledge Handbook of Human-Animal Studies, London, Routledge.

Massaro A., Fossati P. (2015). SUMMER SCHOOL – Cibo: la vita condivisa. Academia.edu.

McHugh, S. (Winter 2016). Critical Animal Studies: Thinking the Unthinkable. Critical Inquiry 42, no. 2: 414-415.

O'Sullivan, S. and Bennison, R. (2011). Riding the crest of a human-animal studies wave. Society & Animals 19(4): 333-336.

Shapiro, K., DeMello, M. (2010). The state of human-animal studies. Society & Animals, 18(3): 307-318. Available at: https://www.animalsandsociety.org/wp-content/uploads/2016/04/shapiro-1.pdf.

Taylor, N. (2013). Humans, Animals, and Society: An Introduction to Human-Animal Studies, New York, Lantern Books.

Section 9.
Media, transparency and trust

51. Tracing trust – on tracking technologies and consumer trust in food production

*S.G. Carson and B.K. Myskja**
Department of Philosophy and Religious Studies, Norwegian University of Science and Technology – NTNU, 7491 Trondheim, Norway; bjorn.myskja@ntnu.no

Abstract

Food production has been an arena for asymmetrical trust, i.e. a trust based in an uneven power-relation where the trustor is dependent on the trustee. Over the past decades, this has changed dramatically with a stronger emphasis on the rights of the consumer. This is expressed in the increased focus on labelling, product tracing and consumer choice. This paper explores how the use of tracking technology in food production affects consumer choice and trust. Previously, the consumers were left to passively, or even blindly, trust the food delivery chain. In the emerging food delivery system one can talk about a new level, maybe even a new form, of autonomy through increased access to information, and by that, increased empowerment of the individual. It is reasonable to assume that the new technology based product information provides a better, more detailed picture, but also more complicated picture, disturbed by information 'noise'. How does this affect trust – or how should our trust in the food delivery system and expertise be affected? Consumers need help to interpret information in order to make informed choices and the basis for trust will be altered, due to the distribution of responsibilities and use of technology. This development requires a turn from 'blind' to reflexive trust. While it may lead to information overflow and thus to a *de facto* decrease in consumer choice, increasing risk and development of new technologies arguably imposes a moral responsibility on the side of the consumer to collect and assess the available information.

Keywords: food tracking, reflexive trust, responsibility, consumer choice

Introduction

In 2006, a 4-year old, Norwegian boy died from *E. coli* infection, and several others fell seriously ill. The source of the outbreak turned out to be infected sausage from a most prominent and well-reputed Norwegian meat producer (Folkehelseinstituttet, 2006). In the aftermath of tragedy and scandal, the Norwegian government invested heavily in an ambitious electronic food traceability project. After seven years and 62 million NOK, the project shipwrecked due to technical and administrative problems, but it was argued that the state-led project had lasting effects on the traceability of food products in terms of improved knowledge and raised consciousness (Ulstein *et al.*, 2014), and food-tracking technologies continues to be developed by the private sector in Norway. Meanwhile, in the same period the focus on traceability of food products was increasing globally because of several food scandals and in general the industrialization of food production and the resulting complexities and anonymity of modern supply chains.

The demand for food traceability leads to the development of a number of different tracking technologies involving the screening of food products throughout production, processing and distribution. Thus, these two concepts – tracking and tracing – indicates the two directions one can register the movement through the food production chain: '[T]racing and tracking have been interpreted as exploration of an entity (e.g. food product) under consideration (in the supply chain) in the upstream direction and downstream direction, respectively.' (Bosona and Gebresenbet, 2013, 34). Modern food production consists of various stages including the sourcing of seeds, eggs, and fertilizers, the farming and harvesting

EurSafe 2018 – DOI 10.3920/978-90-8686-869-8_51, © Wageningen Academic Publishers 2018

of fish or produce, processing, storage, transportation and retail sales. On all of these stages, there are risks involved, for example the risk of contamination. ISO (International Organization for Standardization), which develops voluntary international standards for products and services, defines traceability as the 'ability to trace the history, application, or location of that which is under consideration' (ISO 2011), or, specifically for food production, the 'ability to follow the movement of a feed or food through specified stage(s) of production, processing and distribution' (ISO 2007). While the primary purpose of food tracking is risk management and food safety, there are several other important objectives of traceability. Better surveillance, more cost-effective management of the supply-chain and quality assurance are other key objectives. Finally, increasing the transparency of the supply chain through traceability may empower consumers in terms of facilitating informed and ethical consumer choices regarding issues such as animal welfare and environmental concerns (Coff *et al.*, 2008).

There are a number of new tracking technologies such as RFID, GPS, electronic tags, nuclear techniques and biometrics (Bosona and Gebresenbet, 2013). They could potentially enable increased consumer control through improved information flow and more detailed knowledge about how, when and where the food is produced. However, the technology based product information may also provide a more complicated picture, disturbed by information overload and 'noise'. Currently, the existing technologies are unable to provide consistent, reliable information. However, the systems are improving, and novel 'internet of things' – technologies, make it likely that more reliable information will be available in the near future (Badia-Melis *et al.*, 2015). However, more information and traceability does not always increase the trust of consumers, but may rather create, under certain circumstances, uncertainty and suspicion (Dulsrud *et al.*, 2006). Arguably, consumers must go from 'blind' to reflexive trust, and seek to be enlightened and empowered rather than become sceptical and overwhelmed by the increased access to information. This paper discusses how the use of tracking technology in food production affects consumer choice and trust. *Gladlaks*, a branding program for Norwegian farmed salmon is used as an example. It is argued that new technologies for food traceability changes the basis for consumer trust and furthermore, that consumers' responsibility to seek out information about how their food is produced increases in light of these new technologies.

The happy salmon

In 2015, the Norwegian seafood producer Lerøy joined forces with NorgesGruppen, the largest player in Norwegian grocery, in launching the brand *Gladlaks*, which translates to 'happy salmon' and is a common phrase for being carefree and unconcerned. The idea behind the brand name is that consumers should not have to worry about anything concerning this product. On the brand web page, under the heading 'Full transparency – full enjoyment', they state: 'Our customers are increasingly concerned about health and wellbeing. We can tell that by the demand for fresh fish products. More and more also want to make sure that the food they eat is safe and that it is produced in a responsible way.' (*Gladlaks*; authors' translation) Further, it is claimed that *Gladlaks* is good news for these concerned consumers, since they offer full transparency in all regards. The brand guarantees that the fish is antibiotics free and fed with palm oil free feed, and it also signifies traceability: '*Gladlaks* is traceable from egg to finished product. Through this tracking you as a consumer get to know everything that has happened to your particular salmon: What it has eaten, where it has swum, which vaccines it has been given – and so much more' (*Gladlaks*; authors' translation).

Needless to say, this information is a lot to handle, and the producer might not literally encourage the consumers to seek out every tedious detail of the life of the fish on their plates. Rather, the intention is to let the consumers know that all the information is available, in case they ever do get an urge or a need to seek it out. At the top of the web page is a box where consumers may type the code printed on the packaging of the fish, and from there they can access the information they want.

Based on the increasing focus on traceability nationally and globally, one might expect this initiative to be wished welcome on the Norwegian market. However, upon first launching the brand, *Lerøy* and *NorgesGruppen* received a lot of criticism, among others from the Norwegian Consumer Council, who claimed that this was misleading marketing. Specifically, the Consumer Council reacted to the way *Gladlaks* specified that their salmon was antibiotics free, since actually all Norwegian salmon is antibiotics free. Thus, it was argued that this was 'too much information' and misleading in the sense that it led consumers to assume that other, competing products may contain antibiotics. Similarly, it was argued that stressing how their feed is palm oil free is misleading given that feed containing palm oil is more or less non-existing in Norwegian salmon farming. The Norwegian Food Safety Authority reviewed the case and ordered *Lerøy* and *NorgesGruppen* to change their marketing of the fish.

The case illustrates how traceability and increased product information may not always improve the decision-making ability of consumers. *Lerøy* and *NorgesGruppen* disagreed in the Food Safety Authority's decision, claiming that they were only trying to do their part in correcting some widespread misconceptions regarding farmed salmon. The question in this case was not about whether the information given was correct – this was not disputed – but whether the consumers were able to process the information in the appropriate way. Part of the question might arguably have been whether Norwegian consumers were prepared for this level of transparency.

Challenges to consumer trust in Norwegian food production

Tracking technology increases the available information on how, when and where food is produced. As the *Gladlaks* case illustrates, the significance of information for consumer trust is disputed. While the right amount and kind of information might build trust, too much or irrelevant information could rather be counterproductive and lead to distrust. According to a 2006 study of consumers trust in seafood, Norwegian consumers had an 'undisputed trust' in the Norwegian Food Safety Authority as a guarantor of food safety, compared to German consumers who to a larger extent sought information from retailers and producers but also to a larger extent distrusted the available information (Dulsrud *et al.*, 2006, 226). Norwegian consumers traditionally have a high trust in food safety resulting from a high trust in the institutions safeguarding it, but arguably, they too are affected by food scandals and the globalization and industrialization of food production, although to a lesser extent than many other consumer groups (Berg, 2003). New systems of traceability and quality control imply a gradual redirection of the responsibility for food safety from the direct control of the authorities to the industry actors themselves (e.g. by help of quality standards such as ISO). The safety of food products gradually becomes a matter of product differentiation, in other words something the producers use in order to distinguish certain products from other, similar products, e.g. through explicit branding, while it used to be secured by authorities and simply trusted by the consumers. In addition to food safety, ethical, environmental and social values are increasingly used for product differentiation, first in imported goods such as fair trade coffee and fruit, and gradually for Norwegian food products as well.

Self-regulation and increasing international competition creates a new situation where Norwegian food producers communicate more explicitly on issues of quality, health and safety (Ursin *et al.*, 2016). The development towards self-regulation challenges the 'blind' trust in the Norwegian food sector, since this trust traditionally was closely linked to the comprehensive external control systems (Terragni, 2004).

From blind to reflexive consumer trust

According to Luhmann (1988), trust is a way to manage uncertainty by reducing the complexity of decision-making situations. Thus, trust is not a purely rational construct, but rather what Zinn calls an 'in between strategy' based on contextual knowledge, for example personal experience (Zinn, 2008, 442-

443). Luhmann argues that the modern conception of risk – indicating that unexpected consequences might result from our decisions rather than simply from fortune – alters the human condition and calls for a distinction between trust and confidence. While confidence is the natural condition of 'blind' trust – simply presuming that everything is in order without questioning it – to trust is to take an active stand, to make a 'leap of faith'. Thus, when (blind, unquestioning) confidence in the safety of food is challenged, consumers must reflect on the available knowledge, and based on this make a more or less informed decision to trust or distrust. Giddens (1991) and Beck (1992) argues that the shift towards the 'reflexive modernity' of a 'risk society' changes the nature of trust, from unconditional to reflexive trust (cf. Dulsrud *et al.*, 2006, 214). Reflexive trust, as opposed to confidence, requires an active stand to the available facts.

There are perceived as well as real risks following from globalization and industrialization of food production, increasing self-regulation and consumer choice – concerning present and future food safety as well as environmental and social risks. Consumers must arguably face these challenges by taking a more active stand and seek out information about the food they consume. Following Hans Jonas' imperative of responsibility, the risk associated with technological development brings about changes in our moral duties, and an important element of this is the duty to seek knowledge about possible outcomes, and to take precautions according to a 'heuristics of fear' (Jonas, 1979, 63). Thus, there is an internal connection between the turn from blind to reflexive trust and Jonas' analysis of the new dimensions of responsibility. The power of technology creates a sphere of collective action, where everyone shares responsibilities for the outcome, and everyone is required to seek knowledge about the outcome of these collective acts (Jonas, 1979, 26-28).

The duty of individual consumers to act as concerned citizens is, however, not possible to live up to without proper information provided by food producers and food authorities. A guiding principle for a proper information flow could be to reduce 'noise' and provide the information consumers actually need to make informed choices. In the *Gladlaks* case, the declared intention of the producer was 'full transparency', but the result seemed to be 'too much information'. At the time it was launched, the information and traceability seems to have contributed to increasing complexity rather than reducing it in order to establish trust. In a different context, however, the level of information might have been proper. Research show that Norwegian consumers in general are positive to more labelling and increasing product information and consumer control, but at the same time that they have problems assessing the information (Roos *et al.*, 2010, 58). This might suggest that, given the right kind of training and information, consumers are ready to take on more responsibility in order to acquire a reflexive rather than blind trust in food.

References

Badia-Melis, R., Mishra, P and Ruiz-García, L. (2015). Food tracability: New trends and recent advances. A review. Food Control 57: 393-401.

Beck, U. (1992). Risk society: Towards a new modernity. Sage, London, UK, 260 pp.

Berg, L. (2004). Trust in food in the age of mad cow disease: a comparative study of consumers' evaluation of food safety in Belgium, Britain and Norway. Appetite 42(1): 21-32.

Bosona, T. and Gebresenbet, G. (2013). Food tracability as an integral part of logistics management in food and agricultural supply chain. Food Control 33: 32-48.

Coff, C., Korthals, M. and Barling, D. (2008). Ethical traceability and informed food choice. In Coff, C., Korthals, M., Barling, D. and Nielsen, T. (eds.) Ethical traceability and communicating food. Springer, Dordrecht, the Netherlands, pp. 1-18.

Dulsrud, A., Norberg, H. M. and Lenz, T. (2006). Too much or too little information? The importance of origin and traceability for consumer trust in seafood in Norway and Germany. In Luten, J.B., Jacobsen, C., Bekaert, K., Saebo, A. and Oehlenschlager, J. (eds.) *Seafood research from fish to dish: quality, safety and processing of wild and farmed fish.* Wageningen Academic Publishers, Wageningen, the Netherlands, pp. 213-228.

Folkehelseinstituttet. 2006. *E. coli* O103 – Utbruddet 2006 Oppsummering AV Spekepølsesporet, Arbeidsrapport 09-06-2006.

Giddens, A. 1991. Modernity and self-identity: Self and society in the late modern age. Stanford University Press, Stanford, California, USA, 256 pp.

Gladlaks (?). Full åpenhet, full matglede. Available at: http://www.gladlaks.no/full-apenhet-full-matglede. Accessed 28 January 2018.

ISO 22005 (2007). Traceability in the feed and food chain – General principles and basic requirements for system design and implementation.

ISO 12877 (2011). Traceability of finfish products – Specification on the information to be recorded in farmed finfish distribution chains.

Jonas, Hans. 1979. Das Prinzip Verantwortung. Suhrkamp, Frankfurt am Main, Germany, 414 pp.

Luhmann, N. (1988). Familiarity, confidence, trust: Problems and alternatives. In Gambetta, D. (ed.) Trust: Making and breaking cooperative relations. Blackwell, Oxford, UK, pp. 94-107.

Roos, G., Kjærnes, U. and Ose, T. (2010). Warning labels on food from the point of view of consumers. Report. National Institute for Consumer Research, Oslo, Norway. Available at: https://www.mattilsynet.no/mat_og_vann/merking_av_mat/allergener/sifo_warning_labels_onfood_from_the_point_of_view_of_the_consumers_2010.1551/binary/SIFO:%20Warning%20labels%20on%20food%20from%20the%20point%20of%20view%20of%20the%20consumers%202010. Accessed 31 January 2018.

Terragni, L. (2004). Institutional strategies for the production of trust in food in Norway. Report. National Institute for Consumer Research, Oslo, Norway. Available at: http://www.hioa.no/extension/hioa/design/hioa/images/sifo/files/file54070_fagrapport_2004-08.pdf. Accessed 29 January 2018.

Ulstein, H., Wifstad, K., Loe, J.S.P. and Skogli, E. (2014). Etterevaluering av eSporingsprosjektet. Report. Menon Business Economics, Oslo, Norway, 28 pp.

Ursin, L., Myskja, B. K., and Carson, S. G. (2016). Think Global, Buy National: CSR, Cooperatives and Consumer Concerns in the Norwegian Food Value Chain. Journal of Agricultural and Environmental Ethics, 29(3): 387-405.

52. Achieving effective animal protection under the threat of 'Ag-gag' laws

A.S. Whitfort
The University of Hong Kong, Faculty of Law, 10/F Cheng Yu Tung Building, Pokfulam Rd Pokfualm, Hong Kong; whitfort@hku.hk

Abstract

Significant improvements to legislation protecting farm animals in the past two decades can be directly attributed to the influence of animal welfare science on regulation and policy in many parts of the world, most notably in the European Union. Outside of the EU, these improvements are now under very serious threat. In ten states in the USA, and three in Australia, new legislation, coined 'Ag-gag' law, has been enacted prohibiting public dissemination of material depicting on farm animal use. In both countries, media corporations and private citizens are liable to up to three years' imprisonment for publishing photographs or recordings depicting the conditions of animals on farms or at slaughter. Controls on the publication of information documenting animal use have compromised transparency in the food chain, eroded the accountability of those involved in the management of animals and undermined the case for enhanced legislation and policy reform. This paper describes recent challenges in the US states of Idaho and Utah to the constitutionality of 'Ag-gag' laws and evaluates the impact such laws may have on the transparency of animal use in agricultural facilities in the USA and Australia.

Keywords: farm, legislation, transparency, welfare, criminal

Introduction

Undercover surveillance of farm animal conditions has long proved an effective tool in animal protection, not only in exposing animal cruelty cases that might otherwise have gone undetected, but in inciting and informing public discourse on the appropriate and acceptable use of animals. In Europe, the importance of public discourse to effective change for animals, is well illustrated by the work of Ruth Harrison in the United Kingdom (Harrison, 1964). The publication of Harrison's book, 'Animal machines', led to the appointment of the UK Parliamentary Committee which produced the 'Brambell Report; in 1965. As noted by Webster (2013), Harrison's visual images of hens in battery cages, which were reproduced in her book, and later published in a major London newspaper, were able to do more to influence public opinion than a thousand diligent scientific studies of the welfare of a laying hen might ever have achieved. It is significant to note that today the publication of the images, reproduced by Ruth Harrison in 'Animal machines', would render her criminally liable in several jurisdictions in the USA and Australia.

In the USA, undercover investigations have also led to law reforms to better protect animals. Most famously, federal laws were amended to protect cattle after undercover video was recorded by the Humane Society of the United States showing downed cattle being fork lifted by meat workers in California in 2007 (Flaccus, 2009). Such investigative work is now paying the price of success. In recent years, both the USA and Australian governments have enacted new laws effectively undermining the enforcement of animal protection statutes on farms. Since 1990, ten states in the USA and three in Australia have passed so called 'Ag-gag' laws. These laws criminalise unauthorized access to farms and the recording and publication of animal conditions found there. Offenders are liable to up to three years' imprisonment.

Svenja Springer and Herwig Grimm (eds.) ***Professionals in food chains***
EurSafe 2018 – DOI 10.3920/978-90-8686-869-8_52, © Wageningen Academic Publishers 2018

'Ag-gag' laws are the result of an increasing concern, within the agricultural industry, that undercover investigations of farm animal conditions are undermining public confidence in animal welfare and food safety. Idaho's 'Ag-gag law' was drafted and sponsored by the Idaho Dairymen's Association. It was enacted in 2014, after a Los Angeles based animal rights group, Mercy for Animals, posted a secretly-filmed expose of operations at an Idaho dairy farm on the internet. The video showed, amongst other abuses, workers using a moving tractor to drag a downed cow, chained by the neck, across the floor of the farm. In Iowa and Utah, similar laws criminalising the publication of images obtained by undercover investigators on private farms had already been enacted in 2012. The Idaho law followed their model.

The Iowa, Utah and Idaho laws are not the earliest examples of 'Ag-gag' laws in the USA. Such laws have been enacted since the 1990's, however their passage has become more prevalent in recent years. As noted in a report prepared by the John Hopkins University; the increasing number of 'Ag-gag' laws in the USA represents a growing threat to transparency within the American agricultural industry (John Hopkins Centre for a Livable Future, 2013).

The scope of the threat in the USA

In the USA, 'Ag-gag' laws were passed in Arkansas in 2017, Wyoming and North Carolina in 2015, Idaho in 2014, Utah, Iowa and Missouri in 2012, Montana and North Dakota in 1991 and Kansas in 1990. Most State's laws criminalise the making of video or audio recordings depicting agricultural facility operations, without the express consent of the owner, but the laws of Idaho, Iowa and Utah also permit undercover investigators who take jobs on private farms to be prosecuted for misrepresenting their intention to cause their employers economic loss or other kinds of injury.

The 'Ag-gag' laws of Idaho and Utah have recently faced constitutional challenge. In early 2015, the District Court for Idaho heard an action against the State's 'Ag-gag' law, Idaho Code § 18-7042. A group of animal advocates, headed by the Animal Legal Defence Fund, sought an injunction against the use of the law claiming it was inconsistent with the First and Fourteenth Amendments to the US Constitution. The court agreed and found that in seeking to suppress criticism of agricultural operations, the law was in breach of the Constitution's First Amendment right to free speech. Giving judgment in *Animal Legal Defense Fund and Ors v Otter*, the Chief Judge of the District Court for Idaho, B Lynn Winmill, found that the law was not only unconstitutional but also unnecessary, stating: 'Food and worker safety are matters of public concern. Moreover, laws against trespass, fraud, theft, and defamation already exist. These types of laws serve the property and privacy interests the State professes to protect through the passage of § 18-7042, but without infringing on free speech rights'.

The judge also ruled that the law was in violation of the Constitution's Fourteenth Amendment which provides for equal treatment for all parties under the law. He found that the State's 'Ag-gag' law was motivated by animus towards animal welfare groups and undercover investigators, targeting them unfairly.

Emboldened by the decision of the District Court for Idaho, animal advocates challenged the constitutional validity of the State of Utah's 'Ag-gag' law (Utah Code § 76-6-112). In 2017, after hearing legal arguments in *Animal Legal Defense Fund and Ors v Herbert,* the District Court for Utah declared the State's law to be in breach of the First Amendment right to free speech. While the State of Utah did not appeal the striking down of their 'Ag-gag' legislation, in Idaho, an appeal against the Idaho District Court's decision was filed with the US Court of Appeals for the Ninth Circuit, in late 2015 (*Animal Legal Defense Fund and Ors v Wasden*). The Court of Appeals handed down its judgment on that appeal in early January 2018, affirming the lower court's ruling in part and reversing it in part. The Court of Appeals judgement declared that while the filming of animal use on farms is form of protected free speech, and cannot be legislated against, the State of Idaho has a legitimate right to prosecute those who lie to gain employment in agricultural facilities, with the intent to cause economic harm.

While lawyers for the Animal Legal Defence Fund declared the Court of Appeals judgment a landmark ruling protecting the right to record animal abuses on farms and other private property (Stempel, 2018), the judgment reveals very differing views within the judiciary of the validity of 'Ag-gag' laws. Even within the Court of Appeals, the judges were not unanimous in their opinions. Dissenting in part with the majority, Circuit Judge Bea, took the view that clause (1)(a) was constitutionally valid. Citing the common law right to seek damages for trespass to land, regardless of whether any actual harm had been done on the land, the judge rejected his fellow judges' reasoning and held that a misrepresentation used to gain entry to an agricultural facility, without the owner's true consent, permitted a landowner to claim legally cognizable harm.

The same question of whether a misrepresentation to gain access to an agricultural facility amounted to a legally cognizable harm was also posed in the challenge to Utah's 'Ag-gag' law: *ALDF and Ors v Herbert*. In that case the District Court for Utah relied on two earlier Court of Appeals' decisions (*Desnick v American Broadcasting Companies Inc.* (Seventh Circuit, 1995) and *Food Lion Inc. v Capital Cities/ABC Inc.* (Fourth Circuit, 1999) to support its finding that a person who tells lies in order to gain access to private premises is not a trespasser unless or until he causes interference with ownership or possession of the land. Citing the same two cases in *Animal Legal Defense Fund and Ors v Wasden*, Judge Bea found nothing in the cases prohibited a state court or legislature from establishing a law which would vitiate consent to enter land, where that consent had been procured by fraud and ruled that the State of Idaho could legitimately enact § 18-7042 (1)(a) of the Idaho Code.

The scope of the threat in Australia

Three states in Australia have passed laws which may be used to stifle undercover investigation and publication of animal misuse on farms. In 2015, the state of New South Wales passed the Biosecurity Act 2015 (New South Wales). The purpose of the Act is to minimize biosecurity risks and the law criminalises, with up to three year's imprisonment, any action which might create a biosecurity risk (such as an unauthorized access to a farm) or a failure to report a risk having occurred. During the debate on its passage, legislators opposing the new law expressed concern that it might be used to suppress undercover investigations of animal abuses on farms. While the law purports to target biosecurity risks, rather than animal activists, it has only been in force since July 2017, making its effect not yet clear. A similar law imposing duties on the public to prevent and report biosecurity risks was passed in the State of Queensland in 2016. The Biosecurity Act 2014 (Queensland) imposes liability for biosecurity breaches with up to 3 years' imprisonment.

While it is common for states in Australia to prohibit the unauthorized recording of private conversations and activities, the Surveillance Devices Act 2007 of New South Wales was recently used in a novel way to prosecute an animal activist for recording and publicizing animal welfare breaches on pig farms. While the case was dismissed on technical grounds, it is feared that a precedent has been set for future prosecutions (Bettles, 2017).

Most concerning, from the Australian perspective, is the federal Criminal Code Amendment (Animal Protection) Bill 2015 which is currently under discussion in the Australian Senate. The draft law would directly criminalise the unauthorised recording of animal abuses at federally regulated farms, laboratories or other animal enterprises in Australia. The law purports to have been drafted to protect animals, by requiring those who record abuses to pass their information to authorities within 24 hours of recording. However, political opponents to the law and Australian animal welfare organisations have observed that the short time allowed for reporting would undermine the opportunity for any long term investigation of systematic abuse.

Conclusions

In recent years, governments have become increasingly concerned with the threats to public safety posed by the criminal activities of some members of the animal protection movement (Garner, 2004). Some activists have resorted to violence against persons and property to demonstrate their condemnation of animal use. Unfortunately, the use of criminal means to effect change, has allowed those condemning animal activists to blur the line between legitimate and non-legitimate forms of protest. It has been argued that the 9-11 attacks have given the American government an excuse to crack down on dissent, and demonise animal activists (Sorenson, 2009). Certainly, the moves towards the criminalisation of those who seek merely to publish records of animal mistreatment is alarming. It is important to recognize that direct action involving criminal activity was illegal in the USA and Australia long before the introduction of 'Ag-gag' laws and the law already provides effective sanctions for those convicted of such acts. Further laws are not necessary to address what is already illegal. Conversely, undercover investigations, which cause no damage to people or property, are not generally regarded as criminal activities under the law. Even where the persons involved in such activities trespass on private property, it is remarkable for legislation to permit what has long been regarded as a civil transgression to be prosecuted, by the state, as a criminal act. Introducing a law which converts a civil wrong into an act of terrorism is an extraordinary step for any legislature to take, and the policies behind it should be carefully examined.

The state of the law in regard to the legitimacy of 'Ag-gag' is still in flux. Constitutional challenges to the US laws are at a very early stage and while the legitimacy of 'Ag-gag' legislation remains uncertain, its future impact on undercover investigations is difficult to determine. What is clear, however, is that in the USA and Australia the introduction of 'Ag-gag' legislation has served to threaten public trust in the food industry. The passing of each new law undermines the accountability of those responsible for animal management and erodes transparency of animal use on farms. For those seeking to improve the responsibility and accountability of professionals working within the food chain, 'Ag-gag' laws present an ongoing and serious threat. Urgent action, in the form of constitutional and political challenges, is necessary to halt their continuing passage.

References

Animal Legal Defence Fund and Others v Herbert No 2.13-cv-00679-RJS, United States District Court for Utah, 7 July 2017.

Animal Legal Defence Fund and Others v Otter 45 ELR 20146, No. 1:14-cv-00104, United States District Court for Idaho, 3 August 2015.

Animal Legal Defence Fund v Reynolds Case No 4:1-cv-362, District Court for the Southern District of Iowa, filed 10 October 2017.

Animal Legal Defence Fund and Others v Wasden No 15-35960, Unites States Court of Appeals for Ninth Circuit, 4 January 2018.

Arkansas Code § 16-118-113. Available at: http://www.arkleg.state.ar.us/assembly/2017/2017R/Acts/Act606.pdf. Accessed 23 January 2018.

Biosecurity Act (Queensland) (2014). Available at: https://www.legislation.qld.gov.au/view/html/inforce/current/act-2014-007. Accessed 23 January 2018.

Biosecurity Act (New South Wales) (2015). Available at: http://www.austlii.edu.au/au/legis/nsw/consol_act/ba2015156. Accessed 23 January 2018.

Bettles C 'Animal Activists 'Let Off' Charges Under NSW Surveillance Devices Act Due to Technicality' The Daily Advertiser August 9 (2017). Available at: http://www.dailyadvertiser.com.au/story/4842942/animal-activists-let-off-charges-due-to-technicality. Accessed 15 March 2018.

Brambell F W R (1965). Report of the Technical Committee to Enquire into the Welfare of Animals kept under Intensive Livestock Husbandry Systems. Command Paper 2836. London: Her Majesty's Stationary Office.

Criminal Code Amendment (Animal Protection) Bill (Commonwealth of Australia) (2015). Available at: https://www.legislation.gov.au/Details/C2015B00002. Accessed 23 January 2018.

Desnick v American Broadcasting Companies Inc. No 99-3715, Unites States Court of Appeals for the Seventh Circuit, January 10 1995.

Flaccus G 'Suit: Meat packer Used 'Downer' Cows for Four Years Associated Press September 24 2009. Available at: http://www.sandiegouniontribune.com/sdut-us-slaughterhouse-abuse-092409-2009sep24-story.html. Accessed 15 March 2018.

Food Lion Inc. v Capital Cities/ABC Inc. No 97-2492 and 97-2564, United States Court of Appeals for the Fourth Circuit, October 20 1999.

Garner, R (2004). Animals, Politics and Morality. 2nd Ed. UK: Manchester University Press, pp. 236-242.

Harrison, R (1964). Animal Machines London, United Kingdom: Vincent Stuart Publishers Ltd.

Idaho Code Title 18, § 7042. Available at: https://legislature.idaho.gov/idstat/Title18/T18CH70SECT18-7042.htm. Accessed 23 January 2018.

Iowa Code Title XVI, § 717A.3A. Available at: https://www.legis.iowa.gov/law/iowaCode/sections?codeChapter=717A&year=2016. Accessed 23 January 2018.

John Hopkins Centre for a Livable Future (2013), Industrial Food Animal Production in America. Available at: http://scalar.usc.edu/works/field-guides-to-food/industrial-food-animal-production-in-america-examining-the-impact-of-the-pew-commissions-priority-recommendations-pdf. Accessed 23 January 2018.

Kansas Statute § 47-1827. Available at: http://www.ksrevisor.org/statutes/chapters/ch47/047_018_0027.html. Accessed 23 January 2018.

Missouri Revised Statute Title XXXVIII, § 578. 013. Available at: http://www.moga.mo.gov/mostatutes/stathtml/57800000131.html. Accessed 23 January 2018.

Montana Code Title 81, § 30-103. Available at: http://leg.mt.gov/bills/mca/81/30/81-30-103.htm. Accessed 23 January 2018.

North Carolina Civil Remedies for Criminal Actions § 99A-2. Available at: http://www.ncga.state.nc.us/EnactedLegislation/Statutes/pdf/ByChapter/Chapter_99A.pdf. Accessed 23 January 2018.

North Dakota Century Code Title 12.1, § 21.1. Available at: http://www.legis.nd.gov/cencode/t12-1c21-1.pdf#nameddest=12p1-21p1-01. Accessed 23 January 2018.

Stempel J 'Court voids Idaho ban on secret farm videos, revives some curbs on probes' Reuters January 5 2018. Available at: https://www.reuters.com/article/us-idaho-animalabuse/court-voids-idaho-ban-on-secret-farm-videos-revives-some-curbs-on-probes-idUSKBN1ET2C5. Accessed 23 January 2018.

Sorenson J (2009) Constructing terrorists: Propaganda about animal rights Critical Studies on Terrorism 2(2): 237-256.

Surveillance Devices Act (New South Wales) 2007 Available at: http://www8.austlii.edu.au/cgi-bin/viewdb/au/legis/nsw/consol_act/sda2007210. Accessed 23 January 2018.

United States v Alvarez No 11-210, United States Supreme Court, June 28 2012.

Utah Code Title 76, § 6-112 Available at: http://codes.findlaw.com/ut/title-76-utah-criminal-code/ut-code-sect-76-6-112.html. Accessed 23 January 2018.

Webster J (2013). Ruth Harrison-Tribute to an Inspirational Friend. In R Harrison (1964) Animal Machines. 2nd Ed. United Kingdom: CAB International p. 8.

Wyoming Statute §6-3-414. Available at: http://legisweb.state.wy.us/NXT/gateway.dll?f=templates&fn=default.htm. Accessed 23 January 2018.

53. The GMO debate reloaded – a survey on genome editing in agriculture

S.N. Bechtold[1], S. Schleissing[1] and C. Dürnberger[2]*
[1]Institute Technology-Theology-Natural sciences (TTN) at the Ludwig-Maximilians-University Munich, Geschwister-Scholl-Platz 1, 80539 Munich, Germany; [2]Unit of Ethics and Human-Animal-Studies, Messerli Research Institute, Veterinaerplatz 1, 1210 Vienna, Austria; sarah.n.bechtold@gmail.com

Abstract

Applications of genome editing (GE) technologies in plant breeding for agricultural use recently became a remarkable field of innovation, fascinating basic research scientists as well as commercial researchers. It is just now, that the debate about GE and its possible use in food and feed production transcends the scientific circle towards a political discussion. Although a general public debate in Germany is yet to come, it very likely will be influenced by the existing controversy about food products modified by classical genetic engineering, i.e. genetically modified organisms (GMOs). The Institute Technology-Theology-Natural Science (TTN) investigates potential ethical aspects of the public debate with a focus on the question of freedom of choice and labelling of GE products. An expert online survey we conducted revealed that GE technologies have the potential to change the stand-off situation in the debate about genetical engineering in agriculture as they were positively assessed by the majority of the experts. But therefore, questions of risk and questions of values need to be discussed in a comprehensive debate. We find that a voluntary label for GE-free as well as approved GE products will be highly effective to ascertain freedom of choice and to improve the knowledge basis for a value-based deliberation of GE technologies.

Keywords: freedom of choice, value conflicts, product labelling

Procedure and evaluation of the online survey

In spring 2017 we conducted an online survey on the ethical purport of freedom of choice and labelling of genome edited food products. We primarily addressed scientists from the natural sciences, arts and social sciences, but also selected stakeholders. All participants were known to be familiar either with the technique in particular or with assessment of innovative technologies in general. Additionally, they were provided with a summary about GE and its opportunities and problems. We received 27 fully completed questionnaires, composed of multiple choice and free text questions. The analysis presented here additionally relies on a filter profiling, which allowed us to correlate between answers to detailed questions and more general ones.

Ethical assessment of GE technologies in agriculture

One of the general questions of our survey asked for an overall ethical evaluation of GE applications in the agricultural sector. As shown in Figure 1, experts throughout all disciplines had a predominantly positive attitude towards GE.

This result is in strong contrast to the disapproval of genetic engineered food dominating the public view in Germany, lately confirmed by a focus group study of the German Federal Institute for Risk Assessment (The German Federal Institute for Risk Assessment, 2017). The negative public attitude towards GE was also anticipated by the experts we consulted. Where does that remarkable difference originate from? Can it be compensated by communicating scientific knowledge about GE? The free text contributions

EurSafe 2018 – DOI 10.3920/978-90-8686-869-8_53, © Wageningen Academic Publishers 2018

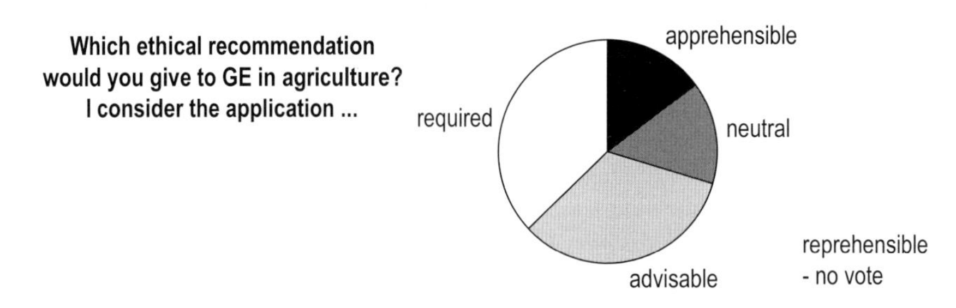

Figure 1. Asked for their general ethical appraisal of GE technologies in agriculture, most experts showed a positive attitude.

of the experts could indicate that divergent attitudes rather ground on non-scientific deliberations and aspects that cannot be addressed in a purely scientific debate: 'When it comes to fundamental decisions, it is mostly about values and practical knowledge.' Moreover, one participant of our survey argued, that 'knowledge should not be conceived as ticket to the debate about genome editing in agriculture.' She is certainly right that specialist knowledge cannot be a necessary precondition for an opinion on food, as food is relevant to everyone. Nevertheless, one possible explanation for the difference between the expert opinion and the public view could be, that a fact-based and rational deliberation – in other words a scientific approach to the new breeding technologies (NBTs) – indeed fosters an open and affirmative stance regarding the use of GE in agriculture. We wanted to find out whether scientific insight and value-based attitudes depend on each other – and if so, how they relate. Therefore, we focused on two different lines of arguments distinguishable in the responses to our survey.

Elements of the debate about genome edited food

Intuitive arguments based on emotional attitudes

As documented in Figure 2, the majority of experts we consulted agreed on the assumption that consumers are guided predominantly by non-rational criteria when buying food.

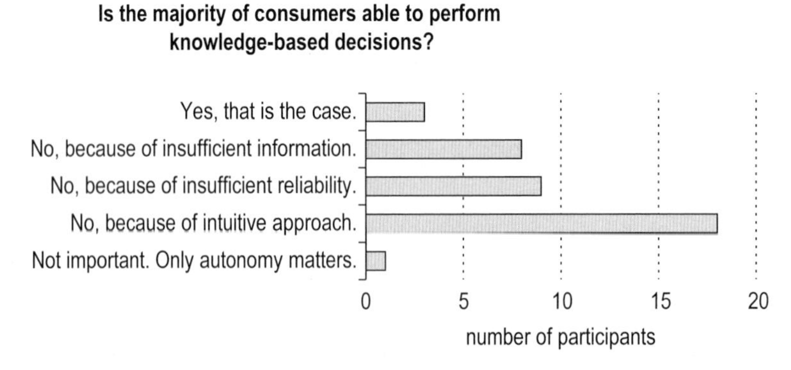

Figure 2. Intuitive emotional consumer choices were expected to be most prominent. Only few experts consider the current consumer knowledge sufficient (multiple selection was possible).

The virtue of naturalness is a good example of such a criterium (for an overview of nature conceptions in the bio-technological context: Siipi, 2008). Although it is far from clear what a natural food product is and whether it is generally better than an unnatural one, our experts identified unnaturalness as the major concern that will be brought up against GE products. But what is the purview of such intuitive choices based on positive and negative feelings towards a product? Some experts argued that the right of free choice, fundamental to liberal societies, basically implies that no reason has to be instanced for a particular decision. Those participants conceive every consumer decision as a basic act of self-determination, which has to be regarded and acknowledged as an ethically relevant action: 'If someone has a bad feeling with GE products, she should be able to choose alternatives.' However, those choices have limited validity as they do not allow for the deduction of general rules for dealing with GE products. While they have to be respected as significant expressions of personal sovereignty, we doubt that they are sufficient as a basis for general regulation of GE products.

Rational arguments based on scientific knowledge

Very few participants of the survey refused to respect individual decisions as an act of personal self-determination. But many insisted on the notion that choices need to ground on scientific knowledge in order to be compulsory. They argued that 'scientific knowledge is the only possibility to reach an objective statement.' Actually, however, science does not produce facts but hypotheses and opportunities. In the case of GE plants and animals for agricultural purposes the concrete impact of the technique is finally determined rather by its commercial application than by scientific research. Hence, in our view the scientific perspective constitutes an important but non-exhaustive approach to GE technologies. Furthermore, participants who take up the scientific stance also referred to overarching values, especially to the value of food security, when justifying the need for new breeding procedures: 'Resigning GE technologies is irresponsible regarding the global demand for food.' And indeed, GE technologies hold the promise to improve food supply especially under difficult growth conditions – but only under the assumption of the appropriate societal and economic conditions. Hence, although scientific risk assessment of the last decade was able to show that genetically engineered food products as such do not constitute a risk to human health or the environment (The German Reference Centre for Ethics in the Life Sciences, 2018), we find it falls short in determining the consumers right to choose or reject GE foods. Choices according to factual circumstances, values and personal preferences have to be possible and respected.

Value conflicts at the basis of the controversy?

Analysing our survey, we identified two groups of experts guided by two different core values. One group focused on the value of consumer sovereignty, while the other group considered food security the major goal. Both values are connected to freedom of choice, but in different ways: A free choice between two or more objects can only be made if a person is able to discriminate those objects and is allowed to reject one or the other (negative freedom). But free choice also requires that there is a variety to choose from at the first place (positive freedom) (on the relation between negative and positive freedom e.g.: MacCallum, 1967). However, we found a strong tendency towards the notion that, concerning GE products, the values of consumer sovereignty and food security are mutually exclusive: 'My freedom of choice just means that other people do not have a choice at all.' Although we would argue that this interpretation is misled, the ostensible value conflict appears to be fundamental for the difficulties arising in the public debate about genetically engineered foods. We find that an ethically acceptable regulation of GE in the agricultural sector should take into account positive and negative freedom – not only of consumers, but also of producers. At this point the discussion often turns towards the question of product labelling.

Communicating via product labels

Risk communication

Product labelling is supposed to facilitate autonomous consumers decisions based on relevant product features (The German Ministry of Food, Agriculture and Consumer Protection, 2011). Figure 3 shows that 40% of the experts regarded the use of GE techniques in food production *per se* as a relevant information for the consumer.

About the same percentage held GE products as equivalent to products of conventional breeding techniques, such as mutagenesis, as long as the modification achieved by GE technologies does not exceed those that can be obtained by conventional breeding, i.e. no transgene product. They argue that '...concerning possible risks for human health and the environment, GE products are comparable to conventionally bred plants or animals.' If, however, GE products are to be labelled, proponents and criticists of GE labelling agreed that the aim of labelling is achieved best by a state-run and -controlled label. This is interesting as the German GMO label, a state-controlled, mandatory label, was in general judged rather critically, especially with respect to its reliability. Another frequently mentioned objection is that state-run, mandatory labels are associated with possible safety and health concerns as those concerns constitute the main arguments for governmental intervention into a free market economy (for an elaborate deliberation of reasonable governmental food policy: Rippe, 2000). It is highly probable that this interpretation of the GMO label contributed to the situation that no production and distribution of GMO products takes place in Germany. A GE label modelled on the GMO prototype, may support negative freedom of consumers, but it might severely impair positive freedom. Opposing an obscure public good to potential individual risks, a mandatory GE label could aggravate value-based consumer decisions, impede the farmers freedom of choice and reduce the supply consumers can choose from.

Communicating virtues

Labels can, however, also communicate virtues of food products. In case of GE this may be by informing the consumer about the benefits that were achieved by use of the new technique, i.e. absence of specific substances such as allergens or ecological benefits (positive labelling). Alternatively, a product may be labelled as 'GE-free' in order to satisfy consumers rejecting GE foods (negative labelling) (for a comprehensive comparison of labelling strategies: Patterson and Josling, 2002). As illustrated in Figure 4, the possibility to identify non-GE products was clearly the most important requirement for a free consumer decision, according to the experts.

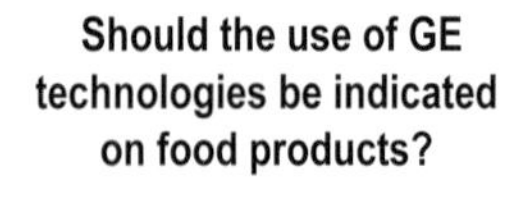

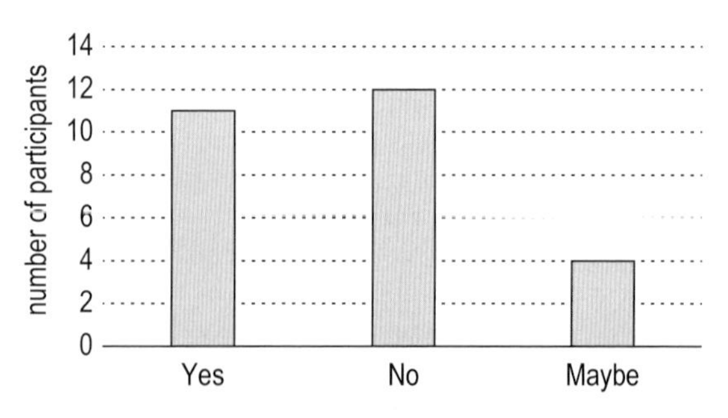

Figure 3. The question of whether GE products are to be labelled is highly controversial throughout the experts.

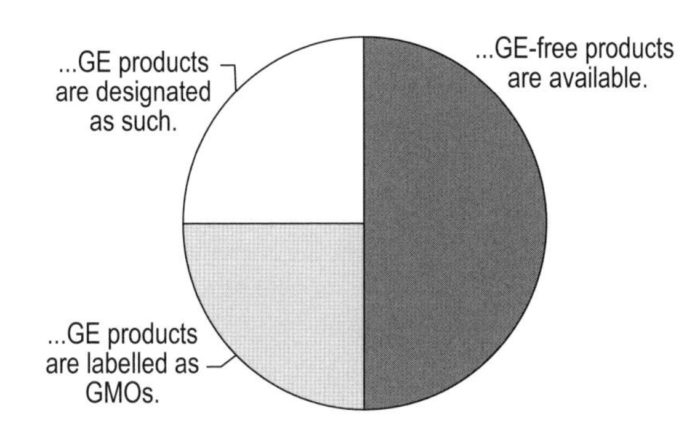

Figure 4. Asked for the appropriate type of label for GE products, 50% of the experts voted for an implicit non-GE label.

Again, in a liberal society communication about product amenities is not a governmental measure, but is supposed to happen between producers, vendors and consumers. Hence, in order to communicate virtues a voluntary labelling is indicated. Positive voluntary food labelling has the advantage that it can address consumers according to their individual interests, preferences and values, thereby possibly improving the efficiency of information transfer. Moreover, superordinate values like sustainability can be reified by pointing out particular advantages of specific GE applications. Thereby a label communicating contextualized and comprehensive knowledge can help to overcome the obstacles currently impeding a socio-ethical deliberation of GE. However, it will only reach people that are interested in knowledge about food and its production. Hence, it is only appropriate, if consumer knowledge about the use of GE is defined as being desired, but not essential.

Communication partners

The strong preference of state controlled labelling prompts the conclusion that the communication between food producers and consumers is seriously troubled by distrust and mutual scepticism (Meijboom, 2007). We think that transparency in terms of the scientific foundation of GE technologies and the fact of its application alone is insufficient to build up trust, as trust is a necessary precondition to rely on the provided information. Given the experiences with the German GMO label, we doubt that a state-controlled GE label is able to improve that problem. Moreover, it might further prevent the establishment of direct communication channels between food producers and consumers and the joint elaboration of our future agricultural system. Comprehensive transparency also takes into account the reasons and goals of the use of GE in agriculture, thereby transcending the risk-dominated scientific debate towards a debate about values. Those values provide a field of discussion relevant and open to food producers and consumers, while sustaining reasonable freedom of choice for each civic group as to how these goals are achieved in particular. Moreover, in the course of finding an accepted way to deal with GE technologies in agriculture it may be possible to address the general ethical question of how to deal with freedom of choice in situations where risk assessment is uncontentious, but benefits are assessed controversially.

Conclusion

We find that the potential benefits of using GE technologies in agriculture, such as reduced application of pesticides or drought resistance, are ethically relevant. Therefore, the regulation of GE products should encourage further research and should not rule out the practical application of the technique.

However, we also find that the argument of global food demand, frequently brought up by scientists, is not sufficient to deny consumer choices against GE foods. In order to guarantee freedom of choice, the consumer must at least be able to identify foods abstaining from GE technologies. As a third aspect the future dealing with GE products should foster social trust between food producers and consumers and enable a discussion about overall objectives and responsibilities in the agricultural sector. As knowledge was shown to support an unbiased discussion, detailed information, including the purpose of GE in the production process, should be accessible to the interested consumer in order to support knowledgeable value decisions. However, specific biotechnological knowledge is unlikely to facilitate competent consumer decisions unless it can be linked to societal values and individual preferences. Therefore, we find that a future labelling strategy for genome edited food products should also be benchmarked against its ability to promote meaningful connections between knowledge and values in the upcoming debate. Our preliminary work indicates that voluntary labelling of GE-free products and products specifically improved by the use of GE technologies could be able to meet the demands of proper consumer information about GE.

References

MacCallum, G.C.Jr. (1967). Negative and Positive Freedom. The Philosophical Review, 76(3), 312-334.

Meijboom, F.L.B. (2007). Trust, Food, and Health. Questions of Trust at the Interface between Food and Health. Journal of Agricultural and Environmental Ethics, 20, 231-245.

Patterson, L.A. and Josling, T.E. (2002). Regulating Biotechnology: Comparing EU and US Approaches. Paper presented at the Western Economic Association International 76th annual conference, San Francisco, July 8, 2001.

Rippe, K.P. (2000). Novel Foods and Consumer Rights: Concerning Food Policy in a Liberal State. Journal of Agricultural and Environmental Ethics, 12, 71-80.

Siipi, H. (2008). Dimensions of Naturalness. Ethics and the Environment, 13(1), 71-103.

The German Federal Institute for Risk Assessment (BfR) (2017). Conduct of Focus Groups Concerning the Perception of Genome Editing (CRISPR/Cas9) [German language]. Available at: http://www.bfr.bund.de/cm/343/fokusgruppen-zur-erhebung-der-oeffentlichen-meinung-zum-genome-editing.pdf. Accessed 26 December 2017.

The German Federal Ministry of Food, Agriculture and Consumer Protection (BMEL) (2011). Joint Statement of the Scientific Advisory Boards on Consumer and Food Policy and on Agricultural Policy on the Political Strategy for Food Labelling. Available at: https://www.bmel.de/SharedDocs/Downloads/EN/Ministry/Scientific_Advisory_Board-Food_Labelling.pdf?__blob=publicationFile. Accessed 18 March 2018.

The German Reference Centre for Ethics in the Life Sciences (DRZE), Summary on studies investigating health risks of genetically modified foods. Available at: http://www.drze.de/in-focus/genetically-modified-foods/modules/health-risks-posed-by-genetically-modified-foods. Accessed 18 March 2018.

54. Public opinion on dogs as a first step for solving dog welfare problems

C.S. Ophorst[1,3], M.N.C. Aarts[1], B. Bovenkerk[2] and H. Hopster[2,3]*
[1]Radboud University, Comeniuslaan 4, 6525 HP Nijmegen, the Netherlands; [2]Wageningen University & Research, Droevendaalsesteeg 4, 6708 PB Wageningen, the Netherlands; [3]Van Hall Larenstein University of Applied Science, Agora 1, 8934 CJ Leeuwarden; susan.ophorst@hvhl.nl

Abstract

The Dutch are dog lovers. Nevertheless numerous problems regarding dogs occur. There are concerns about risk of zoonoses due to (illegal) imports of puppies and rescue dogs; hereditary health problems in (pedigree) dogs; dog bite incidents and welfare issues related to dogs in animal shelters. Although problems vary largely, they all may originate from the moment at which a specific owner decides to purchase a particular dog. While government and animal protection organizations already target (potential) dog owners by providing lots of information, preventing said problems by changing purchase behaviour remains challenging. According to Kahneman (2012) the quintessence is that people take emotional, fast decisions (system 1), while information campaigns target the rational and slow system 2. As behaviour is largely influenced by the behaviour and opinion of others (Smith and Christakis, 2008; Pompe *et al.*, 2012) this is a starting point for determining factors in the decisions taken with system 1. We therefore study the interaction between the opinion of others and the person's mindset when purchasing a dog as well as how people deal with possible inconsistencies. As a first step we will map the public's opinion on dogs, qualitatively as well as quantitatively. The complexity of this public opinion is shown through frame analysis (Gray, 2003; Aarts and Van Woerkum, 2006) typifying the various aspects of the public opinion. Dutch media will be analysed to retrieve these frames. Ultimately these results will provide the basis for more effective methods to influence the purchase of dogs, having a greater chance of preventing risks for dogs, owners and society.

Keywords: purchase behaviour, behaviour modification, social influence

Introduction

Dog welfare issues, public health risks, severe biting incidents, illegal dog trafficking, for an important part boil down to that one moment where a specific person decides to purchase a particular dog. A decision that can cause tremendous misery to the dog, its owner and society.

Well-considered purchasing decisions are at the start of risk prevention. Several measures already aim to accomplish this. Since 2013 all 'new' dogs in the Netherlands must be RFID-chipped and registered to prevent illegal and malicious trafficking in dogs. In addition, effective rabies vaccination for dogs entering the Netherlands is obligatory. Unfortunately, morbidity due to other illnesses and illegal trade still exists (Van Rijt *et al.*, 2015). For a long time, information campaigns have aimed at increasing awareness in (potential) dog owners concerning the risks associated with purchasing and owning a dog. Parties regarded providing knowledge as the crucial step in empowering people making better decisions. However, as the problems continue (NVWA, 2016), the effectiveness of the current approach needs to be improved. The extent to which the purchase of a dog is guided by conscious processes and the autonomy of the decision making is, however, largely unknown.

We therefore need insight in the way people decide to purchase a dog, especially in the unconscious and impulsive behaviour that is involved. Careless choices entail the risk that the knowledge, skills, needs,

EurSafe 2018 – DOI 10.3920/978-90-8686-869-8_54, © Wageningen Academic Publishers 2018

and home situations of dog owners do not match the needs and characteristics of the dog (Jagoe and Serpell, 1996). Careless choices can, for example, be triggered by the appearance of the dog, his tough image or his sad eyes. The social environment of the (potential) dog owner, including prevailing values and norms, is likely to play a significant role (Smith and Christakis, 2008; Pompe *et al.*, 2012). This can include the direct environment, but also virtual networks (social media) and society at large. The focus of this study is therefore to gain more knowledge of the public opinion on dogs in the Netherlands. Subsequent studies will focus respectively on the coping strategies of dog buyers with this public opinion and living lab settings to track possibilities for influencing the behaviour of people in buying dogs.

To define the public opinion regarding dogs in the Netherlands, the following research question has been formulated:

> Which values and norms exist in Dutch society regarding dogs and how have these changed over time?

This study aims at discovering coping strategies in dealing with the values and norms in society when purchasing a dog. Coping strategies reflect people's reactions in situations that cause cognitive dissonance (Festinger, 1957). According to Kahneman (2012), people tend to rationalize decisions after the purchase has been made instead of beforehand. The actual decision has by then already been made automatically. This phenomenon is illustrated by the study of Pompe *et al.* (2012) concerning the purchase of 'high-risk' pets. A stunning 95% of dog owners considered buying a high-risk dog again, knowing the health and welfare consequences. An increase in knowledge (like experience provided) has apparently not influenced the decision. The search for values and norms and the way people cope with these values and norms may open up possibilities to influence the purchasing behaviour more effectively.

Research design

In agreement with Kahneman (2012) that people rationalize their decisions afterwards, we will not focus on methods as surveys because these could mask system 1 influence on the actual purchasing decision. As ultimately we want to design methods to be able to influence exactly this moment, we need to use a more indirect and contextual approach. We hypothesize that public opinion influences purchasing decisions and media discourse is argued to be 'an essential context' for understanding formation of public opinion on topics (Gamson and Modigliani, 1989). It is either viewed a reflection of public opinion (McCombs, 2014; Gamson and Modigliani, 1989) or as of crucial influence on the formation of public opinion (Terkildsen and Schnell, 1997). Both options provide a valuable contribution to this study, without the necessity to distinguish between one or the other.

This study is interpretative; we assume we live in a world that is interpreted in different ways (Yanow and Schwarts-Shea, 2013). The keyword 'framing' is crucial in this process: when people frame situations and experiences they emphasize certain aspects, at the expense of others (Dewulf *et al.*, 2009). Frames illustrate how we give meaning to facts and figures and which goals we strive for in interaction. These goals can entail the justification of a purchase, but also the way we view ourselves and others, for example as a dog lover. We use frame analysis to discover frames in Dutch society regarding dogs.

Messages including the word 'hond' (Dutch for dog) in the following selection of media were analysed:

- the three largest national daily newspapers: Telegraaf, AD and Volkskrant;
- two regional daily newspapers: Dagblad van het Noorden and De Stentor/Gelders Dagblad;
- social media: Facebook and Twitter.

This selection provides enough data enabling both quantitative and qualitative analysis and covers differences in opinion throughout society. To discover the variety in public opinion, not only a variety of resources was used, but messages were also analysed over time.

Time frame media: Newspapers (2007, 2012, 2017). Within these years January, April, July and October were selected to correct for or discover seasonal effects; Facebook and Twitter (2012, 2017).

All hits containing 'hond' or 'hondje' (dog or little dog) were coded in a database, except hits where 'hond' had nothing to do with the animal or the way it is viewed (when a name contained 'hond' or the expression 'geen hond'). All hits were allotted to a frame. In the first two-hundred hits new frames where created when no suitable frame existed. After this, no hits were found that couldn't be allotted to an existing frame.

All frames where ranked according to frequencies. Inter- and intra-year frequencies of frames were compared as well as their distribution over different media.

This quantitative step will guide further methodological choices regarding qualitative analysis since outcomes can point to certain frames, specific media or particular time frames.

Appropriate qualitative methods will be applied and selected from the following methods: frame analysis, category analysis (Yanow, 1999) and narrative analysis (Macnaghten *et al.*, 2015). Two options will be elaborated on by the following examples of possible outcomes:

- *Frame on nuisance caused by dogs*. This frame illustrates an underlying conflict with regard to dogs and/or their owners. A further frame analysis, based on the division of frames by Gray (2009) in identity, characterization, power and problem can elaborate on the conflict and directions for solutions.
- *Frame in which proverbs or metaphors with dogs feature*. When this frame occurs consistently over time it illustrates that certain stories/phrases/sayings/expressions involving dogs are part of Dutch culture. By means of a narrative analysis (Macnaghten *et al.*, 2015) the origin of these narratives can be tracked and give direction to the way people can be addressed with regards to the topic of dogs.

Based on the frames and trends found, articles including again the term 'hond' will be analysed subsequently. These articles are selected by the news service function in LexisNexis. In this selection dogs now feature as a topic, where this was not necessarily the case in the selection used for quantitative analysis.

Together, the quantitative trend analysis and the qualitative frame analysis provide insight in the way dogs are portrayed and viewed in Dutch society. We will use these insights in in-depth interviews with dog-owners to reveal their coping strategies (Festinger, 1957). Based on the coping strategies found, influencing methods will be developed and tested in living lab circumstances.

Preliminary results

The first and preliminary results of our research endorse the approach that we aim for. The quick scan consisted of 606 hits in five different newspapers. The months used for this quick scan were: January 2007, January 2017 and April 2017.

Frames that were found are:

1. *Dog as a metaphor for a state of normalcy*. The word dog is used to emphasize how normal a situation or a person is or seems to be. Typical phrases found with this frame are 'just walking the dog' or

'taking the dog out for a stroll' or 'petting the dog'. Example: 'Since then I'm really feeling better. I'm walking the dog in the forest again. Lovely.' (AD, 15 January 2007)

2. *Specific dog with a specific relation with a person or a specific place in society*. The word dog is combined with the name of the dog and/or specific characteristics, such as breed or appearance, but also character traits of the dog. In these cases the relationship with the dog is mentioned, the meaning of the dog to the person in the article. Example: 'Her little dog Bus lies in a guitar case. She uses him conveniently to fill up the silences between the songs with some chitter chatter.' (Telegraaf, 16 January 2007)
3. *Dog in proverbs or used as metaphors, with a negative connotation*. Proverbs with dogs are part of the Dutch language. Found proverbs combine the use of dog with a negative connotation, such as cowardly ('laffe hond'). Example: 'I long for not having to get up in time and go home like a scared dog.'(AD, 25 April 2017)
4. *Dog in proverbs or metaphors, with a positive connotation*. Found proverbs combine the use of dog with a positive connotation, such as loyal as a dog ('trouwe hond'). Peculiar is the use of positive connotations as loyal for situations where this quality is seen as negative. In these instances loyal is regarded as willingly following a unsuitable leader. Example: 'Until a friend points out that Kea is for Hans what a guiding dog is for a blind man. De dog helps and protect, but the blind man stays in control.' (de Volkskrant, 22 April 2017)
5. *Dog related to special accomplishments*. Special accomplishments by dogs include unusual performances, finding missing persons, saving someone ('s life) by warning others, playing a part in solving crimes and assisting people in need. Mostly these referrals are made to specific dogs, sometimes in official capacities such as army dogs, police dogs, assistance dogs. Example: 'Agata isn't just a dog. She's one of the best drug sleuths in Colombia.'(De Stentor/Gelders Dagblad, 25 January 2007)
6. *Dog related to nuisance*. Dogs are seen as a source of nuisance, including straying dogs on railways, airports and highways, barking, biting people and animals. Example: ' The dog has attacked two women and an employee of the animal rescue service.' (AD, 5 January 2007)
7. *Dog abuse*. This frame tells the story of dogs that are victimized. Example: 'He stabbed the dog multiple times with a letter opener and hit him on the head with a hammer.' (Telegraaf, 14 January 2007)

Table 1 shows the frequencies of the frames in the quick scan. The distribution over time of the frame of the dog as a metaphor for a state of normalcy, seems quite consistent. In every year (and month, data not shown) it is the most frequent occurring frame (65-77). In different newspapers, however, some variation is found. In De Volkskrant it was the most dominant frame by far (65 times against 34 for the

Table 1. Frequencies of frames in quick scan.

Frame	Frequency				Percentage
	January 2007	January 2017	April 2017	total	
State of normalcy	77	77	65	219	36.1
Relationship specific dog	64	69	51	184	30.4
Proverb – negative connotation	20	28	16	64	10.6
Nuisance	21	15	10	46	7.6
Special accomplishments	14	8	13	35	5.8
Dog abuse	16	7	11	34	5.6
Proverb – positive connotation	7	12	5	24	4
Total				606	100

runner up, relationship with a specific dog), in the AD it was the other way around with 27 times the normalcy frame and 38 times for the relationship frame.

Conclusion and discussion

The quick scan already provides us interesting data on the public opinion of dogs in the Netherlands. The normalcy frame and the use of dogs in proverbs and metaphors indicate how ordinary dog are in the Netherlands. Remarkably frames related to many of the problems that are signalled in relation to dogs are virtually lacking. In the quick scan hereditary diseases related to breeding dogs only occur once, zoonotic diseases are absent and illegal trade comes up rarely.

As these are severe problems for all kinds of stakeholders, ranging from individual dog lovers to animal rights groups, political parties and the Ministry of Agriculture, Nature and Food Quality, one can question if Dutch society is aware of or interested in the problems. As the ongoing efforts to educate people on dogs doesn't lead to less problems, could this be a sign that 'ordinary people' view dogs so much as an ordinary part of life, that they don't pay specific attention to them? Does this then also apply to the way they purchase dogs, not giving it a second thought at all?

Further analysis of the public opinion, including the way people cope with public opinion when acquiring a dog, will provide insight in the significance of the frames found and their potential as an agent of influence.

References

Aarts, N. and Woerkum, van C. (2006). Frame construction in interaction. In Engagement. Proceedings of the 12th MOPA International Conference, ed. Nicholas gould. Pontypridd, IK: University of Glamorgan, 29-37.

Dewulf, A., Gray, B., Putnam, L., Lewicki, R., Aarts, N., Bouwen, R., and Van Woerkum, C. (2009). Disentangling approaches to framing in conflict and negotiation research: A meta-paradigmatic perspective. Human relations, 62(2), 155-193.

Festinger, L. (1957). A Theory of Cognitive Dissonance. Stanford, CA: Stanford University Press.

Gamson, W. A., and Modigliani, A. (1989). Media discourse and public opinion on nuclear power: A constructionist approach. American journal of sociology, 95(1), 1-37.

Jagoe, A. and Serpell, J. (1996). Owner characteristics and interactions and the prevalence of canine behaviour problems. Applied Animal Behaviour Science, 47, 31-42.

Kahneman, D. (2012). Thinking, fast and slow. Penguin Books, Mann.

McCombs, M. (2014). Setting the agenda: Mass media and public opinion. John Wiley & Sons.

Macnaghten, P., Davies, S. R., and Kearnes, M. (2015). Understanding public responses to emerging technologies: a narrative approach. Journal of Environmental Policy & Planning, 1-19.

NVWA (2016). Illegale hondenhandel. Available at: https://www.nvwa.nl/onderwerpen/dieren-dierlijke-producten/dossier/honden-en-katten1/illegale-hondenhandel. Accessed 16 December 2016.

Pompe, V., Hopster, H. and Dieren, M. van (2013). Liefde maakt blind? Onderzoek naar waardenoriëntaties en waardenafwegingen van kopers/houders van 'risicovolle'dieren.

Rijt, van W., Verhoeven, W. and Kok, R. (2016). Beleid hondenfokkerij en -handel in Nederland: beleidsdoorlichting en evaluatie I&R hond. Panteia.

Smith, K. and Christakis, N. (2008). Social networks and health. Annual Review of Sociology.

Schwartz-Shea, P., and Yanow, D. (2013). Interpretive research design: Concepts and processes. Routledge.

Terkildsen, N., and Schnell, F. (1997). How media frames move public opinion: An analysis of the women's movement. Political research quarterly, 50(4), 879-900.

Yanow, D. (1999). Conducting interpretive policy analysis (Vol. 47). Sage Publications.

55. Portraying animals to children: the potential, role, and responsibility of picture books

J. Benz-Schwarzburg
Messerli Research Institute Vienna, Department of Ethics and Human-Animal Studies, Veterinaerplatz 1, 1210 Vienna, Austria; judith.benz-schwarzburg@vetmeduni.ac.at

Abstract

Recently, Agrarmarkt Austria (the main body of Austrian agribusiness marketing), pulled back an almost launched project after massive critique from the public. They had cooperated with a children's book author to do a picture book on meat production. The book was accused of unashamedly whitewashing the killing of animals and using the genre to influence children's moral intuitions. Picture books indeed come with an ageless tradition of conveying ethical messages. But it has been scarcely addressed so far how children make sense of pictures and narratives or by what processes they learn to respond to them. This is especially true with regard to normative messages concerned with nature in general and animals in specific. By reference to the AMA book as a case example I will identify ethically relevant elements of pictures and narrative. Their discussion will be set against the background of current empirical studies on anthropomorphism and anthropocentrism in picture books. The analysis will reveal the animal ethics potential of picture books which are 'usually more complicated and occasionally more potent than they seem at first glance'.

Keywords: children's books, advertisement, slaughter, anthropomorphism, anthropocentrism

Introduction

Picture books come with an ageless tradition of conveying ethically relevant messages. Some of them, like Sarah Trimmer's 'Fabulous histories' published as early as 1796, directly convey messages about the human-animal relationship as a normatively shaped one. Many more picture books indirectly transport a specific idea of this relationship and its ethics. Marriott speaks of 'the moral imperative' of picture books. With this he means, that 'consciously or unconsciously, overtly or covertly, picture books provide through the combination of images and words, themes and ideas, texts and subtexts, a representation not only of how the world is but also of how it ought to be' (Mariott, 1998: 5-6). Given the weight of moral messages in picture books and the often didactical or pedagogical intent of the authors, it is surprising that it has scarcely been addressed so far how children make sense of pictures or by what processes they learn to respond to the narratives and messages of picture books (Mariott 1998: 2, emphasis by JBS). This gap in research, combined with the fact that anthropomorphised animals have always been a staple of children's literature (Geertz, 2014; Marriott, 2002), can easily lead to the hypothesis that the animal ethics content of picture books might have been largely overlooked. For the purpose of this paper the animal ethics content of picture books shall be defined as the messages about the human-animal relationship which are conveyed by text and pictures alike. Mariott argues rightfully, that picture books are 'inescapably plural', meaning that there is a 'peculiar relationship between words and pictures' (Mariott, 1998: 4). Often, it is the illustrations that provide 'a starting point from which the reader gets meaning and to which the reader gives meaning' (Evans, 1998: xv). Reading picture books more specifically requires to focus on the gap between what the texts says and what seems to be in the pictures. Reading, then, ideally becomes an engaging and 'highly creative process' (Mariott, 1998: 4), also, because the readers' relationship with the book is 'personally meaningful' to them (Evans, 1998: xvi). 'In days gone by', as Evans states, authors and illustrators used this to 'position' readers 'in certain roles' (Evans, 1998: xiv). Given that picture books still bear the potential of the moral imperatives and

Svenja Springer and Herwig Grimm (eds.) ***Professionals in food chains***
EurSafe 2018 – DOI 10.3920/978-90-8686-869-8_55, © Wageningen Academic Publishers 2018

that they can be seen as 'inherently ideological', they will always come with the danger of conveying 'pervasive' ideologies, like sexism, racism (Marriott, 1998: 5) or, one might argue, also speciesism and anthropocentrism. This means that any analysis of the animal ethics content of picture books should pay attention to possible (sublime or blunt) ideologies transported by text and pictures, notably speciesist or anthropocentric ideas. They could for example be prevalent in the character or dynamics as well as in the power relations or hierarchies of the portrayed human-animal-relationship. Indeed, uncovering these aspects and their implementation in the media isn't just an interest of ethicists but also a key objective of Critical Animal and Media Studies in general (Almirón *et al.*, 2016).

Anthropomorphism vs anthropocentrism

In the past few decades, psychologists have embarked on answering the question if anthropomorphism or anthropocentrism in children's books affects the young readers' learning and conception of animals. I can only highlight a few of the most relevant studies here.

Ganea *et al.* confronted 3-, 4-, and 5-year old children with books featuring realistic drawings of a novel animal. Half of the children saw these pictures accompanied by a factual, realistic language describing the animals, while the other half heard an anthropomorphized language. In a second study they replicated this paradigm but used anthropomorphic illustrations of real animals. Their combined results show that the different kind of languages had an effect 'on the children's tendency to attribute human-like traits to animals'. They impact how much (or how little) children learn about animals and influence their conceptual knowledge of animals.

The question is if this is a 'manipulation' of any kind. It wouldn't be, if children universally began with an anthropocentric perspective on the world, maturing to a somewhat biological perspective in a basic conceptual change (cf. Carey, 1985). But recent research reveals that an anthropocentric perspective, which we can find in studies of urban 5-year olds, is not evident in 3-year-olds already (Herrmann *et al.*, 2010). This indicates that the anthropocentric perspective is not an obligatory first step in children's reasoning about biological phenomena. They are rather influenced in adopting this stance. Waxman *et al.* (2014) could show that urban 5 year-olds can indeed be manipulated into either an anthropomorphic or an animal-centred perspective by books that we read to them. Priming children with the one or the other perspective 'had a dramatic effect. Children primed with Berenstain Bears [an anthropomorph portrayal of animals] revealed the standard anthropocentric pattern. In contrast, children primed with Animal Encyclopedia [an animal-centred, biological portrayal] adopted a biological reasoning pattern.' The authors conclude that 'children's books and other media are double-edged swords. Media may (inadvertently) support human-centred reasoning in young children, but may also be instrumental in redirecting children's attention to a biological model in which humans are one among the animal kinds' (Waxman *et al.*, 2014: 7, insertions JBS).

If these authors are right that 'anthropocentrism is not an initial step in conceptual development, but is instead an acquired perspective, one that emerges between 3 and 5 years of age in children raised in urban environments' (Hermann *et al.*, 2010) animal ethicists might indeed want to have a closer look at the literature targeting children of that age. Ethicists could support the debate by providing a discussion of the normative content and impact of such a manipulation in children. What normative messages do children learn from anthropocentric picture books? And how is anthropomorphism employed as a tool to support an anthropocentric moral imperative?

For the purpose of this paper I will define anthropomorphism as a portrayal of an animal character that restores to human characteristics (behaviour, appearance, mental processes/states, interests and alike) as we can find it in animal characters who can speak, wear cloths, play the piano, or engage in other interactions typically for humans and human culture. Anthropocentrism in contrast to this is a normative evaluation targeting a (solely) human-centred perspective of a representation of animals and/or the human-animal relationship, wherein the actually prevalent (but supposedly divergent) perspective of the animal(s) is being ignored or erased.

While anthropomorphism is first of all a descriptive category, in the sense that it is a description of how animal characters look like and behave, it can be normatively relevant if the characteristics or interest depicted substantially lead away, disguise, or contradict the actual interests of the animals present in the scenery or narrative. This concern is at the heart of the psychologists' criticism on anthropomorphism. Understood like this, anthropomorphism can merge into anthropocentrism. It can be argued that what happens here is then a normatively relevant step: the description of the animal becomes ethically relevant because something we should recognize and respect is left-out: the animal's own perspective, given by its own species-specific and individual behaviour, appearance, mental processes/states, interests and alike. However, anthropomorphism also has a positive side: its power, which is surely attractive to children's books authors, comes from the fact that it invites for identification: after all the animal character seems to be so much like us. But this identification again comes with an ethical danger: it makes the reader forget that who is represented here is actually someone else.

Based on the introduced conceptual clarifications we can say that anthropomorphism serves to connect the reader with the message. It also can be employed as a tool, or a pedagogical means to convey moral norms concerned with anthropocentric interests. In such constellations, the real animal behind the anthropomorphised animal character becomes an absent referent (Adams, ([1990]2010), in fact, it becomes absent to such a degree that we might even argue the story is not about an animal at all but solely about humans. The animal's role is then to serve an instrumental aim within and for the story told by the author. This is where the ethical relevance of such a portrayal is located. How problematic this is, however, also depends on the content of the messages which are brought forward by such books. This means we need to focus on anthropomorph characters but also on the anthropocentrism in their stories. The main question of the following analysis of a case example will be: What overall moral imperative does the combination of an anthropomorph character with an anthropocentric message generate? Does the author leave some gaps for the young reader to unfold her/his creativity with regard to recognizing a normative meaning in the book? Or does he rather deliver a moral ideology to 'position' the child with regard to specific power and hierarchy related roles in the human-animal relationship?

Tegetthoff (2017): 'Meat where are you coming from?!'

AMA Austria (responsible for the marketing of agricultural products from Austria) recently cooperated with a very renowned Austrian author and story teller, Folke Tegetthoff, and the illustrator Jera Kokovnik on a book called 'Meat where are you coming from?! How the meat comes to your plate' (German: 'Fleisch, woher kommst denn Du?! Wie das Fleisch auf Deinen Teller kommt'. All translations from Tegetthoff (2017) in the following by JBS).

The first four scenes of the book tell us that Lena, Max, and Lukas are on farm holidays. Every morning, they go to meet their favourite animals. At lunch they tell their family what the animals did and how they reacted to them (for example how happy the cow seemed when seeing Max, flipping her tail, winking at him, jumping out of joy). The children's' descriptions of these encounters include fairly biological descriptions about the animals, also references to their behaviour, (supposed) interests or desires, emotions, and intelligence. Scenes five to seven move the story to a different, more fictional level: the children go to bed and the animals talk to them in their dreams. The cow says to Max: 'Like you, I won't be here anymore tomorrow. You have to go back home to school to learn how to count and write. And I, too, have a duty to fulfil.' She explains: 'to become a good piece of meat we cows have to stand on the pasture, eat good food, drink clear water and breath fresh air. And now, me and my colleagues set off.' Calming the child's emotions she argues: 'you don't have to be sad, because we will see each other again'. When the story proceeds with the farewell of the other animal protagonists it becomes clear that the cow actually means they will see each other again on the child's dinner plate.

In scene six the chicken tells Lukas that tomorrow his 'big journey' begins: 'Team chicken is ready. Right at the beginning (of our trip) we meet humans who fly us from HERE to THERE. This happens so fast that we even don't realize it. Like on a rocket. And do you know where we are going to land? With YOU. Because the THERE, that's YOU'. With this he gives Lukas the reason, namely the child's appetite for chicken wings, for himself (the chicken) to be slaughtered.

In scene seven the pig lies down besides Lena and tells her: 'I am out now. The food was really cool. The farmers took great care of me. But I have to think about my future'. While her first utterance repeats the rather harmonious and romantic image of organic farming given already by the cow, the second utterance causes the reader to expect that the following rationale is in the considerate interest of the pig. She continues: 'For me, the best specialists are waiting. They work with me so that I become a Schnitzel Star. A Chop Wonder. A Mega-Super-Douper-Filet'.

The final condition of the animals as a piece of meat is described by the typical superlatives of meat advertisement. The chosen explanations put into their mouths indicate that they are not only cooperating consensually with their butchers but are also totally fine with, even excited about, their destiny (semantically disguised as a journey, an adventurous trip on a rocket, a flight from HERE to THERE). Thus, the step from being a living being to a piece of meat is covered up by their excitement, and immediately rendered into a language of consumption while their perspective (especially the one of the pig) solely and strongly emphasises consensus, even mutual cooperation, within a system that does not name the killing as killing but as a professional processing by qualified specialists. This kind of processing is stripped of any subjective emotions, notions of discomfort, fear, loss or harm. The perspective of the animals is significantly reduced to this view on their fate. The subjective side of the animal is virtually non-existent but collapses into the anthropocentric interest of the consumer which the animals happily consent to. The pig concludes by asking: 'And do you know why I am doing all of this?' She gives the answer herself: 'To make YOU happy. See you! So long!' Note, that in this wording the pig claims that she is doing something, instead of an outside force doing something to her. This erases the responsibility of the butcher. The phrase 'do you know why I am doing all of this' indicates that someone is wilfully choosing to do something as a sacrifice to someone else. The pig's cheerful exit thus again emphasizes that while the animal perceives her choice to be butchered as a sacrifice to make the consumers happy, she is very much positive about this decision.

The remaining scenes describe the next morning when the children (well prepared for what happens with the animals) serenely tell their parents that they do not need to say good-bye to them: 'they are already gone'. The farmer told them, they will 'see them again', easily recognizable on the supermarket shelves 'because they will all wear a medal [the AMA label (German: AMA Gütesiegel)]'. The children display full accordance with the story told by the animals in what they say and in how they behave. They have completely adopted the animal's relaxed attitude and the logic of the perspectives and reasons given by them. The medal, Lena explains, is for the 'great work they [the animals] have accomplished' – transforming themselves wilfully and happily into a piece of meat.

The book is oscillating between picture book and meat advertisement. In the first part (scenes one to four) it only employs little signs of anthropomorphism in text and pictures (exceptions are for example the Barbie-like eyelashes of the animals). This changes profoundly, when the animals tell the story of their 'journey'. The talking creatures restore in almost all utterances to human cultural knowledge, human beliefs and human interests. This is perfectly mirrored by anthropomorph illustrations in which they pull a trolley, carry a bag, wear a baseball cap, or skate on a longboard. This change towards obvious anthropomorphism nicely shows where the message of the story turns from a description of animals on a farm into an attempt to rationalize and moralize the purpose of their life and death. The text is clearly not normatively neutral but 'inherently ideological'. The relaxed happiness of the animals, their wilfulness to enter in a supposedly mutual understanding of sacrifice for the sake of consumer demands, and finally the awarded medal for their 'great work' are all value-assigning expressions and pictures that are in agreement with the (AMA) idea of romantic, welfare-friendly organic farming, but also with the hierarchies and power-relations inherent to it including the instrumental role of the animals.

The anthropomorphism thereby stretches to the entire rationale reproduced by the animals: like children who have to go to school it is the 'cultural' role of animals to go to the butcher and end on our plate. The fact that it is our culture, not theirs where human interests in consumption shape this role is hidden. Their own interest in a continued life is completely erased. Anthropomorphism invites the reader to identify with the figure. But the interests installed in the figures are, by way of their anthropocentric messages, human interests. Thus, ultimately, the identification taking place positions the reader along the lines of the message. This is the specific problematic we can find in the combination of an anthropomorph animal character conveying an anthropocentric message. The serenity of the children in the last scene, where they explain the situation as they have come to see it, is almost scary, given the actual knowledge they have gained. But it exercises and displays right there, in the book, that the envisioned positioning of children regarding the role of the animal and their own role as a consumer has been executed successfully. Their attitude comes as an inbuilt consequence of the anthropocentric story told by the anthropomorphized animals in their dreams. The same attitude is likely to arise in the young readers.

The book wasn't distributed to kindergarten children and didn't enter the book market because of a complaint by an animal welfare organisation to the Austrian Advertising Council. The Council argued that the book indeed conflicts with several points in their Advertising Industry Ethics Code (Austrian Advertising Council, 2012). This code, though self-regulatory, addresses ideas of honesty, truthfulness, social responsibility, and pedagogical responsibility (forbidding for example any abuse of the imaginative limitations of children for advertisement and causing them physical, psychological, and moral harm). The fact that 'moral harm' is named here is especially interesting as the message conveyed by the AMA book clearly targets the normative beliefs of children which potentially stand in the way of meat consumption. The Media Council recommended an immediate stop of the book or a complete change in subject (naming text and pictures both as sources of their objection). They stressed in their decision that the book is 'trivializing' the topic of meat production in general and slaughtering in specific which is not only misleading but also 'overburdening the imagination of children' (Austrian Advertising Council, 2017). What might be overburdening, however, is not so much the trivialization itself, but the promotion of a rationale concerning meat production which is diametrically opposed to every scientific and common sense perception of animals and to the values associated with the children's encounter of the animals (including the encounter of the human protagonists with their favourite animals in the first scenes of the book). To achieve the implementation of this far-fetched rationale the book in some way cannot but restore to heavy anthropomorphism and anthropocentrism.

Conclusions

The example shows that whenever authors (aim to) convey an anthropocentric moral message the traditional moralizing of picture books and the tool of anthropomorphism become problematic from an animal ethics perspective. The 'moral imperative' of picture books is central in their potential to be indeed a double-edged sword (Waxman *et al.*, 2014). The AMA example shows an extreme case to the one side where a picture book is ultimately reduced to a merchandising tool. But as Waxman *et al.* also remark, the sword can as much cut to the other side and become 'instrumental in redirecting children's attention to a biological model in which humans are one among the animal kinds' (Waxman *et al.*, 2014: 7).

It may be tempting to animal ethicists to argue that the double-edged sword should be used to work into this direction. However, we should not forget that it is questionable if the moral imperative comes as blunt (and actually outdated) moralizing. Such moralizing prevents the books from working as a real piece of literature because it doesn't allow for a creative de-coding of text, pictures, and the gaps between. The book or the author directly convey normative meaning instead of leaving the derivation of such to a well-developed story, well-developed animal characters, and ultimately to the young reader engaging with both. This leaves us with an interesting question: How rare are picture books about animals that are not moralizing right away? Given the potential of the picture book genre it should be possible to illustrate animals as clever and sentient actors to whom it matters how humans treat them. For the derivation of any ethical implications authors can trust in the ability of children to develop their non-anthropocentric stance into what constitutes their first animal ethics.

References

Adams, C. ([1990]2010). The sexual politics of meat: a feminist-vegetarian critical theory. Continuum, New York, London.

Almíron, N., Cole, M. and Freeman, C.P. (eds.) (2016). Critical Animal and Media Studies: Communication for Nonhuman Animal Advocacy. Routledge, N.Y, London.

Austrian Advertising Council (2017). AMA Kinderwerbung. Council decision published on 21.06.2017. Available at: https://www.werberat.at/beschwerdedetail.aspx?id=5214. Accessed 20 January 2018.

Austrian Advertising Council (2012): Advertising Industry Ethics Code. Vienna, 1st June 2012. Available at: https://www.werberat.at/layout/ETHIK_KODEX_6_2012_EN.pdf. Accessed 20 January 2018.

Evans, J. (ed.) (1998). Introduction. In: Evans, J. (ed.) What's in the picture?: responding to illustrations in picture books. Paul Chapman, London, 1-24.

Geerds, M. (2014): Anthropomorphic media and children's biological knowledge. Dissertation submitted to the Graduate School – Newark Rutgers, The State University of New Jersey. Available at: https://rucore.libraries.rutgers.edu/rutgers-lib/43766/PDF/1/play. Accessed 19 March 2018.

Ganea, P.A., Cranfield, C.F., Simons-Ghafari, K. and Chou, T. (2014). Do cavies talk? The effect of anthropomorphic picture books on children's knowledge about animals. Frontiers in Psychology 5 (2014), Article 283.

Herrmann, P. Waxman, S. R. and Medin, D. L. (2010). Anthropocentrism is not the first step in children's reasoning about the natural world. Proceedings of the National Academy of Sciences US 107(22): 9979-9984.

Marriott, S. (1998). Picture books and the moral imperative. In: Evans, J. (ed) What's in the Picture? Paul Chapman, London, pp. 1-24.

Marriott, S. (2002): Red in tooth and claw? Images of nature in modern picture books. Children's Literature in Education 33 (3): 175-183.

Tegetthoff, F. (2017): Fleisch, woher kommst Du? In Kooperation mit AMA, No longer online available.

Waxman S., Herrmann P., Woodring J. and Medin D. (2014). Humans (really) are animals: picture-book reading influences 5-year-old urban children's construal of the relation between humans and non-human animals. Frontiers in Psychology 5 (Article 172): 1-8.

Section 10.
Animal ethics

56. Personalism as a ground for moderate anthropocentrism

S. Aerts
Odisee University College, Ethology and Animal Welfare, Hospitaalstraat 23, 9100 Sint-Niklaas, Belgium; stef.aerts@odisee.be

Abstract

One of the main characteristics of personalism is an emphasis on the exceptional position of humans in relation to animals and the environment. However, one could also argue that the personalist view of humans as connected entities should not only be interpreted as connectedness to other humans. More than ever, there are good reasons to also consider the bonds between humans and animals, and even the environment. This line of reasoning gives us a unique opportunity to value animals, and nature, without having to sacrifice the focus on humans. By that, personalism provides a realistic framework to engage with the pressing issues in animal and environmental ethics in the 21st century. One of the main problems with a classic anthropocentric ethic is that it tends to focus on short-term and narrow interpretations of human interests. That environmental problems can have an impact on humans is quite evident, but similar connections can also be found between animals and humans (individually and collectively). In 2015, even Pope Francis has alluded to this in Laudato Si'. Recognising the interconnectedness of all human and non-human life forms in the biosphere is a major step towards a personalism that awards dignity or intrinsic value to animals and the environment. Then, stewardship – another important concept – becomes important, but no longer only because of the importance of animals and the environment for humans. Following these arguments in depth, we are confronted with a moral framework that implies (or facilitates) fundamental changes in our mentality, not only for the animals and the environment themselves, but also because it is important for humans. A value and person based philosophy such as personalism guides us to a moderate anthropocentrism that allows for concrete steps forward for animals and the environment.

Keywords: animals, environment, stewardship

Introduction

Personalism as a philosophical and moral tradition is currently under severe pressure. In the secular Western European societies, it is almost only known – if at all – as the underpinning of Christian democratic political positions, and the philosophical ground for the Catholic Church teachings. One of the main characteristics of those is the heavy emphasis on the exceptional position of members of the human species. In that sense, it is certainly an anthropocentric position.

Looking for the roots of some of the most pressing issues of the 21st century, we often stumble upon our own species. This is rather evident when discussing the problems associated with the use of animals in animal husbandry, animal experimentation, etc. But also with regard to the ecological crisis (crises?), most authors have been compelled to point to (different aspects of) human society. For example, Lynn White (1967) considers the Judeo-Christian vision of human dominance over nature a fundamental problem, while John Passmore (1974) refers to the Greek idea of rationality (and anthropocentrism), but in both the underlying problem is the 'objectification' of nature. Others have added other explanations, such as the well-known demographic explanation by Malthus (1798), or the classic Tragedy of the Commons by Hardin (1968).

The traditional ethical systems that have evolved over the centuries (virtue ethics, deontology, and consequentialism alike) have proven to be ill equipped to deal with the emerging ethical issues of the

EurSafe 2018 – DOI 10.3920/978-90-8686-869-8_56, © Wageningen Academic Publishers 2018

latter part of the 20th century. They were in fact different forms of 'human ethics' (implicitly, or explicitly anthropocentric), while the new problems dealt with different non-human entities (animals, ecosystems, etc.). The solution was to develop new, 'modern' ethical systems, opposing the 'classic', or old systems. We've seen the development of a zoocentric ethics (Singer, 1975; Regan, 1983), and of an ecocentric ethics (see e.g. Naess, 2010).

Problems with non-anthropocentric ethics

The development of these new ethics has had some influence on the way humans behave towards animals (and the environment), and certainly more than any of the writings that conveyed essentially the same messages decades (or centuries) earlier (e.g. Salt, 1894). On the other hand, they have not resulted on the great shift away from the systems and situations that are creating these problems. At appears that, on a purely practical level, a call for a radical change in behaviour and mentality including a shift away from humans as a focal point of morality (however urgent, or well-argued), is not the fastest way forward.

On a conceptual level, we encounter a more profound critique; animal and environmental ethics cannot be 'dehumanised'. Not only are we (as already stated) the source of most of the problems, but – more fundamentally – because ethics itself is a human activity. In a sense, the problem only exists from our perspective. This is especially clear with regard to the ecological crisis: we only consider the current ecological changes as problematic (and not the catastrophical changes form the past that resulted in new geo-biological balances), because this is the only one we have been implicated in. A similar line of argument can be followed with regards to animals, but in both cases it is clear that any ethics should be a human ethics; (1) value appropriation is by definition a human activity, and (2) ethical rules or principles can never be inferred from biological situations.

Therefore, citing William Grey (1993), it is not the anthropocentrism itself that is the problem, but the 'short term and narrow conception of human interests and concerns'. He states 'a suitably enriched (non-atomistic) conception of humans as an integral part of larger systems – that is, correcting the misconception of humanity as distinct and separate from the natural world – means that anthropocentric concern for our own well-being naturally flows on to concern for the nonhuman world.' Although his analysis focuses on ecology, there is arguably no reason why it should not apply to animal ethics as well.

Personalism and animals

Connectedness and spiritualism are two major concepts in personalism. Historically, in the majority of cases those have been connected to humans, almost exclusively. It seems, however, that there is no fundamental reason to do so.

With regard to connectedness: it should need no explanation that we, humans, are intimately connected to many, if not all, of the animals on this world. That is especially true for those animals that we share our houses and lives with, but there are many other types of human-animal relations (see e.g. Aerts, 2015). These relations are reciprocal: in many the animals are the dependent party, relying on humans for food and shelter (e.g. farm animals, pets), but the animals also provide something to humans (food, companionship, etc.). In a single relation (that of assistant animals) the balance shifts and humans can even be said to be the dependent party.

When John H. Lavely (1986) argues personalism provides a basis for the concept of 'dignity of nature', he clearly refers to the interdependency of human and non-human entities. For him, the concept of personality or 'person' should not be applicable to humans alone, but must – at least partially – refer to all creatures. He does not support the mind-body dualism, and is thereby able to assign intrinsic value to all entities. He was certainly not the first to argue for a departure from the mind-body dichotomy, as it was already argued by Karol Wojtyla (1979) (who would become pope John-Paul II).

The importance of connectedness resonates well with what pope Francis has recently written on ecosystems and animals in his encyclical Laudato Si' (Francis, 2015): 'if we feel intimately united with all that exists, then sobriety and care will well up spontaneously'. He clearly assigns intrinsic value to non-humans as well: 'It is not enough, however, to think of different species merely as potential 'resources' to be exploited, while overlooking the fact that they have value in themselves.' That Grey's flow of concern goes both ways, is also clear to Francis: 'Moreover, when our hearts are authentically open to universal communion, this sense of fraternity excludes nothing and no one. It follows that our indifference or cruelty towards fellow creatures of this world sooner or later affects the treatment we mete out to other human beings.' The latter is, of course, something that was already argued by Immanuel Kant. By this, we have also shown that the spiritual dimension that is important in personalism can easily be extended to animals (and ecosystems).

Then, the concepts of '*bonum commune*' and stewardship – other important concepts in personalism – become important, but no longer only because of the importance of animals and the environment for humans. Both are in need of upgrades. The concept of *bonum commune* can no longer refer to the human society alone, but needs to be extended to all entities that are within our 'community'. Stewardship is often understood quite economically as the duty to use and develop resources, without compromising the resource capital. Houtepen (1995) warned that real stewardship is more than that: a good steward should also be a watcher who warns for long-term dangers. Again, the different aspects of personalism combine to develop an ethics that urges to act for animals, not only because they are important for humans, but also because of their intrinsic value.

Personalism as moderate anthropocentrism

We are thus confronted with a moral framework that implies (or facilitates) fundamental changes in our mentality, not only for the animals (or the environment) themselves, but also because it is important for humans. In this sense, personalism could possibly be the philosophy that makes certain 'that anthropocentric concern for our own well-being naturally flows on to concern for the nonhuman world.' The personalist anthropocentric focus then becomes no longer a hindrance to progress, but an opportunity. A value and person based philosophy such as personalism guides us to a moderate anthropocentrism that allows for concrete steps forward for animals and the environment.

References

Aerts, S. (2015). Named, numbered or anonymous: how the Human-Animal Relation affects the naming of individual animals. Beiträge zur Namenforschung, 50(3/4), 309-318.

Francis. (2015). Encyclical Letter Laudato si' of the Holy Father Francis on care for our common home. Rome: Libreria Editrice Vaticana.

Grey, W. (1993). Anthropocentrism and Deep Ecology. Australasian Journal of Philosophy, 71(4), 463-475.

Hardin, G. (1968). The Tragedy of the Commons. Science, 162(3859), 1243-1248.

Houtepen, A. (1995). Christelijk geloof, humanisme en intrinsieke waarde milieu. Christen democratische verkenningen, pp. 83-83.

Lavely, J.H. (1986). Personalism Supports the Dignity of Nature. Personalist Forum, 2(1), 29-37.

Malthus, T.R. (1798). An Essay on the Principle of Population. London: J. Johnson.

Naess, A. (2010). The Ecology of Wisdom: Writings by Arne Naess. Berkeley: Counterpoint.

Passmore, J.A. (1974). Man's Responsibility for Nature: Ecological Problems and Western Tradition. Scribner.

Regan, T. (1983). The case for animal rights. Berkeley: University of California Press.

Salt, H.S. (1894). Animals' rights considered in relation to social progress. London: MacMillan & co.
Singer, P. (1975). Animal Liberation. A new ethics for our treatment of animals. New York: New York Review/Random House.
White, L.J. (1967). The Historical Roots of Our Ecologic Crisis. Science, 155(3767), 1203-1207.
Wojtyla, K. (1979). The Acting Person (Analecta Husserliana – The Yearbook of Phenomenological Research, Vol. X). D. Reidel.

57. Objectification and its relation to Kant's moral philosophy

S. Camenzind
Unit of Ethics and Human-Animal Studies, Messerli Research Institute, Veterinaerplatz 1, 1210 Vienna, Austria; samuel.camenzind@vetmeduni.ac.at

Abstract

Within the context of bioethics, feminism and animal ethics the notion of 'objectification' is used pejoratively to express a morally impermissible action. Objectification is roughly defined as treating a human being or another animal as a thing. Concerning contemporary theories on objectification, it is remarkable that explanatory approaches refer prominently to Kant's ethic, especially to the second formula of the Categorical Imperative (Formula of Humanity, FUL). It states: 'So act that you use humanity, in your own person as well as in the person of any other, always at the same time as an end, never merely as a means' (GMS, IV: 429). The aim of this paper is to examine how objectification can be considered in a Kantian moral framework in general and how it can be based on the FUL in particular. It will be argued that although there are overlaps between contemporary concepts of objectification and Kantian ethics, they differ in various aspects, especially concerning objectification of animals. This concerns first of all the fact, that Kant didn't use the term 'objectification' (*Verdinglichung*) at all. Secondly FUL can only be applied to persons, i.e. morally autonomous beings. Against the concept of objectification of many contemporary scholars, according to Kant's moral philosophy objectification of animals is not possible. Therefore FUL is smaller concerning the scope of morally relevant entities. At the same time FUL is wider regarding objectifying actions, including various forms of (self-) objectification, like gluttony, drunkness and suicide, that are not normally part of contemporary debates on objectification.

Keywords: objectification, animal ethics, Kantian ethics, formula of humanity

Introduction

Within the context of bioethics (e.g. Fiester, 2005: 339ff; Camenzind, 2015), feminism (cf. Adams, 2013: 4) and animal ethics (cf. Petrus, 2013), the notion of 'objectification' is used pejoratively to express a morally impermissible action. Concerning contemporary theories on objectification it is remarkable that many explanatory approaches refer prominently to Kant's ethic. Especially Kant's 'Formula of humanity' (FUL), which states: 'So act that you use humanity, in your own person as well as in the person of any other, always at the same time as an end, never merely as a means' (GMS, IV: 429)[1], seems to get to the heart of the idea of objectification (cf. Nussbaum, 1999: 218, 224; Radin, 2001: 118; Wilkinson, 2003: 28ff.).

In the following paper, four different aspects will be discussed regarding the relation between the contemporary use of the term 'objectification' and Kant's ethics, and how they differ: (1) Based on the fact that FUL deals with the problem of unpermitted total instrumentalisation (use someone merely as a means), a comparison between objectification and unpermitted instrumentalisation is made first. It will be argued that *prima facie* the grounding of the wrongness of objectification with FUL is possible. Yet, the underpinning of objectification with Kant's ethic should not be overstated for two reasons (aspect 2 and 3). (2) First of all, it is to mention that Kant didn't use the term 'objectification' at all. (3) The third aspect concerns Kant's dichotomy between 'persons' and 'things'. It will be shown that the group

[1] Kant's work is cited by the standard abbreviations of the Kantian Society, referring to the volume and page number of *Kants gesammelte Schriften*, published by *Preussische Akademie der Wissenschaften*.

EurSafe 2018 – DOI 10.3920/978-90-8686-869-8_57, © Wageningen Academic Publishers 2018

of possible subjects of 'objectification' is limited to persons only. According to Kant, animals belong to the category of 'things' by definition. Therefore, it isn't possible to objectify animals in a morally relevant way. (4) Further, FUL also condemns various forms of (self-)objectification, like gluttony, drunkenness and suicide, that are not normally part of contemporary debates on objectification.

Instrumentalisation and objectification

The core of objectification may be defined as transformation of a living entity into a mere object. This transformation can happen (1) literally, (2) verbally, (3) by way of a presentation, and (4) within a certain modus operandi of treatment:

1. The literal transformation occurs, for instance, when animals are physically fragmented into different body parts and transformed to food products.
2. Verbal objectification often comes along with figurative language (e.g. abstraction, renaming, metonymy) to transform an animal or human subject into an object (e.g. Adams, 2013: 13).
3. In advertisements or art, someone (paradigmatically women) can be presented in a manner that reduces them to their body or appearance (cf. Langton, 2009: 228f.)
4. A human being or an animal can also be treated in a manner that is characteristic of our treatment of things. In her well-received essay 'Objectification', Martha Nussbaum (1999) locates 'seven ways to treat a person as a thing'. Besides denial of subjectivity, denial of autonomy, denial of violability, inertness and ownership, Nussbaum mentions instrumentality as one notion of objectification (cf. Nussbaum, 1999: 218). Her intuition that objectification and instrumentalisation are interlinked phenomena is right, because many objectifying acts are identified as acts of instrumentalisation, which can be defined as the (mis)use of an entity (means) in a specific mode to a certain end (cf. Camenzind, 2017).

According to my suggestion (Camenzind, forth.), the hierarchy that Nussbaum identifies is rather the other way round: objectification is one particular mode of unpermitted instrumentalisation, because there are (morally permitted) modes of instrumentalisation that do not entail objectification at all. I just mention two examples here: if Peter uses a cabdriver (with his consent) as means to bring him to the airport, neither one of the other notions of objectification (denial of subjectivity, denial of autonomy, denial of violability, inertness and ownership) are present, nor is it obvious why the driving service provided by the cabdriver should be stated as objectification. Nevertheless, he is clearly instrumentalised. The second example concerns the fact that actual, non-living things can be instrumentalised, but not objectified. If Peter uses a non-living thing (a book or a pen) as a means for various ends, the category of objectification cannot be applied to this case (cf. also Nussbaum, 1999: 218).

The connection between Kant's expression in FUL 'to use someone as a mere means' (instrumentalisation) and the concept of 'treating someone as an object' (objectification) is very tempting because of two reasons. One concerns Kant's terminology. In the derivation of the FUL (GMS IV: 428-429), Kant distinguishes sharply between things (*Sachen*) and persons (*Personen*), and mentions later that to use a person as a mere means is morally wrong: 'So act that you use humanity [...] never merely as a means' (GMS, IV: 429). In Kantian terms, to (mis)use someone as a mere means is the same thing as using someone as mere thing, tool or instrument. The second reason concerns the fact that both concepts are used pejoratively.

Despite these overlaps between contemporary concepts of objectification and Kant's moral philosophy, I will argue in the remaining part of this essay as follows: an underpinning of objectification with Kant's ethic, especially regarding FUL, should not be overplayed, because FUL has implications in different aspects that differ with respect to the use of the term 'objectification' by contemporary philosophers. These differences will be outlined in the following paragraphs.

Objectification as *Versachlichung*

A first minor remark concerns Kant's terminology. Kant didn't use the term 'objectification' (*Verdinglichung*) at all. He knew the '*Verdingungsvertrag*' (*locatio conductio*) and the common parlance of '*verdingen*', which means to enter into service of someone as employee (cf. TL VI: 330). This terminology was used in a descriptive way and distinguished from morally unpermitted forms of misuse of the service of others like serfdom or slavery.

Nevertheless, a suitable term for objectification that is closely connected to Kant's own terminology would be '*Versachlichung*', which would be another word for an unpermitted instrumentalisation, the use of a person as a mere tool or thing (*Sache*).

Animals as things

The sharp distinction between persons and things is another reason why a contemporary concept of objectification cannot fully rely on Kant's moral philosophy. It has been demonstrated sufficiently that the ethical analysis of objectification concerning animals in various contexts, e.g. animals for food production (Petrus, 2013), animal experimentation (Camenzind, 2015) or the media portrayal of animals in the context of organic food production (Leitsberger *et al.*, 2016), is possible and fruitful. Most of these studies rely on Nussbaum's approach and are in line with her assessment that objectification constitutes a morally wrong practice.

But assessments that argue that objectification does a moral wrong to animals cannot be founded on Kantian grounds. In contrast to persons, who have absolute inherent value and therefore a dignity that has to be respected (GMS, IV: 428, 435), animals belong to the category of 'things', which do not have a moral status. 'Things' is a *terminus technicus* borrowed from the Roman law that comprises all non-rational entities: animals, plants and (non-living) objects. In Kant's paradigm, all 'things' have only a relative value, a market price (cf. GMS, IV: 435) and are not part of the moral community. According to Kant's autonomocentrism,[2] they are '[...] means and instruments to be used at will for the attainment of whatever ends he [man] pleased' (Anfang, VIII:114). Therefore, the group of possible subjects of objectification is limited to persons only.

A Kantian interpretation of objectification

Concerning persons, two aspects of the Formula of Humanity should be mentioned here that differ from contemporary approaches to objectification. Firstly, the distinction between persons and things opens up further forms of objectification that are in line with Kantian ethics, but which are at odds with contemporary concepts of objectification. Since the category of things consists, besides of objects, also of plants and animals, it would not only be a form of objectification if a person is treated like a non-living object, but also if a person is treated like a plant or an animal. This could be the case if someone is reduced to functions like self-maintenance, survival and reproduction (plant), or physical pleasures (animals). This interpretation is at odds with contemporary concepts of objectification, which only deal with the problem of treating someone as a non-living object.

Secondly, Kant's FUL doesn't only refer to interpersonal interactions, but also to the relation of a person to herself – a relation that is neglected in contemporary discussions on objectification. As the FUL states,

[2] It is common within the field of animal ethics to categorize Kant as anthropocentrist. But because Kant's moral community includes all morally autonomous beings – beside human beings, also angels, god and aliens on other planets (cf. Anth, VII 331) – I do not think it is accurate. For the term 'autonomocentrism' I am thankful to Jens Timmermann.

one should never use others as mere means, but in the same way one should never use oneself as a mere means. Kant mentions here the duties to oneself, which 'take first place, and are most important of all' (V-Mo/Collins, XXVII: 341). Since the sharp dualism between persons and animals is reflected in the predispositions personality and animality within a person, certain actions, like gluttony, drunkenness and suicide (cf. TL, VI: 417ff.), are examples of self-objectifications. According to Kant, those practices are wrong because, by engaging in them, one degrades oneself to a non-autonomous 'thing'. My main point is here that those duties to oneself, the core of Kant's late moral philosophy, are not mentioned when referring to his FUL by contemporary theorists of objectification.

To sum up: the reference to Kant's FUL by contemporary objectification theorists is consistent insofar as objectification is one particular mode of unpermitted instrumentalisation concerning persons. But the connection between objectification and Kant's ethic should be made carefully. First of all, according to Kant's theory, a morally relevant objectification of animals isn't possible. This means that the scope of FUL is smaller compared to contemporary notions of objectification. Secondly, the FUL is wider regarding actions of objectification, because it covers practices that are at odds with or at least a blind spot of contemporary notions of objectification.

References

Adams, C.J. (2013 [1990]). The Sexual Politics of Meat. A Feminist-Vegetarian Critical Theory (20th Anniversary edition), Bloomsbury Academic, New York/London.

Camenzind, S. (2015). On Clone as Genetic Copy: Critique of a Metaphor. Nanoethics 9 (1): 23-37.

Camenzind, S. (2017). Instrumentalisation as Evaluation Criterion in Animal Ethics. Presented at the 3rd Minding Animals Symposium, Erlangen, June 24-26, 2016, Erlangen, Germany.

Camenzind, S. (forth.). Instrumentalisierung. Zur Transformation einer kantianischen Grundkategorie der Moral in der Ethik der Mensch-Tier-Beziehung. Unpublished doctoral dissertation, University of Veterinarymedicine, Vienna.

Fiester A. (2005). Ethical Issues in Animal Cloning. Perspectives in Biology and Medicine 48 (3): 328-343.

Langton, R. (2009). Sexual Solipsism. Philosophical Essays on Pornography and Objectification, Oxford University Press, Oxford/New York.

Leitsberger, M., Benz-Schwarzburg, J. and Grimm, H. (2016). A Speaking Piglet Advertises Beef: An Ethical Analysis on Objectification and Anthropomorphism. Journal of Agricultural and Environmental Ethics 2016; 29: pp. 1003-1019.

Nussbaum, M.C. (1999). Objectification. In: Sex and Social Justice, Oxford University Press, New York *et al.*: 213-239.

Petrus, K. (2013). Die Verdinglichung der Tiere. In: Chimaira – Arbeitskreis für Human-Animal Studies (eds.): Tiere, Bilder, Ökonomien. Aktuelle Forschungsfragen der Human-Animal Studies, Transcript, Bielefeld: 43-62.

Radin, M.J. (2001 [1996]). Contested Commodities. The Trouble With Trade in Sex, Children, Body Parts, and Other Things, Harvard University Press, Cambridge/London:.

Wilkinson, St. (2003). Bodies for Sale. Ethics and Exploitation in the Human Body Trade, Routledge, New York.

58. Legal protection of animal intrinsic value – mere words?

*M.F. Trøite and B.K. Myskja**
Department of Philosophy and Religious Studies, Norwegian University of Science and Technology – NTNU, 7491 Trondheim, Norway; bjorn.myskja@ntnu.no

Abstract

The Norwegian Animal Welfare Act states that animals have intrinsic value independent of their 'usable value' for humans. Despite this explicit statement, Norwegian animal husbandry involves numerous generally accepted practices that appears to be contrary to an everyday understanding of this concept. By analysing the moral terms used in the Act in light of central contributions to the animal ethics debate, we aim to bring some clarification to a certain use of this otherwise abstract and complex ethical concept of intrinsic value, and discuss its possible importance in future regulation of animal husbandry. According to the former Minister of Agriculture, the use of 'intrinsic value' is intended to be of symbolic value only and should not have consequences for animal welfare. We will argue that such symbolic use of moral concepts in animal welfare legislation could have two opposite effects, regarded from an animal rights or welfare perspective. It may undermine the meaning of the concepts and make them less valuable in promotion of animal rights, or they may be valuable as tools for long-term political change, despite their mere symbolic function. Although the argument is referring to Norwegian law and animal husbandry, it is also relevant for countries with laws using more or less similar expressions, such as Germany, Switzerland and Austria.

Keywords: animal welfare, animal husbandry, moral status, respect

Introduction

The purpose of the Norwegian Animal Welfare Act (AWA) is to promote good animal welfare and respect for animals, and it states that animals have intrinsic value independent of their use-value for humans (AWA, 2009, §§ 1 and 3). The AWA replaced the Animal Protection Act in 2009, and the government justified the need for a new law by referring to the significant development in knowledge about animal abilities and needs, adding that the population has in general become more concerned about the treatment of animals (Næringskomiteen, 2009). While the previous Act emphasized protection from suffering, the aim of the new Act, according to the preparatory works, is to contribute to the thriving of animals by focusing on the interests of animals (Landbruks- og matdepartementet, 2008).

Despite these explicit statements of intrinsic value, respect and protection of interests, and the positive aims behind the formulations of the AWA, Norwegian animal husbandry involves numerous generally accepted practices that appear to be contrary to these ideas and goals. One example is the maceration of newly hatched male chickens. Immediately after determining the sex, male chickens are dropped live into a high-speed grinder and cut to pieces. Although regulations require that death is immediate, this is a common practice that does not only seem contrary to the purpose of the Act, but also to an everyday understanding of 'intrinsic value'. These animals are killed due to their lack of 'usable value' for humans, meaning that their treatment is justified merely according to instrumental considerations. Similar considerations underlie other practices in Norwegian animal husbandry. According to the former Minister of Agriculture the term 'intrinsic value' is meant to have merely a symbolic function and should not have consequences for animal welfare (Næringskomiteen, 2009). By analysing the moral terms used in the Act in light of central contributions to the animal ethics debate, we aim to bring some clarification to a certain use of this otherwise abstract and complex ethical concept of intrinsic value, and discuss its possible importance in future regulation of animal husbandry. We will discuss two possible effects on the public debate and animal husbandry of such a symbolic use of moral concepts in animal welfare legislation.

Legislating intrinsic value

The AWA (§3) states that: 'Animals have an intrinsic value which is irrespective of the usable value they may have for man. Animals shall be treated well and be protected from danger of unnecessary stress and strains.' According to the preparatory works, this means that the treatment of animals should be justifiable and respectful independent of economic or other utility for humans (Næringskomiteen, 2009). The purpose of ascribing intrinsic value to animals in law was to clarify their status and rights. In regard to their legal status, animals still have status as property, and hence, they are not legal subjects, i.e. entities with rights and duties. Still, the value ascription in the AWA must be regarded as an elevation of their status and in a commentary on the Act, Stenevik and Mejdell (2011) suggest that their legal status should be described as that of protected subjects. We could therefore say that the Act gives animals rights *prima facie*, i.e. a right to have their interests taken into account and to be treated well and with respect as long as this is not overruled by other necessary considerations.

By using the term 'intrinsic value', the act apparently accords moral status to animals. A common definition of moral status is the following: 'An entity has moral status if and only if it or its interests morally matter to some degree for the entity's own sake, such that it can be wronged' (Jaworska and Tannenbaum, 2017). The preparatory works states that animals should be protected for their own sake where considerations shall be based on the interests of the animal, including its species-specific and individual needs (Landbruks- og matdepartementet, 2008). Thus, according intrinsic value to animals implies that they should be treated well and with respect for their own sake, and not merely as a means to prevent injuries and to promote production and public legitimacy. Still, the value ascription does not amount to independent rights for animals, as was made clear by the former Minister of Agriculture (see above).

Interpretations of 'intrinsic value'

Intrinsic value is usually understood as a value an entity has 'in itself' or 'in its own right', and is generally contrasted with instrumental value – a value an entity has in virtue of being a means to someone else's end (Zimmerman, 2015). Because inherent value can be given the same description, inherent and intrinsic value are sometimes used synonymously. It is, however, useful to distinguish between intrinsic and inherent value as two different accounts of non-instrumental value. In the first case, the entity is a carrier of value, in the latter it is a possessor of value. The first understanding is associated with utilitarian approaches to animal ethics. As the AWA emphasizes animal interests, a preference utilitarian approach is the relevant framework here. According to this approach, preferences have intrinsic value. This means that sentience is a necessary and sufficient condition for moral status. The assumption, according to Singer, is that sentience is a condition for having preferences, as all sentient beings are interested in avoiding pain (Singer, 1990). The right act is therefore that which maximizes the overall preferences of all affected sentient beings after equal consideration has been given to each and every preference. On this account then, the intrinsic value of chickens is respected in maceration if death is painless and their only preference is to avoid pain. Certainly, many other husbandry practices that are currently accepted are contrary to these principles of utility and equal consideration. Most of the stress and strains that animals are subjected to and which are viewed as necessary from an anthropocentric perspective concerned with economy, would be unacceptable speciesism on Singer's account.

Regardless of this, the logic of the AWA is utilitarian to the extent that it states that animals should be protected from 'unnecessary stress and strains', implying a weighing of interests. It is the weight of different interests, however, that needs adjustment. If we assume this to be the case, there still seems to be something of moral importance missing in an utilitarian approach. When living beings can be treated as mere waste and destroyed only on condition that it is done painlessly, it goes against the intention of the Act, which is 'to promote good animal welfare and respect for animals' (§1). The preparatory work states that 'respect' signifies consideration of how animals ought to be treated beyond mere welfare considerations (Landbruks- og matdepartementet 2008), and should be based on the animal's distinctive character and intrinsic value (Landbruksdepartementet, 2002). The use of 'respect' in the AWA as explained in the preparatory works, indicates therefore a deontological understanding of intrinsic value. On this account, the entity is not a carrier of value but the possessor of value, where respect for intrinsic value entails never to regard or treat such an entity as a mere means to someone else's end.

All beings that have the capacity to experience well-being have equal inherent value and are due respect, according to Regan. He calls such beings 'subjects-of-a-life', and it is the entity itself, rather than its experiences, that is the source of value (Regan, 2004). On this perspective, the question is not whether the 'stress and strains' are necessary, but whether the treatment is showing respect for beings that are subjects-of-a-life. Macerating newly hatched male chickens will therefore be wrong on this perspective– even if it is painless – because killing shows total lack of respect for the inherent value of these beings. They are destroyed because they are considered merely in instrumental terms, and their commercial value alive is considered too small.

Regan's deontological concept of inherent value captures moral considerations implicit in the AWA's conceptual apparatus that is lost in a utilitarian approach, namely respect as something more than mere welfare considerations. However, as this deontological value concept requires unconditional duties and inviolable rights, it is not applicable as an interpretation of the AWA concepts of intrinsic value and respect. One possible solution is to interpret the Act's 'intrinsic value' as a hybrid concept, drawing on a combination of modified utilitarian and deontological principles (Parfit, 2011) for one attempt to combine these approaches). Some may object that this is combining contradictory and mutually exclusive theories, but there is a crucial similarity between the two. Both challenge anthropocentric ethics by presenting rational arguments for the moral relevance of the animals' own experience of something as good or desirable, providing a non-human source of evaluation. This change of perspective may be required to fulfil the legal shift from avoiding suffering to ensure that animals thrive or prosper. However, when these moral principles are used as basis for rational deduction to right conduct, as is common in moral philosophy, they become irrelevant in this legal context.

The moral significance of attitude

We should not conclude that the term 'intrinsic value' in the AWA is useless in practical matters, although it falls short of the rigorous moral standards proposed by Singer and Regan. The turn from the welfarist approach in the previous Act is combined with a seemingly virtue ethical emphasis on displaying the proper attitudes. This is evident by the combination of 'intrinsic value' with 'respect', as well as the prohibition of killing animals as 'an independent form of entertainment or competition' (§12). As Hursthouse makes clear, as long as one avoids the moral status-question, 'intrinsic value' can also be a useful concept within virtue ethics. 'We can recast talk of things having intrinsic value as talk about their being worth our pursuing or having or preserving (...) for their own sake, and there is no reason why virtue ethicists shouldn't agree with Regan that the good of the other animals is such a thing and thereby has intrinsic value in that sense' (Hursthouse, 2006: 151-152). Support for a virtue ethics aspect of the AWA is also found in the preparatory works, which states that the exhibition of animals should be done in a way that inspires admiration and respect in the spectators (Landbruksdepartementet, 2002). It is emphasized that the circumstances can undermine these attitudes, for example in animal husbandry where large stocks and low stimulation environment make the individual animal invisible and prevent them from developing individual characteristics. This may result in loss of respect and in a treatment of them as production units rather than individual beings. Thus, the right attitudes are taken as essential for proper treatment, which opens for an interpretation of the AWA as concerned with the way we regard animals. This may be a more appropriate way to interpret the Act compared to attempts at specifying the moral status we ascribe to them. Saying that animals have intrinsic value is then an expression of the kind of attitude we should have towards animals which should be displayed through our dealings with them.

To sum up: the AWA represents a change in how we legally regulate the treatment of animals by stating that protecting animals from unnecessary suffering is not sufficient. Our treatment of them should display certain attitudes that show that they are more than mere things or usable products, where respect for intrinsic value entails that animals ought to be regarded and treated as individuals that are valuable for their own sake. Put this way, it may sound as if the purpose of the Act is to serve as a tool for moral education. That is certainly not the intention of laws. The law is not primarily directed towards people's thoughts or feelings, motivations or attitudes, but concerned with regulating human behaviour. This means that the AWA's virtue approach must be understood as directed towards acts as expressions of respect for animals. This is clearly problematic when several accepted practices are far from showing respect. We have mentioned maceration of male chickens, but the general living conditions provided for chickens are hardly much more respectful, if placing them 'in an environment which is consistent with good welfare, and which meets the animals' needs which are specific for both the species and the individual' (§ 23), is understood as an expression of respect for the animal itself. A month-long life with an unnaturally rapid growth in a barn with thousands of other chickens and no flock containing adult birds, is meeting neither species-specific nor individual needs. Failing to meet such needs can be said about other husbandry practices in Norway as well. Even with a virtue or attitude oriented interpretation of the AWA 'intrinsic value' term, we have an unsurmountable gap between the letter of the law and the accepted animal related practices. As long as it is intended to have a symbolic effect only, there seems to be a morally problematic division between what is said and what is done.

Moral effects of the symbolic use of 'intrinsic value'

Legislating respect for the intrinsic value of animals, while allowing practices that are clearly not respectful, seems at first glance to be morally wrong, at least from a deontological or virtue ethical perspective. It is a case of lying or deceit, and lying is not acceptable under any circumstances within deontology, while deceit is a vice in virtue ethics. The question is whether it actually is a case of deceit. Who is deceived, and who are the liars? People understand that terms such as intrinsic value have a symbolic function in the law. The moral wrong is not because people wrongly believe what is said, but rather in the devaluation of language and the source of the statement. This is particularly so in the context of law. If the law is not taken seriously by the law-makers and the law-enforcers, the citizens will also lose respect for the law. That is, unless there are good reasons for this kind of 'symbolic' language. Such a reason could be that the language of the Act is intended to have an indirect effect, changing our appreciation of animals with consequences for how we behave. Then the Act is effective even if failing to treat animals according to the words of the Act does not result in any legal action.

Part of the symbolic function of intrinsic value might therefore be that it presents an ideal we should strive for, making it valuable for long-term political change, even if it has no direct effect on legal practice. This change can take at least two, closely related forms. The first is described in Sunstein's analysis of how laws can have the expressive function of 'making statements' rather than directly regulating practices. The effect of these statements, he writes, is 'to change social norms' (Sunstein 1996: 2025). We will not go into a detailed analysis of the AWA in light of Sunstein´s complex discussion, but point at a simple way the law can contribute to such change of norms. Legislating intrinsic value and respect for animals can be used as a reference in moral debates as an expression of how we, the citizens of the country, think that animals ought to be treated, given that the law is made by our elected representatives. Thus, it may be referred to as a norm everyone ought to live up to in our treatment of animals. Not adhering to generally accepted norms is something most want to avoid as it is connected to shame. The second form the law can be used to effect change, is a less individualistic way. The formulations in the law can be used to raise discussions whether concrete practices are actually acceptable when regarded in light of the law. Then the words become tools for change in practices. Referring to statements in law strengthens arguments for altering practices by giving them a generalized, universal basis.

Conversely, the mere symbolic use of moral terms without any impact on practice may undermine their meaning and leave them without significance. If one claims something is valuable in itself, while allowing it to be treated as a mere commodity, the value ascription is worthless. Stating something without having any intention of living up to it, is a way to say that one's statements are not worth pursuing. This lack of concern with verity has similarities with the rhetoric of bullshit, in Frankfurt's (2005) analysis. When we talk bullshit, the aim of the talk is to give the audience a certain impression of us, and the facts of the matter are by and large irrelevant. Similarly, when we can read the AWA as saying that animals have intrinsic value, but this should not have decisive impact on our treatment of them, it is a kind of bullshit. As Frankfurt points out, the serious forms of bullshit are a more severe threat to truth than direct lying is, as it undermines the capacity to distinguish truth from falseness. Within moral discourses, saying that our moral language should not lead to practical consequences, could be a form of bullshit that undermines the truth claims in morality. The practical consequence might be that it becomes more problematic using terms like 'intrinsic value' and 'respect' in the protection of animals also outside of the context of the law.

Conclusions

We have pointed to two opposite consequences of using a certain kind of symbolic moral language in the legal protection of animals. It is an empirical question to what extent the letter of the law actually has these consequences, but it is not unlikely that both phenomena are taking place. On the one hand, the AWA displays a double standard which might result in a loss of respect for the law and devaluation of the moral terms themselves. On the other hand, the symbolic moral language in the Act can be a useful tool for those who actually think that we ought to change how we behave towards animals, where they can challenge the rest of us to take seriously the moral standard we proclaim in our legislation when discussing concrete cases of treatment of animals and future regulation. There is also a well-known socio-psychological effect of pretending to be better than you are. You are influenced by your words and behaviour, even if you use them to give a certain impression and do not care about their truth content. What starts out as bullshit, gradually becomes truth.

References

Animal Welfare Act (2009). Act of June 19, 2009 no. 97 on animal welfare. Available at: https://www.regjeringen.no/en/dokumenter/animal-welfare-act/id571188. Accessed 21 January 2018.

Dillon, R.S. (2016). Respect. The Stanford Encyclopedia of Philosophy. Stanford University. Available at: https://plato.stanford.edu/archives/win2016/entries/respect. Accessed 28 May 2017.

Frankfurt, H. (2005). On Bullshit. Princeton University Press, Princeton, New Jersey. 77 pp.

Jaworska, A. and Tannenbaum, J. (2017). The Grounds of Moral Status. The Stanford Encyclopedia of Philosophy. Stanford University. Available at: https://plato.stanford.edu/archives/fall2017/entries/grounds-moral-status. Accessed 15 December 2017.

Landbruksdepartementet (2002). Om dyrehold og dyrevelferd. Regjeringen, Oslo, Norway. Available at: https://www.regjeringen.no/no/dokumenter/stmeld-nr-12-2002-2003-/id196533. Accessed 23 January 2018.

Landbruks- og matdepartementet (2008). Om lov om dyrevelferd. Regjeringen, Oslo, Norway. Available at: https://www.regjeringen.no/contentassets/4c83935a183e45ea92761d8b864383dd/no/pdfs/otp200820090015000dddpdfs.pdf. Accessed 23 January 2018.
Næringskomiteen (2009). Innstilling fra næringskomiteen om lov om dyrevelferd. Stortinget, Oslo, Norway. Available at: https://www.stortinget.no/globalassets/pdf/innstillinger/odelstinget/2008-2009/inno-200809-056.pdf. Accessed 21 January 2018.
Parfit, D. (2011). On What Matters: Volume One. Oxford University Press, Oxford, United Kingdom, 540 pp.
Regan, T. (2004). The Case For Animal Rights. University of California Press, Berkeley, USA, 425 pp.
Singer, P. (1990). Animal Liberation. Avon Books, New York, New York, 320 pp.
Stenevik, I.H. and Mejdell, C. (2011). Dyrevelferdsloven Kommentarutgave. Universitetsforlaget, Oslo, Norway, 472 pp.
Sunstein, C.R. (1996). On the Expressive Function of Law. University of Pennsylvania Law Review, 144: 2021-2052.
Zimmerman MJ. (2015). Intrinsic vs. Extrinsic Value. The Stanford Encyclopedia of Philosophy. Stanford University. Available at: https://plato.stanford.edu/archives/spr2015/entries/value-intrinsic-extrinsic. Accessed 15 December 2017.

59. Why insect sentience might not matter very much

S. Monsó
Institute of Philosophy, Karl-Franzens-Universität Graz, Attemsgasse 25/II, 8010 Graz, Austria; susanamonso@gmail.com

Abstract

The last decade has seen an emergence of evidence that suggests that insects may be sentient, which raises the question of the sort of treatment that these beings are owed. In this paper I focus on the case of edible insects and put forward some reasons why the sentience of these insects, assuming they do indeed possess it, might not matter very much from an ethical perspective. I will offer four reasons to support this claim. First, I will argue that insect sentience does not automatically imply a right to life. Second, I will offer some reasons to believe that insect consciousness may matter less than that of other animals, who might have a stronger claim to moral protection. Third, I will argue that a commitment not to harm other types of animals, such as mammals, may actually imply an obligation to consume insects. Lastly, I will offer some considerations that speak against the implementation of welfare measures in insect rearing.

Keywords: farm animals, pain, moral status, food production

Introduction

In recent years, the environmental costs of traditional sources of meat have motivated organisms such as the FAO to call for the increasing consumption of insects, which are a much more sustainable source of protein (FAO 2017; Halloran and Vantomme, 2013). At the same time, recent years have also seen the emergence of evidence that suggests that insects may possess consciousness, and so may be capable of experiencing pain (Barron and Klein, 2016; Klein and Barron, 2016; Yarali *et al.*, 2008; Bateson *et al.*, 2011). Given that sentience, or the capacity to feel pain and suffer, is usually considered the threshold condition for being worthy of moral consideration, this raises the question of whether it matters, morally speaking, if insects are sentient, and whether there is anything ethically problematic about raising them for food.

In this paper, I will start from the assumption that (at least some) insects possess some degree of sentience and I will offer several reasons why insect sentience might not matter very much. The terms in italics mark two important qualifications. Firstly, due to the limited space available, in this paper I can only offer some tentative reasons to remain sceptical regarding the ethical relevance of insect sentience. I do not intend to settle the matter, and it is entirely possible that the reasons offered are trumped by other philosophical considerations or by the appearance of new empirical evidence. My only aim is to put forward some reasons why the discovery of insect sentience would not warrant jumping to conclusions regarding the moral status of insects. At the same time, however, I do not want to imply that insect sentience would not matter *tout court* and that we would be allowed to do with insects as we please. Rather, I want to argue that it might not matter very much, that is, that insect sentience, while not entirely insignificant, should probably not be at the top of our moral worries.

I will offer four reasons to support this claim. First, I will argue that insect sentience would not automatically imply a right to life. Second, I will offer some reasons to believe that insect pain may matter less than that of other animals, who would then have a stronger claim to moral protection. Third, I will argue that a commitment not to harm other types of animals, such as mammals, may actually imply an obligation to consume insects. Lastly, I will offer some considerations that speak against the implementation of welfare measures in insect rearing.

EurSafe 2018 – DOI 10.3920/978-90-8686-869-8_59, © Wageningen Academic Publishers 2018

Would insect sentience imply a right to life?

Discussing the moral status of insects means debating the sort of treatment they are owed. With respect to some of the animal species that have traditionally been raised for food, certain theorists and activists argue that they have an interest in continued life that must be respected, which means they cannot be legitimately raised and slaughtered for food (e.g. Regan 2004, Francione 2009). Assuming that insects are sentient, we need to ask whether it is plausible to consider that they too have an interest in continued life that can ground a right to life.

From a practical perspective, we can note, following Fischer (2016b), that ascribing a right to life to insects would have many undesirable consequences. It might imply, for instance, that we have a moral obligation not to drive our cars for non-essential purposes, since a huge number of insects are killed whenever we go on a car trip. We might also have to stop using insecticides, refrain from stepping or sitting on the grass, and so on. Of course, slave-owners also had to face a radical change in their lifestyles when slavery was abolished, so the fact that it would make our lives less comfortable is not, in and of itself, enough to conclude that insects do not have a right to life.

A second consideration that can be offered is not practical, but theoretical. An interest in continued life, and the corresponding right to life, only makes sense in those individuals that can be harmed by death. Currently, the issue of whether and in what sense death harms nonhuman animals is a hotly debated topic with no signs of resolution in the near future (Višak and Garner, 2015; Meijboom and Stassen, 2015). There seems to be a widespread consensus, however, that the mere capacity to feel pain is not enough to ground an interest in continued life. Most scholars would argue that, at the very least, a being needs the capacity to experience pleasure and joy in order for death to constitute a harm (by depriving the individual of future positive experiences). While we tend to think of pain and pleasure as two sides of the same coin, and of sentience as a pack that necessarily includes both, in principle it is possible for insects to be capable of feeling pain, but not of feeling pleasure, which would mean that most ethicists would regard the painless death of an insect as harmless.

Even if insects are capable of experiencing pleasure and enjoyment, a significant number of scholars would argue that this is not enough for death in itself to be harmful to them. It has been postulated that for death to be harmful the individual in question needs to have a psychological unity (McMahan, 2002), a concept of death (Cigman, 1981), the capacity to form plans for the future (Belshaw, 2015), the capacity to hold a preference to go on living (Singer, 2011), and so on. While insects may have the ability to feel pain, it is questionable that they will fulfil these further conditions for death to be harmful to them. We are far from being able to rigorously conclude that insects lack a right to life, because the criteria listed have not been defended and the capabilities of insects are merely being assumed, but these considerations are enough to make my point: given the substantial philosophical disagreement on what grants a right to life, the discovery that insects are sentient would not immediately secure them a right not to be raised and slaughtered for food.

Does insect pain matter as much as the pain of other animals?

There is a very big difference between the size of insect and mammalian brains. The brain of the honeybee, which is notoriously large for an insect, possesses less than one million neurons. In comparison, the brains of mice (6.8 million neurons), rhesus monkeys (6.4 billion neurons), and humans (86 billion neurons) are enormous (Klein and Barron, 2016). Some authors have used this difference to argue that insect brains may simply not be big enough to support sentience (Feinberg and Mallat, 2016), while others are confident that number of neurons is not so important, since the structures relevant for sentience may be implemented on vastly different scales (Merker, 2016). Assuming that insect brains do support sentience, their smaller size might mean that their experiences are less fine-grained, less complex, than those of animals with much bigger brains, like mammals. This idea is supported by the fact that insects are known to use their limbs as usual when they are damaged and to continue feeding or mating despite severe injuries or even while being consumed by another insect (Eisemann *et al.*, 1984). If they do feel pain, these considerations suggest that it may somehow 'feel less painful' than it does to other animals, such as mammals. This argument presupposes that the 'painfulness' of pain can be measured by the reactions of an animal being similar to our own, which can of course be brought into question. Nevertheless, these reactions in insects do suggest that the pain they feel is less incapacitating than in the case of mammals, since for the latter pain comes with an inability to use the affected body parts, a loss of appetite, etc., which in turn diminishes the range of possibilities for thriving or pursuing enjoyment. If pain for insects is indeed less incapacitating, we could conclude that it is less harmful, if we are willing to accept the claim that the blighting of one's capacities is a form of harm. Given the difficulties involved in rearing and slaughtering conventional livestock without inflicting significant amounts of pain and suffering (cf. Singer 2009: 145ff.), the fact that pain may feel less painful or be less harmful to insects would speak in favour of consuming the latter rather than the former.

A reason that might speak against the consumption of insects would be the numbers involved. Indeed, we would need a huge number of, say, crickets to provide us with the same nourishment as one pig. So, it might be preferable to raise and slaughter one pig rather than a thousand crickets. Scholars of certain philosophical traditions, such as utilitarianism, may find this a powerful argument. However, it is also important to note that the numbers involved are only relevant within certain normative frameworks. For some scholars of the deontological tradition, for instance, numbers do not matter. Taurek (1977) famously argued that faced with the choice of killing one individual or killing several, one should simply flip a coin, for each particular individual whose life is at stake would be equally concerned about her own fate, and all should be given an equal chance to survive.

Admittedly, not everyone will find Taurek's argument convincing enough to allow us to kill one thousand individuals with a clear conscience when we could instead kill just one. So to the idea that numbers only matter within some normative frameworks we should add that the harm inflicted on a pig that is reared for food in a factory farm might not be comparable to that inflicted on a cricket under the same circumstances. This is because pigs are highly intelligent animals with very complex ethological needs. Studies performed on pigs suggest that they are not only sentient, but that they also possess a wide range of emotions, personalities, self-awareness, long-term memory, the ability to engage in perspective-taking and complex social relationships (see Marino and Colvin, 2015 for a comprehensive review). Following Nussbaum (2004: 309), we can note that '[m]ore complex forms of life have more and more complex capabilities to be blighted, so they can suffer more and different types of harm.' Raising pigs for food does not just harm them because of the pain and suffering it inflicts on them, but also because they have, in Nussbaum's terminology, many other capabilities that are thwarted when they are treated this way. This might mean that when we raise and kill a pig for food we are generating an amount of harm that cannot be compared on a 1:1 basis with the harm involved in raising and killing a cricket for food. If this were true, it is possible that cognitively complex animals, such as pigs, would have to be placed higher on the moral hierarchy than insects, and their interests favoured in cases of conflict, regardless of the numbers involved.

Are we morally obliged to eat insects?

A possible question that emerges at this point is why we should eat insects at all. Why not opt for a purely plant-based diet instead, and ensure that no suffering is involved in our food production? While this may indeed be the best option (although see Calvo, 2017 for a defence of the possibility of plant sentience), a case can also be made that incorporating some insects into our diet is actually preferable than a purely plant-based one. Meyers (2013) has made this claim on environmental grounds, by emphasising the fact that insects, in contrast to protein-rich plants like soybeans, have a much bigger capacity to recycle our waste, since they can feed on things such as cardboard or industrial refuse. So, complementing our vegan diets with some insects may actually result in a better output for the environment.

More relevant for the present discussion is the case made by Fischer (2016a). He has argued that it's not just good to eat insects, but that we actually have a moral obligation to do so. He bases his argument on a precautionary principle that establishes that we should treat animals that are perhaps conscious (a group where he locates insects) as though they are definitely conscious only in those cases where it wouldn't prevent us from fulfilling any obligation towards beings that are definitely or probably conscious. He points out that the techniques used for planting and harvesting fruits and vegetables routinely harm or kill animals that happen to be in the field and that are much more likely to be conscious than insects: animals such as mice or rabbits. Since a harm-free diet is not possible, he argues that we have an obligation to opt for the diet that minimises the harm inflicted on animals that are probably or definitely conscious.

We are departing from the assumption that insects are definitely conscious, so Fischer's precautionary principle would not require us to consume insects. However, if we are right to suppose that pain may be less painful or less harmful to insects, then we could say that combining a vegan diet with the occasional consumption of insects might actually result in less harm being delivered to animals whose pain matters in a full-blown sense than a strictly vegan diet. If we were to favour a hierarchical account of moral status, where the interest of mammals in not being harmed ranks higher than the interest of insects in not being raised for food, then we might have a moral obligation to consume some insects and ensure that the collateral damage of our farming techniques is minimised. On the other hand, not favouring such a hierarchical account could have highly problematic consequences wherever there are conflicts of interests. It might mean, for instance, that we would not be allowed to treat our dog when he has fleas, or our child when she has lice, since the interests of these insects in not being harmed would presumably rank higher than the dog or the child's interest in being itch-free.

Should we adopt welfare measures in insect farming?

Assuming that we are morally obliged to eat insects, their capacity to feel pain might nevertheless mean that we are required to adopt welfare measures to ensure that they suffer as little as possible during the production process. In principle, this seems like a sound solution. However, it is important to bear in mind that these welfare measures will not come for free. As De Goede *et al.* (2013) point out, '[i]nsect biodiversity is too large to generalize upon welfare standards,' and there are currently over 2,000 species of insects that are being eaten by humans (Röcklinsberg *et al.*, 2017). This means that a large amount of money will have to be invested into studying the welfare of each of these species. Implementing the resulting welfare measures will also likely come with significant costs. We thus have to determine, not only what level of protection we want to grant these animals, but also the price we are willing to pay in exchange for implementing protection measures (Knutsson and Munthe, 2017). Realistically speaking, and seeing the difficulties we are already facing in getting consumers to care about the welfare of animals that are much more similar to us than insects (such as those that make up conventional livestock), the price humans will be willing to pay to adopt welfare measures for insects will probably not be very high. Meyers (2013), however, has offered a reason why this might not be too much of a problem. He has pointed out that the characteristics of industrial farming that are so detrimental to the welfare of pigs, chickens, cattle, and so on may not cause insects that much suffering, due to the fact that they actually tend to prefer over-crowded, hot, dirty, and unlit spaces. One cannot generalise this point to all insects, precisely due to the aforementioned biodiversity that characterises them, so I wouldn't go as far as asserting that 'there is no reason to worry about the suffering problem for insect farming' (Meyers, 2013: 124). However, if Meyers is partially right and at least some insect species prefer these living conditions, then perhaps an alternative to demanding the implementation of welfare measures would be to encourage the farming of these species.

Conclusion

If we were to establish that insects can feel pain, it would be a huge scientific discovery. What might follow in ethical terms, however, is less clear. In this paper I have offered some tentative reasons why insect sentience might not matter too much from an ethical perspective. I have argued that, at least with respect to edible insects, learning that they are sentient might not have drastic implications for the sort of treatment they are owed. Regardless of whether these considerations are successful, it is important to reiterate that I began from the assumption that insects are indeed conscious, which is far from proven. If we factor in the uncertainty surrounding the issue of insect sentience, things get even muddier. For this reason, I would be inclined to assert that we should focus our efforts on protecting those animals who are known to not only feel pain, but also have complex subjective experiences and ethological needs, and who are currently suffering immeasurably at the mercy of our food production systems. However, I acknowledge that this may be just a vertebrate-centric bias.

References

Barron, A.B. and Klein, C. (2016). What insects can tell us about the origins of consciousness. Proceedings of the National Academy of Sciences 113 (18):4900-4908.

Bateson, M., Desire, S., Gartside, S.E. and Wright, G.A. (2011). Agitated honeybees exhibit pessimistic cognitive biases. Current Biology 21 (12):1070-73.

Belshaw, C. (2015). Death, pain, and animal life. In Višak, T. and Garner, R. (eds.) The ethics of killing animals. Oxford University Press, New York, USA, pp. 32-50.

Calvo, P. (2017). What is it like to be a plant? Journal of Consciousness Studies, 24 (9-10):205-227.

Cigman, R. (1981). Death, misfortune and species inequality. Philosophy and Public Affairs 10 (1):47-64.

De Goede, D.M., Erens, J., Kapsomenou, E. and Peters, M. (2013). Large scale insect rearing and animal welfare. In The ethics of consumption. Wageningen Academic Publishers, Wageningen, The Netherlands, pp. 236-42.

FAO (2017). Insects for food and feed. Available at: http://www.fao.org/edible-insects/en. Accessed 22 March 2018.

Feinberg, T.E. and Mallat, J.M. (2016). The ancient origins of consciousness: How the brain created experience. The MIT Press, Cambridge, USA.

Fischer, B. (2016a). Bugging the strict vegan. Journal of Agricultural and Environmental Ethics 29 (2):255-63.

Fischer, B. (2016b). What if Klein & Barron are right about insect sentience? Animal Sentience: An Interdisciplinary Journal on Animal Feeling 1 (9). Available at: http://animalstudiesrepository.org/animsent/vol1/iss9/8. Accessed 22 March 2018.

Francione, G. (2009). Animals as persons: Essays on the abolition of animal exploitation. Columbia University Press, New York, USA.

Halloran, A. and Vantomme, P. (2013). The contribution of insects to food security, livelihoods and the environment. Available at: http://www.fao.org/docrep/018/i3264e/i3264e00.pdf. Accessed 22 March 2018.

Klein, C. and Barron, A. (2016). Insects have the capacity for subjective experience. Animal Sentience: An Interdisciplinary Journal on Animal Feeling 1 (9). Available at: http://animalstudiesrepository.org/animsent/vol1/iss9/1. Accessed 22 March 2018.

Knutsson, S. and Munthe, C. 2017. A virtue of precaution regarding the moral status of animals with uncertain sentience. Journal of Agricultural and Environmental Ethics, 30 (2):213-224.

Marino, L. and Colvin, C.M. (2015). Thinking pigs: A comparative review of cognition, emotion, and personality in Sus domesticus. International Journal of Comparative Psychology, 28(1). Available at: https://escholarship.org/uc/item/8sx4s79c. Accessed 22 March 2018.

McMahan, J. (2002). The ethics of killing: Problems at the margins of life. Oxford University Press, New York, USA.

Meijboom, F.L.B. and Stassen, E.S. (eds.). (2015). The end of animal life: A start for ethical debate. Wageningen Academic Publishers, Wageningen, The Netherlands.
Merker, B. (2016). Insects join the consciousness fray. Animal Sentience: An Interdisciplinary Journal on Animal Feeling 1 (9). Available at: http://animalstudiesrepository.org/animsent/vol1/iss9/4. Accessed 22 March 2018.
Meyers, C.D. (2013). Why it is morally good to eat (certain kinds of) meat: The case for entomophagy. Southwest Philosophy Review 29 (1):119-26.
Nussbaum, M.C. (2004). Beyond 'compassion and humanity': Justice for nonhuman animals. In Sunstein, C.S. and Nussbaum, M.C. (eds.), Animal rights: Current debates and new directions, Oxford University Press, New York, USA, pp. 299-320.
Regan, T. 2004. The case for animal rights (updated with a new preface edition). University of California Press, Berkeley, USA.
Röcklinsberg, H., Gamborg, C. and Gjerris, M. (2017). Ethical issues in insect production. In van Huis, A. and Tomberlin, J.K. (eds.) Insects as food and feed: From production to consumption, Wageningen Academic Publishes, Wageningen, The Netherlands, pp. 364-79.
Singer, P. (2009). Animal liberation: The definitive classic of the animal movement. Reissue edition. Harper Perennial Modern Classics, New York, USA.
Singer, P. (2011). Practical ethics, third edition. Cambridge University Press, New York, USA.
Taurek, J.M. (1977). Should the numbers count? Philosophy and Public Affairs 6 (4):293-316.
Višak, T. and Garner, R. (eds.) (2015). The ethics of killing animals. Oxford University Press, New York, USA.
Yarali, A., Niewalda, T., Chen, Y., Tanimoto, H., Duerrnagel, S. and Gerber, B. (2008). 'Pain relief' learning in fruit flies. Animal Behaviour 76 (4):1173-85.

60. Animal protection vs species conservation: can the relational approach solve the conundrum?

B. Kliesspiess and H. Grimm*
Unit of Ethics and Human-Animal-Studies, Messerli Research Institute, Veterinaerplatz 1, 1210 Vienna, Austria; 1001090@students.vetmeduni.ac.at

Abstract

The management of invasive alien species gives rise to a moral conflict between animal ethicists, who generally take an individual-based approach to animal protection, and environmental ethicists, who generally hold species- and ecosystem-based accounts. Management programs that require the killing of sentient animals to save a species have been the subject of extensive criticism in animal ethicists. In contrast, environmental ethicists who attribute value to species and their conservation normally argue that eradication programs are legitimized if they contribute to saving a species. So far, this moral conflict between animal ethics and species conservation has been insurmountable, although many philosophers have tried to reconcile the two approaches. Palmer argues for a relational approach that addresses the moral status of wild animals. Unlike most classical individualistic views, Palmer bases her claims not solely on the animals' capacities, but also on the relations between animals and humans. The potential of her account regarding this issue will be analysed in the context of the management of alien sentient animals in New Zealand. We will clarify whether her relational approach can be used to minimize the conflict between environmental and animal ethics positions, that is, whether it can offer a way to preserve a species through the protection of individual animals, without attaching intrinsic value to the species itself. This analysis of Palmer's view will make it obvious that, in situations in which the rights of two sentient animals are in conflict, negative rights outweigh positive relation-based rights, resulting in a relapse into a classical animal ethics approach. We will therefore conclude that the potential of the relational approach to minimize the conflict is limited and that it cannot overcome the gap between the diverging values and goals of environmental ethics and animal ethics.

Keywords: alien species management, value conflict, positive and negative duties

Introduction

The desire to care about the environment and at the same time care about sentient animals is common in society. A person who abolishes the use of animals for food, clothing or entertainment might be also in favour of reforesting a rainforest or in keeping the ocean free from plastic and toxic substances. There might be even the same enemies detected that pollute the environment (Jamieson, 1998: 42), pose a threat to species in various ways and inflict major animal suffering. In such cases, animal liberationists and environmentalists often share common thoughts about how to solve those problems (Paquet and Darimont, 2010: 178). But in other cases, their shared view falls apart. The striving to conserve values like '[...] the integrity, stability, and beauty of the biotic community [...]' (Leopold, 1968: 224-225) is therefore a valid reason for sacrificing individuals if this practice contributes to the conservation of an intact environment. When sentient animals become invasive and have negative effects on the environment and other species, environmentalists would ordinarily support the reduction of those animal populations. This moral stance is quite contrary to individualistic accounts in animal ethics. Classical animal ethics approaches focus on domesticated animals and mostly advocate the argument from ignorance claiming that if we don't know the exact consequences regarding actions of wildlife management, humans shall not intervene e.g. Peter Singer (Singer, 1983: 251). For this reason, Palmer's approach regarding wild animals is a suitable animal ethics position to examine the possibility of

EurSafe 2018 – DOI 10.3920/978-90-8686-869-8_60, © Wageningen Academic Publishers 2018

individual-based species conservation, as she claims for obligation to 'assist' individual sentient wild animals. This moral community also includes sentient animals which are members of endangered species, like Palmer mentioned in her polar bear example, regarding the obligation to assist them due to their suffering from human-induced climate change and poaching (Palmer, 2010: 142). Environmental ethics deal with a comprehensive amount of interlinked aspects like biotic and abiotic environmental factors and the conservation of species including beside sentient and non-sentient animals also plants and fungi. For the purpose of a clear representation of the analysis of Palmer's approach, we will focus on one aspect of species conservation, namely the management of invasive alien species. As invasive alien species are one of the leading cause for species extinction (including animal suffering) (Clavero and García-Berthou, 2005) it is also crucial to investigate this topic from an animal ethics perspective, that values sentient animals such as mammals and birds, like Palmer does.

Even if there is a shift to more humane methods in wildlife management (Dubois *et al.*, 2017; Marks, 1999), at least when dealing with invasive alien species[1], that has massively detrimental effects on other species and pushes them to the edge of extinction, the usually applied management method is the killing of this 'pest' (Bellingham *et al.*, 2010). A vivid example for such a case is the alien predator problematic on New Zealand. The native avifauna evolved without any predators and therefore shows very limited flight reaction. Invasive alien species reached the islands of New Zealand through assistance of humans. Since the introduction during the 19th century the numbers of the native animals rapidly decline as a result of the growing alien predator populations, which put major predation pressure particularly on birds (Clout, 2006; Clout and Merton, 1998). The need of an effective management measure in order to minimize the invasive alien animal population became an urgent concern, therefore extensive eradication projects were implemented. The result of this 'pest' management measures, involving the killing of millions of (potential) predators (Bellingham *et al.*, 2010) like rats, cats and stoats, is for example the conservation and slight increase of native bird populations e.g. the kakapo (*Strigops habroptilus*) – currently 157 individuals are left (IUCN, 2016).

In view of the rapid progress of species loss (and the interlinked animal suffering) caused by habitat loss or the toxification of various habitats, the question arises: is there an all-embracing reconciling theory which allows both sentientistic and holistic values? Clare Palmer tried a step in the direction of such a potentially reconciling theory.

Relational approach

The book 'Animal ethics in Context' (2010) written by Clare Palmer is a context-sensitive animal ethics approach regarding the moral consideration of sentient wild animals. Palmer claims that her view can be accepted by both animal ethicists and environmentalists (Palmer, 2010: 166). This is a very promising claim, which shall be investigated further. In her ethical approach, Palmer includes capacities of non-human animals as well as relations to animals in certain situations and contexts. Negative obligations are based on capacities, while positive obligations can be derived from contexts and human relations (Palmer, 2010: 96).Bossert as well states that Palmer's approach could allow for holistic values (Bossert, 2015: 111), and hence establishes a basis for a reconciliation of the introduced ethical approaches. The statement that some positions familiar from the animal ethics literature could constitute a convincing baseline for environmental ethics (Bossert, 2015: 145) implies that Palmer has accomplished her aim of closing the gap between animal ethics and environmental ethics.

[1] Alien species (synonym: exotic, introduced or non-native species) are '[...] species that are foreign to an ecological assemblage in the sense that they have not significantly adopted with the biota constituting that assemblage or to the local abiotic conditions' (Hettinger, 2001: 193).

Consequentially, the hypothesis of this paper is the following: 'Because she presents a convincing theoretical framework for the moral consideration of wild animals, Clare Palmer's relational approach holds the potential to justify conserving a species by protecting individual sentient animals without directly valuing entities such as species, and thereby minimizes the conflict between environmental and animal ethics'.

Negative and positive duties

Palmer proposes a novel ethical approach to the consideration of wild sentient animals. Most animal ethics approaches focus on domesticated sentient animals and argue that humans should not intervene in the lives of wild animals. Palmer, on the contrary, argues that we indeed have the duty to intervene in the lives of wild animals, particularly and exclusively when humans are the reason for the animal suffering in nature. Sentient domesticated animals are always dependent on humans and therefore have the right to assistance in every situation, wild animals have the right to assistance only if humans make wild animals vulnerable or dependent, or if humans directly harm them. Negative duties are granted to all sentient animals equally; neither domesticated nor wild animal shall be harmed or killed without a powerful reason. Positive duties or in Palmer's words special obligations are generated after an animal is harmed, which must be implemented by the responsible person, the person who did or benefitted from the harm. In nearly all situations, negative duties weigh more, morally speaking, than positive ones. Some exceptions along the lines of Tom Regan's miniride principle (Regan, 2004: 305) are mentioned by Palmer. These cases involve overriding the right of a few individuals in order to protect the greater amount in cases where an intervention would minimize the human caused suffering of most individuals substantially.

Palmer's core aim is to establish a convincing theoretical conception, which permits assistance in certain wild animal-human encounters, namely those cases in which wild animals are harmed directly or indirectly by humans, which she calls causal relations[2], without undermining the LFI (laissez faire intuition). Derived from the claim that beings who possess rights (in the case of Regan, the subjects-of-a-life) deserve some kind of 'compensatory justice' after past rights infringement (Regan 2004: XI). Palmer claims that non-human animals who possess similar capacities to humans (e.g. socio-cognitive abilities) may also be able to raise a claim of compensatory justice (Palmer 2010: 100).

Application of Palmer's relation approach in an invasive animas species management case

In transposing the notion behind the relational approach into a practical example considering 'harmful' invasive alien animals, one could argue that both the alien and the endangered native species should be granted negative and positive rights. There are no values in Palmer's view that justify the conservation of a species only on the basis of their endangered status or being valued due to properties like native or alien. Palmer considers all animals that share the ability to suffer in her approach, which includes at least birds and mammals (Palmer, 2010: 4). As with other individualistic animal ethics approaches, Palmer too rejects the prima facie harm (eradication) of the alien animals.

Invasive alien species reach their new habitat by human assistance. That means humans are responsible for the detrimental effects alien predators pose to the native fauna. Human-caused harms can be detected: (1) towards native and endangered animals as they are indirectly harmed (by humans) due to the predation pressure of introduced species, and (2) toward the alien animal which in turn are pursued and killed to protect the native fauna. This constitutes a kind of vicious circle. Due to the harm induced

[2] Palmer defines a 'causal relation' as '[...] cases where human beings have caused, or partially caused, animals to be in a particular situation and contexts in which they are' (Palmer, 2010: 54).

either to the endangered sentient animals or the sentient invasive alien species, special obligations toward both animal groups are generated.

Killing in Palmer's approach is a permitted option just in case most of the animals could be saved from the harm constituted by humans (Palmer, 2010: 146-148). In the invasive alien animal example, if more animals must be killed to save the endangered (far less in number) animals, killing would not be a reasonable justification, following Palmer's approach to 'save' the individuals of the endangered species.

Conclusion

To conclude, Palmer's relational approach cannot overcome the gap between the diverging values of species- and environmental-orientated wildlife management and animal ethics when killing sentient (alien) animals is the 'prerequisite' to save a species. In other words, the promising hypothesis is false at least in its generality and regarding the specific treatment of sentient animals. Of course, Palmer is limited from an environmental ethics point of view in not considering animals that lack evidence of being sentient and abiotic factors, which are irrelevant considering the wellbeing of sentient animals. Beside this fundamental-value-based conflict, there are moral problems when the integrity of a sentient animals is impaired by another sentient animal. Wildlife management per se is not contradicting Palmer's approach, as there are many cases, which do not entail sentient animal suffering or killing. Following Palmer's approach, we have obligations towards the native sentient animals to assist them to survive and not getting harmed, as humans are responsible for the introduction of alien species. Although Palmer allows the killing of animals as a form of special obligation, if there are a minimal number of sentient animals harmed, the killing of millions of sentient alien predators in order to save much less native sentient animals from being preyed on is not justified. On the other side humans have also special obligations towards alien sentient predators, as humans are responsible for their situation of being dependent on prey that is considered as endangered and native by humans. A moral agent is not able to follow those contradicting obligations. As negative duties weigh more, if it comes to killing some animals to assist others, Palmer's theory collapses into a classical animal ethics view and special obligations become irrelevant.

References

Bellingham, P.J., Towns, D.R., Cameron, E.K., Davis, J.J., Wardle, D.A., Wilmshurst, J.M., and Mulder, C.P.H. (2010). New Zealand island restoration: seabirds, predators, and the importance of history. New Zealand Journal of Ecology, 34(1), 115-136.

Bossert, L. (2015). Wildtierethik: Verpflichtunge gegeüber wildlebenden Tieren (1st ed.). Baden-Baden: Nomos.

Clavero, M., and García-Berthou, E. (2005). Invasive species are a leading cause of animal extinctions. Trends in Ecology and Evolution, 20(3), 110. http://doi.org/10.1016/j.tree.2005.01.003.

Clout, M.N. (2006). A celebration of kakapo: progress in the conservation of an enigmatic parrot, 53(1), 1-2.

Clout, M.N., and Merton, D.V. (1998). Saving the Kakapo: the conservation of the world's most peculiar parrot. BirdLife International. http://doi.org/10.1017/S0959270900001933.

Dubois, S., Fenwick, N., Ryan, E.A., Baker, L., Baker, S.E., Beausoleil, N.J., ... Fraser, D. (2017). International consensus principles for ethical wildlife control, 31(4), 753-760. http://doi.org/10.1111/cobi.12896.

Hettinger, N. (2001). Exotic species, naturalisation, and biological nativism. Environmental Research Letters, 10(2), 193-224. http://doi.org/10.3197/096327101129340804.

IUCN. (2016). The IUCN Red List of Threatened Species.

Jamieson, D. (1998). Animal Liberation is an Environmental Ethic. Environental Values, 7(1), 41-57.

Leopold, A. (1968). A Sand County Almanac: And Sketches Here and There (1st ed.). Oxford University Press.

Marks, C.A. (1999). Ethical issues in vertebrate pest management: can we balance the welfare of individuals and ecosystems?

Palmer, C. (2010). Animal Ethics in Context. New York: Columbia University Press.

Paquet, P., and Darimont, C. (2010). Wildlife conservation and animal welfare: two sides of the same coin? Animal Welfare, 19(2), 177-190.
Regan, T. (2004). The Case for Animal Rights (3rd ed.). Los Angeles: University of California Press.
Singer, P. (1983). Animal liberation. Wellingborough: Thorsons.

61. Ethical dilemmas of fertility control in wildlife – the case of white-tailed deer

C. Gamborg[1*], *P. Sandøe*[1,2] *and C. Palmer*[3]
[1]*Dept. of Food and Resource Economics, University of Copenhagen, Rolighedsvej 25, 1958 Frederiksberg C, Denmark;* [2]*Dept. of Veterinary and Animal Sciences, University of Copenhagen, Grønnegårdsvej 8, 1870 Frederiksberg C, Denmark;* [3]*Dept. of Philosophy, Texas A&M University 4237 TAMU College Station, TX 77843, USA; chg@ifro.ku.dk*

Abstract

This paper explores ethical issues raised by the use of non-surgical, pharmaceutical fertility control to manage reproduction of white-tailed deer (WTD). A high density of WTD, especially in suburban areas, has led to human-deer conflicts, conflicts traditionally solved by hunting, Recently, however, there has been a push towards non-lethal control, especially fertility control. While the scientific and technical aspects are beginning to be well understood, the ethical issues raised require further exploration. The paper begins by discussing the challenges of high-density WTD populations, and the possibility of using fertility control as a response to these. Then the paper identifies major ethical issues raised, as viewed from the perspectives of animal rights, utilitarianism and concern for wildness. Our conclusion is that changes in human behaviour, rather than pharmaceutical fertility control to reduce deer populations, seem to ethically preferable from all three perspectives. However, it is less clear how pharmaceutical population control compares with hunting in ethical terms.

Keywords: non-lethal control, rights, welfare, wildlife management, wildness

Introduction: development of human/deer conflicts

The control of white tailed deer (WTD) is a vexing question for wildlife managers and professionals across the Eastern US. Effective control is elusive, and there are heated public debates about available strategies. Ethical analysis of human-deer conflict may assist in understanding these debates and the reasons for their apparent intractability, help to see possible resolutions more clearly, and cast light on possible management issues relating to other problematic wildlife species.

WTD are native to the Americas. Historically abundant, their numbers had declined significantly by the early 20th century. As important game animals, population growth was encouraged by reintroductions of WTD, imposing bag limits on hunters, and habitat management by creating forest clearings. These measures coincided with a general growth in habitat that, as ecological generalists, WTD could exploit: leafy suburbs and patchy reforestation of formerly agricultural land, with relatively sparse numbers of predators. However, as WTD populations expanded, so did concerns about their presence, especially in urban areas. Deer-motor accidents were a major concern, leading to 150-200 human deaths a year. A questionnaire survey in upstate New York showed that while 91% of respondents 'enjoyed deer to some extent', 54% of respondents were also worried about them. 83% mentioned deer-car accidents, 57% Lyme disease, 26% damage to yard plantings, 24% damage to garden vegetables and 16% damage to farm crops (Stout *et al.*, 1993). There were also worries about effects of deer browse on ecosystems. These factors, taken together, led to popular calls across the eastern USA to manage and control growing populations of WTD.

EurSafe 2018 – DOI 10.3920/978-90-8686-869-8_61,

Possible options for addressing human/deer conflicts

The three main options for addressing problems caused by WTD populations are (1) attitudinal and behavioural changes in humans: trying to keep deer and humans apart, adopting risk-avoiding behaviour, or trying to change attitudes to deer; and/or (2) actual changes in the deer population by means of lethal control (e.g. culling/sport hunting); or (3) changes by non-lethal control, primarily pharmaceutical methods of fertility control.

Option (1) has significant difficulties. Deer fencing may be effective, but at considerable aesthetic, practical and economic cost. Changing attitudes to WTD by increasing human tolerance for co-operative living, and changing behaviours to make human environments less attractive to deer have been difficult to achieve on a large enough scale to reduce perceptions of conflict. Option (2) – primarily in the form of game hunting – has been the primary way, historically, of managing burgeoning deer populations. This does reduce populations, but game hunter goals don't necessarily align with the goal of population control. Game hunters prefer to hunt antlered bucks to does; and since one buck can service many does, killing bucks is unlikely effectively to reduce population size. Even targeted culls by professional wildlife managers for the purpose of population management are not always the solution. First, in some suburban areas, human population density is such that hunting is either not logistically possible, or not safe. And second, there's increasing ethical resistance to killing healthy sentient mammals in order to control populations. So, neither sport hunting nor targeted culls alone are likely to resolve WTD population problems.

These difficulties have increased interest in option (3), non-lethal population reduction through fertility control. One possible solution is surgical sterilisation, but there has been concern about lack of evidence of efficacy in population reduction (Boulanger and Curtis, 2016); and the necessary involvement of professional veterinarians makes this expensive. Pharmaceutical contraception, then, offers alternatives to culling and surgical sterilization. There are two main methods: PZP and GonaCon. PZP is licensed in many states in the US and can be used for deer population control. In female deer, it prevents egg fertilization; it does not otherwise change mating behaviour. However, since female deer continue to have fertility cycles until they become pregnant, it does have the effect of causing additional cycling and thereby lengthens the mating season. Recent studies suggest that PZP can be administered in timed-release pellets that mean deer can go several years without boosters. It can be hand injected or remote darts may be used (Rutberg *et al.*, 2013). GonaCon is a form of vaccine that was recently registered for use on free-ranging deer herds by the US Environmental Protection Agency (EPA). A single dose can make a doe infertile for a minimum of one year and up to four years (Miller *et al.*, 2008). Currently it may only be hand-administered (though not necessarily by a veterinarian) since there's a risk of humans or other animals being stabbed with stray darts and becoming infertile. GonaCon represses all mating behaviour; and after administration, female deer do not undergo fertility cycles.

Both PZP and GonaCon seem to have reasonably predictable and consistent effects. In principle, both are reversible, though PZP seems to cause permanent infertility after about 5 years of use. However, administering pharmaceutical fertility control to a population of wild animals faces practical challenges. First, animals have to be found and to be placed in a position where contraception can be applied, especially difficult where application must be by hand. Second, animals need to be tagged to avoid multiple applications. Third, the application has to target a very high proportion of the female population – around 90% – to be effective in population reduction. Fertility control in WTD also raises ethical concerns.

Ethical issues raised by fertility control

Pharmaceutical contraception as here described, has not yet been much used in non-human animals. Recently, it has been used in zoo animals and among companion and farm animals – the latter two mostly with the aim of changing unwanted behaviours in male animals (such as horses, dogs and pigs) – see Palmer *et al.* (2018). Therefore, the case of WTD differs by having a focus on population control in free-ranging wildlife. In addressing the ethical issues raised by this case, mainly from the perspective of animal ethics, we'll first consider *rights* arguments. Then we'll consider the possible welfare consequences of using fertility control in WTD, taking a broadly utilitarian perspective. Discussing welfare consequences, though, requires considering fertility control in relation to other options for managing deer-human conflicts: hunting/culling and intervention in human attitudes and behaviour. Finally, we'll consider one further value that may be at stake with WTD rather than in domesticated populations: wildness.

It's possible to extrapolate from some existing animal rights views that using fertility control in WTD would be problematic. Regan (1984), for instance, argues that individual animals that are 'subject of a life', that is, who have the appropriate kind of psychophysical unity over time, should not be treated as mere instruments; they should not be reduced to means to benefit others, human or animal. Clearly fertility control in WTD is not meant to benefit individual deer; it is mainly intended to benefit humans, who think that there are too many deer. Even if fertility control were being used to reduce population size with the goal of making the population as a whole healthier (by increasing the resource base) it's unlikely that Regan would consider this sufficient justification, unless there was clear evidence that the individual deer to whom birth control was administered would be made better-off. And even if this applies, it's not part of Regan's position that we should improve wild animal welfare; wild animals have rights not to be harmed, but not rights to be assisted. Thus Regan would probably consider fertility control unacceptable, as even if it is not harmful, it still uses individual deer for the benefit of humans.

However, it may also be argued that these forms of fertility control can be seen as harmful, though the harms vary in relation to different forms of fertility control. First, their application may be seen as harmful; less so in the case of PZP delivered by a dart than PZP or GonaCon injected by hand, which requires capturing and restraining deer. Second, the effects of the treatment may be harmful. Some harms relate to individual deer subjective experiences, in particular with PZP, where female deer continue to cycle, and male deer continue to mate with them (Miller *et al.*, 2001). The extended rutting season (an extra two months) may be exhausting for male and female deer, and – presumably – may produce negative experiences. But harms may also concern deprivation of positive experiences. If female deer receive GonaCon they do not cycle at all; there is no mating behaviour (except for individuals that are not vaccinated). So, deer may be harmed by deprivation of rewarding mating experiences (assuming it is rewarding); while female deer lose plausibly pleasurable experiences from tending to offspring. Further, on some views, harms from fertility control may not concern subjective experience at all, but rather the loss of 'natural behaviours,' independently of whether those behaviours are experienced. Do such harms make the use of fertility control in WTD impermissible from rights perspectives less restrictive in terms of instrumentality than Regan's? Boonin (2003) argues that surgically sterilizing animals would be consistent with a rights view if it imposed 'relatively minor harms on animals (and relevantly analogous humans) ...where this produces great benefits for others, and that is not only consistent with the attribution of rights to animals, but is motivated by the same sorts of considerations that justify such attribution'. Boonin's main concern about surgical sterilization is the invasive procedure itself, rather than the experiences or natural behaviours foregone. So in this respect, fertility control looks better, and may be permissible. But on the other hand, the motivation for WTD fertility control is human benefit, and it's unlikely that these benefits could be called 'great'. So, although the procedure is less invasive, benefits are also less urgent; it's therefore unclear whether Boonin's account would permit fertility control.

Donaldson and Kymlicka (2011) have developed another version of animal rights theory, including wide-ranging negative and positive rights. They argue that sentient animals have a right not to be coercively sterilized. However, pharmaceutical fertility control in companion animals is 'relatively non-invasive,' and may, they suggest, be permissible once animals have had the opportunity to have one family. Obviously, though, it would be near impossible to target fertility control on free-ranging female deer that have reproduced once; and in any case, for Donaldson and Kymlicka, there's likely to be insufficient justification for acting at all to limit deer populations. After all, we can live alongside them, even if this requires behavioural change (as option (1) above). It isn't reasonable, on this view, for WTD to give up reproductive rights so that (for instance) we can enjoy our gardens more.

We don't want to say that no approach to animal rights could accept pharmaceutical fertility control in WTD; but many existing versions, at least, will be highly sceptical that it could be compatible with individual deer's rights. However, it's worth noting here, though, that if forced to choose, fertility control would clearly be preferred to culling. The right not to be killed is the primary right extended to animals in almost all animal rights theory; fertility control would be the lesser of two evils.

We'll now consider a utilitarian approach to fertility control in WTD. Different utilitarian approaches to animal ethics exist; while all aim at bringing about best consequences, they can have direct and indirect forms, and may adopt different ideas of the good. Here we'll take the good to be welfare. However, welfare can be interpreted in different ways (Fraser, 2008). We will focus on welfare understood both as subjective experience and as performing natural behaviour. Since we're considering best consequences overall, this utilitarian approach (unlike the rights approach) requires us to consider the use of pharmaceutical fertility control alongside the other two most plausible approaches: hunting/culling and human behavioural change.

As already noted, pharmaceutical fertility control raises welfare concerns for the deer themselves. This includes the application of the treatment; negative subjective experiences from long-lasting fertility cycling (using PZP); deprivation of potentially positive experiences from mating and reproduction; and changes in natural behaviours. Alongside this, though, when considering best outcomes, we also need to take into account negative experiences that might be averted, such as from pregnancy, birth and lactation in wild deer, which are draining and can be risky. A welfarist, utilitarian view would also need to take into account other factors. First, fertility control means that many deer will not come into existence that otherwise would have, probably meaning fewer net positive experiences and natural behaviours. Second, the absence of mating (with GonaCon) and perhaps the persistence of mating (with PZP) will result in changes in relations between male and female deer, perhaps with welfare consequences. Third, precisely because young deer are not born, deer population structures are likely to be weighted towards older rather than younger deer, which may also have welfare impacts. However, assuming that the fertility treatment succeeds in reducing populations, the reduction in numbers of WTD is likely to result in an improvement in human welfare.

Lethal control has different impacts on welfare than fertility control. Obviously, there's no need to capture and restrain deer for treatment, and deer can continue mating and reproducing as normal. Assuming the balance of experience from these activities is positive, there's no loss of positive reproductive experience. In addition, there's no necessary impact on the structure of deer populations, especially if hunting bags are set to ensure this is the case. In all these ways, the outcomes from hunting look better than fertility control. But, of course, hunting potentially has other negative impacts. If hunted deer are not killed cleanly and quickly, there may be extremely negative subjective consequences for the deer concerned; worse than the effects of fertility control. Using professional sharpshooters makes clean kills more likely; but frequently control of deer population is left to game hunters who may not be so effective.

In addition, of course, hunting kills deer. This may or may not be regarded as a welfare issue (Kasperbauer and Sandøe, 2015), but it means that deer lives, and so their experiences/natural behaviours, are prevented from continuing. Hunting means more deer are born, but their lives are shorter because they are killed before they would die naturally. For most welfarist utilitarians, there are no grounds to distinguish between fewer longer lives and more shorter lives, provided that the deer experience that exists is of similar quality. So, if the effects of fertility control would reduce the quality of deer' lives, while leaving the number of deer in existence unchanged, then culling would be preferable. However, empirical uncertainties remain here, requiring considerable further research into the effects of fertility control on deer welfare. In practice, though, utilitarians may agree with rights theorists here: unless deer populations get so large that they suffer from malnutrition or starvation, the best outcomes overall are likely to follow from changes in human behaviour without hunting or fertility control.

Finally, there are concerns that the use of fertility control in wild populations may, in some way, undermine their wildness. This value doesn't fit comfortably into standard rights or utilitarian views, but many people nonetheless highly value the wildness of wild animals. However, wildness in this context may be interpreted in different ways; for instance as 'not tame' or as 'not domesticated' or as 'being free from human control' (see Palmer, 2016 for a systematic discussion). The key question here is whether pharmaceutical fertility control is likely to compromise wildness in one or more of these senses.

First, pharmaceutical fertility control is unlikely to make deer less wild in the sense that they will be tamer (less fearful of people) – perhaps the reverse. Second, it's unlikely to make deer less wild in the sense of being more domesticated, at least if domestication refers to selective breeding; fertility control may determine which deer actually come into existence, but it doesn't determine particular traits. However, if wildness is taken to mean something more like 'being free from human control in terms of performing natural behaviour' (resembling the 'natural behaviour' interpretation of welfare, above), clearly fertility control does affect wildness, since on both methods of fertility control, deer no longer perform some natural behaviours (in the case of GonaCon, this includes fertility cycles, rutting, mating and reproduction; while in the case of PZP, rutting and mating is 'unnaturally' extended, while reproduction is absent). Relatedly, a broader concern is that applying fertility control compromises wildness in the sense that it increases the pervasiveness of human deliberate intervention into the non-human world. Here fertility control may compare unfavourably with hunting. Hunting could be seen as an episodic intervention into deer communities, removing some adult deer, but otherwise allowing deer to continue to live and reproduce as deer normally do. Fertility control, on the other hand, exerts significant influence over deer lives all the time, not just episodically, including when the deer are nowhere near people. Even where deer have moved into human communities, and may be partially dependent on human provision, these are still actions determined by the deer themselves. Fertility control involves human intentional management of beings and behaviours from which human intention was previously absent. Compared to a population that is not managed at all, or managed only by hunting, then, the use of fertility control might be seen as a further loss of wildness.

Concluding discussion

This paper takes some first steps towards a systematic analysis of the ethical issues raised by different methods for managing high density WTD populations by focusing on pharmaceutical fertility control. WTD management methods are of high significance for wildlife professionals, veterinarians, farmers, hunters and residents. One conclusion here is that we lack information about some impacts of fertility control on deer welfare and population structures, and that further research is needed. We also suggest that stakeholders with different ethical concerns and deeply held values are likely to understand the nature of the problem very differently and to defend different ways of resolving it (resembling a classic 'wicked problem').

Our analysis, drawing on animal ethics, suggests that culling and pharmaceutical fertility control raise ethical concerns from both utilitarian and rights perspectives. From a rights perspective, pharmaceutical birth control is likely to be preferable to culling; but this judgment is much less clear from a utilitarian perspective; and fertility control probably compromises wildness values more than culling. From the ethical perspectives considered here – rights views, utilitarian views and views emphasizing the protection of wildness – changes in human behaviour, rather than lethal or non-lethal efforts to reduce deer populations, may well be ethically preferable either to pharmaceutical fertility control or to culling. We recommend, therefore, that research continues into human behavioural solutions.

References

Boonin, D. (2003). Robbing PETA to spay Paul: do animal rights include reproductive rights? Between The Species 13(3), 1.

Boulanger J.R., and Curtis, P.D. (2016). Efficacy of surgical sterilization for managing overabundant suburban white-tailed deer. Wildlife Society Bulletin 40: 727-735.

Donaldson, S. and Kymlicka, W. (2011.) Zoopolis: A political theory of animal rights. Oxford University Press, Oxford, UK, 338 pp.

Fraser, D. (2008). Understanding animal welfare. Acta Veterinaria Scandinavica, 50 (S1).

Kasperbauer, T. J., and Sandøe, P. (2016). Killing as a welfare issue. In Visak, T. and Garner, R. (eds.) The ethics of killing animals. Oxford University Press, Oxford, pp. 17-31.

Miller, L., Gionfriddo, J., Fagerstone, K., Rhyan, J., and Killian, G. (2008). The single-shot GnRH immunocontraceptive vaccine (GonaCon™) in white-tailed deer: Comparison of several GnRH preparations. American Journal of Reproductive Immunology, 60: 214-223.

Palmer, C. (2016). Climate change, ethics, and the wildness of wild animals. In Bovenverk, B. and Keulartz, J. (eds.) Animal ethics in the age of humans. Springer, Cham, pp. 131-150.

Palmer, C., Pedersen, H.G., and Sandøe, P. (2018). Beyond Castration and Culling: Should we use non-surgical, pharmacological methods to control the sexual behavior and reproduction of animals? Journal of Agricultural and Environmental Ethics 31/1 2018.

Regan. T. 1984. The Case for Animal Rights. University of California Press, Los Angeles.

Rutberg, A. T. (2013). Managing wildlife with contraception: why is it taking so long? Journal of Zoo and Wildlife Medicine 44: S38-S46.

Stout, R. J., Stedman, R. C., Decker, D. J., and Knuth, B. A. (1993). Perceptions of risk from deer-related vehicle accidents: implications for public preferences for deer herd size. Wildlife Society Bulletin 21: 237-249.

62. The black box of rodents perceived as pests: on inconsistencies, lack of knowledge and a moral mirror

M.A.A.M. van Gerwen[1*] *and F.L.B. Meijboom*[1,2]
[1] *Centre for Sustainable Animal Stewardship, Faculty of Veterinary Medicine, Utrecht University, Yalelaan 2, 3584 CM Utrecht, the Netherlands;* [2] *Ethics Institute, Faculty of Humanities, Utrecht University, Janskerkhof 13, 3512 BL Utrecht, the Netherlands; m.a.a.m.vangerwen@uu.nl*

Abstract

Discussions about food production and animals often focus on the welfare and treatment of production animals. However, more animals are involved. All over the world, unspecified, but large numbers of wild living rats, mice and other animals are killed because they are perceived as pests. If labelled as a pest animal, discussions on moral status and welfare seem to disappear from the public debate. This seems rather strange, if we take into account that these rats and mice do not differ in their capabilities to suffer compared to rats and mice in other contexts. In this paper, we present the first results of a study on this topic by the Centre for Sustainable Animal Stewardship. It includes the current status of the treatment of rodents perceived as pests, an overview of stakeholder attitudes and first reflections on its moral dimensions. Participating stakeholders feel the need to improve the treatment of mice and rats perceives as pests in terms of animal welfare, humanness and effectiveness of control. They urge for a better application of preventive measures by all people involved in pest control, including private persons. More attention should be paid to the humanness of control measures. National coordination and monitoring under the responsibility of the government is necessary for better implementation of IPM and more responsible rodent control. The outcomes of the study seem promising for the start of a dialogue about the treatment of rodents perceived as pests and improvements to make. Furthermore, they may be an inspiration for and start of a broader dialogue about ways to achieve a responsible and sustainable human-animal relationship.

Keywords: rodent control, animal welfare, ethics, pest control, human-animal relationship

Introduction

The combination of food production and animals often leads to ethical discussions on production animals, such as cows, pigs or poultry. However, in practice both on a farm level and in the processing and storage of food many other animals play a role. We are pointing at the unknown large numbers of wild living rats, mice and other animals that are killed because they are perceived as pests spreading diseases, causing losses in food production, damaging human property, etc. (cf. Littin *et al.*, 2004, 2014). In literature, no exact data can be found about the numbers of rats and mice, the nuisance they cause and the numbers of animals killed. Only estimations can be found (Mason and Littin, 2003). This also holds for the impact on human health. Rats and mice are capable of transferring the diseases either directly to other animals (including humans) by biting them or contaminating food. Or indirectly, for instance humans can be infected by swimming in water contaminated by rats. Next to the transfer routes, a number of animal and human diseases can be carried and transferred by rodents, such as Weil's disease (or leptospirosis), Hanta virus, Salmonella, Campylobacter (Meerburg *et al.*, 2009). Nonetheless, exact numbers and impact are less clear.

Svenja Springer and Herwig Grimm (eds.) ***Professionals in food chains***
EurSafe 2018 – DOI 10.3920/978-90-8686-869-8_62, © Wageningen Academic Publishers 2018

Legal aspects

The uncertainty is not restricted to precise negative impact of these animals, also on a legal level there is much unclarity. Most wild living animals in the Netherlands are protected by the Dutch nature protection law (*Wet natuurbescherming*), which is predominately an implementation of the European Birds and Habitats Directives (Directive 2009/147/EC and 92/43/EEC respectively). According to this law, it is prohibited to disturb, catch or kill many species on purpose and without a proper reason. Rats and house mice, however, are not protected by this legal frame. A general duty of care applies to them though, based on both nature and animal protection legislation. This duty of care means that everyone needs to provide a certain amount of care for (wild living) animals. It is prohibited to inflict pain or damage to an animal or to impair the welfare or health without a proper reason. Negative consequences for animals need to be minimized. However, terms as 'duty of care' and 'proper reasons' for harming animals, are not further specified. Consequently, this leaves a lot of space for different interpretations. People have a lot of freedom to decide how to deal with these animals. Like in many other countries with comparable legal frames, this often results in situation in which rodents are routinely exposed to cruelty and their treatment and methods for killing do not get much attention (Mason and Littin, 2003).

In the Netherlands, the use of rodenticides (anticoagulants) is regulated for the control of rats though. Since 2017, a certification for Integrated Pest Management (IPM) is a requirement for the use of anticoagulants for the control of rats outside buildings (Ctgb, 2018). According to the IPM principles, preventive and non-chemical methods should be taken before rodenticides can be used (see also Meerburg *et al.*, 2008 for a description of IPM). Everyone who wants to use rodenticides outside buildings, needs to obtain a certificate. Otherwise, persons should contract a professional. The underlying arguments for this requirement relate to environmental concerns and the increasing resistance in rats against the poison, rather than any animal welfare concern.

Animal welfare

The silence on animal welfare is not restricted to legal documents. There is a relatively small number of papers available about the welfare and treatment of animals that are perceived as pests. In the small amount of studies, the authors focus on humanness and welfare implications of pest control by looking at the amount of pain and distress, the duration, the methods for killing and the effects on animals that either escape from the control or non-target species (Broom, 1999; Mason and Littin, 2003; Littin, 2010). Broom (1999) states that when animals lose consciousness and/or sense of pain before they are killed, methods can be classified as humane since there is no real welfare impairment. Examples of these methods are certain fast-acting poisons or killing traps. Fast-acting poisons are however, not used very often because of their negative effects on non-target species. Most poisons which are used, are slow-acting anticoagulants with severe welfare implications and therefore not considered humane (Broom, 1999; Mason and Littin, 2003). This also holds for glue traps.

Consistency and moral concerns

It is striking that once labelled as a pest animal, discussions on moral status and welfare seem to disappear from the debate. In papers about the treatment of animals perceived as pests, authors point out the inconsistency in the treatment of different animals depending on context (Mason and Littin, 2003; Meerburg *et al.*, 2008). For instance, within the current legislation, people have a lot of freedom to decide how to deal with these rats and mice that do not differ in their capabilities to suffer compared to rats and mice in other practices, such as the laboratory or as pets (Littin, 2010). While, a rat in scientific research for the production of rat poison, is never allowed to die without pain relief over a period of days and will be euthanized in due time. A wild rat that is perceived as a pest, on the contrary, often dies slowly and painfully after it has ingested the same poison (Meerburg *et al.*, 2008). It seems this is not only the result of indifference, but even by an attitude of public denial: there is a tendency to be involved as least as possible in pest control and to outsource it to professionals. Consequently, the actual control measures become a magical black box (Jackson, 1980 in: Meerburg *et al.*, 2008).

In order to integrate ethical considerations in pest control, several authors suggest to apply the 3Rs principle (replacement, reduction and refinement) from laboratory animal science to pest control, thereby starting with a justification and cost-benefits analysis of the measures to be taken (Littin *et al.*, 2004; Meerburg *et al.*, 2008; Yeates, 2010; Littin, 2010). Killing of animals should only be allowed when no other methods serve the desired outcome. When animals need to be killed or harmed, as few animals (including non-target animals) as possible should be involved and the methods need to be as humane as possible (refinement). In this way, unnecessary suffering is prevented or minimalized.

Taking all the above-mentioned information into account, the current situation in rodent control is problematic for three reasons: (1) the mentioned problem of inconsistency, (2) the lack of clarity on the need and effect of current control measures and (3) climate change, increasing world population and destruction of natural habitat for these animals have an impact on the increase of animals perceived as pests (Littin *et al.*, 2014). It asks for better evidence-based views on the risks that come with rodents, reflection on moral dimensions of human interaction with pest animals and a start of a dialogue in order to bring this subject into society. In this paper, we present the first results of a stakeholder investigation on this topic by the Centre for Sustainable Animal Stewardship (CenSAS, www.censas.org). The study is based on the Dutch situation, but is also relevant for other countries facing similar issues regarding pest control and animal welfare.

Methods

For this study, next to an extensive desk study, semi-structured interviews were conducted. In total 15 stakeholders (19 persons) were interviewed between August 2017 and March 2018. Stakeholders represented governments, animal protection NGO's, pest controllers, food retail and the agricultural sector or were researchers or advisors in the field of pest control and/or wildlife management. Inclusion in the study was based on willingness and availability to cooperate. During the interviews, which lasted each 1.5 to 2 hours, stakeholders were asked about (1) their own experiences with rats and mice as pests, (2) their opinion about current rodent pest control in the Netherlands, (3) welfare of the animals, (4) moral aspects of control and responsibilities and (5) possibilities for future improvements.

Summaries of the interviews were sent to the stakeholders for feedback afterwards. After approval, the summaries were merged together and included in the report and this paper.

Results

Own experiences

Most stakeholders experience no direct nuisance themselves, but rather indirect due to their profession as pest controller, researcher, advisor or working for (local) government. None of the stakeholders is able to quantify the real damage, costs or effects for public health. Thereby indicating that there is a lack of data and proper monitoring and registration. The term 'pest animal' or especially 'vermin' is not favoured by most of the stakeholders. They rather speak about 'animal plagues', indicating that not the animal in itself is the problem, but rather the circumstances and perceptions of humans (whether humans experience a plague). 'In pest control, the focus is and should be on controlling the plague, not the animal itself', the stakeholders tell. Some stakeholders point at the positive role of rats and mice.

They are, for example, very useful animals that clean our environment. One stakeholder even calls them 'super animals', since they are evolutionary champions and excellent survivors. Rats and mice may also be seen as a good indicator for bad hygiene, e.g. presence of garbage, in the city. In that way, humans facilitate the animals in their survival success.

Current pest control

All stakeholders think that IPM with a focus on preventive measures is a good development towards a more responsible pest control. Stakeholders differ, however, in their opinion about the effectiveness of IPM. Most people think that professional pest controllers apply IPM quite well. Outside the professional context, the focus on prevention should be stronger though. Furthermore, stakeholders indicate that IPM should also form a requirement for non-chemical methods.

Stakeholders find it problematic that (private) persons are not aware of their own role. Since people (mostly unconsciously) create favourable circumstances for rats and mice, e.g. by providing infrastructure, hiding places and food, they often cause animal plagues themselves. Many people wait for the situation to become problematic, before they start thinking of rodent control. Consequently, more animals need to be killed.

Finally, stakeholders signal that responsibilities are divided among different ministries, provinces and local governments in the Netherlands. Proper coordination and monitoring are lacking, according to stakeholders. For example, every province can make its own considerations for allowing certain methods for killing pest animals, making the treatment inconsistent. Due to the lack of monitoring, it is hard to say whether current control measures are effective. This in combination with the large number of animals killed and the type of methods used, makes the current pest control irresponsible.

Animal welfare

All stakeholders recognize the value of animal welfare and see the importance of a minimal harm principle. This is reflected in the shared view that it is important to treat rats and mice in pest control as humane as possible and suffering should be minimized. In order to achieve this, the focus should be on preventive measures. When lethal control methods are necessary, the animal should be killed fast. Stakeholders therefore, prefer mechanical killing traps over rodenticides and glue traps. However, sometimes more humane methods may be less effective to control a plague. Development of new, more humane, methods is desired by most stakeholders. In spite of the shared understanding of the importance of welfare, there was no consensus on the definition by the stakeholders. Furthermore, some stakeholders used context-dependent definitions of animal welfare in which they differentiated between pest animals and domestic animals. For example, stakeholders used the five freedoms as a definition for welfare in the case of domestic species, while for pest animals they indicated that welfare mainly refers to a fast and painless (e.g. humane) killing of animals. Others used a single definition for all contexts, but indicated that an impaired welfare is easier to justify in the case of pest animals than domestic animals.

Moral aspects

With regard to the treatment of rats and mice, most stakeholders detect certain degrees of inconsistency in the human behaviour towards and legal status of pest animals compared to other animals, for example other wild living animals or rodents in scientific research. They say: 'it seems that people feel sorry for some animals, but not for others' and 'people seem to feel less moral responsibility towards these rats and mice, compared to 'beautiful' or 'useful' animals'. The stakeholders who are active in the field of animal protection formulated direct moral concerns and described the current treatment of animals as 'frightful' or 'disrespectful'.

Some stakeholders worry about the intentions of pest controllers. They are concerned that pest controllers might have an interest in not solving a plague completely, since their job is dependent on the presence of pest animals and the nuisance caused by them.

Stakeholders indicated that it would be a good idea to apply the 3R principles from laboratory animal science as part of taking moral justification and responsibility in the context of pest management. Thereby indicating that the principles of IPM resemble the 3Rs to some extent. Refinement of the methods used deserves more attention however and the practical implementation of the 3Rs remains complex, especially for private persons.

Suggestions for improvements

All stakeholders see opportunities or feel the need for improvements in the treatment of rats and mice perceived as pests. During the interviews, the following suggestions for improvements were given.

1. *Education of the general public.* Since people (mostly unconsciously) actively create favourable circumstances for rats and mice, good education, starting at schools, is essential. Municipalities are assigned great responsibility for this. This is not restricted to prevention of problems. Some stakeholders also argue that more knowledge about the biology of rats and mice will lead to more respect of animals and better treatment.
2. *Improved implementation of IPM.* According to stakeholders, IPM principles should be implemented better. First, this is a matter of attitude. Professionals in the field of pest control need to have an intrinsic motivation to actively strive for prevention and if necessary responsible control measures. Second, it implies expanding the application of IMP outside the professional field, for example in the agricultural sector. Furthermore, it implies even more attention to prevention. For example, companies hiring pest controllers have a responsibility to ask and pay for preventive methods. Also in the construction of buildings and design of (green) cities, attention should be paid to the presence of rats and mice. Finally, stakeholders think that IPM should be a requirement for the use of non-chemical methods. In all cases, not only for chemical methods, prevention should be the first step. 'Animals we do not have, do not need to be killed', one of the stakeholders says.
3. *Improved legal protection.* Most stakeholders claim that rats and mice are entitled to have better legal status and protection. This should be comparable to the position of other wild animals. Furthermore, certain ethically inspired conditions for killing rats and mice should be set in order to prevent unnecessary harm. The Dutch government should take responsibility for this.
4. *Refinement of methods.* According to stakeholders, it is important to know the welfare effects of all control methods. Humaneness of methods and development of methods with less severe welfare implications deserve more attention. For prevention measures, knowledge about natural mice and rat behaviour and ecology should be integrated more. A few stakeholders suggest that rodenticides should not be available for individual citizens and that only professionals are assigned to their use.
5. *National coordination and monitoring.* For proper implementation of IPM including attention to animal welfare, national coordination, registration and monitoring are needed. The Dutch government should take the responsibility for this.

Discussion and conclusions

This study presents the first outcomes of a stakeholder assessment about the treatment of rats and mice perceived as pests. Since this is a first exploration stakeholder opinions, it is not possible to draw general conclusions about the public attitudes towards rodent control. Ways for improvements as mentioned by the participating stakeholders, are not necessarily the same as those of other stakeholders and/or the general public. Nonetheless, we can conclude that the professional stakeholders in this study share their concerns and ideas for improvements, despite the different interests and views they have. The interviews clearly show that stakeholders feel the need to improve the treatment of mice and rats perceived as pests. This results in attention to the humanness and effectiveness of control methods and more explicit attention to animal welfare. Generally, stakeholders stress (1) the importance of the use of preventive measures and consider this a shared responsibility for both professionals and the general public. (2) The need of more systematized attention to the animal welfare impact of control measures. And (3) the importance of national coordination and monitoring as a necessary condition for better implementation of IPM and a more responsible rodent control.

The findings of the study show that animals perceived as pests seem to be the black box of the human-animal relationship. On the one hand, they are a black box because relatively little is known about the number of rodents, the actual risk to human health and food safety and the efficacy of pest control. On the other hand, they function as a black box like an airplane's flight recorder: they tell us something important about the human-animal relationship depending on context. From this perspective, it is striking that from the stakeholder analysis we can conclude that the attention to animal welfare is – although still limited – increasing and that the need for preventive measures are also mentioned with reference to animals' interests. This is an interesting development and is an indication of the dynamic character of the moral dimensions on the human-animal relationship. Therefore, we aim to explore in further research how animals in this specific context can help and inspire us to start a broader dialogue about ways to achieve a responsible, consistent and sustainable human-animal relationship.

Acknowledgements

We would like to thank all stakeholders for their willingness to cooperate, their openness during the interviews and their useful input. We thank Triodos Foundation for the financial support of CenSAS, which makes this type of research possible.

References

Broom, D.M. (1999). The welfare of vertebrate pests in relation to their management. In: Advances in Vertebrate Pest Management, ed. P.D. Cowan and C.J. Feare, 309-329.

Ctgb (Dutch Board for the Authorisation of Plant Protection Products and Biocides) (2018). Outdoor control of rats with anticoagulants. Available at: https://english.ctgb.nl/topics/rodenticides/outdoor-control-of-rats-with-anticoagulants. Accessed 23 January 2018.

Littin, K.E. (2010). Animal welfare and pest control: meeting both conservation and animal welfare goals. Animal Welfare 19: 171-176.

Littin, K.E., Mellor, D.J., Warburton, B. and Eason, C.T. (2004). Animal Welfare and ethical issues relevant to the humane control of vertebrate pests. New Zealand Veterinary Journal 52:1, 1-10.

Littin, K.E., Fisher, P.M. and Beausoleil, N.J. (2014). Welfare aspects of vertebrate pest control and culling: Ranking control techniques for humaneness. Revue scientifique et technique (International Office of Epizootics) 33 (1), 281-289.

Mason, G. and Littin, K.E. (2003). The humaneness of rodent pest control. Animal Welfare 12: 1-37.

Meerburg, B.G., Brom, F.W.A. and Kijlstra, A. (2008). Perspective, The ethics of rodent control. Pest Management Science 64: 1205-1211.

Meerburg, B.G., Singleton, G.R. and Kijlstra, A. (2009). Rodent-borne diseases and their risks for public health. Critical Reviews in Microbiology 35 (3): 221-270.

Yeates, J. (2010). What can pest management learn from laboratory animal ethics? Pest Management Science 66: 231-237.

Section 11.
Animal research

63. Prosocial animals showing human morality – on normative concepts in natural scientific studies

A. Huber[1,2], H.B. Schmid[3] and H. Grimm[4]*
[1]ABC Welfare Group, School of Life Sciences, University of Lincoln, Lincoln LN6 7DL, United Kingdom; [2]Division of Animal Welfare, Vetsuisse Faculty, University of Bern, 3012 Bern, Switzerland; [3]Institute for Philosophy, University of Vienna, 1010 Vienna, Austria; [4]Unit of Ethics and Human-Animal Studies, Messerli Research Institute, Veterinaerplatz 1, 1210 Vienna, Austria; annika.huber@vetsuisse.unibe.ch

Abstract

Traditionally, morality has mostly been researched on in the humanities and the social sciences. In recent times, however, the topic has received increasing attention from the natural sciences. Natural scientific disciplines such as ethology, that studies animal behaviour, increasingly conduct research on morally-relevant behavioural capacities, e.g. prosociality, empathy, or inequity aversion in various animal species. However, the resulting scientific publications often lack an explicit and reflected examination of the underlying concept of morality. This is problematic because as is evident from philosophical research, 'morality' is a highly contested concept. We hypothesise that the study paradigms and experimental designs are informed by a very specific conception of morality which is implicit in the experimental design; being able to make explicit through critically analysing the latter. We argue that the concept of morality applied represents a particular core idea of the morally good and bad and is strongly oriented along egalitarian lines. However, as we will argue, philosophy can provide alternative conceptions. We will first identify how this hidden concept of morality is at work in a sample of relevant publications in the field by highlighting a number of thick ethical concepts (morally evaluative descriptive terms; cf. Williams, 1985) which reveal the moral foundation underlying these studies. This is further elucidated by, second, turning to an examination of the study paradigms, e.g. through identifying and analysing the criteria to classify the study subject as behaving prosocial (i.e. morally good) or not prosocial (i.e. morally bad). Besides providing an important – so far ignored – perspective on studies dealing with interdisciplinary research topics like morality, we strengthen the importance of a mutual exchange of knowledge between scientific disciplines in order to conduct reasoned science. Furthermore, we argue that natural scientific studies like the ones discussed can be considered as a tool to identify what moral or immoral behaviour is for ourselves, thereby addressing human morality through animal studies.

Keywords: animal morality, morality concept, natural sciences, prosociality

When morality entered natural scientific studies

Studying morality has a long tradition in the humanities and social sciences with both mainly focussing on conducting research on the concept of morality in the human domain. Addressing the definition of morality, the Stanford Encyclopaedia of Philosophy (Gert, 2016) states that there not one single definition of morality applicable to all moral discussions but definitions of morality should broadly be differentiated into two directions: (1) normative and (2) descriptive or empirical. In the normative sense, morality deals with questions about `What ought to be?` and is consequently tied to rationality (Fitzpatrick, 2017) and a collection of societal norms that ought to be followed (Fitzpatrick, 2016). Second, morality can also be defined from a descriptive or empirical sense (Fitzpatrick, 2017; Gert, 2016). In its pure sense, descriptive definitions of morality are independent of normative claims and evaluations (Fitzpatrick, 2017; Gert, 2016). Instead, in the descriptive or empirical way, morality is – somewhat circularly – defined as requiring the possession of 'psychological capacities that enable the holding of various beliefs and attitudes about moral issues', referring to an individual's moral cognition

EurSafe 2018 – DOI 10.3920/978-90-8686-869-8_63, © Wageningen Academic Publishers 2018

and action independent of normative evaluations(Fitzpatrick 2017), making morality an empirically assessable and observable phenomenon (Fitzpatrick 2016). More recently, debates on morality beyond the human species have been increased, focussing on moral capabilities of nonhuman animals (e.g. Fitzpatrick, 2017; Monsó, 2015; Rowlands, 2012). This recent debate oftentimes especially considers and refers to empirical findings from natural scientific studies on morally-relevant behaviour in animals and takes them as basic premise for developing the argument on animal morality.

In the natural sciences, animal morality has received increasing attention in recent times. However, especially in the context of this research, differentiating between the normative and the empirical or descriptive definition of morality is essential. Failure to distentangle these two definitions can significantly affect and impede empirical research on animal morality if the existence of moral behaviours in animals is considered reliant on what we as humans might consider as morally praiseworthy (Fitzpatrick, 2017). This has not always been sufficiently avoided in the past although it should, when the aim is to study moral capabilities of animals as 'whether or not animals behave in ways that we might judge to be right or good according to a particular normative standard is [...] irrelevant to the question of whether they are capable of moral cognition or action' (Fitzpatrick, 2017). With regard to an empirical approach on the topic, Konrad Lorenz, one of the main founders of ethology as a natural scientific biological discipline (Brigandt, 2005), was one of the first who associated certain observed behaviours of animals with morality, labelling them as morality-analogous (Lorenz 1954; 1964). Specifically, he defined mechanisms that inhibit lethal aggression against conspecifics in social animals as morality-analogous and described observations of this behaviour in the context of ritualized fights of several animal species ranging from fish to birds to game to carnivores (Lorenz 1954, 1964). Furthermore, Lorenz considered caring behaviours as analogies of moral behaviours in humans by describing animals that do not harm their offspring although they would have the capacity to do so as showing moral-analogous behaviours. By introducing these examples, Lorenz refers to an aspect of human morality that may be less obvious and may not be the focus in the recent empirical research on features of animal morality – that a significant part of human morality is considered to consist of omissions like the principle of not harming others (*neminem laedere*).

The two sides of the morality coin

Morality, as considered in philosophy, incorporates two principles, each representing one side of the same coin. First, it incorporates an omission-based part which is to refrain from doing bad to others with examples from the animal kingdom as for instance described by Lorenz (1954, 1964). Also Müller (2004) referred to the ideas of Lorenz. According to her, all 'higher' social living animals possess some form of morality and she builds on the approach of Lorenz arguing that we can differentiate between moral-analogous behaviour and moral behaviour in animals. The criterion for this differentiation is the freedom of decision in the animal. Hence, a bee that acts in favour of its colony only performs moral-analogous actions as it cannot do otherwise. However, a kitten that inhibits to chase the tail of another cat despite its drive to do so acts morally as this is according to the social rules. The criterion whether an animal shows moral behaviour or moral-analogous behaviour is, according to her, the freedom of decision to act so. If an animal inhibits an inclination due to social rules, it acts moral (Müller 2004: 152).

The second part of the morality coin is an activity-based one – incorporating actions of doing good to others (Müller 2004: 153). This activity-based part of morality can be referred to behavioural or psychological concepts that are oftentimes linked to moral behaviours in animals in the more recent debate on animal morality such as prosociality or empathy (see for some examples of empirical studies on further concepts linked to animal morality e.g. Fitzpatrick, 2017). There are natural scientists who intensively scrutinized the theoretical conception of animal morality with integrating potentially morally-relevant behavioural, cognitive and psychological concepts observed in the animal kingdom, especially mentioning de work of Frans de Waal (e.g. de Waal, 1996). However, usually experimental studies on morally-relevant concepts in animals lack an explicit and reflected examination of the underlying concept of morality. Addressing this aspect, Sarah Brosnan, an ethologist and co-author of several publications with de Waal, argued that working definitions of morality are few and far between, a fact that she assumed to be based on the concept of morality being already well-understood and simultaneously such a difficult concept to define (Brosnan, 2006: 168). However, such a statement is problematic because as is known from philosophical research, morality is a highly contested concept.

Besides the lack of an explicit appraisal of the concept of morality underlying the investigation, natural scientific studies focussing on morally-relevant behavioural concepts such as prosociality as well as empathy or inequity aversion oftentimes do not specifically indicate that those behaviours studied can also be considered to be of normative value. However, by looking closer at the study paradigms and experimental designs applied as well as on the terms used in the relevant study publications, the implicitly underlying moral conception as well as a normative evaluation can be made explicit and thereby, can be scrutinized in more detail. Our aim is an approach to the philosophy of science. By philosophical reflection, we indicate that in empirical research on morally-relevant behaviours of animals normative values can be identified.

Scrutinizing an example study on prosociality – making the implicit morality concept explicit

In the following, a published study (Tennie *et al.*, 2016) on prosocial behaviours of chimpanzees is summarized in order to later discuss the implicit conception of morality underlying this study.

Tennie and colleagues (2016) aimed to investigate aspects of intended helping – which belongs to actively-performed prosocial behaviours, thus, behaviours that intend to benefit others (Jensen, 2016) – in captive chimpanzees. In chimpanzees, helping of others has been reported in a range of studies in an experimental context (e.g. Yamamoto *et al.*, 2012). However, contradictorily, chimpanzees oftentimes do not willingly share food with others in experiments that require the delivery of food to a conspecific even if there would be no cost in doing so (see e.g. Vonk *et al.*, 2008). In the study of Tennie *et al.*, (2016) this paradox should be addressed by examining whether the delivery of food to one individual (the recipient chimpanzee) in experiments following the manipulation of a certain stimulus like a lever peg by another individual (the deliverer chimpanzee) can be considered as actively intended prosocial helping behaviour or, alternatively, whether the delivery of food to the recipient can be considered to be happened rather coincidentally as a by-product of the testing and thus cannot be considered to be truly prosocially motivated. Furthermore, this study examined the deliverer individuals` behaviour when they are able to actively prevent the recipients from accessing the food by the manipulation of the lever peg; and in case they would do so, according to Tennie *et al.* (2016), they would behave spiteful.

For investigating these research questions, two different experiments have been conducted in this published study. In the first experiment, the two individuals (deliverer and receiver chimpanzee) were in two different cages that were separated by a gangway. Visual contact between both individuals was possible. In the receiver's cage a box was mounted that contained food. This box was attached to a string that went into the deliverer's cage with a release peg at the end. By pulling this release peg, the deliverers could either open the box and thereby provide access for the receiver chimpanzees to the food in the box (this was the case for half of the tested animals) or, for the other half of the tested animals, pulling the releasing peg resulted in locking the food box and thereby preventing the receiver chimpanzee from accessing the food in the box. Tennie and colleagues (2016) predicted that if the chimpanzees behaved helpful (i.e. prosocial), then the deliverer chimpanzees for which pulling the peg resulted in the delivery of food for the receiver should pull the peg more often compared to the deliverer chimpanzees in the other group where the pulling of the peg resulted in the receiver individual having the access to the food prevented as a consequence. In the reversed case, so their words, chimpanzees would behave spiteful (i.e. anti-social; Tennie *et al.*, 2016: 2). Intriguingly, the researchers could not find a difference in pulling the peg between groups and therefore draw the conclusion that chimpanzees neither behaved prosocial nor anti-social but were just personally motivated to pull the peg, regardless of its consequences to the recipient individual.

That this study is not free from normative evaluations of the animal's behaviour and that morally-relevant features are addressed, even if not explicitly discussed by the authors, can be identified by the use of thick ethical concepts, thus, morally evaluative descriptive terms (cf. Williams, 1985). Thick ethical concepts are concepts or terms used, respectively, that have an evaluative aspect and appraise a certain situation. In doing so, thick ethical concepts are considered to be different from scientific concepts that are argued to enjoy a kind of objectivity (Moore, 2006: xvi), thus, they are supposed to lack normatively-evaluative features. What, however, if we can now detect such evaluative terms in publications of natural scientific studies as in the one by Tennie and colleagues (2016)? We argue that by doing so, we can reveal the moral foundation that underlie these studies. In the publication of the above summarized study, we have detected a number of thick ethical concepts, such as the terms helpful and spiteful of which especially the latter is frequently used throughout the text. Helpful and spiteful are both terms that have an evaluative aspect and can be used to appraise a situation or the behavioural response of an individual, respectively. Consequently, they can be regarded as thick ethical concepts and thereby identify a moral foundation underlying this study on prosocial behaviour of chimpanzees. Based on the moral foundation contained, the chimpanzees` behaviours can also be evaluated as morally good or morally bad. Consequently, prosocial helpful behaviour can be considered to be regarded as morally good whereas if an individual is not behaving prosocially, thus, spiteful, it is behaving morally bad.

Having a closer look at the study paradigm and experimental design of this publication can reveal a second interesting aspect which is the very specific conception of morality that is implicitly underlying this study. Therefore, we addressed the criteria to classify the study subject as behaving prosocially (i.e. morally good) or not prosocially (i.e. morally bad)? In Tennie *et al.* (2016) the test subjects (deliverer chimpanzees) behaved prosocially when they provided the recipients with access to the food reward (by pulling the peg to open the box) or avoided to prevent the access to the food reward for the recipient (in case the pulling of the peg would result in the food box to be locked). Consequently, the chimpanzees behaved prosocially (morally good) whenever their behaviour allowed the other individual to access a food reward and they behaved not prosocially (morally bad) when they behaved otherwise.

The moral concept underlying this study can be argued to be oriented along egalitarian lines. The core of egalitarianism is that it advocates equality of some sort; in the human domain it advocates that all humans are equal and should be treated as equal concerning the rights granted as well as the resources allocated (Gibson, 2008). With regard to the study published by Tennie *et al.* (2016) the study subjects are only then classified as being prosocially helpful and thereby as behaving morally good if they provided the other with access to food, primarily regardless of the social relationships or hierarchy between the individuals (however, the researchers noted that they took the possibility of personal relationships between the deliverer and receiver subject into account by stating that 'To minimize the effects of personal relationships, three male chimpanzees of a similar age range were chosen to be recipients' (Tennie *et al.*, 2016: 3)). Consequently, the underlying moral conception can be argued to primarily be based on egalitarian lines. Such a moral conception is not without alternatives – in Nietzsche for instance we can find a concept of morality that does not follow egalitarian lines but one with a hierarchical view on morality, ranking the value of human beings (Peery, 2010); without advocating for human equality or the equal distribution of resources to be considered as the morally *good*. Following this hierarchical moral conception, the morally good act – in the presented study the provision of access to food –

would be much more evaluated in the light of personal and hierarchical relationships between the two individuals where the morally good behaviour would not be to grant every recipient chimpanzee with access to the food but that there should rather be a state of inequality based on the social ranks of the individuals involved.

When in empirical research on morally-relevant behaviours in animals normative evaluations can be detected, the latter could provide us not only with indications about the moral context of the study, but furthermore with insights into human judgments of right and wrong; thus, the particular normative standard followed by the researchers. By defining morally good behaviours in animals and differentiating them from those behaviours defined as morally bad, no matter whether these attributions were done explicitly and implicitly, we argue that inferences could be discussed about what is regarded as moral or immoral amongst humankind. Detecting a rather egalitarian than hierarchical approach in the exemplified study on prosociality in chimpanzees (Tennie *et al.*, 2016), could give us some basis for discussing a priorisation of an egalitarian compared to a hierarchical moral conception amongst humankind. In that sense, the prosocial act of providing food to another individual regardless of its status, that has been shown to be implicitly evaluated as morally good in the chimpanzee study, could be inferred to be an act regarded as morally good as well amongst humans and the same holds true for the morally bad act. Consequently, we argue that our approach, a philosophical analysis of the underlying conceptions of morality in animal behaviour studies, might provide us with even more insights beyond the nonhuman animal kingdom – namely with what moral or immoral behaviour is for ourselves, for us as humans.

Conclusion

Searching for empirical evidence of morally-relevant behaviour in animals – either explicitly addressed such as by Konrad Lorenz or implicitly as done in the presented study by Tennie and colleagues (2016) – is facing increasing progress in the recent times. Usually natural scientific study publications mainly lack an explicit and reflected examination of the conception of morality underlying the reported study. Stating that morality is a concept that is already well understood but difficult to define (Brosnan, 2006) should not necessarily be considered as valid but questioned. Our aim was to indicate how such hidden moral conceptions are nevertheless at work in natural scientific studies. By critically analysing the study paradigms and experimental designs of natural scientific studies, we can, firstly, show that the behavioural concepts studied are not value-free but critical from a normative point of view and, secondly, we can make the implicit moral conceptions underlying these natural scientific studies explicit in order to further discuss them. Moral conceptions are not without alternatives as we can see in the history of research in the philosophical domain. A mutual exchange of theoretical approaches and practical applications between scientific disciplines such as philosophy and ethology would be important in order to advance informed and reasoned science and to provide a broadened, interdisciplinary research perspective on topics of importance across research disciplines such as morality.

References

Brigandt, I. (2005). The instinct concept of the early Konrad Lorenz. Journal of the History of Biology 38: 571-608.

Brosnan, S. (2006). Nonhuman species′ reactions to inequity and their implications for fairness. Social Justice Research, 19: 153-185.

De Waal, F.B.M. (1996). Good natured. The origins of right and wrong in humans and other animals. Cambridge, Harvard University Press.

Fitzpatrick, W. (2016). Morality and evolutionary biology. In: Zalta, E.N. (ed.) *The stanford encyclopedia of philosophy*. https://plato.stanford.edu/archives/spr2016/entries/morality-biology.

Fitzpatrick, S. (2017). Animal morality: What is the debate about?. Biology & Philosophy. doi: 10.1007/s10539-017-9599-6.
Gert, B. (2016). The definition of morality. In: Zalta, E.N. (ed.) The stanford encyclopedia of philosophy. Available at: http://plato.stanford.edu/entries/morality-definition.
Gibson, D.C. (2008). Egalitarianism. In: Kolb, R.W. (ed.) Encyclopedia of business ethics. Thousand Oaks, Sage Publications, pp. 661-664.
Jensen, K. (2016). Prosociality. Current Biology 26: R748-R752.
Lorenz, K. (1954). Moral-analoges Verhalten geselliger Tiere. Forschung und Wirtschaft, 4: 1-23.
Lorenz, K. (1964). Moral-analoges Verhalten von Tieren – Erkenntnisse der Verhaltensforschung. Universitas, 19: 43-54.
Monsó, S. (2015). Empathy and morality in behaviour readers. Biology & Philosophy 30: 671-690.
Moore, A.W. (2006). Introduction. In: Williams, B. Philosophy as a humanistic discipline. Princeton, Princeton University Press, pp. xi-xx.
Müller, S. (2004). Programm für eine neue Wissenschaftstheorie. Würzburg, Königshausen & Neumann Verlag.
Peery, R.S. (2010). Nietzsche for the 21 century. New York, Algora Publishing.
Rowlands, M. (2012) Can animals be moral?. Oxford, Oxford University Press.
Tennie, C., Jensen, K. and Call, J. (2016). The nature of prosociality in chimpanzees. Nature Communications 7: 13915. doi: 10.1038/ncomms13915.
Vonk, J., Brosnan, S.F., Silk, J.B., Henrich, J., Richardson, A.S., Lambeth, S.P., Schapiro, S.J. and Povinelli, D.J. (2008). Chimpanzees do not take advantage of very low cost opportunities to deliver food to unrelated group members. Animal Behaviour 75: 1757-1770.
Williams, B. (1985). Ethics and the limits of philosophy. Cambridge, Harvard University Press.
Yamamoto, S., Humle, T. and Tanaka, M. (2012). Chimpanzees′ flexible targeted helping based on an understanding of conspecifics′ goals. Proceedings of the National Academy of Sciences USA 109: 3588-3592.

64. The logic, methodological and practical flaws of the harm-benefit-analysis in Directive 2010/63/EU

M. Eggel[1,2] and H. Grimm[1]*
[1]Unit of Ethics and Human-Animal-Studies, Messerli Research Institute, Veterinaerplatz 1, 1210, Vienna, Austria; [2]Institute for Biomedical Ethics and History of medicine, University of Zurich, Winterthurerstrasse 30, 8006 Zurich, Switzerland; matthias.eggel@ibme.uzh.ch

Abstract

Directive 2010/63/EU regulates the use of animals for scientific purposes in EU member states and mandates that every project proposal involving procedures on living non-human vertebrates and cephalopods has to be approved in a review process that includes a Harm-Benefit-Analysis, to assess 'whether the harm to the animals in terms of suffering, pain and distress is justified by the expected outcome taking into account ethical consideration and may ultimately benefit human beings, animals or the environment (EU Directive 2010/63, Art 38d). The aim of this paper is (1) to summarize recent criticism on the epistemic and practical limitations of the prospective benefit assessment in the HBA in its current form and on the focus on tangible societal benefits in project evaluation and (2) as a proof of principle, demonstrate the argumentation of these papers on two concrete examples, namely the insulin inhalator *Exubera* and the cancer drug *Ipilimumab*. First, we show that the HBA suffers from a logical and methodological flaw. The outcome of an experiment is per definition uncertain. If it wasn't, the experiment would not generate new knowledge and would therefore be illegal. Moreover, as long as animals are used as models for humans there will always be uncertainty regarding the translatability of knowledge from model to target species. Second, we show that practical flaws further complicate prospective benefit assessment. There are non-scientific factors, such as market potential, lobbying, patient compliance, etc., that are impossible to predict and yet, are important parameters in prospective benefit assessment. Together, these uncertainties make a prospective benefit assessment implausible. Also, the requirement to demonstrate societal benefits might incentivize researchers to overstate the tangible benefits of their research in project proposals, thereby making prospective benefit assessment in project evaluation more difficult for committees. Overstating potential societal benefits that are eventually not realized might also be detrimental to the credibility of science. In light of these flaws we think it necessary to develop an alternative model for project evaluation that focuses on potential knowledge gains as outcome of a project rather than prospective assessment of potential societal benefits.

Keywords: harm-benefit-analysis, animal research ethics, project evaluation, credibility of research, benefit concept

Introduction

Directive 2010/63/EU mandates that every research proposal entailing experiments on living non-human vertebrates and cephalopods is evaluated in a 'harm-benefit-analysis' (HBA) to assess 'whether the harm to the animals in terms of suffering, pain and distress is justified by the expected outcome taking into account ethical consideration and may ultimately benefit human beings, animals or the environment (EU Directive 2010/63, Art 38d). From this it follows, that societal benefit (in terms of practical benefits for human, animals or the environment) and animal harm are competing interests that ought to be and can be weighed against each other. The HBA thus functions as legal as well as moral evaluator of research projects. Due to moral concerns regarding animal suffering, we agree that animal research should be regulated very strictly. However, we claim that the HBA as the current evaluation method to decide on what is scientifically and morally justified research has conceptual and practical flaws that make it unsuitable for its purpose. These flaws are connected to the nature of the relation between 'outcome' and 'benefit' and to the focus on societal benefits within the HBA (Bateson, 1986; Bout *et al.*, 2014; Brönstad *et al.*, 2016; Hirt *et al.*, 2015; Laber *et al.*, 2016; Scharman and Teutsch, 1994; Stafleu *et al.*, 1999; EWG for project evaluation. 2013). The outcome of an experiment (i.e. knowledge) has been attributed little justifying power in a HBA compared to societal benefits. The potential of the knowledge to turn into societal benefit is considered to be a more valid benefit able to outweigh inflicted harm on animals. But exactly here lies the problem; prospective assessment of potential societal benefits of research projects in a HBA is, due to epistemic and practical flaws that are connected to the inherent uncertainties of the scientific process, vague speculation at best and thus, unsuitable for legal and moral project evaluation. In the following, we will (1) summarize the argumentation of recent publications that critizise the current HBA in project evaluation due to logic, methodological and practical flaws (Grimm *et al.*, 2017; Grimm and Eggel, 2017) and (2), as a proof of principal, we demonstrate the argumentation of these papers on two concrete examples, namely the insulin inhalator *Exubera* (Heinemann *et al.*, 2008) and the anti-cancer drug *Ipilimumab* (Williams *et al.*, 2015).

It is important to note, that we do not question on a principle level whether animals ought to be used for research. Our argumentation starts from a 'real world' assumption, where as a matter of fact animals are used in research under particular legal requirements. Also, our argumentation is based on the moral premise of the Directive, which is based in consequentialist moral theory. We will show, that based on that premise, the HBA in its current form fails to meet its goal (i.e. ethical evaluation of research proposals) due conceptual and practical flaws and should be replaced by a different model that replaces 'HBA' with a 'harm-knowledge-analysis'.

Epistemic flaws of the HBA in project evaluation

In hypothesis-driven research, or any research for that matter, the goal is to answer an unsolved question, i.e. to produce new knowledge. Thus, the outcome of research is per definition unknown; before performing an experiment, one does not know whether a hypothesis will be verified or falsified or whether unintended knowledge will be produced. If the outcome of an experiment was clear, there would be no knowledge gained and thus, there would be no reason to perform the experiment and it would ethically as well as legally be wrong to do so. The implication of this for the HBA is not trivial. The HBA clearly favours societal benefits over knowledge. Here, we argue that for prospective benefit assessment only verification and only in exceptional cases the falsification of a hypothesis will contribute to the potential generation of societal benefit. After all, drug discoveries for example, usually happen on the back of positive results and to a far lesser extent on negative results. However, as mentioned above, whether the hypothesis will be verified or falsified is unknown. This inherent uncertainty of hypothesis-driven research makes prospective evaluation of societal benefit implausible.

This logical flaw is exacerbated further by the methodological flaw that is associated with the uncertainty of translatability. Animals are used as model organisms to study human diseases. A model organism in this context is per definition never the same as the target species, i.e. humans. The knowledge gained in animals thus can't always in its entirety be translated to human application. In general, one can say that the translatability increases with greater similarity. However, since there will always be differences between model and target species, there will also always be uncertainty regarding the translatability of the knowledge gained. Since prospective benefit assessment is dependent on translatability, it is strongly affected by the uncertainty of translatability.

Practical flaws

So far, we have elaborated on science-internal factors, e.g. uncertainty of outcome and uncertainty of translatability. Next to these logic and methodological flaws there are also extra-scientific, practical flaws that further complicate prospective assessment of societal benefits. Whether societal benefit is realized does not only depend on knowledge gained in research. The knowledge gained in research can only ever be a necessary but never sufficient condition for the generation of societal benefits. This is due to the fact, that extra-scientific factors, such as market potential (Qureshi *et al.*, 2011; Heinemann, 2008), patient compliance (Hunter *et al.*, 1999), lobbyism (Miller *et al.*, 2007; Ventola *et al.*, 2011), among others play a crucial role for the human application of knowledge gained research. This is illustrated by the fact that drugs that are approved for application never make it to the market or are discontinued (importantly, we are not talking about drugs that are withdrawn for safety or efficacy reasons but rather because they showed limited market potential or failed due to low patient compliance) (Qureshi *et al.*, 2011; Heinemann, 2008). Orphan diseases (Aronson *et al.*, 2006) represent an example, where a genuine medical need does not translate into sufficient market potential. Furthermore, the nonsteroidal anti-inflammatory drug *Duract* (Hunter *et al.*, 1999), had to be withdrawn from the market because it caused kidney failure because patients took it longer than indicated – the drug was effective and safe, however it failed due to patient compliance. Also, approved drugs first need to be integrated into treatment regimens of doctors and hospitals. This might sound trivial, however, these decisions are not only based on medical merits of drugs, e.g. they are also influenced by lobbying (Miller *et al.*, 2007; Ventola *et al.*, 2011). These are only but a few 'proof of principle' examples, that illustrate that there are non-controllable factors outside the scope of research, that further complicate a prospective benefit assessment in project evaluation.

The American association for laboratory animal science ('Aalas') and the European federation of laboratory animal science association ('Felasa') working group on the HBA states that 'Since HBA drives ethical reflection and discussion on current practices, it is important for building public support...' (Brönstad *et al.*, 2016). While we agree, that the Directive and the HBA were formulated with good intentions we think that they might actually not build but rather erode public support in the long term. In our opinion this is due the fact that scientists are incentivized to speculate on the potential societal benefits of their research in project proposals and thus might be incentivized to promise too much. This might be a problem for two reasons. First, overstating societal benefits makes project evaluation more difficult for competent national authorities. Furthermore, 'non-technical project summaries' are a mandatory requirement in writing a project proposal and they are open to the public. Thus, repeated failure of science to meet self-proclaimed goals (i.e. societal benefits) might erode the trustworthiness of science in the long term.

The cancer drug *Ipilimumab* and the insulin inhalator *Exubera* as 'proof of principle' examples

In the following we will use *Exubera* and *Ipilimumab* as two 'proof of principle' examples to demonstrate the logic, methodological and practical flaws of the HBA in current project evaluation. It was shown that the discovery and market approval of the cancer drug *Ipilimumab* was mainly contingent on 433 basic and applied research publications over 46 years (Williams *et al.*, 2015). However, the HBA with its focus on prospective assessment of potential societal benefits implies a rather direct causality between a research project and the generation of a new therapeutic drug. When looking at the above example, the nature of this causality becomes rather indirect, i.e. not a single research project generates societal benefit, but rather literally hundreds of projects over many decades. If one acknowledges the logic and methodological limitations of prospective assessment, i.e. in each of these studies the outcome of the research and the translatability of the results are uncertain, then it becomes salient that a prospective benefit assessment of single research projects in animal research that goes beyond the assessment of knowledge is not plausible.

Furthermore, the financial failure of the Insulin inhalator Exubera is a good example to demonstrate the practical limitations of prospective benefit assessment. Exubera, although effective and safe, never really generated a societal benefit, since it was taken off the market already after one year of its approval due to low sales numbers. The main reason for the low sales numbers were not limited efficacy or safety issues, but is rather comical and simple – the inhalator was just too big (Heinemann, 2018). The device was, from a scientific point of view smartly designed to optimize insulin application into the deep lung, but did not accommodate for the patient's wish for discretion. If you used this inhalator in a public space you would draw a lot of attention. Also, inhalation of a large dose of insulin was time-consuming and the teaching efforts necessary to use the inhalator were underestimated by the manufacturer (Heinemann, 2018). This example illustrates that even in cases, where one assumes that all the speculations regarding potential benefits gained from a research project will hold true, i.e. generation of Insulin therapeutic; it is still uncertain what actual societal benefit might arise from it, due to uncontrollable, extra-scientific factors.

Another striking example is the discovery of CRISPR-Cas9. This obscure microbial system, discovered 20 years ago in a salt marsh at the Costa Blanca in Spain is revoluzioning agriculture, research and medicine (Lander, 2016). Interestingly, the early research on CRISPR was not a quest to edit DNA, nor a study of human diseases. It was a hypothesis-free study of odd DNA sequences with unknown biological function without any foreseeable benefit to society. This is not to say that 'hypothesis-driven' research is not very important but rather an acknowledgement that big scientific breakthroughs are often serendipitous and completely unpredictable. Since the early work on CRISPR was done in prokaryotes, these research proposals did not have to pass a HBA. However, it is quite obvious that these proposals would have had a hard time demonstrating potential societal benefits.

Conclusion

Summarizing arguments from recent publications (Grimm *et al.*, 2017; Grimm and Eggel, 2017) we have argued that the HBA suffers from a logic flaw connected to the uncertainty of the outcome of an experiment and a methodological flaw connected to the uncertainty associated with the translatability of knowledge gained in model organisms. Furthermore, we have shown that a practical flaw regarding non-controllable, extra-scientific factors further exacerbates the problem of prospective benefit assessment.

Also, we believe that scientists might be incentivized to overstate the potential societal benefits of their research, which could further complicate project evaluation for competent authorities from a practical perspective. Next, we have argued that the focus on societal benefits might not only have implications for the HBA, but also might have consequences regarding the credibility of research in the long term (Grimm and Eggel, 2017; Grimm *et al.*, 2017).

We have argued that prospectively assessing societal benefits is vague speculation at best and is thus not well-suited for legal and ethical evaluation of project proposals. We believe that knowledge can be more accurately assessed prospectively compared to societal benefits. Thus, we propose to replace the current HBA with a harm-knowledge-analysis which evaluates the expected knowledge contribution of a project to a research field rather than societal benefits. To maximize epistemic benefit, project evaluation would focus on scientific factors, e.g. scientific merit, strength of hypothesis, experimental setup, among others. The parameters described are not new to scientists. These are parameters that are already used in peer-review and thus one can expect researchers to have little trouble complying with them. By focusing on knowledge, rather than benefit, our proposal would also overcome the uncertainty regarding translatability and the problem associated with uncontrollable extra-scientific factors. Also, scientists would no longer be incentivized to promise too much and thus, the HKA might be beneficial for the credibility of science. Importantly, our model separates scientific from ethical assessment in project evaluation. We believe that the decision on the ethical acceptability of animal research should not be decided by competent authorities but should be decided in the political arena.

References

Aronson, J.K. (2006). Rare diseases and orphan drugs. *Br. J. Clin. Pharmacol.* 61, 243-245.

Bateson, P. (1986). When to experiment on animals. *New Sci.* 30-32.

Bout, H.J., van Vlissingen, J.M.F. and Karssing, E.D. (2014). Evaluating the ethical acceptability of animal research. *Lab Anim.*

Brønstad, A. *et al.* (2016). Current concepts of Harm-Benefit Analysis of Animal Experiments. Report from the AALAS-FELASA Working Group on Harm-Benefit Analysis – Part 1. *Lab.Anim.* 50, 1-20.

European Parliament. Directive 2010/63/EU Available at: http://eurlex.europa.eu/LexUriServ/LexUriServ.do?uri=OJ:L:2010:276:0033:0079:en:PDF. Accessed 19 January 2018.

European Commission Expert Working Group. Working document on Project Evaluation and Retrospective Assessment. Brussels: European Commission, 2013.

Grimm, H. and Eggel, M. (2017). White Paper and Colourful Language: Toward a Realistic View of Animal Research. *Altern Lab Anim* 101-103.

Grimm, H., Eggel, M., Deplazes-Zemp, A. and Biller-Andorno, N. (2017). The Road to Hell Is Paved with Good Intentions: Why Harm-Benefit Analysis and Its Emphasis on Practical Benefit Jeopardizes the Credibility of Research. *Animals* 7.

Heinemann, L. (2008). The Failure of Exubera: Are We Beating a Dead Horse? *J. Diabetes Sci.Technol. Online* 2, 518-529.

Hirt, A. Maisack, C. and Moritz, J. (2016). *Tierschutzgesetz: TierSchG.* (Verlag Frans Vahlen).

Hunter, E. B., Johnston, P. E., Tanner, G., Pinson, C. W. and Awad, J.A.B. (1999). (Duract) associated hepatic failure requiring liver transplantation. *Am J Gastroenterol.*

Laber, K. *et al.* (2016). Recommendations for Addressing Harm-Benefit Analysis and Implementation in Ethical Evaluation – Report from the AALAS-FELASA Working Group on Harm-Benefit Analysis – Part 2. *Lab. Anim.* 50, 21-42.

Lander, E.S. (2016). The heroes of CRISPR. Cell. 2016;164(1):18-28.

Miller, J.D. (2007). Study affirms Pharma's Influence on Physicians. *J Natl Cancer Inst.*

Qureshi, Z.P., Seoane-Vazquez, E., Rodriguez-Monguio, R., Stevenson, K.B. and Szeinbach, S.L. (2011) Market withdrawal of new molecular entities approved in the United States from 1980 to 2009. *Pharmacoepidemiol. Drug Saf.* May 2011 doi:10.1002/pds.2155

Scharmann, W. and Teutsch, G.M. (1994). Zur ethischen Abwägung von Tierversuchen. *ALTEX 11*, 191-198 (1994).

Stafleu, F.R., Tramper, R., Vorstenbosch, J. and Joles, J.A. (1999). The ethical acceptability of animal experiments: a proposal for a system to support decision-making. *Lab. Anim.*

Ventola, C.L. (2011). Direct-to-Consumer Pharmaceutical Advertising. *Pharm. Ther.* 36, 669-684.

Williams, S.R., Lotia, S., Holloway, A.K. and Pico, A.R. (2015). From scientific discovery to cures: bright stars within a galaxy. *Cell.*

65. Raising the stakes in the stakeholder theory: should animals be considered stakeholders by businesses that affect them?

A. Molavi and F.L.B. Meijboom*
Utrecht University, Ethics Institute, Janskerkhof 13, 3512 BL Utrecht, the Netherlands; a.molavi@uu.nl

Abstract

Within business ethics literature, a spectrum of definitions for the term 'stakeholder' has been put forward to serve different theories of corporate social responsibility. Stakeholder theories often come with the normative idea that a business has, at least, a fiduciary duty, and at most, a moral duty, to protect the interests of its stakeholders. Therefore, how a stakeholder is defined usually shapes a stakeholder theory. In this paper, I show that there has been a clear move from acceptance of stakeholdership under the minimum criterion of 'actively participating in and benefiting from the system' towards 'being affected by the system'. I will argue why the reasoning behind this move should not necessarily stay human-centred. What elevates the minimum criterion for humans can be applied to animals used in businesses such as ones in biomedical research and the food industry. This will of course pose challenges as to the limitations on defining what constitutes having a stake in a corporate system in non-human stakeholders. I will respond to these challenges by showing that the above-mentioned move on minimum criterion has only become possible by taking into account what I will call 'outside attributes'. Outside attributes are those attributes that manifest themselves outside the primary function of a business. For example, private life interests of an employee count as an outside attribute with respect her workplace. My hypothesis is that looking at outside attributes of laboratory animal models or animals raised for food can qualify them for a status of stakeholdership. The consequence of this change in our attitude would be that our scope of understanding animal welfare in these areas will have to reach beyond attributes observed within the confines of research and food chain.

Keywords: stakeholder theory, corporate responsibility, moral status of animals

Background on the stakeholder theory

There is abundant literature on stakeholder theory and there are many versions of it in organizational management and business ethics literature. Before answering the question whether the stakeholder theory should be expanded to include animals as stakeholders, we have to identify some common grounds in lines of arguments developed in the past three decades. Then, based on a set of common grounds, I will attend to the following questions: (1) should animals affected by businesses be considered a stakeholder group; (2) do businesses have a moral duty to protect animals' interests as stakeholders. A common way of making sense of where the stakeholder theory came from and what it has meant to business ethics is too look at its early development, particularly in Edward Freeman's works.

The stakeholder theory was developed in the 1980s to critique how the power within corporations are organized in the modern capitalist system. The problem lied in the traditional capitalist format that would usually subscribe to a 'shareholder view'. This view states that the sole fiduciary duty of a corporation's manager is to protect and further the interest of her shareholders. The stakeholder theory, in response, argues for giving due consideration to the interests of all stakeholders affected by the corporation. Stakeholders are people who have a stake or claim on the corporation. Besides the shareholders, they include employees, customers, suppliers and the local community. Edward Freeman's formulation of the theory in his 'Strategic management: a stakeholder approach' (1984) is widely considered to have started the debate that spanned the next few decades:

EurSafe 2018 – DOI 10.3920/978-90-8686-869-8_65, © Wageningen Academic Publishers 2018

> My thesis is that I can revitalize the concept of managerial capitalism by replacing the notion that managers have a duty to stockholders with the concept that managers bear a fiduciary relationship to stakeholders (...). Each of these stakeholder groups has a right not to be treated as a means to some end, and therefore must participate in determining the future direction of the firm in which they have a stake.

Freeman's move from 'instrumental 'to 'deontological' position

Freeman's motivation for 'Strategic management' was to offer a practical framework for business management strategies, which sets out to extend corporate responsibility beyond shareholders' interests. His approach in this book to business is primarily an 'instrumental' one. However, as he proclaims, it has a normative component to it. But it is not normative in the sense that it advocates a particular moral position over others. It, instead, triggers a much-needed discussion about opposing moral views on what constitutes ethical management. For example, the proposition that managers bear 'fiduciary relationship' towards stakeholders has been interpreted in different ways (Gibson, 2000). The reason is that this relationship can either be framed in purely strategic terms which can work without a moral component; or it can be seen as entailing a strong moral duty which would beg the following question: On what grounds should we accept a firm's relationship with its stakeholders as being morally charged and not of a solely fiduciary nature? Freeman does not subscribe to an ethical framework until much later in another paper (1997) where he defends a moral duty on deontological grounds inspired by John Rawls's 'Theory of justice'.

According to Rawls (1971), through an abstract thought experiment, in which all stakeholders in an organized community are stripped of their socially constructed attributes, such as gender, ethnicity, wealth, etc., we can create a moral paradigm for fair treatment of all stakeholders. He calls this thought experiment the 'original position'. The basic idea is that in an organized community, the only way to ensure a social contract free from prejudices is to act on basic facts of the world. In the context of a corporation's relationship with its stakeholders, the basic facts will be how and to what extent each stakeholder group or individual is affected by the business. To some business ethicists, including Evan and Freeman (1988), This might mean that facts, such as an stakeholder's capacity to benefit a business in profits or her ownership of stocks, should be rendered irrelevant in determining a fair fiduciary relationship between the management and the stakeholders.

Inspired by a Rawlsian moral paradigm, Freeman moves from a 'instrumental' position in 'Strategic management' to a deontological one where he focuses specifically on conglomerate forms of corporations whose influence can be compared to those of governments. A contemporary example for such corporation is Apple which is set to become the first trillion-dollar company in value this year (Neate, 2018). Freeman (1997) argues that, in each stakeholder theory, there is a normative core that is either explicit in the literature or implicit. The power of corporations today coupled by growing influence of stakeholder groups, including ones largely ignored by majority of stakeholder theories, he thinks, calls for provision of a clear normative framework (Phillips, 2003). This normative framework is meant to support the following argument:

> P1. There are stakeholders that are in a fiduciary relationship with a firm's management by virtue of having shares in the company. The firm, by definition, has a clear duty towards these stakeholders because the function of a firm centres around reciprocal relationship between the firm and its shareholders based on self-interest and towards the shared goal of making profit.

P2. There are, however, stakeholders who do not stand in the P1 relationship to the firm. These could include employees, customers and the local community. The firm, by definition, does not have a clear duty towards these stakeholders on the basis of self-interest and a shared goal.

P3. Stakeholders in P2 are still affected by the actions of the firm. This effect could be beneficial or harmful. Moreover, this effect could be based on a consensual or non-consensual interaction between the firm and its stakeholders.

P4. All members of society have rights and interests, and these rights and interests are protected by the law of the country.

P5. By virtue of being affected by businesses, stakeholder individuals and groups have legitimate claims and obligations towards businesses.

Therefore,

C1. Businesses have fiduciary duties towards all stakeholders.

Freeman's Rawlsian approach applied to the above argument amounts to the following conclusion:

C2. The normative core of this redesigned contractual theory will capture the liberal idea of fairness if it ensures a basic equality among stakeholders in terms of their moral rights as these are realized in the firm (Freeman, 1997).

An objection to the Rawlsian approach is that there is an argumentative flaw to automatically assume there is a normative core to any stakeholder theory that affirms some sort of fiduciary duties towards non-shareholder stakeholders. Marcoux (2003) argues that shareholder-manager relation should hold a special moral status that is not expandable to other stakeholders based on the following premises:

- if some relations morally require fiduciary duties, and
- the shareholder-manager relation possesses the features that make fiduciary duties morally necessary to those relations, then
- stakeholder theory is morally lacking.

This objection does not necessarily dispute the idea that a firm's manager has moral duty to protect the interest of people affected by her company who are not shareholders. But, it highlights a difficulty in reaching from:

P4. All members of society have rights and interests, and these rights and interests are protected by the law of the country.

To

P5. By virtue of being affected by businesses, stakeholder individuals and groups have legitimate claims and obligations towards businesses.

This difficulty has been addressed by making a moral distinction between different types of stakeholders: Primary and Secondary. Primary stakeholder are those who have a reciprocal relationship with the firm manager, usually based on a contract, such as stockholders, customers, suppliers, creditors and employees. Whereas, secondary stakeholders are those who are not in a direct relationship with the firm, yet, they do affect the firm and are affected by the firm themselves, such as the general public and activist groups and the media (Carroll, 1993). But this response seems to create even more problems, as there are stakeholders that fall in between the direct and indirect types. Children or severely mentally challenged among the general public for example could only be passively affected by actions of a firm and not engage with the form in the sense the above types do. Moreover, for the purposes of this paper, going beyond an anthropocentric view of the stakeholder theory also renders animals for commercial use as possible contender for the in-between stakeholder group.

Animals as stakeholders

So far, stakeholder theories have only been concerned with human stakeholders. The contribution of animals to the businesses that use them for profit is usually considered to be merely of an instrumental value. However, in what follows, I argue that the same line of reasoning used by Freeman (1984, 1997, 2010) or Donaldson (1993, 2002) can be applied to raising the status of animals to stakeholders. What triggered a shift in attitude in business management from a purely fiduciary relationship to a morally charged relationship seems to be taking into accounts stakeholder group attributes that do not strictly fall under a reciprocal profit scheme.

Ensuring adherence to fiduciary duties in shareholder-manager relationship certainly does not suffice to maintain a corporate system on its own. This much has always been clear to the defenders of the traditional capitalist format of management. However, the practical implications of adding a normative component to management strategies threaten a formula that drives profit in the traditional format.

Freeman, along with other major contributors to the field, initially thought this formula has to change on grounds that it simply does not work, hence the need for the stakeholder theory. They realized that there are other stakeholders worthy of consideration for a managerial strategy to survive in long-term. Because non-shareholder stakeholders could help a firm prosper, expand and flourish; but they also had equal power to discredit or bankrupt a firm.

An instrumental account of the stakeholder theory described in 'Strategic management', holds firms accountable towards stakeholders and offers ways to fix stakeholder relationship through a new formula: care for stakeholders leads to steady and secure profit pattern. However, a normative account questions the very raison d'être of businesses as profit-making machines. Consideration of attributes that do not serve the purpose of making profit enables a normative account to be compelling. These attributes are crucial part of stakeholder's existence and their expression could be protected by law. For example, an employee's right to privacy after working hours, or a customer's safety are attributes that might hinder the process of profit-making. Nevertheless, they are protected by virtue of these stakeholders being citizens of a jurisdiction and entitled to certain rights.

A similar line of argument can be developed for non-human stakeholders. The reason such attempt is important is that animals are widely used for food, research, clothing, entertainment, etc. They make businesses that rely on them profitable. Yet, they do not have any stake in those businesses and exploitative treatment of animals in these areas is, more often than not, deemed morally objectionable. Even though there have been considerable development on animal welfare front, the majority of people view animals' value as being purely instrumental and this view is inevitably reflected in corporate behaviour too. The idea that animals should be given a higher moral status which would mandate ethical treatment by humans is, of course, not new (Donaldson and Kymlicka, 2013). However, it is something that has been missing in the stakeholder theory debate. My hypothesis is that, by using Freeman's Rawlsian approach to the stakeholder theory and including animals in the marginal stakeholder group that falls in between the direct and indirect types, it is possible to extend the scope of the stakeholder theory to non-human stakeholders. This hypothesis could be implemented in Freeman's deontological argument as follows:

> P1. There are stakeholders that affect the firm and are affected on a reciprocal basis by virtue of attributes that contribute to profit-making.
>
> P2. There are also stakeholders that affect the firm and are affected by the firm in a non-reciprocal basis by virtue of attributes make them morally significant outside the profit-making paradigm.
>
> P3. Stakeholders, as far as they are being affected by the firm or affect the firm, are worthy of moral consideration, regardless of reciprocity.
>
> P4. In commercial use of Animals, animals are stakeholders by virtue of affecting and being affected in a non-reciprocal basis.
>
> P5. Animals' outside attributes such as ability to feel pain, willingness to live a good life make them morally significant outside the profit-making paradigm.

Therefore,

> C. Animals are stakeholders and businesses that affect them have a moral duty to treat them as such.

References

Carroll, A. (1993). Business and Society: Ethics and Stakeholder Management. Cincinnati: South-Western Publishing.

Donaldson, T. (1993). The Language of International Corporate Ethics. In Business Ethics: Japan and the Global Economy (Vol. 5, pp. 115-131). Dordrecht: Springer. Available at: http://doi.org/10.1007/978-94-015-8183-7_6.

Donaldson, T. (2002). The Stakeholder Revolution and the Clarkson Principles. Business Ethics Quarterly, 12(2), 107-111.

Donaldson, S. and Kymlicka, W. (2013). Zoopolis: a political theory of animal rights. Oxford New York: Oxford University Press.

Evan, W. and Freeman, E. (1988). 'A Stakeholder Theory of the Modern Corporation: Kantian Capitalism', Beauchamp, T. and Bowie, N. (eds.), Ethical Theory and Business, Englewood Cliffs, New Jersey, Prentice Hall pp. 97-106.

Freeman, R.E. (1984). Strategic Management: A Stakeholder Approach, Boston: Ballinger.

Freeman, R.E. (1994). 'The Politics of Stakeholder Theory: Some Future Directions, Business Ethics Quarterly, 4(4), 409-421

Freeman, R. E. (1997). A Stakeholder Theory of the Modern Corporation. In M. Clarkson (Ed.), The Corporation and Its Stakeholders. Toronto: University of Toronto Press. http://doi.org/10.3138/9781442673496-009.

Freeman, R.E. (2015). The Politics of Stakeholder Theory: Some Future Directions. Business Ethics Quarterly, 4(04), 409-421.

Freeman, R.E. (2010). Stakeholder theory: the state of the art. Cambridge New York: Cambridge University Press.

Gibson, K. (2000). The Moral Basis of Stakeholder Theory. Journal of Business Ethics, 26(3), 245-257.

Marcoux, A. (2003). A Fiduciary Argument against Stakeholder Theory. Business Ethics Quarterly, 13(1), 1-24.

Neate, R. (2018). 'Apple Leads Race to Become World's First $1tn Company'. The Guardian. Available at: http://www.jstor.org/stable/3857856.

Phillips, R. (2003). Stakeholder theory and organizational ethics. San Francisco: Berrett-Koehler.

Rawls, J. (1971). A Theory of Justice. Harvard University Press.

Section 12.
Biotechnology

66. The ethical dilemma with governing CRISP/Cas genome editing

*F. Pirscher** *and I. Theesfeld*
Department of Agricultural, Environmental and Food Policy, Faculty of Natural Science III, Martin Luther University Halle-Wittenberg, 06099 Halle (Saale), Germany; frauke.pirscher@landw.uni-halle.de

Abstract

CRISPR/Cas is one of the newly developed genome editing techniques, that is expected to have great innovative potential in the very early stages of the food change, i.e. plant breeding. CRISPR/Cas remains within species boundaries. The cisgenetic plants could also be the outcome of conventional breeding or natural evolution. Compared to former genetic modification (GM) techniques, this system is considered as relatively easy to apply, more precise, quicker and much cheaper. This new technological development has induced a debate on the adequacy of the current governance system within the EU, due to the current non-traceability of the modifications. A controversy point is whether cisgenic plants should be viewed as GMOs and thus fall under the current EU law for GMO approval, which was developed at a time when GM meant introducing transgenic DNA into a genome. The current debate on regulation is very much science based and focuses mainly on risk management aspects. However, there are a number of fundamental trade-offs to be considered in the context of governing CRSPR/Cas genome editing. Certain characteristics of the new technology set the key-requirements for governance options. Because it is easy to apply, cheaper and much quicker and non-traceability its outreach will be immense, therefore governance has to reflect the balance between the diverging interests and values and requires a continuous discourse between science and society. We will shed light on the ethical dilemma that, on the one hand, fast regulative action is needed to catch up with the developments and, on the other hand, the required societal debate needs time.

Keywords: CRISPR/Cas, genome editing, governance, ethical dilemma, agricultural ethics

Introduction

CRISPR/Cas is one of the newly developed genome editing techniques, that is expected to have great innovative potential in plant breeding by speeding up breeding, increasing yields and allowing plant production under less favourable conditions. It allows for modifications of genes by adding, cutting out or suppressing certain gene sequences of the DNA. Compared to former genetic modification (GM) the technology is considered as relatively easy to apply, more precise, quicker and much cheaper (Baker, 2014).

In its current application in agriculture CRISPR/Cas remains within species boundaries. These so called cisgenic plants could also be the outcome of conventional breeding or natural evolution. In this case they are called nature-identical genetically modified organisms (nGMOs). The possible molecular equivalence of nGMOs and the fact the CRISPR/Cas is based on a defence mechanism naturally occurring in bacterial cells has raised a debate whether CRISPR/Cas modified plants should be viewed as GMOs at all. The decision regarding the classification is currently the basis for governing the respective products. The European law has separate legislative schemes for the approval of GM and non-GM varieties (EU Directive 2001/18/EC; Council Regulation (EC) No 2100/94 on Community plant variety rights). The regulations differ substantially in cost and time needed for product approval. EU jurisprudence on GMOs was developed at a time when GMOs meant introducing transgenic DNA into a genome. Thus,

EurSafe 2018 – DOI 10.3920/978-90-8686-869-8_66, © Wageningen Academic Publishers 2018

the regulation does not distinguish between transgenic and cisgenic plants, and defines GMOs by the process of creation and not its outcomes.

Characteristics of CRSPR/Cas that call for special attention when designing a governance system

Regulation, ban or moratorium, and laissez-faire are three different general approaches to govern a new technology. Due to international trade agreements a ban or a moratorium do not seem to be a viable option. Thus, the decision on how CRISR/Cas modified crops should be governed has to be taken between regulation and laissez-faire. However, in the EU context it means more or less regulation, not purely laissez-faire.

There are particularly five characteristics of CRISPR/Cas and its related transactions that call for special attention when designing a governance system that is based on regulations:

1. The naturalness of the CRISPR/Cas mechanism and the possible natural identity of the outcome: While former GMOs meant introducing transgenic DNA into a genome this is no longer the case.
2. The non-traceability of the modification in the resultant organism: Up to now conventional analytical methods used by inspection bodies (Ahmed, 2002; Waltz, 2016), such as PCR (polymerase chain reaction), cannot detect such traces in the modified product, although it has been argued that CRISPR/Cas leaves small amounts of foreign DNA in the genome (Kim and Kim, 2016).
3. The easy-to-use and inexpensive method: This will lead to a decentralisation of knowledge and use (Araki and Ishii, 2015; Baltimore *et al.*, 2015).
4. The uncertainty about off-target alterations and risk: Some authors stress unintended consequences of genetic modifications, such as horizontal gene transfer, i.e. the naturally occurring transfer of the inserted gene sequences to neighbouring plants, could be reduced by CRISPR/Cas as well (Khlestkina and Shumny, 2016; Paul and Qi, 2016). Critical biologists argue that although it is more precise than former GM techniques it is still not free of errors (Then, 2016; Steinbrecher, 2015).
5. The speed of breeding: In contrast to conventional breeding, CRISPR/Cas genome editing does not require labour-intensive and time-consuming screens to identify the desired plant mutants (Baker, 2014; Belhaj *et al.*, 2015).

In this contribution, we do understand governance as a balancing act between plural values in a society (Vatn, 2016), thus it is more than a technical issue to set regulatory boundaries. Decision are embedded into a general debate on the acceptability of intensive agriculture, including the highly concentrated seed market, the distribution of property rights to genetic resource and the patents to GM plants as well as consumer, seed and food sovereignty. Society will be more involved which brings in a novelty in genome editing governance, i.e. the necessity to consider the time dimension, as it requires time to work out widely agreed rules.

Ethical considerations within the regulatory options

Ad 1. The fact the CRISPR/Cas remains within species barriers and can produce nature-identical outcomes prevents an easy answer to the question, what defines a GMO, the product or process. Some authors (e.g. Schouten *et al.*, 2006; Hartung and Schiemann, 2014; Huang *et al.*, 2016) opt for the product and call to evaluate new plant breeding techniques only according to the new traits and the resulting end-product, which would effectively amount to a laissez-faire approach. Others argue that the intrusion into the cell is a risk in itself because the consequences for the plant are insufficiently known (Then, 2016; Steinbrecher, 2015). In both cases, proponents and opponents hold a consequentialistic perspective. But there are also principal concerns: It is the fundamental critique at human intrusion into natural processes, which further blurs the distinction between 'living' versus 'non-living', 'nature' versus 'artificial' (Ried *et al.*, 2011). Therefore, opponents of genome editing ask whether there is a red line that should not be crossed and how it is defined. As long as there is no clear answer to these open questions, critical voices will call for an urgently needed process-based categorisation of CRISPR/Cas modified products as GMO, which is in line with the precautionary principle.

Against this background, naturalness as a moral category and a baseline for regulatory design no longer provides a clear argument for rejecting the new technology, as it was the case in the former GM debate. Rather, it can be invoked by the technology's proponents as well, due to the supposed nature-identity of CRISPR/Cas-modified products. Therefore, ethical conflicts now arise between different concepts of naturalness; they have moved away from the former focus on risk but also from the general tension between teleological and principle ethics (Pirscher *et al.*, 2018).

Ad 2. If we are not able to determine how the new variety came into being, we cannot start monitoring, and thus governing, from the end-product. Thus, the non-traceability allows only for a process-based regulatory approaches. However, process-based regulation is much more difficult and costly to set up. Especially in case of a centralised regulatory authority like the EFSA the monitoring is more complicated when the number of actors is high. While the low costs of the technology might lead to a diversification of the seed market the high costs of regulation will counteract this development.

An important ethical trade-off to be made is between strict regulation with considerable compliance costs and the widespread use of the technology with associated risks. A strict regulatory framework, particularly process regulation, would lead to a high cost share for companies to meet the various requirements. As a consequence, use of the new technology would most likely be limited to creation of highly profitable crops (Voytas and Gao, 2014). That would limit the benefits of providing a relatively easy-to-use technology, possibly in a partly open-access manner, also to actors who might be interested in increasing the agricultural productivity of minor crops, which lack high profit margins. Such actors might not be able to bear the considerable compliance costs of demanding governmental regulatory packages.

Ad 3. As CRISPR/Cas genome editing is a very easy-to-use and inexpensive method it can be viewed as having the potential to diversify the market for biotechnology, by reducing market entry barriers. Before the advent of CRISPR/Cas, genetic engineering was an enterprise with very high upfront investment costs, which partly explains the highly concentrated market structure of biotech industries. Now, the technique can arguably be applied at very low cost also by small biotech firms, non-profit organisations or public institutions. Moreover, there seems to be a widespread practice of open access to information about this method, such as freeware computer programmes for designing guide RNAs (Khlestkina and Shumny, 2016). This could also lead to a rising role of non-profit organizations and public institutions in the application of GM technology. A decentralized access structure would require a rather diversified assignment of property rights. Current biotechnology regulations are committed to strict protection of intellectual property rights of a small number of large multinational companies. CRISPR/Cas genome editing has the potential to 'open up' this field and add further complexity to the debates about 'patenting life' (Sherkow and Greely, 2015).

More generally, because CRISPR/Cas is likely to result in a surge of innovations, including new varieties of crops and livestock, it will likely trigger a wave of new patents (Webber, 2014). An important and also ethical question, however, is how such patents will be used if their holders are for example non-profit organizations.

Ad 4. Widespread use of CRISPR/Cas, including non-profit organisations and smaller firms, might lead to an increase in biotech-related risks, both because of the diversification of sources of genetic modifications and because of the relative inexperience of some of the actors involved. Even with CRISPR/Cas, genetic engineering is a demanding and complex enterprise, involving large risks and high levels of uncertainty. The current biotech market players have many years of experience in dealing with these issues. The ability of smaller players to handle even existing hazards can be assumed to be smaller than in the case of experienced companies, posing new challenges for regulation. Whether reducing biotech-related risks means hampering decentralization involves ethical questions, too

Ad 5. Finally, other than conventional breeding, CRISPR/Cas genome editing does not require labour-intensive and time-consuming screens to identify the desired plant mutants (Baker, 2014; Belhaj *et al.*, 2015). Additionally, CRISPR/Cas technology is undergoing a rapid evolution itself. This characteristic is extremely relevant for all that has been said above, as it puts the development of adjusted and flexible regulatory governance systems under a time constraint.

Conclusions

The ethical challenge to govern CRISPR/Cas genome editing, goes beyond questions of acceptable consequences and side-effects, but includes fundamental societal questions and moral judgments about the need for and the way of adapting nature for human needs. In fact, this ethical dimension of governance has been overlooked in many recent publications on the topic (Huang *et al.*, 2016; Ishii and Araki, 2016; Pollock, 2016), whose authors effectively advocated for discarding supposedly 'irrational fears' of the wider public and a product-based regulation of CRISPR/Cas – which, in the case of nGMOs, would effectively mean non-regulation or, more precisely, regulation in the same extent as for conventionally bred crops. This leads to circumventing the societal and value debates.

However, there are a number of fundamental trade-offs to be considered in the context of governing CRISPR/Cas genome editing. If nature-identical GMOs are not traceable, their regulation, including patenting, cannot be based on the end-product. If CRISPR/Cas offers a more widespread, even open-access use, it may lead to market diversification, but it likewise brings in new actors without experience in handling risks. If regulators need to act fast to keep up with the technological development, the time required for a thorough normative societal debate may not be available. It is almost a question whether reaction to fast technological development in general denies the option for solid societal debates about its use and expansion. This time constraint leads to the ethical dilemma, which we want to highlight.

To avoid reducing the role of governance to ex-post legitimation of any kind of technological innovations, there is an urgent need for institutionalizing continuous discourse between science and society. Whether this new line of discussion will deepen or blur the division between proponents and opponents of the new technology will strongly depend on whether natural sciences start regarding the ethical dimension intrinsic rather than extrinsic to their task (Bruce, 2002). Thus, on the one hand, it is obvious that CRISPR/Cas technology requires a wide and continuous societal debate about if at all, what and how to regulate and on the other hand, the characteristics of that technology require fast regulatory reactions. Considering the time dimension poses dilemmas to ethical consideration, that currently lack attention.

References

Ahmed, F.E. (2002). Detection of genetically modified organisms in foods. Trends Biotechnology 20: 215-223.

Araki, M. and Ishii, T. (2015). Towards social acceptance of plant breeding by genome editing. Trends Plant Science. 20: 145-149.

Baker, M. (2014). Gene editing at CRISPR speed. Nature. Biotechnology. 32: 309-312.

Baltimore, D., Berg, P., Botchan, M., Carroll, D., Charo, R.A., Church, G., Corn, J.E., Daley, G.Q., Doudna, J.A., Fenner, M., Greely, H.T., Jinek, M., Martin, G.S., Penhoet, E., Puck, J., Sternberg, S.H., Weissman, J.S. and Yamamoto, K.R. (2015). A prudent path forward for genomic engineering and germline gene modification. Science 348: 36-38.

Belhaj, K., Chaparro-Garcia, A., Kamoun, S., Patron, N.J. and Nekrasov, V. (2015). Editing plant genomes with CRISPR/Cas9. Current Opinion in Biotechnolology 32: 76-84.

Bruce, D.M. (2002).A social contract for biotechnology: shared visions for risky technologies? Journal of Agricultural and Environmental Ethics 15: 279-289.

Council Regulation (EC) No 2100/94 of 27 July 1994 on Community plant variety rights. Official Journal of the European Communities No L 227/1:1-30. Available at: http://eur-lex.europa.eu/legal-content/EN/TXT/?uri=CELEX%3A31994R2100.

European Parliament. (2001). 'EU Directive 2001/18/EC of the European Parliament and of the Council of 12 March 2001 on the deliberate release into the environment of genetically modified organisms and repealing Council Directive 90/220/EEC.' Official Journal of the European Communities L 106: 1-39. Available at: http://eur-lex.europa.eu/legal-content/EN/TXT/?uri=CELEX:32001L0018.

Hartung, F. and Schiemann, J. (2014). Precise plant breeding using new genome editing techniques: opportunities, safety and regulation in the EU. The Plant Journal. 78: 742-752.

Huang, S., Weigel, D., Beachy, R.N. and Li, J. (2016). A proposed regulatory framework for genome-edited crops. Nature Genetics. 48: 109-111.

Khlestkina, E.K. and Shumny, V.K. (2016). Prospects for application of breakthrough technologies in breeding: The CRISPR/Cas9 system for plant genome editing. Russian Journal of Genetics. 52: 676-687.

Kim, J. and Kim, J.-S. (2016). Bypassing GMO regulations with CRISPR gene editing. Nature. Biotechnology. 34: 1014-1015.

Paul, J.W. and Qi, Y. (2016). CRISPR/Cas9 for plant genome editing: accomplishments, problems and prospects. Plant Cell Reports 35: 1417-1427.

Pirscher, F. Bartkowski, B., Theesfeld, I. and Timaeus, J. (2018). Nature identical outcomes, artificial processes: Governance of CRISPR/Cas genome editing as an ethical challenge. In: James, H. S. (ed.): Ethical tensions from New Technology: The Case of Agricultural Biotechnology. Wallingford. CABI (forthcoming).

Pollock, C.J. (2016). How should risk-based regulation reflect current public opinion? Trends Biotechnology 34: 604-605.

Ried, J., Braun, M. and Dabrock, P. (2011). Unbehagen und kulturelles Gedächtnis. Beobachtungen zur gesellschaftlichen Deutungsunsicherheit gegenüber Synthetischer Biologie, in: Dabrock, P., Bölker, M., Braun, M. and Ried, J. (eds.), Was Ist Leben – Im Zeitalter Seiner Technischen Machbarkeit? Beiträge Zur Ethik Der Synthetischen Biologie. Verlag Karl Alber, Freiburg; München, pp. 345-368.

Sherkow, J.S. and Greely, H.T. (2015). The history of patenting genetic material. Annual Review of Genetics 49: 161-182.

Schouten, H.J., Krens, F.A. and Jacobsen, E. (2006). Cisgenic plants are similar to traditionally bred plants: International regulations for genetically modified organisms should be altered to exempt cisgenesis. EMBO Reports, 7, 750-753.

Steinbrecher, R.A. (2015). Genetic engineering in plants and the 'New Breeding Techniques (NBTs)'. Inherent risks and the need to regulate (Econexus Briefing).

Then, C. (2016). Synthetic gene technologies applied in plants and animals used for food production: Overview on patent applications on new techniques for genetic engineering and risks associated with these methods. Testbiotech, Munich.

Vatn, A. (2016). Environmental governance: Institutions, policies and actions. Edward Elgar, Cheltenham, UK; Northampton, MA.

Voytas, D.F. and Gao, C. (2014). Precision genome engineering and agriculture: Opportunities and regulatory challenges. PLOS Biolology 12, e1001877.

Waltz, E. (2016). Gene-edited CRISPR mushroom escapes US regulation. Nature 532: 293.

67. Could crispy crickets be CRISPR-Cas9 crickets – ethical aspects of using new breeding technologies in intensive insect-production

M. Gjerris[1*], *C. Gamborg*[1] *and H. Röcklinsberg*[2]
[1]*Department of Food and Resource Economics, University of Copenhagen, Rolighedsvej 25, 1958 Frederiksberg C, Denmark;* [2]*Dept. of Animal Environment and Health, Swedish University of Agricultural Sciences, P.O. Box 7068, 750 07 Uppsala, Sweden; mgj@ifro.ku.dk*

Abstract

The use of biotechnological tools in relation to insects have mainly been discussed in relation with regard to making crops resistant to certain insects or to combatting diseases, e.g. by making genetically modified mosquitoes to reduce malaria. In these cases, typically raised issues are risks to humans, insects becoming resistant to crop traits and to broader environmental consequences. However, novel biotechnological possibilities and growing interest in insects for food and feed could also raise additional ethical issues. In 2013 FAO published a report on edible insects promoting the use of insects in large-scale intensive production systems as an alleged more sustainable source of protein for food and feed than traditional livestock production. So far, a possibility that has received only little attention is the use of biotechnology to modify relevant insect species to speed up domestication to achieve higher productivity or better disease resistance. In this paper we explore some of the ethical aspects related to such a development to show what the discussions are about: E.g. the seemingly legal discussion whether insects produced using biotechnological tools such as CRISPR-Cas9 should warrant different labelling than conventionally bred insects? Would ethical concerns differ from when producing non-altered insects for food and feed? Will knowledge that the insects have been produced through biotechnology make it even more difficult to relate to them as anything other than protein units? Ethical considerations regarding animals often depend on human experiences of empathy for animals to ensure awareness of their welfare. However, with regard to insect ethics the border of both those aspects is challenged. Can insects experience welfare, and if they can, will we care; that is, can we experience any empathy with them? Moreover, issues related to human interference in a being's genome may be ethically significant. In turn, these issues could have ramifications for the social acceptability of GM insects used for food and feed.

Keywords: insect production, CRISPR-Cas9, animal welfare, animal ethics, social acceptability

Introduction

The introduction of CRISPR-Cas9 as a tool to alter living organisms by editing their genome has brought new optimism to those who see biotechnology as a useful tool to address a series of problems facing food production such as climate change, environmental degradation through use of fertilizers, pesticides, herbicides and diminishing resources such as arable land and water. CRISPR-Cas9 is in this perspective the scientific breakthrough that will enable biotechnology to deliver the solutions that have been hyped since the first GM-crops were introduced in the 1990s (Bomgardner, 2017).

Another development within food production that is gaining more and more attention is the use of insects in food and feed production as a source of protein with a more sustainable profile than more conventional protein sources such as meat from cows, pigs, chicken, milk, eggs or soy imported from South America to Europe to be used as animal feed (Van Huis *et al.*, 2013). Insects are not a novelty in human diets, but it is safe to say that they have not been on the menu in Europe and North America

EurSafe 2018 – DOI 10.3920/978-90-8686-869-8_67, © Wageningen Academic Publishers 2018

for centuries. A number of challenges face such a transition to a more insect-based diet in these areas, which we have described elsewhere (Gjerris *et al.*, 2016).

In this article we will provide an ethical analysis of the issues that could arise, if we combine the two above-mentioned developments. What ethical issues would arise, if a company presented a genetically edited cricket to be used as a crispy snack?

We readily admit that we – even though having searched extensively – have found no current projects that attempt to use CRISPR-Cas9 to edit insects in connection with food or feed production. So far the most prominent project directly involving gene-editing insects are attempts to introduce genetic changes coupled with gene drives into mosquitos to help fight malaria (Neves and Druml, 2017). However, it is not always the task of ethicists to function as fire-fighters, only showing up when a new technology is ready to go. The task is just as much to envision future technological developments. After all, our future ambitions have a tendency of shaping the future and – as Hannah Arendt pointed out – technologies are not neutral agents, but should also be seen as political agents (Arendt, 2013). We therefore feel justified to look into the possible combination of biotechnology and insects used for food and feed to uncover the possible ethical issues that such a development could give rise to.

CRISPR-Cas9: something new or more of the same?

For thousands of years animals have been domesticated for food production, clothing, companionship, entertainment, experimentation, etc. Domestication can be defined as '... a process by which a population of animals becomes adapted to man and the captive environment, by some combination of genetic changes occurring over generations and environmentally induced developmental events recurring during each generation' (Price, 1984: 3). During the twentieth century breeding became a more controlled and predictable practice and in the second half of the century new forms of biotechnology, including gene transfer and cloning, emerged (Sandøe and Gamborg, 2008). In recent years a number of gene-editing tools such as CRISPR-cas9 have been developed and applied to animal breeding, especially for research purposes.

Genome editing (note the term 'editing' instead of 'engineering' cf. Nuffield Council of Bioethics (2016)) is the insertion, deletion, or replacement of a segment of DNA in the genome of an organism. It uses so-called site-directed nuclease, which are proteins able to cause DNA-strand breaks at a specific point in the genetic sequence. Most often the term refers to the use of CRISPR/Cas9 which is a genome editing tool modified from a prokaryotic (i.e. single celled organisms such as bacteria) immune system (Østerberg *et al.*, 2017). The huge potential of genome editing is its proposed ability to rapidly modify undesirable traits and hereby speed up the process of domestication understood as genetic changes for e.g. enhanced productivity (*Ibid.*).

Using genome editing technologies to alter insects relevant in food and feed production has so far received only little attention (see Van Huis, 2017 for a rare example) as it is only in the past decade or less that mass production akin to mini-livestock production has been seriously addressed (Gjerris *et al.*, 2016), and current production systems are not in the process of breeding yet. However, it can be envisioned that if insects are to be produced in large scale for food and feed, the industry will be interested in i-accelerating the adopting target insect species, taking advantage of developments within biotechnology.

Within the context of agricultural crops, it is heavily debated whether genome editing should be regulated like genetic modification or whether it should be regulated like conventional breeding – or somewhere in between (The Norwegian Biotechnology Advisory Board, 2017). There is little doubt that the same discussions will arise, when the technology is applied to insects and that the decision will be an important factor with regard to public acceptance of the technology. Including insects in the diet of the average Westerner already faces a range of challenges (House, 2016), which will only increase, if they will be labelled as GM-food.

In conventional animal breeding for food production purposes the genetic change is achieved through selection of existing natural mutants and selective cross-breeding. In genetic modification there is a genetic change that may involve a transfer of a transgene element with a known function (to a certain degree) to a random location in the chromosome, using e.g. bacteria as a vehicle. When it comes to genome editing the change of the original genotype, need not but can involve transgenesis, but is above all editing of the existing genome by e.g. knocking out genes at certain locations (Georges and Ray, 2017).

Relying only on science to regulate the area of biotechnology, however, disregards the ethical dimensions of biotechnology. This relates to two issues: the generic aspects of genome editing (e.g. should the alleged precision, speed and ease of use of the method give rise to ethical concern), and the applied aspects. In relation to the latter, different ethical perspectives would place the emphasis on different issues such as use, risk, justice, naturalness, sustainability, etc. to mention just some of the key issues that have been part of the GM-discussion (Mielby *et al.*, 2013). Whether to regulate gene-edited organisms as GM-organisms or as conventionally bred organisms is a question that has no clear answer, as it depends (too some degree at least) on value-based interpretations of inherently ambiguous concepts.

The use of genome editing to achieve speedy domestication of insects are thus likely to test legal as well as ethical boundaries and will probably give rise to public debate.

What about insects – welfare and beyond

Ethical considerations regarding animals without exemption address their welfare and often touches upon the human capacity to experience empathy with the animal in question. Insect ethics is interesting as it needs to explore the border of both those aspects. If we accept that at least some species of insects as e.g. crickets candidates to have welfare of their own (Horvath *et al.*, 2013), there is, in principle, something to feel empathy with. We have discussed various aspects of insect welfare and ethics elsewhere (Gjerris *et al.*, 2016; Röcklinsberg *et al.*, 2017). Here we will focus on the difficulty of experiencing empathy with insects and discuss whether adding gene-editing to the possible human-insect interactions will increase these difficulties.

It has been suggested that the common disregard for the theme of insect welfare and ethics is due to – among other things – that insects are so 'different' from humans and therefore difficult to relate to in an emphatic way, in the same way as the suffering of farmed fish seems to evoke less compassion than mammals in comparable situations (Horvath *et al.* 2013; Röcklinsberg *et al.*, 2017). Biotechnological tools for breeding purposes might then further estrange us from insects as it will be even easier to perceive them as artefacts, a development witnessed by the term 'animal bioreactor ' rather than the term 'animal' to describe various kinds of GM-animals (Bertolini *et al.*, 2016; Harfeld *et al.*, 2014) and hence clearly different from humans. Another underlying reason for the absence of empathy towards insects is perhaps less connected to the capacities of the insect itself, but rather to expectations of the moral agent and our cultural training, i.e. lack of tradition to *see* the well-being of an individual insect. Even if these aspects might overlap, the ethical concern for insects can thus have its foundation either primarily in the capacities of the insects, or in the capacity of the moral agent to see and relate to the individual insect, independently of its value as resource and regardless of the mental capacities of insects.

Following Cora Diamond (although it should be noted that Diamond does not explicitly apply the idea to insects when writing on animals) and her idea that the moral agent ought to show 'a kind of loving attention' to all beings with whom we share vulnerability (Diamond, 2001), we propose that a fruitful avenue to follow with regard to insects lies in the expectations towards the moral agent. Hence, relating ethically to insects requires a capacity of basic welfare of one's own and the insight that one's own vulnerability is relevant to understand that of the other, setting details about exact experience of a certain mental state aside. Further, Diamond asks the moral agent to see the inherent injustice in how animals are used for our purposes, and suggests loving attention to be an appropriate response. The capacities of the animal are not decisive, but rather the moral agent's attention to their lives and perception of their 'connection with the good' should stop 'us from treating them as props in our show' (Diamond, 2001: 136).

Considering this in relation to applying biotechnological tools such as CRISPR-Cas9 to insects, the phrase 'culture of care' as frequently used in both health care ethics and lab animal science comes to mind. In the former context, care is also reversely made into an abbreviation for e.g. connectedness, awareness, respect and empathy (CARE), as suggested concepts in such a culture (see Gillin *et al.*, 2017 for an overview of approaches). Regarding lab animal science, fostering a culture of care is mentioned as one crucial task of the research community in Recital 31 of the EU Directive 2010/63/EU. Although it does not include insects as research animals, we have reasons to regard some of them as potential candidates given recent knowledge about their capacities (Horvath *et al.*, 2013) and, it could be argued, should at least give them the benefit of the doubt until further welfare research have clarified the issue: absence of proof should not be taken as proof of absence (Gjerris *et al.*, 2016).

From our point of view virtue ethics presents a reasonable framework for such a precautionary approach. As stated in a recent paper on virtue ethics and animals about which there is qualified scientific uncertainty about their sentience: ' It is a requirement of a morally decent (or virtuous) person that she at least pays attention to and is cautious regarding the possibly morally relevant aspects of such animals' (Knutsson and Munthe, 2017: 213). In relation to biotechnological attempts to modify insects, attention to uncertainty of outcomes and consequences is thus called for.

Given the already existing welfare problems in e.g. intensive animal production systems (Harfeld *et al.*, 2014), increased care for individual insects might seem a far-fetched goal. On the other hand, malfunctions in one area should not relativize our responsibility in other contexts. In the case of applying biotechnology to insects produced for food and feed, one might argue that whether one is motivated by the capacities of the insects or by one's own capacity to be an ethical agent based on a thick understanding of shared vulnerability, only makes a small difference. However, in the former case, it can lead to a continuous call for deeper scientific knowledge about each insect species that runs the risk of overshadowing the responsibility for the individual insects one is in charge of, whereas the latter rather opens for 'loving attention' even if one has difficulties feeling empathy. It might, as Lockwood argues when discussing why to anesthetize insects before experimenting on them, 'cultivate an attitude of respect toward other organisms ... We learn the methods of dissection through practices – and we also learn virtues such as compassion through practice' (Lockwood, 2011). Finally, it could increase a sensibility also towards other beings, which is, in our perspective, in itself a desirable side effect.

Conclusions

New technological possibilities such as CRISPR-Cas9 seem to offer hitherto unforeseen opportunities to change the genetic makeup of living being. When seen in combination with a heavily increased interest in industrial, large scale insect production for food and feed, the potential might seem huge, but the (animal) ethical aspects are largely unexplored. One important issue with regard to such a development will be whether such insect products should be labelled as GM or not. As can be seen in current debates about the use of CRISPR-Cas9 and similar technologies in plant breeding, there are no easy answers to this, partly because of underlying value disagreements.

Animal ethics is often based on empathy with the animal leading to considerations of welfare and respect for the capacities of the moral subject (here: insect). Given that developing empathy for insects for a number of reasons seems difficult, it could be argued that a further responsibility of the moral agent, i.e. citizens, producers, retailers, consumers, etc. of insects for food and feed, is called for, if insects' 'interests' should be safeguarded. Evidently, this call depends on the ethical point of departure. In terms of genome editing not only the future purpose or gain for the product is regarded relevant, but also the 'distance' created by using the insect as a mere tool could well be relevant to take into account – especially as applying the new biotechnological tools to the animals could increase the experience of reification of the animals.

It is not necessarily easy to see each single insect in a production chain, but practical difficulties do not by default justify ethical ignorance. If the ethical foundation lies on the capacities of the moral subject consideration of the welfare of insects is called for. If it instead lies on the moral agent, the goal is to develop a sensibility to one's own vulnerability and attention to the good and the vulnerability of other beings – how strange they may initial seem. These two possible approaches seems pertinent in elaborating on ethics of insects used for food and feed. And even more so, if gene editing tools enter the picture.

References

Arendt, H. (2013). *The human condition*. University of Chicago Press.

Bertolini, L.R., Meade, H., Lazzarotto, C.R, Martins, L.T., Tavares, K.C., Bertolini, M. and Murray, J.D. (2016). The transgenic animal platform for biopharmaceutical production. *Transgenic Research* 25(3):329-43.

Bomgardner, M.M. (2017). CRISPR: A new toolbox for better crops. *Chemical and Engineering News*, 95(24):30-4.

Diamond, C. (2001). Injustice and Animals. In Elliott, C. (ed.) *Slow Cures and Bad Philosophers. Essays on Wittgenstein, Medicine and Bioethics*. Duke University Press Books.

Georges, F. and Ray, H. (2017). Genome editing of crops: A renewed opportunity for food security. *GM Crops & Food* 8:1-12.

Gillin, N., Taylor, R. and Walker, S. (2017). Exploring the concept of 'caring cultures' A critical examination of the conceptual, methodological and validity issues with the 'caring cultures' construct. *Journal of Clinical Nursing* 26(23-24):5216-23.

Gjerris, M., Gamborg, C. and Röcklinsberg, H. (2016). Ethical aspects of insect production for food and feed. *Journal of Insects as Food and Feed*, 2(2):101-10.

Harfeld, J.L., Cornue, C., Kornum, A. and Gjerris, M. (2016). Seeing the Animal: On the Ethical Implications of De-animalization in Intensive Animal Production Systems. *Journal of Agricultural and Environmental Ethics* 29(3): 407-423.

Horvath, K., Angeletti, D., Nascetti, G. and Carere, C. (2013). Invertebrate welfare – an overlooked issue. *Annali dell'Istituto Superiore di Sanità* 49(1):9-17.

House, J. (2016). Consumer acceptance of insect-based foods in the Netherlands: Academic and commercial implications. *Appetite* 107:47-58.

Knutsson, S. and Munthe, C. (2017). A Virtue of Precaution Regarding the Moral Status of Animals with Uncertain Sentience. *Journal of Agricultural and Environmental Ethics* 30:213-24.

Lockwood, J.A. (2011). Do bugs feel pain? OUPblog. Available at: http://blog.oup.com/2011/11/bug-pain. Accessed 29 December 2017.

Mielby, H., Sandøe, P. and Lassen, J. (2013). Multiple aspects of unnaturalness: are cisgenic crops perceived as being more natural and more acceptable than transgenic crops? *Agriculture and Human Values*, Volume 30(3):471-80.

Neves, M.P. and Druml, C. (2017). Ethical implications of fighting malaria with CRISPR/Cas9. *BMJ Global Health*, 2(3) DOI:10.1136/bmjgh-2017-000396.

Nuffield Council on Bioethics (2016). *Genome editing. An ethical review*. Nuffield Council on Bioethics. Available at: http://nuffieldbioethics.org/wp-content/uploads/Genome-editing-an-ethical-review.pdf. Accessed 23 January 2018.

Østerberg, J.T., Xiang, W., Olsen, L.I., Edenbrandt, A.K., Vedel, S.E., Christiansen, A., Landes, X., Andersen, M.M., Pagh, P., Sandøe, P., Nielsen, J., Christensen, S.B., Thorsen, B.J., Kappel, K., Gamborg, C. and Palmgren, M. (2017). Accelerating the domestication of new crops: Feasibility and approaches. *Trends in Plant Science* 22(5):373-84.

Price, E.O. (1984). Behavioural aspects of animal domestication. *Quarterly Review of Biology* 59:1-32.

Röcklinsberg, H., Gamborg, C. and Gjerris, M. (2017). Ethical issues in insect production. In: van Huis, A. and Tomberlein, J.K. (eds.): *Insects as food and feed: from production to consumption*. Wageningen: Wageningen Academic Publishers, pp. 365-379.

Sandøe, P. and Gamborg, C. (2008). Animal breeding and biotechnology. In Sandøe, P. and Christiansen, S.B.: *Ethics of animal use*. Oxford: Blackwell, pp. 137-152.

The Norwegian Biotechnology Advisory Board (2017): A Summary: The Gene Technology Act – Invitation to public debate. Available at: http://www.bioteknologiradet.no/filarkiv/2017/12/Genteknologiloven-sammendrag-engelsk-til-web.pdf. Accessed 23 January 2018.

Van Huis, A., van Itterbeeck, J., Klunder, H., Mertens, E., Halloran, A., Muir, G. and Vantomme, P. (2013). *Edible insects: future prospects for food and feed security*. FAO Forestry Paper 171 Rome: Food and Agricultural Organisation of the United Nations.

Van Huis, A. (2017). Edible insects and research needs. *Journal of insects as food and feed*, 3(1):3-5.

68. Potato crisps from CRISPR-Cas9 modification – aspects of autonomy and fairness

H. Röcklinsberg[1]* *and M. Gjerris*[2]
[1]*Dept. of Animal Environment and Health, Swedish University of Agricultural Sciences, P.O. Box 7068, 750 07 Uppsala, Sweden;* [2]*Dept. of Food and Resource Economics, University of Copenhagen, Rolighedsvej 25, 1958 Frederiksberg C, Denmark; helena.rocklinsberg@slu.se*

Abstract

Within the Swedish MISTRA Biotech research programme the quality of starch in potatoes has been changed by use of different technologies such asCRISPR-Cas9. The idea is to increase the level of amylose, both for health reasons and as a mean to investigate possibilities for replacing fossil based oxygen barriers in food packages, thus reducing the climate impact. The goals thus seem laudable to most, but the experience of introducing GMOs on the market shows that even though there might be agreement on the goals, the strategy of using biotechnology to achieve them can be ethically contested. We describe the intentions behind developing the new plants and analyse some of the ethical issues that the development and marketing of the gene-edited potatoes raise. We argue that the concepts of autonomy and fairness are useful tools to understand many of the conflicting ethical values in the discussions relating to gene-editing. From our perspective these concepts are interrelated and relevant in at least two ways: (1) fairness in terms of both financial power and labelling as a means to ensure equal opportunities to make an autonomous decision as an individual ethical consumer and (2) fairness in term of equal market power between autonomous market actors.

Keywords: autonomy, fairness, potato starch, regulation

Introduction

Ever since A.L. Huxley's book Brave New World was published in 1932, reflections on engineered modifications of living beings have engaged a larger public and a wide range of ethical aspects have been scrutinized in both public and academic debates. According to a review of academic papers on ethical aspects of GMO a majority (84%) address consequences of cultivating GMO plants, often in terms of risks, the precautionary principle, sustainability and future generations, whereas 57% discuss the act of modification from a deontological perspective including definitions and the normative role of 'naturalness', and 43% of the philosophical papers focus on the agent along the lines of virtue ethics. They stress ideals such as trustworthiness, responsibility, wisdom, humility and care. In comparison, the review presents that laypersons (as mapped in academic studies) are similarly concerned with consequences, i.e. risks, and with naturalness in their judgement of the act. Although less focus on the actor, trustworthiness is described as an important virtue also by laypersons, and essential for establishing a dialogue between researchers and society (Gregorowius *et al.*, 2012).

The relevance of discussing the above mentioned ethical aspects in relation to genetic modification (GM) and gene editing (GE) is obvious. We do, however, find that there are other important aspects when evaluating genetically modified organisms (GMOs) and gene edited organisms (GEOs) coming out of technologies such as CRISPR-Cas9, which in Gregorowius *et al.* (2012) are classified as 'socio-economic concerns', but which in our perspective also have fundamental ethical dimensions. One example of such an issue concerns the difficulty of making autonomous decisions in a world offering constantly emerging new breeding possibilities and products, as well as the principles for 'fair distribution' of risks and opportunities and goods, among different stakeholders. A Swiss study of lay persons' perception of GM

EurSafe 2018 – DOI 10.3920/978-90-8686-869-8_68,

field trials confirm fairness is important, although the importance differ between outcome and process fairness and relates to whether one finds GM a morally important issue or not (Siegrist *et al.*, 2012).

Within the Swedish MISTRA Biotech research programme a changed quality of potato starch has been developed by use of novel biotechnologies (RNAi and CRISPR-Cas9). One project aims to contribute to lower fossil emissions by developing a potato starch 'plastic' to be used as oxygen barrier in packages (Muneer *et al.*, 2016), another to improve storage qualities of potato starch (Andersson *et al.*, 2017), and a third aims to alter starch qualities to be more beneficial from a health perspective, e.g. through lower glucose levels and insulin responses by higher diet fibre (Zhao *et al.*, 2018). Introducing a GM/GE product on the market, especially for human consumption is, however, a complex undertaking, and neither fair distribution of possibilities or goods, nor requirements for autonomous choices are easily defined (see e.g. Siipi and Uusitalo, 2011). In order to show the relevance of issues such as fairness and autonomy, we begin by briefly describing the new potatoes followed by a description of the regulatory discussions and relate this to the question of the definition of GE. Based on this we analyse two aspects of fairness and autonomy in relation to the development and potential introduction to the market of the potatoes: (1) fairness in terms of financial power and labelling as a means to ensure equal opportunities to make an autonomous decision as an individual ethical consumer and (2) fairness in term of equal market power between autonomous market actors.

The case of gene-modified and gene-edited potatoes

Researchers connected to the Swedish MISTRA Biotech research programme have used CRISPR-Cas9 to combine potato starch with plant protein to create a composite material similar to plastic that can be used as an oxygen barrier in packages, and recycled as compost after usage (Muneer *et al.*, 2016). They have further designed potatoes (*Solanum tuberosum*) with longer storing capacities thanks to a new starch composition (Andersson *et al.*, 2017). Starch consists of two main components, and by eliminating the production of one of them, amylose, by site-directed mutagenesis, the gene edited starch instead consists of increased levels of amylopectin. This is interesting for the industry (which is co-financing the research) as short storing time of potato starch is an obstacle in the food production process. If the method is possible to apply on large scale, it could have an impact on the global production of about 40 million tonnes of starch per year, and according to the industry provide a 'more environmentally sustainable special ingredients to conscious consumers' (Sveriges Lantbruksuniversitet, 2016). Further, by use of classical GM-technology, a slightly different potato has been developed, which – after cooking – takes longer to digest, leading to a number of positive side-effect like slow carbohydrates and benign gut bacteria (Zhao *et al.*, 2018). Contrary to the modified starch product, this potato can be prepared in a traditional way in everyday cooking (Sveriges Lantbruksuniversitet, 2018). This GM-potato, however, would have to be assessed according to EU Directive 2001/18/EC and labelled as a GM-food. It is therefore questionable, if it will reach consumers and whether they in turn will find it acceptable. The researchers are therefore aiming at creating the same change by use of GE through CRISPR-Cas9 instead, as this is regulated as a traditional breeding technology in Sweden (Umeå University, 2015).

These new varieties of potatoes are interesting as they will be among the first CRISPR-Cas9 vegetables for industrial and consumption purposes in Sweden, but also because potato traditionally is the stable vegetable in the Scandinavian and Northern European kitchen and carries many cultural connotations. Considering potato chips and strips as well as the starch used in a large number of products, a wide range of producers and consumers would directly or indirectly come in contact with this new product.

Is CRISPR-Cas9 GMO – a regulatory challenge

When discussing autonomy and fairness in relation to GMOs and GEOs the legal status of CRISPR-Cas9 and other GE-techniques is important, as is the ensuing regulatory marketing framework. The EU Directive 2001/18/EC defines GMO as: 'an organism with the exception of human beings, in which the genetic material has been altered in a way that does not occur naturally by mating and/or natural recombination' (Article 2(2)). Although specified in Annex I A part 1, the definition is regarded to be wide and unclear, but gains some clarity by the explicit exclusion in Annex I B of mutagenesis, cell fusion and certain other methods of exchange of genetic material used in traditional breeding (Zetterberg and Björnberg, 2017).

As new techniques develop the lack of a clear definition has led to different interpretations in different Member States. In Sweden, gene editing through CRISPR-Cas9 is not considered GM but treated as conventional breeding whereas the EU-system still discusses whether the new methods of gene-editing should be considered GM or not (Eriksson *et al.,* 2017). A decision is expected from the Court of Justice of the European Union in 2018. The Norwegian Biotechnology Advisory Board has proposed to differentiate regulation into three levels: changes that could be obtained through conventional breeding (1), intra-species modifications (2), and transgenic changes (3). The goal is to maintain control of GMOs and GEOs, but at the same time have a less demanding approval and control system for e.g. the potato plants described above (The Norwegian Biotechnology Advisory Board, 2017).

Arguments in favour of not defining GE as GM hold that GE involves no transgenesis (introducing a gene form a different species) and that the mutations could be reached by conventional breeding methods just requiring more time (Østerberg *et al.*, 2016). Whether arguments like these will convince a European public highly sceptical towards GMOs remains to be seen (Mielby *et al.*, 2013). Further, since 2015 each Member State in EU has the right to restrict or forbid cultivation of GM-crops within their own country according to the so called 'opt-out' Directive 2015/412. Reasons to activate the Directive may refer to a range of aspects such as environmental, agricultural or public policies, territorial planning and land use, socio-economic impacts or risk of unwanted GMO presence i.e. cross-border contamination. The idea behind the Directive is described as twofold: to revise the often lengthy and non-conclusive decision-making process on GMOs by encouraging Member States to accept a GMO even if they will not cultivate it themselves, and to ensure that aspects other than those related to human or environmental risks can be taken into account (Eriksson *et al.*, 2017). Hence, the Directive can be seen as a way to acknowledge that more than a strictly scientific perspective on what is perceived as a risk plays a role for public opinions on GMO (Gjerris, 2008).

However, some researchers have criticized the Directive for having neither improved the decision-making process, nor created the freedom of choice envisaged. In a 'Corresponcence' in the journal 'Nature Biotechnology' they argue in favour of an 'Opt in'-possibility, i.e. the right for each Member State to accept cultivating GMO-plants (Eriksson *et al.*, 2018). This, it is said, would not change the opt-out possibility, but ensure freedom of act for Member States who are in favour of GMO, but cannot allow it due to the need for a common agreement among all Member States. To what extent neighbouring Member States would have the possibility to veto such a decision based on the risk of GMOs reaching their territories is not discussed by the researchers, but remains to be solved.

On top of this arrives now the debate of whether or not to define new plant breeding technologies as GM-technologies. How the new plants are defined, both with regard to approval procedures and product labelling, will decide where they are grown and how much of a market-share they can obtain, and hence influence what choices farmers are 'allowed' to make. It might also influence the level of citizen trust in the entire food system, as GMOs are covered by assessment and control but GEOs currently are not.

Hence it seems the classical tension between autonomy (here on the level of each Member State's right to decide what to cultivate) and non-malefience will be important the coming years, as well as the issue of how to define the freedom to act (e.g. cultivate what you prefer) or the freedom not to have to act (e.g. not have to protect oneself from other's potentially harmful actions).

Fairness and autonomy

When considering the genetically altered potatoes described above and the unclear regulatory situation, it is relevant to discuss the connection between autonomy and fairness from the perspectives of (1) individual ethical consumers and (2) different market actors. Due to space limitations, we can merely highlight issues of relevance rather than discuss all relevant issues in depth.

Fairness in terms of financial power and labelling as a means to ensure equal opportunities to make an autonomous decision as an individual ethical consumer

It is uncontroversial to state that financially strong consumers have a wider range of purchase options compared to households with a more limited budget, and that one core element in practicing autonomy in everyday life is to choose what to eat, and ideally that food intake is based on one´s values. Given that food prices mirror whether they are produced with or without a certain technology, the ability to express one´s values regarding these technologies thus differ with budget limits. In the case of gene-edited potatoes a question could be, if they would outcompete conventional potatoes on the market. If so, consumers who do not wish to buy gene-edited potatoes (for whatever reason) will be forced to buy organic products which typically come with a higher cost. Further, if GE-potatoes are not legally considered GMOs, organic farmers can begin using GE-plants leaving consumers who do consider GE to be GM with little choice.

Another issue is how the political consumer runs the risk of being left on her own whether the potatoes are labelled or not. In a cacophonic market-place filled with infomercials competing about consumer interest and usually not under-selling the merits of products (Gjerris *et al.*, 2016) one can easily envision that the idea of an informed and autonomous choice by consumers about whether to by GE-potatoes or not will be under a lot of pressure both from those who criticise and those who support the technology. Further, whether one has the ability to gain relevant knowledge, and whether anyone is willing to provide it in a (as far as possible) neutral way, will shape the prerequisites for making autonomous choices. As the familiar discussion about GMOs show there is little agreement on this and as the discussion of e.g. off-targets effects of gene-editing are beginning to take off (Warmflash, 2017), there seems little doubt that ethical issues, including risks and benefits, will also be interpreted in a variety of ways concerning GE-plants.

Finally it is worth noting that attempts to define CRISPR-Cas9 as a conventional breeding tool primarily based on the technical fact that it is not transgenic, carries the risk of misunderstanding the discussion. Transgenesis is not the only reason that GMOs have been opposed and the more proponents of GE seek to avoid substantial ethical discussions about concepts as power and naturalness the easier it is to see the GMO discussion repeat itself.

Fairness in terms of equal market power between autonomous market actors

The first issue in this section concerns the different consequences of whether or not gene-editing such as CRISPR-Cas9 is defined as a GM. Historically, one point of criticism of GMOs points to the unfair competition between market actors, due to different financial strength between small-scale breeding companies and multinational companies, the latter having better chances coping with high regulatory demands (Raybould and Poppy, 2012). One may therefore expect that if CRISPR-Cas9 is regulated as GMO, it may exclude small companies from adopting the technology as the approval procedures will be too time-consuming and expensive. This is relevant not least from a fairness perspective. Creating fair competition is difficult since it raises questions of, for example, what kind of concerns are fair to formulate as assessment criteria for labelling a specific technology as GM or not. Envisaging future marketing of GE potato starch (for industrial purposes) the company developing the product would have an interest in high market shares, which they might lose if defined as GME. Given it is defined as GMO, the suggested 'opt in' legislation would potentially open markets available also for smaller producers, although financial dimensions in handling legal process remain.

The last issue to discuss briefly in this section is that both the suggested opt-in possibility and the definition of GE as non-GM raises an issue with regard to organic farming. Since some of the consumer concerns regarding GMO relate to ideas about naturalness (Gregorowius *et al.*, 2012) as does the choice of organic products, the risk of contamination followed by loss of 'naturalness' would be a serious challenge not least to the autonomy of organic farmers and consumers of their produce. Hence, one may expect that defining GE as conventional breeding and/or choosing a 'liberal' approach for labelling might imply that the GM-free nature of organic farming could be compromised in the public eye.

Conclusion

Recent years have seen the advent of a number of biotechnologies such as CRISPR-Cas9 that carries great promise in relation to changing e.g. crops. Within the Swedish research project MISTRA Biotech, the methods have been used to produce various varieties of potatoes that could result in lower climate impact, higher food security and healthier diets. It is, however, debated, whether such technologies should be defined as GM-technologies or as conventional breeding tools. The choice of definition will make a huge difference both with regards to approval procedures, labelling and consumer acceptance. We show the relevance of linking this issue to the concepts of autonomy and fairness and highlight questions of fairness for the individual consumer in terms of financial strength to choose/de-select the new products and point to the difficulties of obtaining unbiased knowledge to make an autonomous decision for the average consumer in a marketplace filled with infomercials. Further we briefly discuss how different ways of defining the new technologies will have an influence on which companies will be able to utilize them. If GE is not classified as GM, the accessibility of the technique increases and opens to usage by small breeding companies, but might lead to a loss of trust in food regulation among some citizens. We also point to the problems that organic farming might face, if the technologies are accepted as conventional breeding technologies, thus running the risk of undermining the organic labels of GM-free quality in the eye of the part of the public who finds the distinction between conventional and GE important. We conclude that a wide range of issues falls under the heading of autonomy and fairness and therefore find that these concepts are a useful key to further analyse the current discussions of CRISPR-Cas9 and related technologies in connection with breeding of crop plants.

References

Andersson, M., Turesson, H., Nicolia, A., Fält, A., Samuelsson, M. and Hofvander, P. (2017). Efficient targeted multiallelic mutagenesis in tetraploid potato (Solanum tuberosum) by transient CRISPR-Cas9 expression in protoplasts. Plant Cell Reproduction 36(1):117-28.

Eriksson, D., Brich-Pedersen, H., Chawade, A., Holme, I.B., Hvoslef-Eide, T.A.K., Ritala, A., Teeri, T.H. and Thorstensen, T. (2017). Scandinavian perspectives on plant gene technology: applications, policies and progress. Plant Physiology doi:10.1111/ppl.12661 [Epub ahead of print].

Eriksson, D., de Andrade, E., Bohanec, B., Chatzpolou, S., Defez, R., Eriksson, N.L., van de Meer, P., van der Meulen, B., Ritala, A., Sági, L., Schiemann, J., Twardowski, T. and Vaněk, T. (2018). Why the European Union needs a national GMO opt-in mechanism. Nature Biotechnology 36(1):18-19.
Gjerris, M. (2008). The Three Teachings of Biotechnology, in David K. & Thompson P (eds.): What can Nano learn from Bio? Elsevier pp. 91-106.
Gjerris, M., Gamborg, C. and Saxe, H. (2016). What to buy? On the complexity of being a political consumer. Journal of Agricultural and Environmental Ethics 29(1):81-102.
Gregorowius, D., Lindemann-Matthies, P. and Huppenbauer, M. (2012). Ethical Discourse on the Use of Genetically Modified Crops: A Review of Academic Publications in the Fields of Ecology and Environmental Ethics. Journal of Agricultural and Environmental Ethics 25(3):265-93.
Mielby, H., Sandøe, P. and Lassen, J. (2013). Multiple aspects of unnaturalness: are cisgenic crops perceived as being more natural and more acceptable than transgenic crops? Agriculture and Human Values 30(3):471-80.
Muneer, F., Andersson, M., Koch, K., Hedenqvist, M.S., Gällstedt, M., Plivelic, T.S., Menzel, C., Rhazi, L. and Kuktaite, R. (2016). Innovative Gliadin/Glutenin and Modified Potato Starch Green Composites: Chemistry, Structure, and Functionality Induced by Processing. ACS Sustainable Chemistry & Engineering 4(12):6332-43.
Østerberg, J.T., Xiang, W., Olsen, L.I., Edenbrandt, A.K., Vedel, S.E., Christiansen, A., Landes, X., Andersen, M.M., Pagh, P., Sandøe, P., Nielsen, J., Christensen, S.B., Thorsen, B.J., Kappel, K., Gamborg, C. and Palmgren, M.B. (2017). Accelerating the domestication of new crops: feasibility and approaches. Trends in Plant Science 22(5):373-84.
Raybould, A. and Poppy, G.M. (2012). Commercializing genetically modified crops under EU regulations. GM Crops & Foods 3(1):9-20.
Siipi, H. and Uusitalo, S. (2011). Consumer Autonomy and Availability of Gentically Modified Food. Journal of Agricultural and Environmental Ethics 24(2):147-63
Siegrist, M. Connor, M. and Keller, C. (2012). Trust, Confidence, Procedural Fairness, Outcome Fairness, Moral conviction, and Acceptance of GM Field Experiments. Risk Analysis 32(8):1394-1403.
Sveriges Lantbruksuniversitet (SLU) (2016). Specialpotatis för stärkelseindustrin blir Sveriges första 'CRISPR-Cas9-gröda'. Sveriges Lantbruksuniversitet. Available at: https://internt.slu.se/nyheter-originalen/2016/10/specialpotatis-for-starkelseindustrin-blir-sveriges-forsta-crispr-cas9-groda. Accessed 23 January 2018.
Sveriges Lantbruksuniversitet (SLU) (2018). A new potato for everyone who likes slow carbohydrates. Sveriges Lantbruksuniversitet. Available at: https://www.slu.se/en/ew-news/2018/1/a-new-potato-for-everyone-who-likes-slow-carbohydrates. Accessed 23 January 2018.
The Norwegian Biotechnology Advisory Board (2017). A Summary: The Gene Technology Act – Invitation to public debate. Available at: http://www.bioteknologiradet.no/filarkiv/2017/12/Genteknologiloven-sammendrag-engelsk-til-web.pdf. Accessed 23 January 2018.
Umeå University (2015). 'Green light in the tunnel': Opinion of the Swedish Board of Agriculture – a CRISPR-Cas9-mutant but not a GMO. Umeå University. Available at: http://www.umu.se/ViewPage.action?siteNodeId=4510&languageId=1&contentId=259265. Accessed 23 January 2018.
Warmflash, D. (2017). Activists fan concerns about CRISPR 'off target' effects in gene-edited crops and foods. Are they right? The Genetic Literacy Project. Available at: https://geneticliteracyproject.org/2017/06/12/activists-fan-concerns-crispr-off-target-effects-gene-edited-crops-foods-right. Accessed 23 January 2018.
Zetterberg, C. and Edvardsson, B.K. (2017). Time for a New Regulatory Framework for GM Crops? Journal of Agricultural and Environmental Ethics 30(3):325-47.

Section 13.
Aquaculture

69. Aspects of animal welfare in fish husbandry

H. Seibel, L. Weirup and C. Schulz*
Gesellschaft für Marine Aquakultur mbH, Hafentoern 3, 25761 Buesum, Germany; seibel@gma-buesum.de

Abstract

Farming of fish and other aquatic species has increased in recent decades and never before have there been more controversial debates on animal welfare in fish husbandry. The practices used and associated welfare issues are becoming increasingly focused on by scientists, consumers and policy makers. International and national organisations have issued recommendations and guidelines concerning fish welfare but there is still a lot of information lacking. Due to § 2 of the German animal protection law, animals must be adequately nourished, cared for and behaved in a proper manner, and pain or avoidable suffering or damage must be avoided. Keeping fish in accordance with the animal protection laws should be part of a good professional aquaculture practice and is indispensable for ethical, legal and economic reasons. Only a healthy fish growths up adequately. The talk provides and discusses legal aspects of animal protection applied to fish taking the German welfare act and EU regulations as example. First possible indicators for the assessment of well-being will be mentioned. The assessment of fish welfare indicators should be carried out within the frame of daily and regular working routines, e.g. feeding, harvesting, grading or slaughtering. The interpretation of animal welfare in fish is discussed and set into the context of human sensations projected on fish and the debate on consciousness in fish.

Keywords: aquaculture, health indicators, consciousness

Introduction

Fish and its products belong to the richest protein and essential nutrient sources which support a balanced nutrition and good health. In 2013, fish accounted for about 17% of the global population's intake of animal protein and 6.7% of all protein consumed (FAO, 2016). Based on FAO's analysis, 31.4% of marine fish stocks were estimated as fished at a biologically unsustainable level and therefore overfished. One of the world's greatest challenges is to feed more than 9 billion people by 2050, taking climate change, economic and financial uncertainty, and increasing competition for natural resources into account (FAO, 2016). Hence, farming of fish and other aquatic species has strongly increased in recent decades and will still increase in the future (FAO, 2016). The growth and intensification of aquaculture lead to controversial debates on animal welfare in fish husbandry. The farming practices and associated welfare issues are receiving attention from scientists, consumers and consequently from policy makers. Consumers request more and more ecological and animal friendly produced products, but the transparency concerning welfare of fish products is still very incomplete. It even hardly exists for ecological/biological production. Ecological producers, international and national organisations have issued recommendations and guidelines concerning fish welfare, but there is still a lot of information lacking, due to the challenging question: what does welfare in fish husbandry actually mean? In addition, this question needs to be determined for every single aquaculture species and the different rearing systems, because they can be diverse.

For ethical reasons fish welfare needs to be critically determined and put into a measurable way for the consideration of husbandry effects and improvements. The definition of welfare aspects is also essential for the economic success in fish production, because only healthy fish growths up adequately.

Different research programs started recently to further investigate effects of aquaculture procedures on welfare to produce data for best practice and future legislation.

EurSafe 2018 – DOI 10.3920/978-90-8686-869-8_69, © Wageningen Academic Publishers 2018

Legal aspects

The EU, Council Directive 98/58/EC concerning the protection of animals kept for farming gives framework requirements for the protection of animals of all species kept for the production of food, wool, skin or fur or for other farming purposes, including fish, reptiles or amphibians. In Art. 3 it is mentioned that it needs to be ensured that the owners or keepers take all reasonable steps to ensure the welfare of animals under their care and to ensure that those animals are not caused any unnecessary pain with is implemented in § 2 of the German animal welfare act. Some rare recommendations for fish husbandry are given (Full committee of the European convention for the protection of animals in agricultural animal husbandry; recommendation for the keeping of fish in aquaculture) and fish is partially affected by the area of application of the relevant regulations due to transportation, stunning and killing methods in the Directive 98/58/EC.

The German animal welfare act (as amended by the announcement of May 18th 2006; BGBl. I S. 1206, 1313, last amended on March 29th 2017) transposes the Directive 98/58/EC into national German law and the national legislation enunciated specific requirements for fish regarding transport and slaughter (§ 2, § 11 (8) und § 16). Due to § 2 of the German animal welfare act, animals must be adequately nourished, cared for and behaved in a proper manner, and pain or avoidable suffering or damage must be avoided. Equivalent legislation exist as well e.g. in Austria and Switzerland.

Person, who want to keep vertebrates – other than farm animals – professionally, need a permission of the competent authority for that. But there are no requirements for aquaculture farms, holding fish for food production. A proof of expertise or reliability is not required (§ 11 (1) German animal welfare act), but those who keep farm animals for profit, must ensure that the requirements of § 2 (German animal welfare act) are complied, ensured by company internal controls and on the keeper's own responsibility.

In summary, in aquaculture, promising first aspects of animal welfare are already included in some national legislation.

While legislations in European countries are already not very precise, it needs to be kept in mind, that, for example, Asia, including China as the main fish producer and largest exporter of fish and fishery products (FAO, 2016), has no animal welfare regulations at all.

The linchpin of mentioned welfare legislation is the need for verification of requisite qualification and competency in handling and husbandry of fish and an appraisal of species specific fish needs.

Short overview on indicators

First simple and easy applicable characteristics for the assessment of well-being of fish have been published in leaflets or papers (Stien *et al.*, 2013; DLG, 2014; Pettersen *et al.*, 2014; VDFF, 2016; Noble *et al.*, 2017), but the requirements for fish welfare are very diverse and need to be species specific. Given literature includes welfare indicators on the basis of feeding, harvesting, grading or slaughtering and includes normal fish behaviour and potential physiological impacts. The focus of self-inspections in fish populations should be e.g. on mortality, growth, habitus, behaviour status and findings on slaughter and water quality on a regular basis (Good *et al.*, 2009; VDFF, 2016).

To give an example, 'Naturland', an ecological association with the goal to protect the environment and to preserve the natural bases of life by means of organic management in all areas of agriculture (Naturland statutes, 2016), recommends in case of ecological salmon production, that the stocking density of fish shall not exceed 10 kg/m^3. For char (*Salvelinus* sp.) and Coregonen (e.g. whitefish Coregonus) they recommend 15 kg/m^3 and for trout (*Oncorhynchus* sp., *Trutta* sp.) maximum stocking densities of 20 kg/m^3. For water quality Naturland recommends that the conditions comply with the degradation of water quality (in accordance with Directive 2000/60/EC on European Water Framework Directives), as well as an oxygen saturation of minimal 7 mg/l and a minimum inflow of 3 l/s per t fish.

Interpretation of animal welfare in fish

If animals are kept for any kind of human consumption, it is the authors' strong opinion that this needs to be accomplished in the best possible manner to provide the animals welfare.

In the German legislation the ability of vertebrates, including fish to feel pain or to suffer are approved, by claiming § 2 (German animal welfare act). As a consequence to prevent pain and suffering in fish, the German law makes it mandatory to sedate fish before killing or surgery. However the law provides exceptions, if welfare actions are not practicable. As an example a killing method can be faster and more efficient if conducted without prior sedation. Whereas for aquaculture stunning methods for example are strictly determined, the commercial fishery is explicitly excluded in all mentioned acts, although fish welfare regulations are absolutely necessary for this kind of fish production as well. This aspect is one of the main discrepancies in fish welfare. Although, the interpretation of animal welfare in commercial fishery, fishing and aquaculture needs to be defined from each other, the cognition of pain and suffering in fish cannot be ignored only due to the fact that the feasibility of welfare actions is more difficult in fishery.

Evidence of fish perceiving pain are among others: the presence of nociceptors and the transmission of stimuli from the nociceptors to the brain; brain structures that are analogous to the human pain-processing cerebral cortex; opioid receptors and endogenous opioid substances in a nociceptive neural system; the reduction of aversive behaviour and physiological effects after the use of analgesics; as well as fast, inflexible learning of lessening potential painful stimuli (Sneddon, 2014). Beside, fish can have stress (Conte, 2004; Tort 2011).

Despite those indicators some studies doubt or oppose the cognition of pain in fish. Key (2016) for example, states that fish lack the necessary neurocytoarchitecture (e.g. neuron clusters), microcircuitry (connection between the clusters), and structural connectivity for the neural processing required for feeling pain. Rose *et al.* (2012), criticises different studies for not distinguishing unconscious detection of injurious stimuli (nociception) from the conscious experience of pain. After reviewing behavioural and neurobiological evidence Rose *et al.* (2012) claims that fish are not likely to experience pain, especially not in a human-like capacity, even if they were conscious.

Talking about consciousness in fish as well as in other animals can be misleading, as there is a variety of definitions. In a common sense it is often interpreted as an awareness of its own existence and ability to consciously determine and analyse its actions. Other animals might possibly not dispose these abilities; however that is no reason to deny their consciousness for pain.

For a reasonable welfare debate and need for protection, fish must be conscious of at least some basic emotion like subjectively experiences of pain, coldness, comfort or discomfort (Volpato *et al.*, 2007). The ability of experiencing these emotions or states can be termed as being sentient.

To support the idea of fish being sentient it is useful to visualize the benefits that come along with it (Kittilsen, 2013). In general emotions enable us to avoid harm and to seek valuable resources (Panksepp, 1998), or as Cabanac (1992) has argued 'pleasant is useful' and guides human and non-human animals to perform behaviour that serves to enhance their fitness.

However using practical measurements to assess fish welfare in aquaculture systems, the focus should be on objective measurements of biochemical, physiological and behavioural indicators – to evaluate whether human interactions with fish impair the latter's health or prevent them from receiving what they need, if held in captivity (Arlinghaus *et al.*, 2007).

The normative assessment of the artificial intervention in aquaculture is carried out according to a rather emotional one by the consumer (Vanhonacker *et al.*, 2011). While developing guidelines and recommendations for a good fish husbandry practice, we need to account for the fact, that any selection of indicators includes an implicitly value judgments about the interpretation of animal welfare. Humans are on one hand likely to deny feelings and emotions in animals, especially fish, as it compromises their status of being exceptional and separable from the animal kingdom. Also denying sentience in other animals makes the exploitation of them more easy and justifiable. On the other hand humans tend to humanize animals. Being low in a social hierarchy for example can be a more relaxing state than a high position, while humans tend to feel sorry for the ones in the low position. Problems can occur, when human sensations are directly transferred to fish. For example higher stocking densities can be required from the point of animal welfare in fish, which is different for terrestrial animals and often under debate, when animal welfare standards for terrestrial animals are directly transferred to fish. This is not on behalf of fish welfare, because fish have different needs for adequate species specific living conditions.

The subjectivity not only accounts for the fish welfare itself, but also for the perception of rearing systems. Positive aspects of closed land based aquaculture like providing optimal water conditions, as well as regionalism, short transports or a transparent production get lost in the debate, if closed land based aquaculture can be natural like or not. Consumers show fear of technology, because high tech filters and recirculation systems are difficult to understand and estimate, when they are not in touch with them. Therefore, impacts on animal welfare can be over interpreted or misjudged. Thus, artificiality is generally negatively connoted, while naturalness is positive with the renunciation of further pre-misses (Korn *et al.*, 2014). Human understanding of nature can be glorified and forgets about extreme temperatures and oxygen depletion, as well as predators and disease transmission.

Therefore it is important to evaluate welfare indicators for fish in a more proper way on a scientific level, and take the species specific and the kind of production into consideration. Only than an objective statement can be made about the welfare status of a fish, without falling into subjective and humanistic comparisons.

Keeping fish in accordance with the animal protection laws should be part of a good professional aquaculture practice and is indispensable for ethical, legal and economic reasons.

References

Cabanac, M. (1992). Pleasure – the common currency. Journal of Theoretical Biology 155: 173-200.

Conte, F.S. (2004). Stress and the welfare of cultured fish. Applied Animal Behaviour Science 86: 205-223.

Deutsche Landwirtschaft-Gesellschaft, Ausschuss für Aquakultur (2014). Tierwohl in der Aquakultur. DLG Merkblatt 401. Frankfurt am Main. Available at: http://2015.dlg.org/fileadmin/downloads/merkblaetter/dlg-merkblatt_401.pdf. Accessed 26 January 2018.

Food and Agriculture Organization of the United Nations (2016). The state of World fisheries and aquaculture 2016. Contributing to food security and nutrition for all. Rome, 200 pp. Available at: http://www.fao.org/3/a-i5555e.pdf.

Good, C., Davidson, J., Welsh, C., Brazil, B., Snekvik, K. and Summerfelt, S. (2009). The impact of water exchange rate on the health and performance of rainbow trout *Oncorhynchus mykiss* in water recirculation aquaculture systems. Aquaculture 294: 80-85.

Key, B. (2016). Why fish do not feel pain. Animal Sentience, 3.

Kittilsen, S. (2013). Functional aspects of emotions in fish. Behavioural Processes 100: 53-159.

Korn, A., Feucht, Y., Zander, K., Janssen, M. and Hamm, U. (2014). Communication strategy for sustainably produced aquaculture products. University Kassel, report. Available at: http://orgprints.org/28279. Accessed 26 January 2018.

Naturland – Verband für Ökologischen Landbau e. V. – Satzung (2016). Available at: https://www.naturland.de/images/Naturland/Wer_wir_sind/Struktur/QMH_1_Satzung.pdf. Accessed 26 January 2018.

Noble, C., Nilsson, J., Stien, L.H., Iversen, M.H., Kolarevic, J. and Gismervik, K. (2017) Velferdsindikatorer for oppdrettslaks: Hvordan vurdere og dokumentere fiskevelferd. Research report.

Panksepp, J. (1998). Affective Neuroscience: The Foundations of Human and Animal Emotion. OUP, New York.

Pettersen, J.M., Bracke, M.B.M., Midtlyng, P.J., Folkedal, O., Stien, L.H., Steffenak, H. and Kristiansen, T.S. (2014). Salmon welfare index model 2.0: An extended model for overall welfare assessment of caged Atlantic salmon, based on a review of selected welfare indicators and intended for fish health professionals. Reviews in Aquaculture 6: 162-179.

Rose, J.D., Arlinghaus R., Cooke, S.J., Diggles, B.K., Sawynok, W., Stevens, E.D. and Wynne, C.D.L. (2012). Can fish really feel pain? Fish and Fisheries 15: 97-133.

Segner, H., Sundh, H., Buchmann, K., Douxfils, J., Sundell, K.S., Mathieu, C., Ruane, N., Jutfelt, F., Toften, H. and Vaughan, L. (2012). Health of farmed fish: Its relation to fish welfare and its utility as welfare indicator. Fish Physiology and Biochemistry 38: 85-105.

Sneddon, L.U., Elwood, R.W., Adamo, S.A. and Leach, M.C. (2014). Defining and assessing animal pain. Animal Behaviour 97: 201-212.

Stien, L.H., Bracke, M.B.M., Folkedal, O., Nilsson, J., Oppedal, F., Torgersen, T., Kittilsen, S., Midtlyng, P.J., Vindas, M.a., Øverli, Ø. and Kristiansen, T.S. (2013). Salmon Welfare Index Model (SWIM 1.0): A semantic model for overall welfare assessment of caged Atlantic salmon: Review of the selected welfare indicators and model presentation. Reviews in Aquaculture 5: 33-57.

Tort, L. (2011). Stress and immune modulation in fish. Developmental and Comparative Immunology 35: 1366-75.

Vanhonacker, F., Altintzoglou, T., Luten, J. and Verbeke, W. (2011). Does fish origin matter to European consumers? Insights from a consumer survey in Belgium, Norway and Spain. British Food Journal 113: 535-549.

VDFF, Arbeitskreis 'Tierschutzindikatoren' des Verbandes Deutscher Fischereiverwaltungsbeamter und Fischereiwissenschaftler e.V. (2016). Leitfaden Tierschutzindikatoren – mit Empfehlungen für die Durchführung betrieblicher Eigenkontrollen gemäß §11 Absatz 8 des Tierschutzgesetzes in Aquakulturbetrieben. Available at: http://www.vdff-fischerei.de/fileadmin/daten/Leitfaden_Tierschutzindikatoren_Aquakultur_V1_final_Maerz_2016.pdf. Accessed 26 January 2018.

Volpato, G.L., Gonçalves-de-Freitas, E. and Fernandes-de-Castilho, M. (2007). Insights into the concept of fish welfare. Diseases of Aquatic Organisms 75: 165-171.

70. Recirculation aquaculture systems: sustainable innovations in organic food production?

S. Meisch[1*] *and M. Stark*[2]
[1]*International Centre for Ethics in the Sciences and Humanities (University of Tuebingen), Wilhelmstr. 19, 72074 Tuebingen, Germany;* [2]*Seafood Advisory Ltd., Ey 16, 3294 Büren an der Aare, Switzerland; simon.meisch@uni-tuebingen.de*

Abstract

EU regulations explicitly preclude recirculation aquaculture systems (RAS) for aquaculture grow-out from organic certification because they are not close enough to nature (Regulation (EEC) No. 710/2009). Meanwhile, according to another EU regulation, one criterion for organic food production is its contribution to sustainable development (Regulation (EEC) No. 834/2007). Against this background, one might argue that in spite of their distance to nature RAS are innovative solutions to deal with sustainability issues in food production. The paper will deal with the claim that RAS for aquaculture could be one innovative solution to sustainability issues. In this respect, the picture is ambivalent. In the past, the organic movement (OM) has searched for innovative alternatives to industrial forms of agriculture and food production that are non-sustainable. Hence, the majority of the OM cannot warm to industrial RAS, even though one might argue that these systems comply with many of the European OM's founding principles. While there are potential positive effects for a sustainable development, we might still regard these systems as techno-scientific solutions to social problems. This paper discusses innovation narratives related to RAS from the perspective of post-normal innovation critique. It first presents potential contribution to a more sustainable food sector. It then contrasts these arguments within critical assessments of innovation narratives for sustainable development. Finally, it discusses pitfalls that the OM needed to avoid if it wants to lobby for or against organic certification of RAS.

Keywords: recirculation aquaculture systems (RAS), sustainable development, post-normal science, innovation, food systems

Introduction

Recirculation aquaculture systems (RAS) are promoted as innovations in the food sector that will (significantly) contribute to a more sustainable development by providing food security for a growing world population and a more environmentally friendly form of food production (Kloas *et al.*, 2015; Martins *et al.*, 2010). RAS are on-shore 'closed-loop production systems that continuously filter and recycle water, enabling large-scale fish farming that requires a small amount of water and releases little or no pollution' (Jenner, 2010). Aquaponics combines RAS with plant production by using the processed fish excrements to cultivate plants. Amidst recent technological developments (Kloas *et al.*, 2015; Tschirner and Kloas, 2017), the organic movement (OM) faces the challenge whether it should embrace RAS and promote their organic certification. Currently, it is making up its mind. When speaking about the OM, we refer to the global movement of organisations and individuals engaged in the promotion of organic farming and products (Bergleiter *et al.*, 2017; IFOAM, 2017).

From a legal perspective, the case seems clear. The EU regulation on organic aquaculture production explicitly precludes RAS from organic certification. Yet, the regulation binds its exclusion to the discovery of 'further knowledge'. Thus, the OM might embark on producing this evidence, if it was convinced that RAS should be organically certified. Reference to potential contributions to a more sustainable development seems the most promising strategy (Bergleiter *et al.*, 2017; Little *et al.*, 2015).

EurSafe 2018 – DOI 10.3920/978-90-8686-869-8_70, © Wageningen Academic Publishers 2018

There are many ways to engage with RAS, e.g. regarding fish welfare (Seibel *et al.*, 2018). In this paper, we aim to contribute to the process of self-reflexion by the OM. Hence, we question narratives of 'innovation for sustainable development' related to RAS and ask what the OM buys into when subscribing to these narratives. In doing so, we discuss these narratives from the perspective of post-normal innovation critique and focus on the EU context. The paper sketches the legal context within the EU (2). It continues with an overview of the potential contributions of RAS to a sustainable development that might support their organic certification (3). The paper subsequently contrasts these arguments within a critical assessment of innovation narratives for sustainable development (4). It concludes by discussing pitfalls that the organic movement needed to avoid if it ever wanted to lobby for or against organic certification of RAS (5).

The legal perspective

EU regulation on organic aquaculture (Regulation (EEC) No. 710/2009) explicitly precludes RAS from organic certification. Article 11 states that '[due] to the principle that organic production should be *as close as possible to nature* the use of such systems should not be allowed for organic production *until further knowledge is available*' (italics added). Yet, the criterion of 'closeness to nature' is confusing as it leaves implicit what it means by 'nature' and does not provide a scale of what is 'close enough' (Kerr and Potthast, 2018). After all, the term 'nature' is highly contested (Castree, 2017).

Despite the rational given why RAS are excluded from organic certification, EU regulations requires organic food production to contribute to sustainable development and limits RAS 'until further knowledge is available'. What might this mean then? Either other factual knowledge is found showing that RAS are close enough to nature. For many (epistemic) reasons, this is not a very promising strategy (Bergleiter *et al.*, 2017: 47-49). Alternatively, it might be argued that additional values and norms (other than 'closeness to nature') should be considered when evaluating RAS. Advocates of RAS emphasise their positive contributions for sustainable development and point to the EU regulation on organic production and the labelling of organic products (Regulation (EEC) No. 834/2007) according to which organic production 'delivers public goods contributing to the protection of the environment and animal welfare, as well as to rural development'. Against this background and despite their distance to nature, RAS might then be promoted as innovations in the organic sector.

RAS – innovative solutions for a sustainable development

Sustainability issues related to aquaculture

Advocates of aquaculture and RAS stress the potential contributions of these production systems to many urgent challenges of sustainable development (cf. e.g. ECF Farmsystems, 2018; Kloas *et al.*, 2015; Tschirner and Kloas, 2017). In particular, two problems are repeatedly stated. First, it is expected that the global demand for animal proteins will significantly increase because of the growing world population as well as the rising and thus more meat-intensive living standards in many parts of the world. The bottom line is: more people will eat more meat. Fish is expected to play a key role in meeting this increasing demand for animal proteins. Besides, compared to other forms of animal proteins, the ecological impact of fish production is considerably lower. Yet, and here comes the second sustainability challenge, many fish stocks worldwide are already overfished, some close to collapse. Against this background, aquaculture seems a promising way to provide valuable nutrition to a growing world population and to protect marine fish stocks. The aquaculture sector is already one of the fastest growing sectors in the food industry, potentially increasing ecologically and socially damaging effects; this can range from animal welfare in exposed cages and overstocked densities, the abuse of pharmaceuticals resulting in residues in the final product and the natural environment, to the transmission of diseases from farm to wild animals, to the destruction of coastal ecosystems or the neglect of neighbouring stakeholders' interests, to long-distance transport of goods (Bergleiter and Meisch, 2015; Folke *et al.*, 1998). One solution to these problems is organic certification aiming to improve current production (Bergleiter and Meisch, 2015); RAS rate as another one.

Recirculation aquaculture systems

Advocates of RAS claim that this technology has the potential to significantly reduce the shortcomings of other forms of aquaculture. Before discussing RAS, we acknowledge that we would have to differentiate between fish species and production systems, which we unfortunately cannot do within the scope of this paper, and are well aware of this deficit (cf. e.g. Bergleiter *et al.*, 2017).

With regard to fish welfare and biosafety, RAS claim to have advantages over conventional forms of aquaculture. Farmed fish require suitable densities in order to prevent stress from territorial behaviour (too low stock) or overpopulation (too large stock). Given the control options, suitable stocking densities can be achieved in RAS. This controlled environment also favours other important factors for fish welfare, e.g. water quality and the absence of external stressors such as predators. The same applies to the use of pharmaceuticals. The controlled environment makes it easier to keep diseases out and avoid drug applications. In the event of illness, pharmaceuticals can be applied more easily and specifically. As these are on-shore, the spread of diseases to wild fish or the escape of fish, ill or healthy, can be excluded in principle. Finally, if a fish species lives in the water column and does not need a particular environment such as substrate, the closeness to nature becomes a difficult question and ponds are not necessarily more desired than tanks (cf. Seibel *et al.*, 2018; Bergleiter *et al.*, 2017).

Moreover, RAS are expected to have a reduced ecological impact compared to conventional organic aquaculture. As closed on-shore systems, they can be built in areas with low land pressure and thus avoid conflict with land for food and energy production and reduce pressure from precious coastal ecosystems such as mangrove forests. This independence from a particular ground constitutes an advantage over conventional organic aquaculture because there is simply not enough land to substitute the current marine-based fish production with existing forms of organic aquaculture (Bergleiter *et al.*, 2017). RAS are water efficient and can hence be installed in arid areas. Being closed systems, they do (ideally) not pollute water bodies at all. Fish excrements can be filtered out or processed and used for plant production in aquaponics. Recently, an aquaponics system – the so-called 'tomato-fish' – was developed that might even work as a CO_2 sink because it uses more CO_2 than it emits (Kloas *et al.*, 2015; Suhl *et al.*, 2016). The organic concept of closed matter and energy cycles comes close in some RAS such as aquaponics but is largely absent in conventional organic aquaculture.

A consumer review indicated that fish produced in RAS might appeal to the desire of consumers in the Global North as it can allow regional production, which they seem to value more than organic ones (Bergleiter *et al.*, 2017). A prominent example is the 'capital-city perch' ('Hauptstadtbarsch') grown in a RAS in Berlin (Rushe, 2016). Furthermore, regionality might have the additional advantage as consumers can come and see how their fish is grown. Production sites close to consumers thus tend to profit from increased trust and reduced long-distance transportation.

RAS – yet another technological fix?

What can we make of the abovementioned claims that RAS are a step towards an 'environmentally neutral agriculture' (GGN, 2018) and, hence, an innovation for sustainable development under the EU regulation on organic aquaculture? In recent years, post-normal science (PNS) scholarship has critically engaged with narratives of innovation for sustainability (Benessia *et al.*, 2012; Benessia and

Funtowicz, 2016; Rommetveit *et al.*, 2013). Strand (2017: 288) characterises PNS as 'a critical concept originally developed to describe situations in which there are important or controversial public decision problems informed by an incomplete, uncertain or contested knowledge base'. The challenges concerning RAS correspond well to this problem description. PNS scholars have identified different innovation narratives. In their reading, innovation is 'a dynamic system of forces that constantly and necessarily redefine the boundaries between science, technology and the normative sphere of liberal democracy' and 'a phenomenon which is on a path-dependent trajectory, with origins in the scientific revolution and the emergence of the modern state in the 16th and 17th centuries' (Benessia and Funtowicz, 2016: 71). The latest narrative presents innovation, on the one hand, as a way out of various crises in Western capitalist economies and their saturated markets. On the other hand, it evokes an ambiguous sustainability narrative calling for a pause in the human consumption of finite natural resources and at the same time believing that accelerated science-based innovation can manoeuvre humanity out of this crisis by means of smart, marketable technologies (Rommetveit *et al.*, 2013). With this, the 'very same cause of the [...] extreme vulnerability of our life-supporting systems on the one side, and of the massive expropriation and deterioration of natural and cultural systems on the other, is considered as the main and only possible cure' (Benessia *et al.*, 2011: 77, italics in original). This narrative promotes innovation as the pathway to secure the survival of humans and the planet within an unquestioned Capitalist market economy (Leese and Meisch, 2015). So, first, post-normal scholarship critically analyses if innovations really deal with the sustainability challenges themselves or if they simply develop consumer products with the vague message that they would meet these challenges. Second, it looks at the forms of citizenship these narratives create: Can members of society only act in their capacity as consumers or passive recipients of products? Or can they participate actively as citizens in the development of their environments such as food production (Rommetveit *et al.*, 2013; Strand *et al.*, 2017)?

Against this background, RAS appear as a silver bullet to multiple social and environmental predicaments, such as food security, local production, reduced pollution, increased animal welfare or reducing the pressure of destruction of marine habitats. Yet, it remains unclear how they really relate to the solutions of these challenges. Furthermore, their implementation might raise serious issues of justice and democratic citizenship. The stereotypical RAS narrative simply accepts the local desire for global products as well as the global meat consumption continues to increase from its already unsustainable level. At least, advocates of this technology do not make an explicit effort to challenge dominant imaginaries related to meat. With this, they seem to imply that fish products from RAS can fully substitute the current meat consumption. Even if one assumes that humans will never become vegetarians or vegans and continue to eat animal proteins, for RAS to be effective, consumers would have to change their diets radically and eat fish instead of other meat types (poultry, pork, beef, etc.). Convincing people to eat differently is a legitimate aim and constitutes a building block of sufficiency strategies. Yet, this change would have to be a part of democratic deliberations on the good life – and not just the (perhaps desired) side effect of a new food technology. The same applies to the socio-economical context. In order to address sustainability issues, current forms of aquaculture would have to terminate and be replaced by RAS. Again, there are good reasons for a reform of these sectors. Yet, strictly applied, this substitution would turn existing fishing and aquaculture communities upside down by radically altering the ownership structure, knowledge and power base and access rights in the newly designed sector; presumably with a shift from the Global South to the Global North Thus, if RAS were to make its promised contributions to sustainable development, it would by no means be socially neutral and absolutely require democratic deliberation and a social reform that goes beyond the simple installation of a food production facility. Last, but not least, RAS narratives envision members of society as mere recipients or consumers of a technology that might radically change their way of living, working and eating. However, as citizens, they have the capacity and agency to innovate their own food systems (concerning water cf. Meisch, 2014).

RAS: an option for the organic movement?

Proponents usually present RAS as solutions to urgent sustainability challenges. While we do not deny that RAS have the potential to produce marketable food products that are low in residues, produced in a less polluting, more efficient and animal-friendly way, we wonder whether they can really keep their promises. Against this background, should the OM, its collective and individual members then lobby for the organic certification of RAS? On the one hand, it began as a movement against the technologisation, intensification and economization of agriculture and food production. By contrast, it promoted a holistic approach to agriculture, the creation of largely closed matter and energy cycles, the care of soil life, a diversified crop rotation, the adaption of livestock breeding to site conditions, the regard for the needs of the livestock, the production of nutritionally valuable foodstuff, the avoidance of environmental pollution and the lowest possible power consumption of non-renewable resources (Storhas, 1988: 23). The OM's principles relate well to the post-normal critique. Consequently, embracing RAS might feel as a betrayal of core beliefs. On the other hand, the OM originally intended to create new food production systems. Counter-intuitively, RAS conform to OM's aims –particularly, if we concede that many fish species in nature do not live in closeness to sediments and thus references to soil do not apply. In addition, veterinarians suggest controlled RAS to be more in line with animal welfare aims than some of the existing organic systems such as cages. This observation touches on the self-conception of the OM and raises the question what organic certification means after all. Last, but not least, the OM reminds us that agriculture is socially embedded. In this regard, RAS might be tools in order to achieve a more sustainable aquaculture in some places at some times but also tools whose social consequences need to be ethically reflected and democratically deliberated. Yet, this deliberation needs to be embedded in a more encompassing rethinking of current organic systems and (environmentally, socially, globally) valuable innovations.

Acknowledgement

This paper builds on the report of the project *Stakeholder-Studie 'Kreislaufanlagen – Positionen des Ökosektors'* (Bergleiter *et al.*, 2017) funded by German Federal Ministry of Food and Agriculture (project no. 2815OE026).

References

Benessia, A. and Funtowicz, S. (2015). Never late, never lost, never unprepared. In: Saltelli, A. *et al.* (eds.) Science on the Verge. CSPO, Tempe, AZ, pp. 71-113.

Benessia, A., Funtowicz, S., Bradshaw, G., Ferri, F., Ráez-Luna, E., Medina, C. (2012). Hybridizing sustainability: Towards a new praxis for the present human predicament. Sustainability Science 7 (Supplement 1): 75-89.

Bergleiter, S., Böhm, M., Censkowsky, U., Meisch, S., Schulz, C., Seibel, H., Stark, M. and Weirup, L. (2017). Recirculation Aquaculture Systems – Positions of the Organic Sector. Available at: http://www.orgprints.org/32165. Accessed 6 January 2018.

Bergleiter, S. and Meisch, S. (2015). Certification Standards for Aquaculture Products: Bringing Together the Values of Producers and Consumers in Globalised Organic Food Markets. Journal of Agricultural and Environmental Ethics 28(3): 553-569.

Castree, N. (2017). Nature. In: Richardson, D. *et al.* (eds.) The International Encyclopedia of Geography. Chichester, Wiley. DOI: 10.1002/9781118786352.wbieg0522.

ECF Farmsystems (2018). Ein kurzes Wort zum Thema Nachhaltigkeit. Available at: http://www.ecf-farm.de. Accessed 6 January 2018.

Folke, C., Kautsky, N., Berg, H., Jansson, A. and Troell, M. (1998). The Ecological Footprint Concept for Sustainable Seafood Production: A Review. Ecological Applications 8(1) Supplement: S63-S71.

GGN Certified Aquaculture (2018). Recipe suggestion or modern agriculture? Available at: https://aquaculture.ggn.org/en/the-tomato-fish.html. Accessed 6 January 2018.

International Federation of Organic Agriculture Movements (IFOAM) (2017). Consultation on Recirculation Aquaculture Systems (RAS). Available at: https://www.ifoam.bio/en/sector-platforms/ifoam-aquaculture. Accessed 6 January 2018.

Jenner, A. (2010). Recirculating aquaculture systems: The future of fish farming? Available at: https://www.csmonitor.com/Environment/2010/0224/Recirculating-aquaculture-systems-The-future-of-fish-farming. Accessed 6 January 2018.

Kerr, M. and Potthast, T. (2018). 'As close as possible to nature': Possibilities And Constraints For Organic Aquaculture Systems. In: Springer, S. and Grimm, H. (eds.): Professionals in Food Chains: Ethics, Roles and Responsibilities. Wageningen Academic Publishers, Wageningen.

Kloas, W., Groβ, R., Baganz, D., Graupner, J., Monsees, H., Schmidt, U., Staaks, G., Suhl, J., Tschirner, M., Wittstock, B., Wuertz, S., Zikova, A. and Rennert, B. (2015). A new concept for aquaponic systems to improve sustainability, increase productivity, and reduce environmental impacts. Aquaculture Environment Interactions 7: 179-192.

Leese, M. and Meisch, S. (2015). Securitising sustainability? Questioning the ,water, energy and food-security nexus'. Water Alternatives 8: 584-598.

Little, D.C., Newton, R.W and Beveridge, M.C. (2016). Aquaculture: a rapidly growing and significant source of sustainable food? Status, transitions and potential. In: Proceedings of the Nutrition Society 75: 274-286.

Martins, C., Eding, E.H., Verdegem, M.C., Heinsbroek, L.T., Schneider, O., Blancheton, J.P., Roque d'Orbcastel, E. and Verreth, J.A. (2010). New developments in recirculating aquaculture systems in Europe: A perspective on environmental sustainability. Aquacultural Engineering 43: 83-93.

Meisch, S. (2014). The need for a value-reflexive governance of water in the Anthropocene. In: Bhaduri, A., Bogardi, J., Leentvaar, J. and Marx, S. (eds) The Global Water System in the Anthropocene: Challenges for Science and Governance. Springer, Cham, pp. 427-437.

Rommetveit, K., Strand, R., Fjelland, R. and Funtowicz, S. (2013). What can history teach us about the prospects of a European Research Area? JRC Scientific and Policy Report EUR 26120. Publication Office of the European Union, Luxemburg.

Rushe, E. (2016). The Unlikely Fish-Farming Start-Up in the Middle of Berlin. Available at https://psmag.com/news/the-unlikely-fish-farming-start-up-in-the-middle-of-berlin. Accessed 6 January 2018.

Seibel, H., Weirup, L. and Schulz, C. (2018). Aspects of animal welfare in fish husbandry. In: Springer, S. and Grimm, H. (eds.): Professionals in Food Chains: Ethics, Roles and Responsibilities. Wageningen Academic Publishers, Wageningen.

Suhl, J., Dannehl, D., Kloas, W., Baganz, D. Jobs, S., Scheibe, G. and Schmidt, U. (2016). Advanced aquaponics: Evaluation of intensive tomato production in aquaponics vs. conventional hydroponics. Agricultural Water Management 178: 335-344.

Storhas, R. (1988). Grundsätze einer naturgemäβen Landwirtschaft. In: Haiger, A., Storhas, R. and Bartussek, H. (eds.) Naturgemäβe Viehwirtschaft. Ulmer, Stuttgart, pp. 20-27.

Strand, R. (2017). Post-normal science. In: Spash, C.L. (ed.) Routledge Handbook of Ecological Economics: Nature and Society. Routledge, London, pp. 288-297.

Tschirner, M. and Kloas, W. (2017). Increasing the Sustainability of Aquaculture Systems. Insects as Alternative Protein Source for Fish Diets. In: Gaia 26: 332-340.

71. 'As close as possible to nature': possibilities and constraints for organic aquaculture systems

M. Kerr[1] and T. Potthast[1,2]*
[1]International Centre for Ethics in the Sciences and Humanities, Faculty of Mathematics and Science; University of Tübingen, Wilhelmstr. 19, 72074 Tübingen, Germany; [2]Chair for Ethics, Philosophy and History of the Life Sciences, Faculty of Mathematics and Science, University of Tübingen, Wilhelmstr. 19, 72074 Tübingen, Germany; matthias.kerr@posteo.de

Abstract

The Commission Regulation (EC) No 710/2009 is the central reference point for organic aquaculture production in the EU. Accordingly, production has to resemble nature 'as close as possible'. Hence, the regulation explicitly excluded recirculating aquaculture systems (RAS) from organic certification. Yet no definition of the crucial normative criterion 'closeness to nature' is provided. For clarifying the term, we suggest to regard 'closeness to nature' being the outcome of a mixed ethical judgment: a multi-step procedure, which depends on descriptive premises as well as normative judgments. In order to obtain evidence of whether and how RAS may meet the criterion, it is necessary to consider ways on how 'closeness to nature' in (organic) aquaculture can be determined. This paper shall discuss three different perspectives. A first perspective can be assigned from nature conservation. There, concepts of naturalness and 'hemeroby' were developed to measure human impacts on ecosystems. In a second perspective, 'closeness to nature' in animal livestock is associated with indicators for animal welfare. The third perspective deals with existing approaches for 'natural' livestock farming, which were developed throughout the foundation of different organic farming associations. In particular, these approaches stress the importance of closed natural cycles and the resilience of natural systems. We will investigate different possibilities and constraints concerning the 'closeness to nature' criteria focusing on organic aquaculture in the EU. We first clarify the understanding of 'closeness to nature' resulting from a mixed ethical judgment. Then, the different approaches for defining grades of 'closeness to nature' will be contrasted, scrutinising possibilities as well as constraints regarding the application of these criteria. In conclusion, we argue that 'closeness to nature' as such cannot be sufficient to exclude RAS from organic certification completely.

Keywords: recirculation aquaculture systems (RAS), mixed ethical judgment, regulation of aquaculture, naturalness, sustainable production

Introduction: 'closeness to nature' – a mandatory but under-determined condition

Not least due to overfishing of wild stocks in conjunction with increasing demand, aquacul-ture is a fast-growing area of food supply in many countries. This also raises legal and ethical questions for organic production systems (Bergleiter and Meisch, 2015). In the EU, the Commission Regulation (EC) No 710/2009 issues detailed rules on organic aquaculture animal and seaweed production. The regulation explicitly excludes recirculating aquaculture systems (RAS) from organic certification. The reason indicated for this exclusion is the claim to provide 'closeness to nature' in organic production: '[...] Due to the principle that organic production should be as close as possible to nature the use of such systems should not be allowed for organic production [...]' (paragraph 11). However, it is striking that the regulation neither provides a definition of its crucial normative criterion 'closeness to nature' nor states any reference to already established indicators. Instead the regulation enumerates, spread throughout the whole text, aspects like sufficient space, water quality (e.g. oxygen level and temperature), light

EurSafe 2018 – DOI 10.3920/978-90-8686-869-8_71, © Wageningen Academic Publishers 2018

conditions and ground conditions as general aquaculture husbandry rules (*ibid.*, Section 3, Article 25f), but without explicit treatment of the concept of 'nature' and the normative criterion 'closeness to nature'. Considering the complexity and plurality of numerous conceptions of 'nature', an explanation of the specific understanding in this context would be of vital importance in order to ensure a sufficiently valid justification to exclude RAS from organic production. Since the Commission Regulation mentions no approach to measure 'closeness to nature', there is a need for conceptual and ethical analyses, considering different ways on how this normative criterion might be scrutinised. In this paper, we identify 'closeness to nature' as being the result of a mixed ethical judgment, consisting of intertwined empirical as well as normative dimensions. This approach allows us to characterise the components of the conclusion drawn by the Commission Regulation in detail, to discuss the normative aspects and, in this respect, the under-determined content. Then we will outline three perspectives on how 'closeness to nature' can be determined. Perspectives from nature conservation, animal welfare and 'natural' livestock farming provide contexts in which different degrees of 'closeness to nature' already have been defined. The discussion of these different approaches illustrates the problems associated with the attempt to determine certain states or farming methods as near-to-nature or even natural.

'Closeness to nature' – a concept resulting from a mixed ethical judgement

There is no purely descriptive way to determine 'closeness to nature'. 'Nature' is always connected with cultural, religious, social, environmental, economic and aesthetic values. Hence, depending on the underlying conceptions, nature has intrinsic value or crucial instrumental or eudaimonistic value as the basis of life for present and future generations, or a combination of the three value types (Potthast and Eser, 1999). In that sense, 'nature' is always a normative and evaluative reference. Application-oriented ethics and its approaches can increase the transparency and better the understanding of the values, stated or underlying, of the intended preclusion (cf. Ammicht Quinn and Potthast, 2015). From an ethical point of view, the preclusion of RAS from organic certification is a mixed ethical judgement. The EU regulation states certain possibilities and constraints by clarifying what we ought (not) to do (ethical judgment). These statements consist of descriptive and prescriptive premises (mixed ethical judgment), which are essential for a comprehensive consideration and finally to draw a conclusion (recommendation for action). In mixed ethical judgements both ways of reasoning – descriptive as well as prescriptive arguments – are closely connected. Notwithstanding, they can be described independently. Descriptive premises refer to empirically based reasons. The empirical research concerning aquaculture production collects and interprets data. It pays particular attention to the quality of water as well as to other vital measurable parameters for welfare of fish bred in aquaculture and fisheries, such as stress, health, mortality, stereotypies, stocking density, light regimes, feeding, noise, or reproduction according to certain specific indicators (e.g. Ashley, 2007; Huntingford *et al.*, 2006). We already recognise the normative framing of this since animal welfare is not a purely descriptive term. Further, prescriptive premises refer to explicitly normative orientation regarding what we ought to do, i.e. that in aquaculture production animal welfare shall be safeguarded with high priority.

Regarding aquaculture the regulation states 'closeness to nature' as a normative criterion against which the empirically observable properties of RAS appear to be not sufficient for organic production at least based on current knowledge (EC No 710/2009, paragraph 11). To reconsider the ethical judgments, and thereby revise the recommendation for action, one (e.g. potential producer of organic RAS products) may introduce other relevant empirical characteristics and respects, which might change the conclusion of the ethical judgment without altering the prescriptive premise ('closeness to nature'). However, it is still uncertain for production patterns in RAS, whether there is any possibility to realize the compiled empirical standards and at the same time satisfy all the normative requirements or not. An ethical discussion is mandatory to verify approaches on how to determine 'closeness to nature' substantively. 'Closeness to nature' as a condition of organic farming under which animals are kept is obviously not just reducible to (random) environmental facts of respective wild species living in nature, whose lives are often jeopardized by natural circumstances like persecution, predators, stress and diseases. Rather, 'closeness to nature' can be seen as a gradual approach depending on the intensity of human interventions in the living environment of the animals bred in aquaculture, fisheries or livestock farming (Gregorowius, 2008; Wehrli 2013: 52ff). In the following, we discuss three different perspectives on defining 'closeness to nature' by providing gradual concepts to substantively complement the normative premise regarding possibilities and constraints for organic RAS.

'Closeness to nature' – different approaches for defining its dimensions

Nature conservation

To evaluate the 'naturalness' of units (landscapes, ecosystems, communities or populations), two different operable concepts are distinguished in nature conservation: 'closeness to nature' and 'hemeroby'. 'Closeness to nature' stresses a historical perspective by reconstructing the culturally uninfluenced – 'natural' – state of a distinct unit as a reference and contrasts it with the current anthropogenically influenced or induced state (Kowarik, 2006). This procedure allows highlighting deviations stressing the continuity and discontinuity and, in this way, contextualizing the current situation of a unit resulting from a development. The concept of hemeroby, by contrast, is based on a contemporary oriented perspective on naturalness. The term 'hemeroby' is derived from the Greek words *hémeros* (tamed, cultivated) and *bíos* (life) and has at least since the mid-20th century been applied to local ecosystems to measure specific site potentials of vegetation from a contemporary point of view (Sukopp, 1972). In contrast to 'closeness to nature', hemeroby reflects the naturalness of specific land uses regarding human impact by measuring the differences between the current vegetation or population and a constructed non-anthropogenic state, derived from a scenario where the vegetation regulates itself in the complete absence of anthropogenic influences (potential natural vegetation (PNV); Walz and Stein, 2014). Closeness to nature thus refers to the state that existed before human beings became influential, especially sedentary, while PNV describes the vegetation – and, in principle – the whole ecosystem that would appear now without human influences. Hence, the concept of hemeroby takes irreversible changes of land use into account, focusing on actual site potential of the ecosystem to evaluate its naturalness (Kowarik, 2006). As a historical perspective, closeness to nature provides important impulses for culturally sensitive conservation and restoration; but it is not adequate as an operable concept to describe the naturalness of systems that are permanently altered, like cropland, grasslands with livestock or fishponds. The concept of hemeroby, however, emphasises that nature does not cease to exist as a characteristic of a distinct site just by being influenced by humans. Based on the intensity, duration and range of anthropogenic impact, a seven-point scale supplemented by an intersection with the PNV is used to classify land use by degrees of hemeroby (Walz and Stein, 2014; Sukopp, 1969; Stein and Walz, 2012). Landscapes, sites or populations without any human impacts are described as *ahemerob* (culturally unaffected) while overly cultural influenced areas are described as *metahemerob*. The allocation of a land use class to a certain degree of hemeroby is most likely to comply with the idea of naturalness meant by the EU regulation. The above-mentioned approach of nature conservation helps to understand to what extent RAS as a method of aquaculture production is 'distant from nature', i.e. from natural ecosystem conditions. Yet other normative dimensions also have to be taken into account to match the 'closeness to nature' criterion issued by the EU regulation.

Animal welfare

A second evaluative and normative perspective, apparently preferred by the EU regulation, is to focus on the individual animals living in certain production facilities, in this case RAS. The obliged 'closeness to nature' would be satisfied by maximizing animal welfare (EC No 710/2009, Section 3, Article 25f).

For implementing animal welfare in livestock or aquaculture, findings of farm animal biology regarding the vital needs of specific animals must be combined with findings of farm animal ethology regarding innate behavioural dispositions and considerations about behavioural enrichments (Langanke and Voget-Kleschin, 2014; Brydges and Braithwaite, 2009). However, it remains disputable whether a criterion like 'closeness to nature' can be complied with simply by realizing terms of animal welfare, and furthermore if animal welfare is the appropriate scale to constitute a definition of a threshold. This is certainly mandatory to justify the preclusion of RAS from organic farming because there must be a 'tipping point' between organic farming and those production patterns excluded from organic production by reaching or crossing this point and thereby failing on the 'closeness to nature' condition. If 'closeness to nature' can be declared as a principle of organic farming in the EU, it would already have been implemented e.g. in accordingly certified organic livestock, poultry feeds and pig keepings. It is appropriate therefore to take a closer look at the corresponding regulations and standards. The Commission Regulation (EC) No 889/2008 states detailed rules on organic production and conditions for the labelling of organic products by establishing specific rules e.g. on access to open-air areas like pastures, rules regarding stock density, animal health management, mutilations and feed. By applying those rules, the regulation concentrates on physiological, vital and ethological needs and, thereby, on animal welfare. In this regard, 'closeness to nature' would not be classified by degrees of hemeroby, but by maximising the animal-specific welfare. However, the EU regulation (EC) No 710/2009 doesn't seem to encourage this conclusion. If we assume that maximising animal welfare rather than the impact of anthropogenic influences is significant for organic production and thus to match the required 'closeness to nature', the possibility to control the relevant physiological, vital and ethological parameters in livestock farming would be expedient and desirable. Other than already certified organic pond farms, RASs enable operators to fully control water quality, flow conditions, stock density, enrichments, the amount and composition of feed supplied, reproduction, animal health, light regimes, the occurrence of predators and stress due to fisheries management (Wehrli, 2013). In respect of – and abiding by (!) – the parameters relevant for the welfare of fish based on the study of the recent empirical data, RAS would be an acceptable practice for organic aquatic animal farming. This suggests that animal welfare does not equal 'closeness to nature' in the understanding of the EU regulation.

'Natural' livestock farming

A third possible reference to 'closeness to nature' has been proposed by Richard Storhas, co-founder of 'Naturland', an organic farming association operating at international level. One central aim of 'natural' livestock farming is to contribute to global food security without further endangering fertile soils. Land degradation accumulates long-term (environmental) costs on a global scale and in particular for societies or people living in the affected areas. That's why 'natural' livestock farming stresses the importance of closed natural cycles and the resilience of natural systems. Industrial agriculture productions run a linear course: Resources are extracted ('input') and converted into products and services ('transformation') in order to market them as consumer goods ('output') (Storhas, 1988: 21). Due to the lack of recycling processes, this procedure leads to depletion and overexploitation of soils. 'Natural' livestock farming on the contrary emphasizes production patterns oriented on natural cycles ('seeds – growth – reproduction – new seeds', *ibid.*). Just like the concept of hemeroby states, in 'natural' livestock farming anthropogenic influences are not sufficient to categorically preclude agriculture as alienated from nature. 'Closeness to nature' is scrutinized by the intensity, duration and range of human impact and considered by degree, but under principal exclusion of industrialized agriculture. Which animals are to be kept in 'natural' livestock farming depends on site conditions like soil, terrain and climate. 'Closeness to nature' in 'natural' livestock farming is signified by sustainable, environmentally sound agricultural practices, closed natural cycles instead of linear production patterns and scaled by means of human interventions. The overall conception places 'closeness to nature' in a wider context with sustainability. Nevertheless, when considering criteria for naturalness in aquaculture, there are significant conceptual difficulties regarding 'closeness to nature' in 'natural' livestock farming. Although diversified forms of RAS are possible, which especially consider closed energy and material cycles (aquaponics, e.g. Kloas *et al.*, 2015), but there is no equivalency to the meaning of soil. Land use issues in livestock farming cannot be easily compared with the standards in aquaculture. On the contrary, RAS are entirely independent from the quality of the soil. Hence if someone would regard the natural ground of ponds as soil, then RAS would be far from being 'close to nature'. The importance of site conditions to select suitable species for a specific type of farming presents a further valid criterion according to which fishes should be precluded from RAS, if they need structures for their typal behaviour and animal-specific welfare.

Conclusion: 'closeness to nature' – its normative challenges for RAS

In order to develop a better understanding of what 'closeness to nature' might mean and how it could be operationalized also in legal regulation, one should be prepared to enter ethical analysis and discussion. We have pointed out that 'closeness to nature' as a norm results from the combination of descriptive and prescriptive elements, hence a mixed ethical judgment. By investigating three ways to specify 'closeness to nature' and the underlying conceptual framings and problems – hemeroby, animal welfare, natural livestock farming – we have identified various normative dimensions: (1) Hemeroby tends to stress the value of non-anthropogenic states of ecosystems and populations due to the threat of losing such systems and their components by ongoing destruction. Hemeroby does not refer to historical 'originality' and allows for degrees of being 'close to nature'. It might provide some practical advice for creating the artificial environment of RAS mimicking the conditions of natural ponds. Hemeroby would than serve as one indicator, allowing to think about the design of RAS not 'close to nature' but rather artificially constructing a simulation resembling 'natural' conditions (cf. also aquaponics). (2) From an ethical perspective, animal welfare is a necessary condition for RAS anyway, but it is neither a necessary nor a sufficient condition of and for 'closeness to nature'. (3) Natural livestock farming provides hints to design RAS in accord with organic production principles, yet does not give direct advice what 'closeness to nature' exactly means. The three dimensions of 'closeness to nature' might also be understood relating to different recipients: As a principle oriented towards consumer expectations it might emphasize other aspects (e.g. hemeroby) than being oriented towards aiming at animal welfare or environmental protection (e.g. 'natural' livestock farming) protagonists. We conclude that it is not convincing to exclude RAS from organic certification completely only based on the Commission Regulation (EC) No 710/2009. Broader aspects of sustainable production have to be addressed for an informed assessment (cf. also Meisch and Stark, 2018). Whether RAS are sufficiently 'close to nature' or not to meet the principle stated by the EU regulation must be (further) discussed: this comprises animal welfare and the design of RAS at least mimicking environments that are an analogue of being 'close to nature', allowing for their typal behaviour and animal-specific welfare. This might provide a critical perspective on single-species containments of RAS. These more specific questions need to be addressed with further research combining biological and ethical expertise.

Acknowledgement

This paper further develops aspects of the project *Stakeholder-Studie 'Kreislaufanlagen – Positionen des Ökosektors'* (Bergleiter *et al.*, 2017) funded by German Federal Ministry of Food and Agriculture (project no. 2815OE026).

References

Ammicht Quinn, R. and Potthast, T. (eds.) (2015). Ethik in den Wissenschaften. 1 Konzept, 25 Jahre, 50 Perspektiven. Tübingen: IZEW, Materialien zur Ethik in den Wissenschaften, 10.

Ashley P.J. (2007). Fish welfare: Current issues in aquaculture. Applied Animal Behaviour Science, 104: 199-235.

Bergleiter, S. *et al.* (2017). Recirculation Aquaculture Systems – Positions of the Organic Sector. Available at: http://www.orgprints.org/32165. Accessed 26 March 2018.

Bergleiter, S. and Meisch, S. (2015). Certification Standards for Aquaculture Products: Bring-ing Together the Values of Producers and Consumers in Globalised Organic Food Markets. In: Journal of Agricultural and Environmental Ethics 28(3): 553-569.

Brydges, N.M. and Braithwaite, V.A. (2009). Does environmental enrichment affect the behaviour of fish commonly used in laboratory work? In: Applied Animal Behaviours Science 118: 137-143.

European Commission (2008). Commission Regulation (EC) No. 889/2008 of 5 September 2008 laying down detailed rules for the implementation of Council Regulation (EC) No. 834/2007 on organic production and labelling of organic products with regard to organic production, labelling and control. Available at: http://eur-lex.europa.eu/legal-content/EN/TXT/PDF/?uri=CELEX:32008R0889&from=EN. Accessed 26 March 2018.

European Commission (2009). Commission Regulation (EC) No. 710/2009, organic aquaculture animal and seaweed production. Available at: http://eur-lex.europa.eu/legal-content/EN/TXT/PDF/?uri=CELEX:32009R0710&from=EN. Accessed 26 March 2018.

Gregorowius, D. (2008). Landwirtschaft im Spannungsfeld zwischen Natürlichkeit und Künstlichkeit. In: Zeitschrift für ev. Ethik 52(2): 104-118.

Huntingford F.A. *et al.* (2006). Current issues in fish welfare. Journal of Fish Biology, 68: 332-372.

Kloas, W. *et al.* (2015). A new concept for aquaponic systems to improve sustainability, increase productivity, and reduce environmental impacts. In: Aquaculture Environment Interactions 7: 179-192.

Kowarik, I. (2006). Natürlichkeit, Naturnähe und Hemerobie als Bewertungskriterien. In: Fränzle, O. *et al.* (eds.): Handbuch der Umweltwissenschaften – Grundlagen und Anwendung der Ökosystemforschung. Wiley-VCH, Weinheim, VI3.12: 1-18.

Langanke, M. and Voget-Kleschin, L. (2014). Tierethische Maßstäbe zur Beurteilung von landwirtschaftlicher Nutztierhaltung am Beispiel der Haltung von Hühnervögeln – argumentative Möglichkeiten und Grenzen. In: Zeitschrift für ev. Ethik 58(2): 190-202.

Meisch, S. and Stark, M. (2018). Recirculation Aquaculture Systems: Sustainable Innovations in Organic Food Production? (in this volume).

Potthast, T. and Eser, U. (1999). Systematisierungsvorschläge und vier Thesen zum Verhältnis von Naturschutzbegründungen, Ökologie und Ethik. Verhandlungen der Gesellschaft für Ökologie 29: 579-585.

Stein, C. and Walz, U. (2012). Hemerobie als Indikator für das Flächenmonitoring. Methodenentwicklung am Beispiel von Sachsen. In: Naturschutz und Landschaftsplanung 44(9): 261-266.

Storhas, R. (1988). Grundsätze einer naturgemäßen Landwirtschaft. In: Haiger, A. *et al.* (eds.): Naturgemäße Viehwirtschaft. Zucht, Fütterung, Haltung von Rind und Schwein. Stuttgart: Ulmer, 20-27.

Sukopp, H. (1969). Der Einfluss des Menschen auf die Vegetation. Vegetatio, 17, 360-371.

Sukopp, H. (1972). Wandel von Flora und Vegetation in Mitteleuropa unter dem Einfluß des Menschen. In: Bericht über Landwirtschaft, 50(1): 112-139.

Walz, U. and Stein, C. (2014). Indicators of hemeroby for the monitoring of landscapes in Germany. In: Journal of Nature Conservation 22 (2014): 279-289.

Wehrli, S. (Schweitzer Tierschutz STS) (eds.) (2013). Tierwohl in Nutzfischzuchten. Available at: http://www.tierschutz.com/wildtiere/docs/pdf/report_nutzfischzucht.pdf. Accessed 26 January 2018.

Section 14.
Water ethics

72. Water ethics – lessons from post-normal science

S. Meisch
International Centre for Ethics in the Sciences and Humanities (University of Tuebingen), Wilhelmstr. 19, 72074 Tuebingen, Germany; simon.meisch@uni-tuebingen.de

Abstract

Water ethics is an emerging field in application-oriented ethics. It reviews the normative and evaluative implications of human water practices and aims to argue for (more) justified practices. The text corpus on water ethics continues to grow, reflecting an increasing demand for moral orientation. It is time now to take stock and reflect how to address inherent tensions within this corpus, in particular regarding its practice and object. In this vein, the paper does not intend to produce a reductionist and uniform account of the diverse field of water ethics. On the contrary, it perceives itself as a critical and constructive reflection of this rich field. In order to capture the water ethical scholarship, the analysis in this paper addresses ethics primarily as a form of social activity rather than a body of different theories, which it also is. It will first unfold tensions and dividing lines within the water ethical literature. Subsequently, after introducing post-normal sciences (PNS), the paper asks how this approach might help to bridge some of the tensions in the water ethical literature. PNS challenges the dominance of scientific representations in dealing with real-world problems and constitutes an innovative mode of knowledge production for environmental governance under conditions of complexity and uncertainty. The approach of this paper helps assess the potentials and aspirations of water ethics as a field of application-oriented ethics.

Keywords: water, water ethics, ontologies of water, social movements, post-normal science

Introduction

Water is the *sine qua non* for life on Earth (Schmidt and Peppard, 2014). Without it, no human life is possible; without enough of it, living decent lives becomes restricted. Worldwide, a billion people do not have access to safe and secure water and sanitation. This situation may worsen with the expected effects of climate change on precipitation patterns and water bodies. Yet, water shortage results from both physical and social conditions. While water liberalisation was advocated as a means for better and more efficient governance, in effect it has produced more and other forms of water shortage and injustices (Leese and Meisch, 2015).

Water has always been the subject of moral and power struggles, yet only in the last 15 years have we witnessed the emergence of a scholarship within application-oriented ethics that calls itself water ethics. So far, the corpus of literature remains heterogeneous and contradictory. On the one hand, this heterogeneity reflects the actual contentious water reality itself and as such does not need to be an issue of concern. On the other hand, the current water ethical debate might give rise to expectations it cannot easily meet. Against this background, the paper suggests one way of mapping the diverse field of water ethics. In this vein, it unravels inherent tensions and argues for an approach that better grasps the potential of water ethics as a field of application-oriented ethics aspiring to be a practically relevant theory and a theoretically grounded practice. In order to capture the water ethical scholarship, the paper deploys a wide understanding of ethics in two respects. First, generally, ethics is distinguished from morals. Morals comprise the set of beliefs about the evaluatively Good and the normatively Right held by an individual or a group. Ethics is the systematic and methodical philosophical reflection on morals, making morals the research subject of ethics. In principle, the paper sticks with this distinction, but it acknowledges that the terms are used differently within the water ethical debate. Thus, if we only considered scholarship that falls under the abovementioned textbook definition, we would unnecessarily

EurSafe 2018 – DOI 10.3920/978-90-8686-869-8_72, © Wageningen Academic Publishers 2018

restrict the scope of literature and ignore a rich body of normative deliberation on water. Second, the analysis of this paper regards ethics as a form of social activity rather than a body of different theories, which it also is. This activity can become subject of ethical reflection itself – just like any other research activity. Dietrich (2005) described this approach as 'ethics of ethics'. Emphasising the social practices of water ethics contributes to two aims: first, assessing the potentials and aspirations of water ethics as a field of application-oriented ethics and second, providing a more solid basis for water ethical reflection within social contexts and decision-making processes.

Below, I will suggest a way of mapping the water ethical scholarship and discuss some of its inherent dividing lines (2). I will introduce Post-Normal Science (PNS) and argue how it helps bridging some of these tensions (3.). This paper aims to support self-critical reflection within water ethics and improve the process of decision-making on water.

Water ethics

In the last decade, water ethics has emerged as a new field in application-oriented ethics (Grunwald, 2016; Meisch, 2017a). It aims to provide moral orientation in a social field in which the morally right course of action has become uncertain and contested (Meisch, 2014). Water ethics reviews the moral implications of human water practices and aims to argue for (more) justified practices. Currently, there is a growing interest in water ethics within and outside of academia. This interest is driven by challenges related to sustainable development such as water poverty, or the (over-)usage and pollution of water resources. These challenges are expected to worsen due to climate change and population growth. In general, the goal will be to ensure safe access to and a fair distribution of clean water. Meanwhile, global trends of privatising water have spurred further debates on water justice (Feitelson and Chenoweth, 2002; Pahl-Wostl *et al.*, 2013). The text corpus on water ethics continues to grow, reflecting an increasing demand for moral orientation. Different approaches to water ethics might be classified with regard to the moral subject and can be thus subdivided into theories that place either humans only, sentient animals, ecosystems or nature at the centre of their ethical reasoning. Alternatively, a dividing line can be drawn according to ethical theories, such as deontological, utilitarian, eudaimonic or other approaches (Brown and Schmidt, 2010; Kowarsch, 2011; Ziegler and Groenfeld, 2017).

This paper suggests a different way of structuring the domain of water ethics by separating it in two subdomains: (1) practice and (2) epistemology. The practice subdomain refers to the basic orientation of ethical reflection. It focuses on the role of the water ethicist and his or her relation to ethical theories and locates this scholarship in-between the poles of 'academic reflection' and 'social transformation' (Dietrich, 2007). The epistemology domain refers to the object of study and deals with ethicists' perspectives on water (Schmidt, 2014; Yates *et al.*, 2017, Budds and Linton, 2014). Regarding the first (practice) subdomain, the paper uses distinctions suggested by Thompson (2016) and Floyd (2013). In his review of food ethics, Thompson separates academic food ethics from food ethics described as social movement. He discusses what both learn from critically engaging with each other. In this vein, he aims to draw attention to the 'ways in which the very language we use to articulate the nature of an ethical problem in one domain becomes implicated in oppressive or obfuscating measures in another' (Thompson, 2016: 70). In addition, Floyd sorts the environmental security literature according to the role of the analyst and his or her view on security. Based on both approaches, this paper separates the water ethical literature with regard to the ethicist's orientation within the practice of water ethics. This results in a continuum that ranges from general ethics reviewing moral arguments, to water ethics arguing for different water practices, and finally to water activism with various normative agendas. Yet, I expand on Thompson's approach, which seems to imply that food ethics *qua* social movement only consists of NGOs. To me, his view seems too restricted because it underrates commercial economic actors, who also aim for different social water practices (Leese and Meisch, 2015; Schmidt and Matthews, 2018). The second (epistemology) subdomain refers to the lively debate in the water (ethical) community on the ontological status of water i.e. *what* water is. Accordingly, we would analyse whether water ethics takes place within the ambiguous frame of 'modern water' based on the society/nature dualism and reducing water to its predominant scientific representations of H_2O and the hydrological cycle (Schmidt, 2014). Most (natural science) water research aligns itself to this epistemic paradigm, and in general application-oriented water ethics also starts from this perspective: academic ethicists aim to find and define principles for the allocation of and access to water (*qua* H_2O) and social movements call for a different way of dealing with water (*qua* H_2O) (Groenfeldt, 2013; Harremoës, 2002; Ziegler *et al.*, 2017). The opposing view questions the 'modern water'-frame as reductionist, historically contingent and an expression of Western approaches to nature. Related scholars claim that water has 'multiple ontologies' (Yates *et al.*, 2017; Schmidt and Peppard, 2014): academic scholars point to the cultural constructions of water and the embedded value systems. Their disciplinary backgrounds range from political economy to anthropology, cultural and literary studies (Böhme, 1988; Krause and Strang, 2016; Linton and Budds, 2014; Swyngedouw, 2009). In this way, water ethics is also an ethics of water cultures (Meisch, 2017b; Haker 2010). Meanwhile, water ethics qua social movement transcends academic reflection and suggests the recognition of local or indigenous conceptions of water. Again, there are various approaches within each camp turning the fields into continuums instead of oppositions. Against this backdrop, we can create a four-field matrix (Table 1) correlating water ethics (academic reflection, social movement) with the frames (modern water, multiple ontologies).

Scholarship in the different fields (I-IV) has its respective strengths and weaknesses and poses particular ethical challenges. In a textbook understanding of ethics, only scholarship in field I would be considered as water ethics because it is the only approach that explicitly and methodically reviews the moral reasons why we should do (or refrain from doing) something with regard to water. In contrast, social movements in both frameworks (II, IV) argue from particular moral viewpoints that they believe ethically more justified than the status quo, but they vary in the degree to which they make their ethical reasoning transparent. Scholarship in field III critically analyses current water practices and reveals how they embody power relations and value systems. Yet, most of these approaches are inspired by various post-structuralist traditions, which usually tend to reject universal ethical systems (Floyd, 2013: 26; Von Beyme, 2007: 190f). Consequently, they face difficulties in arguing what a better water practice would look like. In this respect, their ethical challenges are inverse to the one in field I which risks losing sight of the different ways of knowing waters and inbuilt power relations. Thus, in a first approximation, we can argue that while the one (I) has epistemological and power blind spots, the other (III) leaves its ethical presuppositions implicit and cannot tell how to proceed from their critical analysis. In addition, there is the challenging question of commensurability: How can we deal with 'other' forms of water knowledge and related norms and values of, for instance, indigenous people? Can we translate and include them in established academic forms of ethical reasoning? If yes, how?

Table 1. Subdomains of water ethics.

Practice	Epistemology	
	Modern water	Multiple ontologies of water
Academic reflection	I Finding and defining principles for the allocation of and access to water (qua H_2O)	III Highlighting the different social and cultural constructions of water and the embedded value systems
Social movement	II Calling for a different ethics with regard to water (qua H_2O)	IV Calling for equal recognition of local or indigenous conceptions of water

The tensions within water ethical scholarship are interesting from a theoretical perspective. They also have consequences for water practices. First, a narrowly defined academic water ethics (I) might overlook hidden epistemological and normative assumptions of the 'modern water'-frame. This could result in theoretical and practical blind spots. Second, water ethics as a social movement (II, IV) aims to reform the water sector, yet by doing so it has to critically consider possible injustices that it might create through new practices, or by reproducing other morally problematic social practices (regarding gender, race, etc.). Third, scholarship in the 'multiple ontologies'-frame (III) raises awareness of the many different water worlds and the injustices embodied by them. While this critical perspective is essential, it needs to be supplemented by (other) ethical approaches in order to be able to adequately argue for more just water practices. This paper suggests PNS as a solution to bridge many of the tensions within the water ethical literature.

A post-normal water ethics

Strand (2017: 288) characterises PNS as 'a critical concept originally developed to describe situations in which there are important or controversial public decision problems informed by an incomplete, uncertain or contested knowledge base'. The challenges that water ethics typically deals with correspond well to this problem description. PNS pursues an epistemic and a social aim. First, it challenges the (growing) dominance of scientific representations in dealing with real-world problems. Second, it constitutes a different mode of knowledge production for environmental governance under conditions of complexity and uncertainty (Benessia *et al.*, 2011; Funtowicz and Ravetz, 1993: 744). In order to classify both perspectives, I take the distinction 'descriptive' for the former and 'normative' for the latter, as recently suggested by Bremer (Bremer, 2017; also Bremer and Meisch, 2017). The descriptive perspective 'helps diagnose the problems of imposing a scientific representation where the science is highly uncertain, values in dispute, stakes high, and action urgent' (Bremer, 2017: 73; also Funtowicz and Ravetz, 1993). It also questions the status of science itself, how the boundaries are drawn between scientific research and other forms of human activity and how this affects the process of democratic decision-making (Benessia and Funtowicz, 2016: 71; Strand *et al.*, 2016). In this way, PNS resembles the science and technology studies aiming for 'the possibility of seeing certain 'hegemonic' forces not as given but as the (co)products of contingent interactions and practices' (Jasanoff, 2004: 36). In this way, PNS can be an analytical, interpretative tool for better seeing certain social phenomena such as 'the hydrological cycle'. It exposes and challenges dominant narratives in water governance (Bremer and Meisch, 2017). The normative perspective 'offers an alternative mode of scientific enquiry; engaging scientists and non-scientists alike as members of an extended peer community, to collaboratively produce new representations and knowledge, and appraise its quality' (Bremer, 2017: 73). As such, it aims for socially robust, accountable and legitimate knowledge in the face of uncertainty (Bremer and Meisch, 2017: 8f.). PNS also deals place-based knowledge and value systems and how they enter debates on sustainable development (Bremer and Funtowicz, 2015; Lam, 2015). In the following, I will discuss three fields in which PNS can bridge some dividing lines in the water ethical literature and thus improve the process of decision-making in water governance.

First, the descriptive perspective of PNS inserts additional layers of self-reflexivity into the ethical research within the 'modern water'-frame. With this, it becomes possible to challenge scientific representations of water and the ways in which water science demarcates itself from other forms of knowledge making. From here, it will also be possible to critically assess the related implications for democratic water governance. This means that PNS challenges the contexts in which water ethics is done. Doing so, it takes a first step to connecting a self-reflexive 'modern water'- to the 'multiple ontologies'-frame. As mentioned above, the latter looks critically at the ways power relations and value systems are embedded in water systems and suggests alternative ways of organising human-water relationships. However, they often need to be supplemented by ethical approaches to indicate how to proceed from a critical assessment of the status quo. Second, in this regard, the normative perspective of PNS can be a useful supplement. It constitutes a different mode of knowledge production that is more sensitive to normative and evaluative issues. Actually, the normative perspective applies to both the 'multiple ontologies'- and to a more self-reflexive 'modern water'-frame. Third, engaging with PNS can help to address two issues related to water ethics *qua* social movement. One concerns the ethics of water activism. As mentioned above, water ethics as a social movement focuses on transforming actual water-related policies. By doing so, it might unintentionally reproduce old or create new injustices. Another concerns the epistemologically difficult question if, and to what degree, place-based value systems with regard to water can be reformulated within the language of academic ethical reasoning. Both, ethics of water activism and of indigenous water cultures, are relevant when claims are taken seriously that stakeholders are to be included in water governance. Thanks to its sensitivity to values and silencing, PNS can inform water ethics as a social movement.

Conclusion

This paper mapped the field of water ethics and unravelled some of the tensions within the field. It pointed to specific challenges and argued how PNS might help to deal with them. One might object that PNS has weak spots when it comes to normative ethics, which can be readily acknowledged. But thanks to its dual character (descriptive, normative), it provides a critical perspective on knowledge production at the science-society interface as well as a constructive approach to producing socially robust knowledge for decision-making. With this, PNS supports self-critical reflection within the field of water ethics and improves the process of decision-making on water.

Acknowledgement

I wish to thank many colleagues for their helpful comments on this paper and in particular Thomas Ammicht, Scott Bremer, Bruna de Marchi, Silvio Funtowitz and Jeremy J. Schmidt.

References

Benessia, A., Funtowicz, S., Bradshaw, G., Ferri, F., Ráez-Luna, E. and Medina, C. (2012). Hybridizing sustainability: Towards a new praxis for the present human predicament. Sustainability Science 7 (Supplement 1): 75-89.

Bremer, S. (2017). Have we given up too much? On yielding climate representation to experts. Futures 91: 72-75.

Bremer, S. and Meisch, S. (2017). Co-production in climate change research: reviewing different perspectives. In: WIREs Climate Change. 8(6); DOI: 10.1002/wcc.482.

Bremer, S. and Funtowicz, S. (2015). Negotiating a place for sustainability science: Narratives from the Waikaraka Estuary in New Zealand. Environmental Science & Policy 53: 47-59.

Brown, P. and Schmidt, J.J. (eds.) (2010). Water Ethics. Foundational Readings for Students and Professionals. Island Press, Washington D.C. *et al.*, 320 pp.

Böhme, H. (ed.) (1988). Kulturgeschichte des Wassers. Suhrkamp, Frankfurt/Main.

Funtowicz, S. and Ravetz, J. (1993). Science for the post-normal age. Futures 25: 739-755.

Groenfeldt, D. (2013). Water Ethics. A Values Approach to Solving the Water Crisis. Routledge, London, 216 pp.

Grunwald, A. (2016). Water ethics – Orientation for water conflicts as part of inter- and transdisciplinary deliberation. In: Hüttl, R., Bens, O., Bismuth, C. and Hoechstetter, S. (eds.) Society – Water – Technology. A critical appraisal of major water engineering projects. Springer, Cham *et al.*, pp. 11-29.

Haker, H. (2010). Narrative Ethik. Zeitschrift für Didaktik der Philosophie und Ethik 2: 74-83.

Harremoës, P. (2002). Water ethics – a substitute for over-regulation of a scarce resource. Water Science and Technology 45: 113-124.

Jasanoff, S. (2004). Ordering knowledge, ordering society. In: Jasanoff, S. (ed.) States of Knowledge: The Co-Production of Science and Social Order. Routledge, London, pp. 13-45.

Kowarsch, M. (2011). Diversity of water ethics – a literature review. 2nd Working Paper prepared for the research project 'Sustainable Water Management in a Globalized World', funded by the German Federal Ministry of Education and Research.
Lam, M. (2015). Reconciling Haida ethics with Pacific herring management. In: Dumitras, D., Jitea, I and Aerts, S. (eds.) Know your food. Food ethics and innovation. Wageningen Academic Publishers, Wageningen, pp. 169-176.
Leese, M. and Meisch, S. (2015). Securitising sustainability? Questioning the 'water, energy and food-security nexus'. Water Alternatives 8: 584-598.
Linton, J. and Budds, J. (2014). The hydrosocial cycle: Defining and mobilizing a relational-dialectical approach to water. Geoforum 57: 170-180.
Meisch, S. (2017a). What is and to what end do we study Water Ethics? Lessons for the Water Ethics Charter. In: Ziegler, R. and Groenfeldt, D. (eds.) Global Water Ethics: Towards a Water Ethics Charter. Routledge, London, pp. 37-55.
Meisch, S. (2017b). Wasserethik als Kulturethik. In: Brand, C., Heesen, J, Kröber, B., Müller, U. and Potthast, T. (eds) Ethik in den Kulturen – Kulturen in der Ethik. Narr Francke, Tübingen, pp. 129-140.
Meisch, S. (2014). The need for a value-reflexive governance of water in the Anthropocene. In: Bhaduri, A., Bogardi, J., Leentvaar, J. and Marx, S. (eds) The Global Water System in the Anthropocene: Challenges for Science and Governance. Springer, Cham, pp. 427-437.
Pahl-Wostl, C., Vörösmarty, C., Bhaduri, A., Bogardi, J, Rockström, J. and Alcamo, J. (2013). Towards a sustainable water future: shaping the next decade of global water research. Current Opinion in Environmental Sustainability 5: 708-714.
Schmidt, J.J. (2014). Historicising the Hydrosocial Cycle. Water Alternatives 7: 220-234.
Schmidt, J.J. and Matthews, N. (2018). From state to system: Financialization and the water-energy-food-climate nexus. Geoforum 91: 151-159.
Schmidt, J.J. and Peppard, C. (2014). Water ethics on a human-dominated planet: rationality, context and values in global governance. WIREs Water 1: 533-547.
Strand, R. (2017). Post-normal science. In: Spash, C.L. (ed.) Routledge Handbook of Ecological Economics: Nature and Society. Routledge, London, pp. 288-297.
Swyngedouw, E. (2009). The Political Economy and Political Ecology of the Hydro-Social Cycle. Journal of Contemporary Water Research and Education 142: 56-60.
Von Beyme, K. (2007). Theorie der Politik im 20. Jahrhundert. Von der Moderne zur Postmoderne. Suhrkamp, Frankfurt/ Main, 450 pp.
Ziegler, R. and Groenfeldt, D. (eds.) (2017). Global Water Ethics. Towards a global ethics charter. Routledge, London, 316 pp.
Ziegler, R., Gerten, D. and Döll, P. (2017). Safe, just and sufficient space: the planetary boundary for human water use in a more-than-human world. In: Ziegler, R. and Groenfeldt, D. (eds.) Global Water Ethics. Towards a global ethics charter. Routledge, London, pp. 109-130.

Posters

73. Food labelling: giving food information to consumers

A. Hrković-Porobija[1], A. Hodžić[1], N. Hadžimusić[2], E. Pašić-Juhas[1], A. Rustempašić[3] and I. Božić[1]*
[1]Department of Chemistry, Biochemistry and Physiology, Faculty of Veterinary Medicine, University of Sarajevo, Bosnia and Herzegovina; [2]Department of Pathophysiology, Faculty of Veterinary Medicine, University of Sarajevo, Bosnia and Herzegovina; [3]Department for Animal Breeding, Faculty of Agriculture, University of Sarajevo, Bosnia and Herzegovina; amina.hrkovic@vfs.unsa.ba

Abstract

This article provides information on food labelling from a consumer perspective and makes recommendations to the food industry and regulators regarding food labelling in order to satisfy consumers' food-labelling needs. A compound ingredient may be included in the list of ingredients, under its own designation in so far as this is laid down by law or established by custom, in terms of its overall weight, and immediately followed by a list of its ingredients. The chosen batch records should contain the following information: the date of production, the vat number, the type of cheese, the formulation name/number or reference, the quantity of milk, partly skimmed milk, skim milk and their ultra-filtered counterparts and cream used in the vat/batch for each milk/cream used, the fat and protein content, the final weight of cheese produced and, if available, its total protein content, % fat and % moisture of the cheese produced. The aim of this paper is to identify the actual composition of the indigenous livestock and herd sheep's cheese in order to protect the consumer and compare it with the qualitative composition shown on the declaration of the same cheese sold on the retail. Total solids were determined in a drying oven at 105±20 °C (Heraeus) according to IDF standard 4A and the fat content in the cheese was determined by Van Gulik-Gerber method (IDF Standard 152A). The determined values were not corresponding in declaration label, considering fat, moisture and dry matter. The food industry and regulators should aim to provide risk communication and intelligible information through food labels and consumer education programs on food labelling. Consumers need to be aware of their right to know what they are purchasing through food labels and take a stand in this regard.

Keywords: food, consumers, declaration

Introduction

Food and nutrition today have completely new dimensions and concepts. In the past, the main role of food was related to survival or hunger, and little attention was paid to its negative or positive effect on health (Rogelj, 1998). Nowadays, with the new nutrition requirements, the focus is directed towards the health of different categories of consumers, as well as on demands for better quality. Therefore, it is quite understandable that we focused this research on the role of food in maintaining and improving the health condition and reducing the possibility of developing various diseases (Rogelj, 1998). Food labels on the products are important source of nutrition information, but is typically underutilized by consumers. Advances in food production and processing have resulted in consumers using more processed food, which makes it more difficult to know the composition of the food they are consuming. With the growing interest among consumers in the link between diet and health and the credence nature of most health attributes, labelling plays a key role in consumers' decision making (Zou and Hobbs, 2010).

Consumers' concerns regarding this, as well as their avoidance of food-borne pathogens, toxins and allergens are increasingly taken into consideration when making food purchasing decisions (Merwe and Venter, 2010). Thus, food determines the behaviour of various subjects in the food industry, such as producers, manufacturers and retailers, and regulators. It is an important concern to consumers, as food directly affects their physical, biological, cultural and social environments (Mepham, 2000).

EurSafe 2018 – DOI 10.3920/978-90-8686-869-8_73, © Wageningen Academic Publishers 2018

Food labels, particularly nutrition labelling, can help consumers identify the nutrient content of foods, compare their foods and make a choice suitable for their individual needs.

Modern food consumers look for quality products which are unique in their characteristics. These products peculiarities can be seen through their characteristics such as physical, chemical, aesthetic, microbiological structure, specific quality of raw material they are made of, use of exactly determined manufacture technology, product presentation and packaging, as well as product identification with its geographical origin. The EU's regulations prescribe that food information should not be misleading, particularly in terms of food characteristics, including its nature, identity, characteristics, composition, country or place of origin and that they must be accurate and clear.

Food labelling is any written, printed or graphic matter that is present on the label, accompanies the food, or is displayed near the food, including that for the purpose of promoting its sale or disposal. Prepackaged food shall not be described or presented on any label or in any labelling in a manner that is false, misleading or deceptive. The following information should appear on the label of prepackaged foods as applicable to the food being labelled which could be different in each country depending on their national food control system and the existing legislation. Today, various animal products of high quality are found on the market, such as traditional products derived from autochthonous breeds. However, it is well known that these products are often falsified by using cheaper raw materials instead of declared, and that the chemical composition is not the traditional sheep cheese (Livno and Travnik) and is easily falsified by the addition of cheaper cow's milk or by lowering it. Bosnia and Herzegovina has a long tradition of family produced dairy products, dominated by 2 to 3 kinds of cheeses. As well-known brands in the dairy industry in B&H are the indigenous Livno and Travnik cheese belonging to a group of cheeses produced from raw thermally unstructured sheep's milk. The production of autochthonous cheeses in Bosnia and Herzegovina is unorganized and volumetric, and is often accompanied by low and uneven quality of milk and cheese.

Materials and methods

For the purpose of determining the correct labelling of cheeses, samples of cheeses produced on family farms sold at the local market in the Travnik and Livno area were sampled. A total of 6 cheese samples were analysed (3 for Livno and 3 for Travnik for each sampling period. The analysis of cheese samples included the determination of physical-chemical characteristics: total solids, moisture and fat. Total solids were determined in a drying oven at 105 ± 20 C (Heraeus) according to IDF standard 4A and the fat content in the cheese was determined by Van Gulik-Gerber method (IDF Standard 152A). All analyses were performed in the Laboratory Milk Technology Faculty of Agricultural in Sarajevo. The statistical data processing was done using the SPSS 15.00 software package.

Results and discussion

Cheese production is an important industry in the food industry. The exact number of all kinds of cheeses in the world is difficult to determine. It is believed that there are more than 2200 different kinds, but that number grows every day. New types of cheese are still produced today, and cheese as a foodstuff is adapted to the wishes and needs of consumers (Lipovcak, 2017). A-cheese belongs to the group of hard cheeses. It is basically made of raw ewes′ milk but nowadays cows′ milk significantly participate in its manufacture. Usual ratio between ewes and cows′ milk in A-cheese manufacture is 80:20 (Table 1). B-cheese is chosen since it is the most popular and consumed cheese among local Bosnian and Herzegovina (B&H) inhabitants. It belongs to the group of white brined cheeses traditionally made of raw ewes milk (Table 1).

Table 1. Physical and chemical characteristics of A and B cheese.

Parameters	A – hard cheese (%)	B – white pickled cheese (%)
Total solids	64.60	46.58
Moisture	35.40	53.42
Fat	32.30	25.00

Acquired results showed that analysed cheeses did not have completely correct declaration. Declarations of cheeses A and B indicated what type of milk was used, milk fat (%), total solids (%) and moisture (%), but without information on minimum durability of the product (Table 1). Analyses of cheeses A and B determined the lower percentage of milk fat, and higher percentage of total solids and moisture than it is stated on the declaration. Consumers have a right to expect that the foods they purchase and consume will be safe and of high quality. They have a right to voice their opinions about the food control procedures, standards and activities that governments and industry use to ascertain that the food supply has these characteristics. Incorrect information on declarations on cheeses that are analysed, do not endanger the human health directly, but it deceiving consumers who are confident that are consuming and paying for high quality products.

The correctness of the declaration is very important, not only to ease the process of placing the product on the market by the producer, but also to prevent consumer misleading. One of the major objections are small letters on declarations, which can be considered as a commercial practice that misleads consumers.

Conclusion

Based on the results obtained, we can conclude that system controls that are in line with European standards, are necessary to monitor labelling of products on the market and to reduce misleading of consumers. Food must be labelled in such way that consumers are not misled with false information of its characteristics, composition, quantity, origin, production process, and minimum durability. Product labelling is essential to gain consumer trust and to improve the accuracy of information relating certain product. In this way, domestic products would be placed on the equal footing with products from the international market.

References

IDF Standard (1982). Cheese-determination of total solids content in cheese and processe of cheese. Volumne 4A.

IDF Standard (1997). Milk and milk products-Determination of fat content. General guidance on the use of butyrometric methods. Volume 152.

Lipovšćak, M. (2017). Provjera ispravnosti označavanja svježih sireva proizvedenih od toplinski obrađenog mlijeka. Diplomski rad. Sveučilište u Zagrebu Agronomski fakultet.

Mepham, T.B. (2000). The role of food ethics in food policy. Proceedings of the Nutrition Society 59: 609-618.

Merwe, M. and Venter, K. (2010). A consumer perspective on food labelling: ethical or not? Koers 75: 405-428.

Rogelj, I. (1998). Istine i zablude o mlijeku i mliječnim proizvodima u prehrani. Mljekarstvo 48: 153-164.

World Health Organisation (2015). Food Safety: What you should know. Available at: http://www.searo.who.int/entity/world_health_day/2015/whd-what-you-should-know/en.

Zou, N.N. and Hobbs, J.E. (2010). The role of labelling in consumers' functional food choices. 'The Economics of Food, Food Choice and Health'-AAEA seminar, Germany.

74. Determination of Travnik's sheep cheese adulteration using the mPCR-method

E. Pasic-Juhas[1], L.C. Czegledi[2], A. Hodzic[1], A. Hrkovic-Porobija[1] and I. Bozic[1]*
[1]Department for chemistry, biochemistry and physiology, Veterinary faculty, University of Sarajevo, Bosnia and Herzegovina; [2]Faculty of Agricultural and Food Science and Environmental Management, University of Debrecen, Hungary; eva.pasic-juhas@vfs.unsa.ba

Abstract

Travnik's sheep cheese is a traditional product in Bosnia and Herzegovina. According to a centuries-old recipe, it is prepared from 100% of sheep milk. Due to high local demand and relatively high prices, some farmers add cow milk as much as 20% or more to increase available quantities for sale. Farmers sell Travnik's sheep cheese at markets, and the farmer's association even exports this cheese to other countries. For the last 20 years there is a sale of Travnik's sheep cheese through large retail chains as product with declaration and enhanced inspection control. The aim of this paper is to determine to what extent farmers respect traditional production (100% sheep cheese), and to what extent they deviate from it. Our goal is also to determine whether the traditional cheeses produced by industrial production in the retail chains correspond to the declaration. In this research PCR-species specific DNA test was used to detect Travnik's sheep cheese (total n=60) adulteration with cow milk. From the farmer we provided 41 pieces of cheese (n=41), and from the store 19 (n=19). We have prepared our own standards for cheese mixing in a different ratio of sheep and cow milk. Multiplex PCR was performed using ovine and bovine 12S and 16S genes fragment in cheese. Our results show that only 46.34% of the farmers respect the traditional recipe, while 53.66% of them verbally declared cheese as pure sheep milk cheese, but they were a mixture of cow-sheep milk. In stores, 63.16% of the cheese packaging corresponds to the declaration, and 36.84% deviates (cheating).

Keywords: traditional cheese, false labelling, small and large producers

Introduction

The origin of the milk used for manufacturing cheese must be declared by the producer, especially in the case of protected denomination of origin cheeses. The adulteration or substitution of different types of milk in dairy products has always been a source of concern for several reasons related to governmental regulation, religion and public health. It was found that most milk proteins, even proteins present at low concentrations, are potential cause of food allergy (Wal, 2004). Cow milk was reported as the main dairy product responsible for human adverse reactions (Darwish *et al.*, 2009).

Pure sheep milk chesses are considered as specialties with characteristic sensory features. The problem with production of sheep milk cheese is the amount of milk per animal and their season supply fluctuation during year. The sheep milk supply depends on the reproductive cycle (Maškova and Pauličkova, 2006). In Bosnia and Herzegovina sheep milking takes place from the end of April to the beginning of September. Calculated average per day is 0.5 l of milk, and around 75 l during season per sheep. This often induces situation where non-declared cow milk is admixed with sheep milk during the manufacture of sheep milk cheese.

For preventing this type of fraud, food labelling regulations requires the species identification of milk in dairy products. Numerous analytical methods such as immunological, electrophoretic, and

Svenja Springer and Herwig Grimm (eds.) ***Professionals in food chains***
EurSafe 2018 – DOI 10.3920/978-90-8686-869-8_74, © Wageningen Academic Publishers 2018

chromatographic have been developed for species identification of milk and dairy products (Zeleňáková *et al.*, 2008; Mayer, 2005; Veloso *et al.*, 2004).

Travnik's sheep cheese is a traditional product in Bosnia and Herzegovina. It is prepared from 100% sheep milk according to an old recipe, but in retail chains there are 100% Travnik's cow milk cheese produced by the same procedure and it is legitimate if it is properly labelled. Increasing market demands have led to changes in cheese production. Because of greater availability and lower price, cow milk is often used as addition for preparing traditional sheep milk cheese. Usually sheep milk cheese products in which more than 1% of cow milk is present are considered to be adulterated if it isn't labelled. In the case of Travnik's sheep cheese tolerance is zero according to an old recipe, but milk mixing is legitimate if it is labelled, as determined by law regulations on food declaration. The adulteration of products creates negative social consequences because it erodes the consumers' trust. The aim of this research, in order to protect manufacturers and consumers from adulterations and imitations, is to (1) to determine whether farmers respect traditional production (100% sheep cheese); (2) to determine whether the composition of cheese from retail chains corresponds to the declaration. Although mixing different types of milk is permitted by law, problems arise when the final product is not properly labelled. There is also a question: Has the desired taste of Travnik's sheep cheese changed over the last few decades, which may have led to a change in the production of cheese?

Materials and methods

The 41 cheese samples were obtained directly from farmers on the Vlasic Mountain, and 19 cheese samples from the large retail chains. The samples were stored at -20 °C until further analysis. We have prepared our own standards for cheese, mixing different amounts of sheep and cow milk, as shown in Figure 1.

In this research we used PCR-species specific DNA test for species identification in Travnik's sheep cheese (total n=60). Multiplex PCR (mPCR) were performed using 12S rRNA and 16S rRNA of

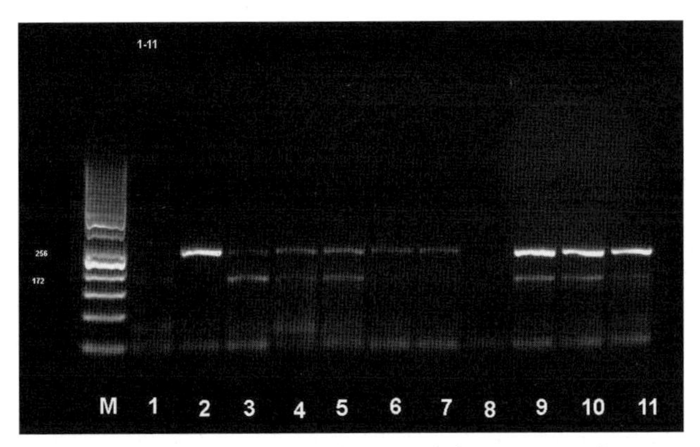

Figure 1. Multiplex PCR of ovine and bovine 12S and 16S genes fragments in cheese made in laboratory. M, 50 bp ladder; lane 1. sheep milk; lane 2. cow milk; lane 3. mixture of sheep and cow milk (99,5% sheep milk); lane 4. mixture of sheep and cow milk (99% sheep milk); lane 5. mixture of sheep and cow milk (95% sheep milk); lane 6. mixture of sheep and cow milk (90% sheep milk); lane 7. mixture of sheep and cow milk (85% sheep milk); lane 8. mixture of sheep and cow milk (80% sheep milk); lane 9. mixture of sheep and cow milk (70% sheep milk); lane 10. mixture of sheep and cow milk (60% sheep milk); lane 11. mixture of sheep and cow milk (50% sheep milk).

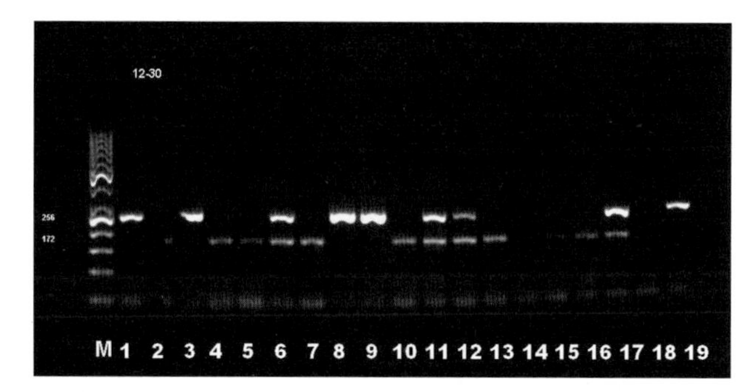

Figure 2. Multiplex PCR results example of ovine and bovine 12S and 16S genes fragments in cheese made by farmers on Vlasic Mountain. Lanes 1-19: every number represent different farmers. M: 50 bp ladder; lanes 2,4,5,7,10,13,14,15 and 16: sheep milk; lanes 1,3,8,9 and 19 cow milk; lanes 6,11, 12 and 17: mixture of sheep and cow milk.

mitochondrial DNA specific primers for ovine and bovine species (Bottero *et al.*, 2003). The DNA from cheese samples was extracted according to De *et al.* (2011). DNA concentration and quality were measured with spectrophotometer. Nucleotide sequence of 12S rRNA of cattle (GQ926965.1) and sheep (JQ622016.1) were used from NCBI GenBank database.

Results

Using the mPCR-method for testing 60 commercial cheeses obtained from retail chains and farmers, the undeclared presence of cow milk in samples was found. The PCR method described enabled the detection of the amplified fragments of cow DNA, size 279 bp and sheep DNA size 282 bp, respectively. The detection limit of mPCR method was 0.5% of cow DNA in cow-sheep milk mixtures (Figure 1). Samples obtained from farmers were not properly labelled, they were verbally declared as a 100% 'sheep milk cheese'.

Table 1. shows presence of cow DNA in 53.66% of 41 cheese samples, and only 46.34% of them consisted of 100% sheep milk. Analysed cheese samples from retail chains labelled as 'sheep milk cheese' proved to contain cow DNA: 36.84% of 19 cheese samples contained undeclared species DNA, while in 63.16% of the cheese samples the composition corresponds to the declaration.

Table 1. An overview of mPCR results for detection of Travnik's sheep cheese adulteration with cow milk in n=60 native samples declared as 100% sheep cheese.

Cheese composition	Origin: farmers		Origin: market	
	N	%	N	%
100% sheep cheese	19	46.34	12	63.16
100% cow cheese	9	21.95	1	5.26
Mix sheep/cow cheese	13	31.71	5	26.32
Not determined	-	-	1	5.26
Total	41	100	19	100

Discussion

Cheese adulteration is a big problem in the dairy industry because it can be easily applied in technology of cheese manufacturing.

Modern molecular techniques based on the polymerase chain reaction (PCR) have found good applicability in detecting such adulterations, and have become effective and reliable also for commercial dairy products. In recent years, an increasing number of PCR-based protocols have been reported for species identification in dairy products, particularly cheese (Feligni *et al.*, 2005; Mafra *et al.*, 2004; Bottero *et al.*, 2003). The higher sensitivity of natural science methods lead to increase in the ethics of the manufacturers, due to absolutely accurate composition identification. The use of this PCR method may therefore ensure the food analyst to enforce legislation and to monitor the illegal use of less expensive cow milk in the manufacture of premium quality sheep milk cheeses and other dairy products. Protection against species substitution or admixture is important for reasons related to religion, conservation, government regulations, and trade (Branciari *et al.*, 2000). Our results demonstrate that mPCR methods are sufficiently susceptible to prove adulteration (the threshold is 0.5%).

Differences in price and seasonal availability might make it attractive for farmers to adulterate expensive sheep milk with cheaper cow milk. In Bosnia and Herzegovina, the amount of produced sheep milk is much smaller than the amount of produced cow milk, but the certain group of consumers demand traditional Travnik's sheep cheese (in original recipe 100% made of sheep milk). This often induces attempts to make these cheeses from raw material where sheep milk is at least partially replaced with undeclared cow milk. Consumers have the right to know the species origin of the cheese, be it for legal, nutritional, health, economic or only force of habit reasons. As a result of this fraudulent practice in the dairy industry and by farmers, adequate control methods are required to evaluate the authenticity of Travnik's sheep cheese. Moreover, our laboratory is challenged not only by the qualitative detection of cow milk, but also by the need for its quantitative estimation in milk mixtures. The question is: Is indeed adulteration in such a large proportion (53 and 32%) present because of economic benefits, or small as well as big producers are trying to accommodate to the consumers claims? Namely, in the 21st century, the consumers demand have changed. Today consumers rather buy cheese made from 80% sheep milk with addition of 20% cow milk, no matter what the price. It is probably due to specific and strong taste of old-recipe cheese. Today, consumers prefer the mild flavour of that cheese, and the cheese manufacturers used this as the possibility of manipulating consumers. False labelling cheese or false verbally declaring is unethical and leads to distrust in customer-manufacturer relationship, which can generate economic problems for manufacturers.

There is a need to adjust to the requirements of consumers and since the Bosnia and Herzegovina legislations don't regulate the composition of 'Travnik's cheese', currently the only obligation for the manufacturers is to proper label product. However, term 'Travnik's cheese' is related to the tradition and authenticity of making 'Travnik's cheese' from sheep milk. We think it isn't ethical to use the name 'Travnik's cheese' for cheese made from some other milk, but there are different opinions on term 'authenticity'. Groves (2001) speaks about authenticity concept: 'individuals either requiring a link with the past or acknowledging that authenticity is a dynamic concept and that the production of products will change over time'.

Conclusions

- On the Bosnia and Herzegovina market there is a large percentage of false declarations (labelling) for Travnik's sheep cheese. In regard to products sold by farmers, 53% were incorrectly verbally declared and 32% of products sold in retail chains were incorrectly labelled.
- There is a need for stricter control of Travnik's sheep cheese production and sales.
- There is a need to adjust to the consumers demands. Should the traditional cheese production be changed, or should a new product be created (cheese with composition of 80% sheep milk and 20% cow milk) on a traditional basis, but with another brand name.

References

Bottero, M.T., Civera, T., Nucera, D., Rosati, S., Sacchi, P. and Turi R.M. (2003). A multiplex polymerase chain reaction for the identification of cows', goats' and sheep's milk in dairy products. International Dairy Journal 13: 277-282.

Branciari, R., Nijman, I.J., Plas, M.E., Eraldo di Antonio and Lenstra, J.A. (2000). Species Origin of Milk in Italian Mozzarella and Greek Feta Cheese. Journal of Food Protection 63: 408-411.

Darwish, S. F., Allam H. A. and Amin, A.S. (2009). Evaluation of PCR Assay for Detection of Cow's Milk in Water Buffalo's Milk. World Applied Sciences Journal 7: 461-467.

De, S., Brahma, B., Polley, S., Mukherjee, A., Banerjee, D., Gohaina, M., Singh, K.P., Singh, R., Datta, T.K. and Goswami, S.L. (2011). Simplex and duplex PCR assays for species specific identification of cattle and buffalo milk and cheese. Food Control 22: 690-696.

Feligini, M., Bonizzi, I., Curik, V.C., Parma, P., Greppi, G. F. and Enne, G. (2005). Detection of adulteration in Italian mozzarella cheese using mitochondrial DNA templates as biomarkers. Food Technology and Biotechnology 43: 91-95.

Groves, A. M. (2001). Authentic British food products: A review of consumer perceptions. International Journal of Consumer Studies 25: 246-254.

Mafra, I., Ferreira, M.P.L.V.O.I., Faria, M.A. and Oliveira, B.P.P. (2004). A novel approach to the quantification of bovine milk in ovine cheeses using a duplex polymerase chain reaction method. Journal of Agricultural and Food Chemistry 52: 4943-4947.

Maškova, E. and Pauličkova, I. (2006). PCR-Based Detection of Cow's Milk in Goat and Sheep Cheeses Marketed in the Czech Republic. Czech Journal of Food Science. 24: 127-132.

Mayer, H.K. (2005): Milk species identification in cheese varieties using electrophoretic, chromatographic and PCR techniques. International Dairy Journal 15: 595-604.

Veloso, A.C.A., Teixeira, N., Peres, A. M., Mendonc, A. and Ferreira, I.M.P.L.V.O. (2004). Evaluation of cheese authenticity and proteolysis by HPLC and urea-polyacrylamide gel electrophoresis. Food Chemistry 87: 289-295.

Wal, J.-M. (2003). Bovine milk allergenicity. Annals of Allergy, Asthma & Immunology 93: 2-11.

Zeleňáková, L., Golian, J. and Zajác, P. (2008). Application of ELISA tests for the detection of goat milk in sheep milk. Milchwissenschaft 63: 137-141.

75. Development and validation of GC-FID method for detection of vegetable oils in dairy products

R. Uzunov[1], Z.H. Musliu[1], M. Arapceska[2], E.D. Stojkovic[1], B.S. Dimzoska[1], D. Jankuloski[1], V. Stojkovski[1] and L. Pendovski[1]*
[1]Faculty of Veterinary Medicine, University Ss. Cyril and Methodius, Skopje, R. Macedonia; [2]Faculty of Biotechnical Science, University St. Kliment Ohridski, Bitola, R. Macedonia; risteuzunov@fvm.ukim.edu.mk

Abstract

Milk and dairy products are frequently included as important components of a healthy diet and are widely consumed by the majority of the human population. Addition of vegetable oils in milk and dairy products is an old and illegal practice that has become increasingly common and complex. The motivation for milk adulteration is economic, as well as the fluctuation in seasonal availability of sheep and goats' milk. Adulteration of milk and dairy products reduces their quality and might be dangerous for human health. Fatty acid composition of milk fat has been used for a long time as a criterion to detect adulteration with vegetable oils, mainly because milk fat is characterized by short-chain fatty acids, whereas vegetable oils have medium to long-chain fatty acids. Based on this knowledge, the aim of this study was development and validation of GC-FID method for detection of vegetable oils in dairy products according to fatty acid composition. FAMEmix standard solution (C4:0-C24:0) was used for identification of individual fatty acids. For detection and quantification of vegetable oils in dairy products the following reference materials were used: butter fat, palm and coconut oil. The linearity of the calibration curves of fatty acid standards showed high correlation coefficients (R^2=0.9845-0.9999). The LOD was from 0.02-0.04 mg/ml and LOQ was from 0.1-0.3 mg/ml. Precision was assessed by determining the coefficient of variation (CV, %) of the areas obtained from the injection of six replicates of the standard solutions and reference materials. The CV was in the range from 0.35-1.91% for standard solution, 0.73-6.50% for butter fat, 0.51-8.19% for palm oil and 0.76-2.16% for coconut oil. The recovery of the method was from 78.93-113.52%. From 60 analysed dairy products samples, purchased from the local market, vegetable oils were detected in 23 samples. Palm oil was detected in 21 samples, and coconut oil was detected in 2 samples. High concentration of C16:0, C18:1n9c and C18:2n6c were found in samples which contain palm oil. Moreover, in the samples which contain coconut oil the content of C12:0 and C14:0 was increased. The results from this study provide valuable information for the consumers and for future studies in this area.

Keywords: dairy products, fatty acids, validation, GC-FID

Introduction

Milk and dairy products are important part of human nutrition. The composition of the milk and dairy varies depending of the animal species, feed, stage of lactation, season of the year. They are a good source of protein, fat, carbohydrates, vitamins and minerals. Because of their nutritional values, milk and dairy products are extensively consumed by large segments of the population, including children, pregnant women and elderly (Azad and Ahmed, 2016; Kim *et al.*, 2016; Poonia *et al.*, 2016; Trbović *et al.*, 2017). Milk fat is one important nutrient because it is a good source of fatty acids and vitamins. Moreover, milk fat plays a very important role in textural and flavour characteristics of milk and cheese, nutritional value and economic role. The adulteration of milk and dairy products with addition of vegetable oils or animal fats is an old and illegal practice, and has become increasingly common and complex (Park *et al.*, 2014; Kim *et al.*, 2015, 2016). The motivation for milk adulteration is economic, as well as the fluctuation in seasonal availability of sheep and goats' milk. On the other hand, adulteration of milk and

EurSafe 2018 – DOI 10.3920/978-90-8686-869-8_75, © Wageningen Academic Publishers 2018

dairy products reduces their quality and may present a risk to human health. Milk fat is characterized by short-chain fatty acids, whereas vegetable oils have medium to long-chain fatty acids; consequently, fatty acid composition of milk fat has been used for a long time as a criterion to detect adulteration with vegetable oils (Azad and Ahmed, 2016; Kim *et al.*, 2016; Hurley *et al.*, 2006). Based on this knowledge, the aim of this study was development and validation of GC-FID (Gas Chromatography with Flame Ionization Detector) method for detection of vegetable oils in dairy products according to fatty acid composition.

Material and methods

Sample collection

A total of 60 samples from dairy products were purchased from the local market. From the total number of samples, 13 samples were sheep cheese (SC), 14 samples were cow cheese (CC), 7 samples were mixed cheese (MC) (from cow and sheep milk), 13 samples were sour cream (SC), and 13 samples were sour cream with peppers (SCP).

Chemicals

N-hexane, HPLC water, ammonium hydroxide, pyrogallic acid, chloroform, ethanol, toluene, diethyl ether, petroleum ether and sodium sulphate anhydrous, hydrochloric acid and boron trifluoride. For detection and quantification of vegetable oils in dairy products were used reference materials (RM): butter fat (RM-BMF), palm oil (RM-PO) and coconut oil (RM-CO). For identification of individual fatty acids was used fatty acids methyl ester mix (FAMEmix) standard solution (SS) (C4:0-C24:0).

Sample preparation

In the first stage the samples were homogenized and extraction of lipid was performed in accordance with the AOAC Official Method 996.06, 2005. After lipid extraction 100 mg of lipids were transferred into dark glass vial (22 ml), and fatty acid methyl esters (FAMEs) were prepared in accordance with the conditions in the AOAC Official Method 996.06, 2005.

Chromatographic analysis

Determinations of FAMEs were carried out on a GC-FID, (GC AgilentTechnologies 7890 GC System, USA). Further, separation of FAMEs was carried out using a column HP88 (60 m × 250 mm × 0.2 mm). The temperature conditions of the column are given in Table 1.

Table 1. Temperature conditions of the column.

	Rate (°C/min)	Value (°C)	Hold time (min)	Run time (min)
	Initial	70	1	1
Ramp 1	5	100	2	9
Ramp 2	10	175	2	18.5
Ramp 3	3	220	5	38.5

Injector and detector temperatures were at 250 and 300 °C, respectively. As a carrier gas was used helium with flow rate of 1.4 ml and split ratio 200:1. Nitrogen was used as a make up gas at a flow rate of 23 ml/min. For separation and determination of the FAMEs, 1 µl of each sample was injected two times into GC-FID (Hajrulai-Musliu *et al.*, 2015; Belichovska *et al.*, 2015).

Individual fatty acids were identified by comparing of retention time of FAMEmix standard. The calculation of results was made with Chemstation software and the results were expressed as percentage (%) of total fatty acid.

Validation of the method

During the validation procedure were determined limit of detection (LOD), limit of quantification (LOQ), linearity, precision and accuracy of the method. The LOD was determined as the lowest concentration that gave a signal-to-noise ratio (S/N)≥3, and the LOQ was determined as the lowest concentration that gave an S/N≥10. The linearity of the method was estimated by performing of 3 replicates of FAMEmix standard solution at three concentration levels, as follows: 1.0, 5.0 and 10 mg/ml. The precision of the method is expressed as the coefficient of variation (CV, %) calculated from repeatability and reproducibility. For repeatability the SS (10 mg/ml), RM-BMF, RM-PO and RM-CO were measured at 6 replicates, whereas for reproducibility the SS, RM-BMF, RM-PO and RM-CO were measured at 6 replicates in two days. Accuracy of the method was verified through recovery (%), calculated from the RM-PO in 6 replicates measured in two different days (Taverniers *et al.*, 2004; Hajrulai-Musliu *et al.*, 2015; Martínez *et al.*, 2012).

Results and discussion

Results of validation

The LOD for the fatty acids was from 0.02 mg/ml to 0.04 mg/ml. The obtained values for LOQ were between 0.10 mg/ml for C11:0 and C17:0, and 0.30 mg/ml for C12:0. The LOD and LOQ values for all fatty acids are presented in Table 2. The coefficients of correlation (r^2) obtained from calibration curves for all fatty acids used in this study were ≥0.9845 (Table 2). As observed from the data, the results were found to be linear over the concentration range studied.

The precision of the method is presented in Table 2. The CV for repeatability ranged from 0.35 to 1.91% for SS, 0.73-6.50% for RM-BMF, 0.51-8.19% for RM-PO, and from 0.76 to 2.16% for RM-CO, while the CV for reproducibility ranged from 0.44 to 1.90% for SS, 0.74-5.81% for RM-BMF, 0.94-7.53% for RM-PO and from 0.78 to 2.04% for RM-CO. The accuracy of the method was expressed through recovery. The recovery was ranged 78.93% for myristic fatty acid to 113.52% for oleic fatty acid (Table 3). On the basis of these results it can be concluded that the method is precise, reproducible and accurate (Taverniers *et al.*, 2004-; Hajrulai-Musliu *et al.*, 2015).

Results of sample analysis

From the 60 dairy products (DP) that were analysed, vegetable oils were detected in 23 samples. Palm oil was detected in 21 samples (DP+PO), and coconut oil was detected in 2 samples (DP+CO) (Table 4). According to previous research, short-chain fatty acids are characteristic for milk fat, whereas medium and long-chain fatty acids are characteristic for vegetable oils (Kim *et al.*, 2016). The fatty acids content of the reference materials and samples are shown in Table 4. The content of the short-chain fatty acids in the RM-BMF and in the milk fat are higher than the content of these fatty acids in the samples with palm oil. According to the research of Park *et al.*, 2013, – C14:0, C16:0, C18:1n9c, C18:0, and C18:2n6c can

Table 2. Limit of detection (LOD), limit of quantification (LOQ) and linearity of the method.[1]

Fatty acids	LOD (mg/ml)	LOQ (mg/ml)	r^2	Precision-repeatability (CV %)				Precision-reproducibility (CV %)			
				SS	RM-BMF	RM-PO	RM-CO	SS	RM-BMF	RM-PO	RM-CO
C4:0 (butyric)	0.03	0.12	0.9964	0.48	4.08	nd	nd	0.57	3.26	nd	nd
C6:0 (caproic)	0.02	0.15	0.9897	0.72	5.12	nd	2.16	0.92	4.78	nd	2.04
C8:0 (caprilic)	0.04	0.15	0.9978	0.35	4.46	nd	1.02	0.46	5.41	nd	1.36
C10:0 (capric)	0.04	0.14	0.9997	1.08	3.21	nd	1.46	1.02	2.17	nd	0.78
C11:0 (undecylic)	0.04	0.10	0.9980	1.13	0.98	nd	Nd	1.90	1.11	nd	nd
C12:0 (lauric)	0.03	0.30	0.9996	0.91	0.93	4.13	2.00	0.98	1.46	2.96	1.15
C13:0 (tridecylic)	0.03	0.12	0.9993	0.93	5.79	nd	nd	0.46	4.41	nd	nd
C14:0 (myristic)	0.02	0.27	0.9982	0.47	6.03	0.51	0.76	0.88	5.78	0.94	1.03
C14:1 (myristoleic)	0.02	0.17	0.9974	1.91	6.50	nd	nd	1.16	5.81	nd	nd
C15:0 (pentadecanoic)	0.04	0.24	0.9981	1.38	3.26	nd	nd	0.92	3.99	nd	nd
C15:1 (cis-10 pentadecenoic)	0.03	0.15	0.9899	1.12	3.34	nd	nd	1.25	3.32	nd	nd
C16:0 (palmitic)	0.02	0.17	0.9962	1.06	4.46	3.28	1.48	0.44	4.08	3.46	1.99
C16:1 (palmitoleic)	0.04	0.21	0.9991	0.52	2.02	8.19	nd	0.76	2.15	7.53	nd
C17:0 (marganic)	0.03	0.10	0.9845	0.67	0.94	2.04	nd	1.88	0.74	2.14	nd
C17:1 (cis-10 heptadecenoic)	0.04	0.22	0.9983	0.77	0.77	nd	nd	0.65	0.92	nd	nd
C18:0 (stearic)	0.02	0.19	0.9997	1.90	1.04	1.13	2.08	1.24	1.17	0.99	1.46
C18:1n9t (elaidic)	0.02	0.23	0.9894	1.78	5.13	nd	nd	1.22	4.74	nd	nd
C18:1n9c (oleic)	0.02	0.12	0.9993	1.43	0.73	3.22	1.44	1.88	3.75	3.16	1.00
C18:2n6t (linolelaidic)	0.03	0.23	0.9987	1.56	1.74	1.17	nd	1.73	2.02	1.84	nd
C18:2n6c (α-linoleic)	0.03	0.22	0.9969	1.12	3.12	5.04	0.93	0.67	3.01	4.28	1.56
C20:0 (arachidic)	0.02	0.27	0.9999	1.17	1.43	0.99	nd	0.92	1.97	1.15	nd
C18:3n3 (α-linolenic)	0.04	0.12	0.9874	1.62	1.78	2.43	nd	1.08	4.01	3.15	nd

[1] nd = not detected.

Table 3. Precision and accuracy of the method.

Fatty acids	RM-PO % of fatty acids	Accuracy			
		Day 1 (n=6)		Day 2 (n=6)	
		% of fatty acids	Recovery %	% of fatty acids	Recovery %
C14:0	1.1	0.87	78.93	0.95	86.36
C16:0	43.5	39.45	90.69	39.41	90.60
C18:0	4.4	4.22	95.91	4.23	96.14
C18:1n9c	38.6	43.82	113.52	43.01	111.42
C18:2n6c	9.8	10.79	110.10	10.77	109.90

use as biomarker to distinguish milk fat from vegetable oils. Table 4 illustrates that the content of the C14:0 in DP is the highest of all samples with palm oil, but the content of these fatty acids in the DP is lower than the samples with coconut oil. The content of the C16:0, which is characteristic fatty acid for palm oil, in DP and samples with coconut oil is lower than in the samples with palm oil. Moreover, C18:0 in samples with palm oil is in low concentration, but the concentration of C18:1n9c and C18:2n6c is higher in samples with palm oil. In the samples with coconut oil, main fatty acid is C12:0 (50.03%). This fatty acid is characteristic for coconut oil and for this reason it is used as biomarker to distinguish milk fat from coconut oil (Wirasnita *et al.*, 2013). Therefore, the content of C16:0 and C18:1n9c in samples with coconut oil is much lower than the DP and samples with palm oil. The concentration of saturated fatty acid (SFA) in dairy samples is from 66.09 to 69.23%, while in samples with palm oil is 51% and in samples with coconut oil is 94.12%. According to Kesenkas *et al.* (2009) the content of SFA can be a good indicator that the dairy product is adulterated with vegetable oils.

Typical chromatograms for DP, DP+PO and DP+CO are given in Figure 1, 2 and 3.

Table 4. Fatty acids content in reference materials and samples.[1,2]

Fatty acids	RM-BMF %	RM-PO %	RM-CO %	CC (n=14) %	SC (n=13) %	MC (n=7) %	SC (n=13) %	SCP (n=13) %	DP+PO (n=21) %	DP+CO (n=2) %
C4:0	2.57	nd	nd	2.62	2.60	2.64	2.68	1.57	0.43	nd
C6:0	1.91	nd	0.59	1.84	2.16	1.90	1.86	1.65	0.33	0.4
C8:0	1.20	nd	7.30	1.13	2.06	1.27	1.14	1.14	0.27	5.41
C10:0	2.75	nd	5.73	2.65	6.39	3.18	2.70	2.64	0.45	4.25
C11:0	0.29	nd	nd	0.12	0.16	0.15	0.12	0.28	0.00	nd
C12:0	3.44	0.24	46.03	3.25	3.78	3.39	3.27	3.28	0.73	50.03
C13:0	0.08	nd	nd	0.09	0.10	0.11	0.10	0.06	0.00	nd
C14:0	10.92	1.00	18.93	11.50	10.57	11.21	11.37	11.40	2.66	15.72
C14:1	1.46	nd	nd	1.69	0.94	1.50	1.50	1.44	0.19	nd
C15:0	1.15	nd	nd	1.29	1.29	1.26	1.22	0.69	0.25	nd
C15:1	0.25	nd	nd	0.24	0.27	0.25	0.29	0.26	nd	nd
C16:0	29.37	39.45	9.74	34.45	25.68	32.31	33.20	33.16	40.21	7.52
C16:1	2.11	0.24	nd	2.15	1.80	2.11	2.15	2.26	0.44	nd
C17:0	0.64	0.09	nd	0.73	0.83	0.71	0.64	0.70	0.17	nd
C17:1	0.50	nd	nd	0.47	0.43	0.49	0.48	0.53	nd	nd
C18:0	11.50	4.22	2.86	9.50	10.94	10.04	9.63	9.35	5.50	10.79
C18:1n9t	1.41	nd	nd	2.54	4.03	2.21	2.20	1.61	0.68	nd
C18:1n9c	24.68	43.16	6.89	20.15	19.82	21.38	21.82	22.71	38.78	3.49
C18:2n6t	0.29	0.15	nd	0.20	0.55	0.25	0.21	0.19	0.09	nd
C18:2n6c	1.78	10.79	1.94	1.98	2.62	2.09	2.13	2.20	8.43	1.88
C20:0	0.18	0.33	nd	0.06	0.24	0.18	0.05	0.17	0.00	nd
C18:3n3	0.90	0.16	nd	0.67	1.29	0.78	0.70	0.69	0.05	nd
SFA	66.00	45.33	91.18	69.23	66.80	68.35	67.98	66.09	51.00	94.12

[1] BMF = butter fat; CC = cow cheese; CO = coconut oil; DP = dairy products; MC = mixed cheese; PO = palm oil; RM= reference material; SC = sheep cheese; SCP = sour cream with peppers.

[2] nd = not detected.

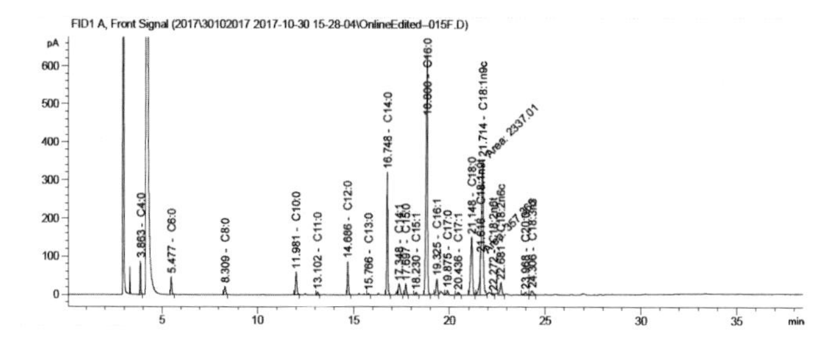

Figure 1. Chromatogram of dairy products.

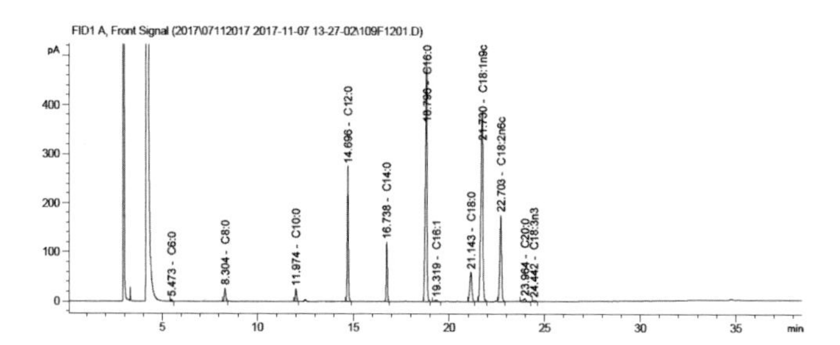

Figure 2. Chromatogram of dairy products + palm oil.

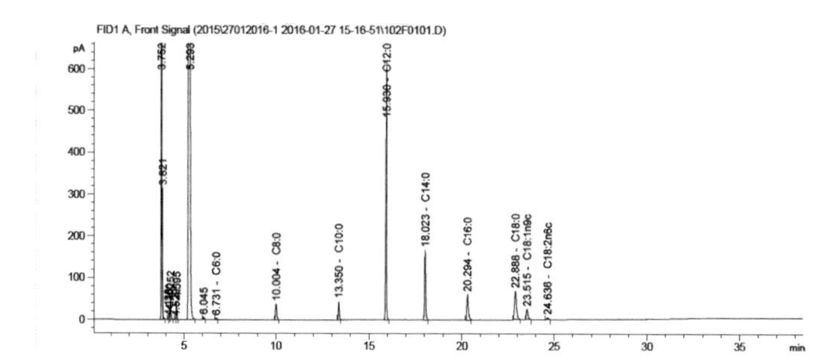

Figure 3. Chromatogram of dairy products + coconut oil.

Conclusion

In summary, the GC-FID method for detection of vegetable oils in dairy products was successfully developed and validated. The C14:0, C16:0, C18:1n9c, C18:0 and C18:2n6c were used as biomarker for detection of palm oil and C12:0 and C14:0 were used as biomarker for detection of coconut oil in DP. From 60 dairy products that were included in this study, palm oil was detected in 21 samples, and coconut oil was detected in 2 samples. In addition, the SFA content can be a good indicator for determination of vegetable oils in DP. The content of SFA in samples with palm oil was 51%, samples with coconut oil contained 94.12%, while the concentration of SFA in dairy products ranged from 66.09 to 69.23%. This study provides the based data for dairy product adulteration. Based on this preliminary research, this method could be useful for detection of vegetable oils in other dairy products in the future.

References

Azad, T. and Ahmed, S. (2016). Common milk adulteration and their detection techniques. International Journal of Food Contamination 3:22.

Belichovska, D., Hajrulai-Musliu, Z., Uzunov, R., Belichovska, K. and Arapcheska, M. (2015). Fatty acid composition of ostrich (Struthiocamelus) abdominal adipose tissue. Macedonian Veterinary Review 38 (1): 53-59.

Hajrulai-Musliu, Z., Uzunov, R., Stojanovska-Dimzoska, B., Stojkovic-Dimitrieska, E., Angeleska, A. and Stojkovski, V. (2015) Determination of fatty acid in asparagus by gas chromatography. Journal of the Faculty of Veterinary Medicine Istanbul University 41: 31-36.

Hurley, I.P., Coleman, R.C., Ireland, H.I. and Williams H.H.J. (2006). Use of sandwich IgG ELISA for the detection and quantification of adulteration of milk and soft cheese. International Dairy Journal 16: 805-812.

Kesenkas, H., Dinkçi, N., Seçkin, A.K., Kinik, Ö. and Gönç, S. (2009). The effect of using a vegetable fat blend on some attributes of kashar cheese. Grasas y Aceites 60 (1): 41-47.

Kim, H.J., Park, J.M., Lee, J.H. and Kim, J.M. (2016).Detection for non-milk fat in dairy product by gas chromatography. Korean Society for Food Science of Animal Resources 36: 206-214.

Kim, J.M., Kim, H.J. and Park, J.M. (2015). Determination of milk fat adulteration with vegetable oils and animal fats by gas chromatographic analysis. Journal of Food Science 80: 1945-1951.

Martínez, B., Miranda, J.M., Franco, C.M., Cepeda, A. and Rodríguez J.L. (2012).Development of a simple method for the quantitative determination of fatty acids in milk with special emphasis on long-chain fatty acidsCyTA – Journal of Food 10: 27-35.

Park, J.M., Kim, N.K., Yang, C.Y., Moon, K.W. and Kim, J.M. (2014). Determination of the authenticity of dairy products on the basisof fatty acids and triacylglycerols content using GC analysis. Korean Society for Food Science of Animal Resources 34:316-324.

Poonia, A., Jha, A., Aharma, R., Singh, H.B., Rai, A.K. and Sharma, N. (2016). Detection of adulteration in milk: A review. International Journal of Dairy Technology 69: 1-20.

Taverniers, I., De Loose, M. and Bockstaele, E.V. (2004). Trends in quality in the analytical laboratory. II. Analytical method validation and quality assurance. Trends in Analytical Chemistry 23: 535-552.

Trbović, D., Retronijević, N. and Đorđević, V. (2017). Chromatography methods and chemometrics for determination of milk fat adulterants. In: IOP publishing (ed). 59th International Meat Industry Conference MEATCON2017, Earth and Environmental Science. October 1-4, 2017. Zlatibor, Serbia, pp. 172-176.

William, H., Latimer, G.W., 2005. AOAC Official Method 996.06, Fat (Total, Saturated and Unsaturated) in Foods.

Wirasnita, R., Hadibarata, T., Novelina, Y.M., Yusoff, A.R.M. and Yusop, Z. (2013). A modified methylation method to determine fatty acid content by gas chromatography. Bulletin of the Korean Chemical Society 34 (11): 3239-3242.

76. Sustainable use of agro-industrial wastes for feeding 10 billion people by 2050

L.F. Călinoiu[1], L. Mitrea[1], G. Precup[1], M. Bindea[1], B. Rusu[1], K. Szabo[1], F.V. Dulf[1,2], B.E. Ştefănescu[1,3] and D.C. Vodnar[1]*
[1]Department of Food Science, Life Science Institute, University of Agricultural Science and Veterinary Medicine, Cluj-Napoca, Romania; [2]Faculty of Agriculture, University of Agricultural Sciences and Veterinary Medicine, Cluj-Napoca, Romania; [3]Department of Pharmaceutical Botany, Iuliu Haţieganu University of Medicine and Pharmacy, Cluj-Napoca, Romania; dan.vodnar@usamvcluj.ro

Abstract

A major problem experienced by agro-based industries in develops and developing countries is the management of wastes. Large quantities of both liquid and solid wastes are produced annually by the food processing industry. The current food wastage in EU27 is 179 kg per capita (89 milion ton per annum). In the context of food waste, paradoxical the food hunger is a growing problem based on a growing population issue, forecasted to 10 billion people by 2050. Therefore, the re-use and valorization of agro-food by-products is a real solution for food waste and hunger. This waste is rich in bioactive compounds and can thus be improved and incorporated into food supplements or used as individual nutrient powders rich in calories and bioactive compounds (European Commission Final Report, 2010). These waste materials contain principally biodegradable organic matter and their disposal creates serious environmental problems (Shyamala and Jamuna, 2010). A variety of processes is being developed towards this direction, aiming at converting the waste materials into bio-fuels, food ingredients and other added-value bio-products. However, in many instances, there is a rather significant lack of appropriate feasibility studies on the exploitation of such wastes, and as a result, their utilization is still in its infancy. For that, the present work proposes an agro-food waste powder formulation and incorporation into food products in order to reduce the waste disposal problem and food hunger by offering a sustainable and cost-effective solution. Is this approach a step forward in the functional food industry by offering an applicative direction to be explored together with a novel route for environmental protection? Does it provide sustainable outcomes for food waste by increasing the economic value of by-products?

Keywords: by-products, hunger, growing population, environmental issues

Introduction

Large quantities of both liquid and solid wastes are produced annually by the food processing industry. The current food wastage in EU27 is 179 kg per capita (89 million ton per annum). In the context of food waste, paradoxical the food hunger is a growing problem based on a growing population issue, forecasted to 10 billion people by 2050. Therefore, the re-use and valorisation of agro-industrial waste is a real solution for food waste and hunger.

Agro-industrial waste is one of the future sector, which must be explore for their bioactive potential. Lately, there is a major issue with respect to the use of agro-industrial waste, industries needing a confirmation of their bioactive potential after thermal processes. This kind of waste is usually produced during the manufacturing processes of the raw materials. Fruit and vegetable juice and pulp industries are generating the highest amount of agro-industrial waste (Dulf *et al.*, 2016, 2017).

Waste products deriving from fruits and vegetable processing industries could be much more valorised mainly for two important reasons: (1) due to their low price and their considerable existing amounts,

EurSafe 2018 – DOI 10.3920/978-90-8686-869-8_76, © Wageningen Academic Publishers 2018

(2) and due to their valuable bioactive potential (Vodnar *et al.*, 2017). Therefore, the following wastes, namely apple peels, carrot pulp, white and red grape peels and red beet peels and pulp, are in focus being massively discarded; however, this waste is rich in bioactive compounds and can thus be improved and incorporated into food supplements (European Commission Final Report, 2010). A variety of processes are being developed towards this direction, aiming at converting the waste materials into bio-fuels, food ingredients and other added-value bio-products. However, in many instances there is a rather significant lack of appropriate feasibility studies on the exploitation of such wastes, and as a result their utilization is still in its infancy.

According to our experimental study (Vodnar *et al.*, 2017), the demonstrated presence of fatty acids and phenolic compounds in above mentioned agro-industrial waste, after exposure to a thermal treatment (10 minutes, 80 °C), enables fruits and vegetable leftovers to become very valuable for the food industry, while providing an extra source of income and a sustainable approach for food waste and hunger. The purpose of the research was to identify the bioactive compounds and to evaluate the antioxidant, antimutagenic and antimicrobial activities of the major agro-industrial wastes (apple peels, carrot pulp, white- and red-grape peels and red-beet peels and pulp) for the purpose of increasing the wastes' value. Each type of waste material was analyzed without (fresh) and with thermal processing (10 minutes, 80 °C). Based on the obtained results, the thermal process enhanced the total phenolic content, but linoleic acid content decreased significantly during thermal processing. The highest antioxidant activity was exhibited by thermally processed red-grape waste followed by thermally processed red-beet waste.

Bioactive potential of waste as fresh and thermally processed matrices and their future applications

In order to underline the bioactive potential of targeted agro-industrial waste, for its further valorization, our latest experimental research studies (Vodnar *et al.*, 2017; Calinoiu *et al.*, 2017) will be the starting point. In Vodnar *et al.* (2017) study, the total phenolic content (via Folin-Ciocalteu method), the total flavonoids content, the antioxidant activity (via DPPH free-radical-scavenging assay) of the waste extracts, both fresh and thermally processed (10 minutes, 80 °C) were examined, as illustrated in Figure 1: Total phenolic content (A), total flavonoids content (B), antioxidant activity (C) of the extracts, both fresh and thermally processed.

Regarding the methodology used, the Folin-Ciocalteu method implied the following: 25 µl of each extract was mixed with 125 µl of Folin-Ciocalteu regent (0.2 N) and 100 µl of 7.5% (w/v) Na_2CO_3 solution; the mixture was incubated for 2 h in the dark at room temperature (25 °C); the absorbance against a methanol blank was recorded at 760 nm; a standard curve was prepared using gallic acid (0.01-1 mg/ml), and the total phenolic content in the extract was expressed as gallic acid equivalents (GAE) in mg/100 g dry weight (DW) of waste.

The total flavonoid content of the extracts was determined using the following method: a 100 µl aliquot of 2% aluminium chloride ethanol solution was added to 100 µl of the extracts and mixed; after incubating for 1 h at room temperature, the absorbance at 510 nm was measured against a prepared regent blank; total flavonoid content was expressed as quercetin equivalents (QE) in mg/g dry weight (DW).

Phenolic extracts of fresh and thermally processed samples were subject to DPPH radical-scavenging activity assessment where the percentage inhibition (I%) was calculated as [1 – (test sample absorbance/ blank sample absorbance)] × 100.

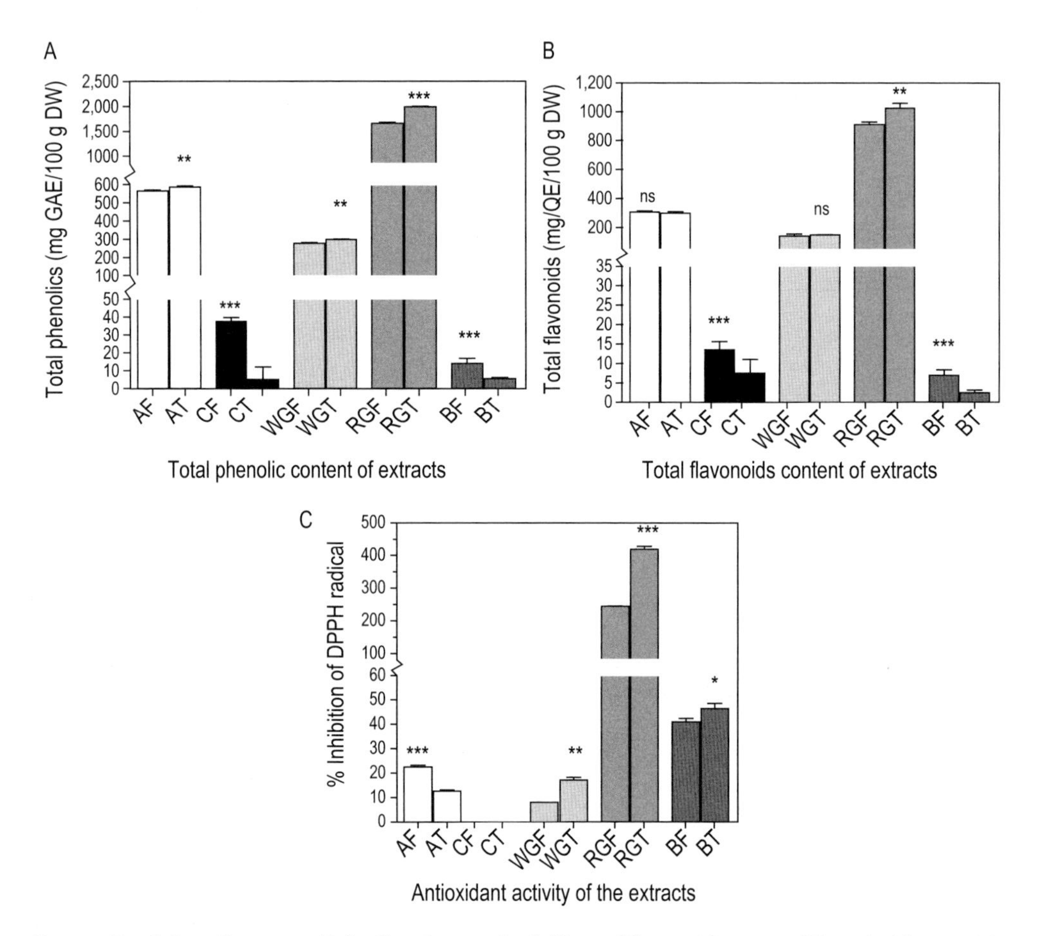

Figure 1. Total phenolic content (Folin-Ciocalteu method) (A), total flavonoids content (B), antioxidant activity (C) of the extracts, both fresh and thermally processed. Total phenolic content of the extract is expressed as gallic acid equivalents (GAE) in mg/100 g dry weight (DW) of waste. Total flavonoid content is expressed as quercetin equivalents (QE) in mg/g dry weight (DW). The percentage inhibition (I%) was calculated as [1 – (test sample absorbance/blank sample absorbance)] × 100. Values are reported as mean ± SD of triplicate determinations and different symbols (, **, ***) indicate significant differences (P<0.05) (paired t-test), while symbol (ns) indicate no significant difference. AF-apple waste fresh; AT-apple waste thermally processed; CF-carrot waste fresh; CT-carrot waste thermally processed; WGF-white-grape waste fresh; WGT-white-grape waste thermally processed; RGF-red-grape waste fresh; RGT-red-grape waste thermally processed; BF-red-beet waste fresh; BT-red-beet waste thermally processed.*

For some of the waste extracts, the thermal process enhanced their bioactive potential, while for others was opposite.

With respect to the total phenolic content, Figure 1A, the thermally processed samples of apple and red grapes had higher total phenolic content, red grape waste had the highest (1,990±52.9 mg GAE/100 g dry weight) while for the carrot and red beet waste extracts the thermal process decreased the phenolic content. The increase of phenolic content was possible due to the intracellular contents extraction enhanced by thermal treatment, where Wang *et al.* (2014) underlined the reason behind it, namely the hydrolysis of polysaccharides. According to the literature, the decrease (in CT and BT) could be attributed to the partial degradation of lignin, responsible for the release of phenolic acids, or to the thermal degradation of the phenolic acids (Morales de la Pena *et al.*, 2011).

Considering further the total flavonoids content, Figure 1B, the thermal process enhanced the red grape waste content (up to 1,050±62.1 mg), while for apple and white grape waste no significant differences among fresh and thermally processed was observed. In accordance with the increased flavonoids content after the process exposure is also the study of Medina-Mezaand and Barbosa-Cánovas (2015), where the exposure to an electric field increased by 96% the flavonoid content of grape peel.

The antioxidant capacity of targeted waste, Figure 1C, was significantly enhanced by the thermal process, for instance red grape waste showed a 59% increase in radical inhibition capacity.

Also, in the same experimental study, the identification of the phenolic compounds in the targeted waste extracts was done in order to validate their bioactive potential via HPLC-DAD-ESI-MS method. In almost all phenolic cases, the fresh and thermally processed samples had significant different values. Therefore, a significant increase in malvidin glucoside (13.68%) was registered after thermal processing for red grape waste; another major example was reported for caffeic acid 4-O-glucoside from apple waste, with a 27% (from 2.492 [mg%] to 3.165 [mg%]) increase after thermal treatment. With respect to red beet waste extract, all the reported betanidin compounds increased being enhanced by the thermal treatment exposure. The same result was reported by the Harivaindaran *et al.* (2008) for dragon fruit where the exposure at 100 °C for 5 minutes increased the yield of betacyanin content.

The fatty acid profile (GC-MS analysis) of each targeted waste sample was reported within the same study. The thermal treatment increased the fatty acids content in almost all the cases, probably due to cell walls breaking and release of fatty acids. In addition, the thermally processed apple waste had the highest content of fatty acids, and the red and white grape waste significantly increased their fatty acids composition/content after exposure. It is known that linoleic acid is very sensitive to the heat; therefore, after exposure to the thermal process, it significantly decreased. In the study of Dulf *et al.* (2016), it was demonstrated that exposure to certain fermentations increased significantly the fatty acids composition and content. Therefore, different treatments were validated as beneficial for exploiting and enhancing the bioactive potential of several waste.

These results offer sustainable solution for utilization of food industry bio-waste that has bioactive potential after thermal treatment (10 minutes at 80 °C). The apple peels, carrot pulp, red and white grape peels, and red beet peels and pulp can be exploited for their bioactive compounds, whose bioavailability increased and can be added in food formulations, like functional beverages, as health promoting products.

For preserving by drying the bioactive materials, freeze and spray drying are established methods. Freeze-drying is a gentle, but long term and expensive method, which is uneconomical. The benefit of spray drying is the formation of free flowing particles in a short time, but it is disadvantageous that the required high temperatures reduce the viability of mesophilic microorganisms, for instance. The fluidized bed drying technology is an alternative, gentle and cost saving method, for the preservation of bioactivity (Vodnar *et al.*, 2015).

Conclusions

The food industry can re-direct their agro-waste via incorporation as powder formulations in different functional products (beverages, food supplements) having bioactive potential after thermal treatments. One possible limitation could come from a decrease of fatty acids content due to the thermal treatment applied. Anyway, the juice industry should become aware of their potential and act vigorously by implementing the several waste practical applications demonstrated in the scientific literature: resistance to different thermal processing (Vodnar *et al.*, 2017); atomization under powder form as functional ingredients (Santos *et al.,* 2017). In this way, this sustainable approach could solve the environmental problem by re-using the waste and food hunger by offering healthy and valuable food ingredients.

The international aims of social and health policies are to improve the quality of life including by ensuring new development in functional food area. In this context, the current approach, suggest formation of new functional food ingredients under powder form using agro food-waste, or new functional products by their incorporation in the final product (e.g. beverages). This fact will on the one hand bring to light data for utilization of agro-food waste for functional beverage design that are lacking from literature, whereas on the other hand will spotlight the potency of these beverages.

References

Călinoiu, L.F, Mitrea, L., Precup, G., Bindea, M., Rusu, B., Dulf, F., Ştefanescu, B. and Vodnar, D.C. (2017). Characterization of grape and apple peel wastes' bioactive compounds and their increased bioavailability after exposure to thermal process. Bulletin of University of Agricultural Sciences and Veterinary Medicine Cluj-Napoca. Food Science and Technology, 74(2): 80-89.

Dulf, F.V., Vodnar, D.C., Dulf, E.H. and Pintea, A. (2017). Phenolic compounds, flavonoids, lipids and antioxidant potential of apricot (*Prunus armeniaca L.)* pomace fermented by two filamentous fungal strains in solid-state system. Chemistry Central Journal, 11: 92.

Dulf, F.V., Vodnar, D.C. and Socaciu, C. (2016). Effects of solid-state fermentation with two filamentous fungi on the total phenolic contents, flavonoids, antioxidant activities and lipid fractions of plum fruits (*Prunus domestica L.)* by-products. Food Chemistry, 209: 27-36.

European Commission. Final Report – Preparatory Study on Food Waste Across EU 27. (2010). Available at: http://ec.europa.eu/environment/eussd/pdf/bio_foodwaste_report.pdf. Accessed 28 September 2017.

Harivaindaran, K.V., Rebecca, O.P.S. and Chandran, S. (2008). Study of optimal temperature, pH and stability of dragon fruit (*Hylocereuspolyrhizus*) peel for use as potential natural colorant. Pakistan journal of biological sciences, 11(18): 2259-2263.

Medina-Meza, I.G. and Barbosa-Cánovas, G.V. (2015). Assisted extraction of bioactive compounds from plum and grape peels by ultrasonics and pulsed electric fields. Journal of Food Engineering, 166: 268-275.

Morales-de la Peña, M., Salvia-Trujillo, L., Rojas-Graü, M.A. and Martín-Belloso, O. (2011). Changes on phenolic and carotenoid composition of high intensity pulsed electric field and thermally treated fruit juice-soymilk beverages during refrigerated storage. Food Chemistry, 129(3): 982-990.

Santos, E., Andrade, R. and Gouveia, E. (2017). Utilization of the pectin and pulp of the passion fruit from Caatinga as probiotic food carriers. Food Bioscience, 20, 56-61.

Shyamala, B.N. and Jamuna, P. (2010). Nutritional Content and Antioxidant Properties of Pulp Waste from *Daucus carota* and *Beta vulgaris*. Malaysian journal of nutrition, 16(3): 397-408.

Vodnar, D.C., Pop, O.L. and Socaciu, C. (2015). Probiotics: Microencapsulation. In: Mishra, M. (ed). Encyclopedia of Biomedical Polymers and Polymer Biomaterials. Taylor & Francis, New York, USA, pp. 4644-4651.

Vodnar, D.C., Călinoiu, L.F., Dulf, F.V., Ştefănescu, B.E., Crişan, G. and Socaciu, C. (2017). Identification of the bioactive compounds and antioxidant, antimutagenic and antimicrobial activities of thermally processed agro-industrial waste. Food Chemistry, 231: 131-140.

Wang, T., He, F. and Chen, G. (2014). Improving bio-accessibility and bioavailability of phenolic compounds in cereal grains through processing technologies: A concise review. Journal of Functional Foods, 7(1): 101-111.

77. Is the consumption of dog´s meat ethical and legal ?

D. Takáčová[*], J. Kottferová, R. Balajty, L. Bodnárová and A. Packová
University of Veterinary Medicine and Pharmacy in Košice, Komenského 73, 04181 Košice, Slovakia; daniela.takacova@uvlf.sk

Abstract

The slaughtering of animals and using their meat for human consumption is a reasonable ground for animal killing according to the legal situation in the Slovak republic. Slaughter treatment for private domestic consumption is regulated by § 23 of the Act on Veterinary Care in the Slovak Republic. This paragraph concerns only bovine and swine. Today we are faced more frequently with cases of dogs being killed for human consumption, across the world, not only in the Slovak Republic. Dogs are being slaughtered and the meat is appearing on the 'menu' of some individuals. The legal rules as well as ethical points of killing dogs for human consumption will be analysed in the Slovak republic and other countries.

Keywords: meat, dog, killing, legislation

Introduction

Animals used for human consumption are farm animals bred in holdings intended for intensive breeding, especially cattle, pigs and poultry. These animals are slaughtered at specialized establishments, slaughterhouses, only by competent personnel. The Slovak legislation does not explicitly require the presence of a competent person at the time of killing of animals intended for the breeder's own consumption. In the case of killing of animals other than those for which the slaughter is legally permitted, there is an obligation to ensure that the animals are rendered unconscious before they are killed. Such a rule of law does not exclude the possibility of killing animals in an inhumane way and consuming products of animal origin from animals other than slaughter animals in the breeder's household (legislation prohibits the marketing of food not obtained from traditional slaughter animals).

Killing of animals in the Slovak legislation

The Slovak Act on Veterinary Care provides a comprehensive list of reasons that are considered a reasonable ground for killing an animal (The Act on Veterinary Care No 39/2007 Coll. as amended; § 22, Section 4). Killing of an animal for any other reason than that included in this list is considered unlawful.

The reasonable ground for killing of an animal includes the following:
- necessary self-defence and extreme necessity;
- killing a slaughter animal or other animal for the purpose of obtaining products of animal origin;
- killing of an animal within the approved procedures;
- painless killing of an animal because of its terminal illness, severe or widespread injury or its age, if its further survival is associated with continuous pain or suffering; painless killing of an animal after the previous loss of consciousness can be carried out only by a veterinarian except for the ending of animal's suffering in urgent cases when it is impossible to quickly secure help of the veterinarian;
- slaughter of an animal at eradication, control, prevention and diagnosis of diseases and rat control;
- killing of unwanted animals, if it is impossible to provide for them alternative care; this does not apply to service animals;
- hunting an animal in a legal manner;
- killing invasive non-native animal species under a special regulation.

EurSafe 2018 – DOI 10.3920/978-90-8686-869-8_77, © Wageningen Academic Publishers 2018

The regulation concerning the killing of animals for private use is laid down in the Section 23 of the Act on Veterinary Care, which applies only to bovine animals and pigs.

As private consumption is considered to be the consumption of the breeder and his/her close persons (the law distinguishes between the group of natural persons that form a household and close relatives (Section 116 of the Civil Code – The close relative is a relative in direct relationship, sibling, husband; other persons in a family or similar relationship are considered to be close to each other if the damage suffered by one of them is reasonably perceived by the other as their own detriment; § 115 Civil Code – The household consists of natural persons who live together and co-finance the costs of their needs). We consider this important due to the fact that the consumption of an animal product for self-consumption should take place in one location and not in different places where close persons can live.

It is generally known that animals of other species are also killed as slaughter animals. If the products of animal origin obtained from such animals are placed on the market, they must be subjected to veterinary inspection.

Although in the Slovak Republic dogs are not considered slaughter animals, some 'breeders' behave towards them as such and increasingly put them on their 'menu'. There is no official procedure for the slaughtering of individuals of this animal species nor of the methods employed in their slaughter/killing. Since the law does not explicitly prohibit consumption of animal products obtained from dogs we believe that such action is not unlawful. However, we know that dogs used for slaughter are often obtained by illegal means. They are often stolen and their killing is done in a way that does not meet veterinary hygiene requirements that ensure the safety of products intended for use in a breeder's household (Bugarsky *et. al.*, 2008). Therefore, unlawfulness cannot be excluded regarding potential danger to the health of consumers. Another aspect is the violation of the Act on Veterinary Care in terms of the protection of animals from the abuse at such killing. There are a number of cases involving economic crime which is the most common motive for the consumption of products derived from this animal species. There are also reasons that can be classified as the traditional consumption of dog meat in some municipalities in the Slovak Republic and it is a well-known fact that citizens who have acquired Slovak citizenship or have legalized long-term residence in the territory of the Slovak Republic have not surrendered the habits that are common in their country of origin.

During the killing of animals, it is forbidden to use a painful method of killing, which can be considered as cruel (the Veterinary Care Act does not directly define the concept of torture to death). Torture to death should be understood as the cause of death resulting from man's infliction of intensive pain and suffering. Killing of animals in an unauthorized way (choking, stabbing, drowning, beating) can also be considered as torture to death. It is forbidden to use poisons and drugs, the administration of which is not associated with bringing the animal into a state of total anaesthesia. In veterinary practice there are means of inducing narcosis which, after overdosing, cause death by stopping the vital functions at a time when the animals lacks perception. Electricity can only be used for killing if its application results in immediate loss of consciousness. Excluded is the use of electric power from a light or socket circuit in houses or blocks of flats.

The Government Regulation (No. 497/2003 Coll., Amending Act No. 315/2003 Coll.) defines terms such as immobilization (use of a method intended to restrict the movement of an animal in order to allow it to be effectively stunned or killed); stunning (a process causing an immediate loss of consciousness of an animal until death); killing (a procedure that makes the animal dead); slaughtering (a procedure that causes death by bleeding).

Dogs are not listed among animals that should be slaughtered in an abattoir. The legal regulations include requirements on slaughtering of animals for fur that are killed by a mechanical device involving brain penetration, by overdose of anaesthetics, electric current followed by cardiac arrest, gaseous carbon monoxide, chloroform or gaseous carbon dioxide. Despite the existence of such regulation, even in these cases, dogs are not mentioned.

Killing/slaughtering of animals in the European Union legislation

Regulation (EC) no. 1099/2009 on the protection of animals at the time of killing introduces welfare rules for the killing or slaughter of animals kept for the production of food and products such as fur and leather as well as killing of animals on farms in other contexts such as disease control situations.

This regulation does not apply to animals killed in the wild, or as part of scientific experiments, hunting, cultural or sporting events and euthanasia practiced by a veterinarian, nor to poultry, rabbits or hares for private domestic consumption. The authority, in laying down procedures in matters of food safety, has adopted two opinions on the welfare aspects of the main systems of stunning and killing. The procedures consider individuals of certain species of animals, namely the welfare aspects of the main systems of stunning and killing the main commercial species of animals (dogs do not belong to this group) and on the welfare aspects of the main systems of stunning and killing applied to commercially farmed deer, goats, rabbits, ostriches, ducks, geese and quail (Council Regulation (EC) No 1099/2009 of 24 September 2009 on the protection of animals at the time of killing). There are definitions involved in chapter 2 of this regulation, e.g.: 'animal' which means any vertebrate animal, excluding reptiles and amphibians; according to the definition for the purposes of this regulation, dogs could be included.

Animals that are killed for private domestic consumption shall be spared any avoidable pain, distress or suffering during their killing and related operations. Animals shall only be killed after stunning in accordance with the methods and specific requirements related to the application of methods regulated by this regulation. The loss of consciousness and sensibility shall be maintained until the death of the animal. Killing and related operations shall only be carried out by persons with the appropriate level of competence to do so without causing the animals any avoidable pain, distress or suffering, or by a person under the responsibility and supervision of the owner (this article applies to the slaughtering of all animals, other than poultry, rabbits, hares, pigs, sheep and goats).

Killing dogs for human consumption in some countries

Case investigations conducted by the US Humane Society (HSUS) and the International Humane Society (HSI), related to the brutal killing of dogs and cats in Asian countries, revealed that 2 million dogs and cats die each year because of their fur. They are killed for making hats, golf gloves, automobile upholstery, boot inserts, and so on (Anonymous 1). Animals are killed by hanging, slow asphyxiation, squeezing (beating), often being skinned alive so that the flayed skin remains intact. This has been carried out by slayers, dealers and various retailers. China is considered to be the main source of cat and dog fur. Farms for breeding of these animals are located mainly in the north of the country, where a cooler climate guarantees a better quality of coat. Many rural settlements have their own fur marketplace where animals are not only sold but also killed (Bugarsky *et al.,* 2008).

Not all dogs are bred for fur. Many of them, packed in sacks, are transported to restaurants that demand fresh meat. Dog meat is increasingly regarded as a delicacy with perceived medicinal properties to improve stamina and strength. Consumption is thought of as a sign of machismo. Men who are impotent eat dog meat to get an erection. Women assume that dog meat is healthy to feed their families when in fact it is not. Unfortunately, it is not yet illegal to eat dog meat in Indonesia. But different Indonesian Criminal Code (KUHP) Articles apply to suppliers, sellers and buyers (BAWA, 2014).

An estimated 30 million dogs across Asia, including stolen family pets, are still killed for human consumption every year, according to the Humane Society International. While not widespread, the charity says the practice is most common in China, South Korea, The Philippines, Thailand, Laos, Vietnam, Cambodia and the region of Nagaland in India (BBC newsbeat, 2017).

In some countries (Canada) it is legal to sell and serve dog meat, provided the dog was killed and his meat was processed under federal inspectors′ supervision (CBC News). If a dog is killed illegally, this is considered as cruelty to the animal, such act violates the provisions of the Canadian Criminal Code, and the offender can be sentenced to imprisonment for up to 5 years (CBC News, 2003). In Switzerland, commercial slaughter of dogs and the sale of their meat is illegal, but farmers may slaughter their dogs for own personal consumption (Savage, 2012).

Because the dog is not included in the list of food animals, the placing on the market and sale of dog meat in the Czech Republic is prohibited by law. Self-consumption of meat is allowed (TV Nova, 2017). Relatively often dog meat is consumed by the Romas, in some cases also by the Vietnamese, while among Czechs the consumption of dogs is very unusual. The police have uncovered several cases of illegal slaughterhouses operated by Vietnamese in recent years (Martinec, 2012). However, reports that dogmeat is being prepared in Chinese and Vietnamese restaurants and bistros and offered to customers like beef or pork are not based on the truth. For Czech Vietnamese, dog meat is a delicacy that is expensive and hard to come by. As such, it is highly unlikely to be used fraudulently in place of other ingredients and served to Czech customers. This is comparable to a customer in a Czech restaurant fearing that instead of a roast chicken he would get a pheasant or instead of beef goulash he would get game (Trávníková, 2014) Such cases were published by various media also in the Slovak Republic, which cited unnamed individuals who praised the positive effects of both dog fat and meat on the human organism. The circumstances under which the dogs got into the pots were not mentioned. Such action could be considered unlawful only in those cases when the dog was unlawfully stolen and killed. There is no doubt that the killing of dogs for the purpose of their consumption takes place in a way not regulated by the legislation.

Ethics of killing dogs for consumption

According to Agnew (1998) animal abuse is defined as 'any act that contributes to the pain or death of an animal or that otherwise threatens the welfare of an animal.' Thus Agnew′s definition of abuse pertains to all animals and includes harm to animals that is normative in our culture, such as use of animals for food, clothing, entertainment, and laboratory research. Not all animals are intended for consumption. Dogs and cats have played a far more intimate role in human lives both during historic agrarian times and the recent past than many other species. They also provide a greater degree of companionship. People often classify animals into those that are eaten and those that are not. If animals are farmed then they have to be eaten. Companion animals in contrast, are not eaten, because they are considered as members of the family, therefore people cannot imagine eating them for any reason. For many this feeling is so strong that the thought of eating them is both taboo and horrifying. Such attitude stems from a person's culture and the way in which they are raised. In Europe it is acceptable to eat animals like cattle, sheep, pigs, but not to eat dogs.

Joy (2010) explores the psychological mechanisms behind *carnism* (Latin *carn*, flesh or body to name and describe the dominant cultural belief system): the ideology according to which eating certain animals is considered ethical and appropriate. Her book starts with a thought experiment: 'Imagine yourself at a dinner party with friends, enjoying a delicious meat stew. When you ask your friend for the recipe, she says: You begin with five pounds of golden retriever meat, well marinated, and then...' The experience of disgust is an important psychological mechanism influencing the decision to eat certain animals. In fact, disgust is considered one of the core moral emotions. According to Joy (2010), generally the more empathy is felt for an animal, the more disgusted you are about the idea of eating it. Because most people feel more empathy towards dogs than cows we are more disgusted at the idea of eating dogs.

Domestication of the dog means many people have developed emotional ties with dogs, but this connection is no grounds with which to consider the consumption of dog as a morally inferior act. Those condemning the consumption of dog as worse than the consumption of pig or cow claim that since dogs are complex, social animals, they are more important animals and thus should be unfit for consumption. However, scientific evidence disproves this. Pigs and dogs are comparable when it comes to loyalty and intelligence, traits for which dogs are prized. Many studies show that pigs demonstrate a higher level of intelligence than do dogs. Even chickens, which are seen as bird-brained creatures, display complex social connections and familial relations with other chickens that merit them as worthy of protection from human consumption as do dogs (Anonymous 2).

According to the ASPCA, 1.2 million dogs are euthanized in the United States alone, (ASPCA 2015-2016). Is putting dogs to death by euthanasia really any better than eating them? Both mean a quick death, with the Chinese situation ('Yulin Dog Meat Festival', an annual celebration in Yulin, China, where festival goers try dog meat, lychees, and other foods) providing more utility as well as being on a much smaller scale than the massive euthanizations conducted in the United States (Anonymous 2).

Conclusion

Inadequate legislation on the breeding of companion animals, the lack of knowledge of existing legislation and a different level of individual legal consciousness of animal breeders, other citizens and, officials in the state administration or self-government bodies, enables behaviour that is often not in line with the legal standards regulating these individuals. Dogs' consumption is not prohibited in some countries. Therefore, the stress must be taken on animal´s welfare as well as on the way of their slaughtering. Some countries have banned the selling and eating of dogs (Taiwan was the first Asian country to crack down on the practice). Ethical views on the consumption of dogs meat differ, which is due mainly to the varying cultural perception of the breeding of various animal species.

References

Agnew, R. (1998). The cause of animal abuse: A social-psychological analysis. *Theoretical Criminology, 2* (2), 177-209

Anonymous 1. Dajte si pozor, čo si obliekate (8 May 2011) Available at: http://waf-kolektiv.blogspot.sk/2011/05/dajte-si-pozor-co-si-obliekate-moze-to.html. Accessed 27 November 2017.

Anonymous 2. The Ethics Behind Human Consumption Of Dogs. Available at: https://www.theodysseyonline.com/ethics-human-consumption-dogs. Accessed 24 March 2018.

ASPCA 2015-2016. Available at: https://www.aspca.org/animal-homelessness/shelter-intake-and-surrender/pet-statistics. Accessed 15 March 2018.

BAWA (Bali Animal Welfare Association): http://www.changeforanimals.org/100000-dogs-eaten-in-bali-press-release. Accessed 30 April 2014.

BBC newsbeat (2017). Available at: http://www.bbc.co.uk/newsbeat/article/39577557/the-countries-where-people-still-eat-cats-and-dogs-for-dinner. Accessed 15 January 2017

Bugarský, A., Buleca, J., Bystrický, P., Hvozdík, A., Kottferová, J. and Takáčová, D. (2008): Pes, chovateľ a právo. 2008, UVLF v Košiciach, s. 229 In Slovak

CBC News. Canine carcasses at Edmonton restaurant were coyotes (11 November 2003). Available at: http://www.cbc.ca/news/canada/canine-carcasses-at-edmonton-restaurant-were-coyotes-1.382176. Accessed 4 December 2017.
Civil Code (Act No. 40/1964 Coll as amended)
Joy, M. (2010): Why We Love Dogs, Eat Pigs, and Wear Cows: An Introduction to Carnism. ISBN: 978-1-57324-461-9; Conary Press, 211 pp.
Martinec, M. (2012). České psy končia na tanieri (In Slovak): Available at: http://www.pluska.sk/plus-7-dni/zo-zahranicia/ceski-psy-koncia-tanieri.html. Accessed 15 March 2018.
Regulation (EC) No. 1099/2009 on the protection of animals at the time of killing.
Savage, J.: Dogs and cats 'still eaten in Switzerland' (27 December 2012). Available at: https://www.thelocal.ch/20121227/dogs-still-eaten-in-switzerland. Accessed 27 November 2017.
The Act on Veterinary Care No. 39/2007 Coll. as amended.
The Government Regulation (No. 497/2003 Coll., Amending Act No. 315/2003 Coll. by which are regulated requirements for animals´ protection at the time of their killing or slaughtering)
Trávníková, B. (2014). Psí pečeně a guláš: když se naši mazlíčci dostanou na jídelníček (In Czech). Available at: http://www.zapnimozek.cz/psi-pecene-a-gulas-kdyz-se-nasi-mazlicci-dostanou-na-jidelnicek. Accessed15 March 2018.
TV Nova (2008). Available at: Je libo psíka? Dejte si nášup, trestné to není. TV Nova 2008-07-22. Available online (In Czech). Accessed 6 December 2017.

78. Sustain or supersede – an exploration of the practices of animal product limiters

S.V. Kondrup
Department of Food and Resource Economics, University of Copenhagen, Rolighedsvej 25, 1958 Frederiksberg C, Denmark; sak@ifro.ku.dk

Abstract

The aim of this paper is to present the context of a current PhD project and the research questions that are guiding the project. To satisfy growing public concerns about the environment, animal welfare, and personal health, animal product limiting (APL), e.g. flexitarianism, vegetarianism, and veganism, provides an alternative to the mainstream animal-based diets in the Western world. However, the exclusion of more or less animal products from the diet is done in a continual movement where the social practices of everyday life and institutionalized eating standards for (in)appropriate eating are thought to play a crucial role. The overall objective of this project is to examine how APL is done among a group of newly practicing (<12 months) flexitarians, vegetarians, and vegans in Denmark and how and why changes in dietary practices among these animal product limiters (APLs) occur over time. A practice theoretical perspective on food and eating is integrated as it opens up for a dynamic relation between the micro-social level and the material and discursive circumstances at the macro-level and emphasizes fundamental questions about how social practices are interconnected and change over time. The project builds on the claim, that more attention should be paid to the embedding of APL in social practices instead of only focusing on individual articulations of attitudes and choices in order to understand why dietary change occur among APLs.

Keywords: vegetarianism, social change, practice theory

Introduction

There is a broad range of motivational factors for changing one's diet in a direction in which less animal product are included, e.g. the perceived benefits for personal health, due to animal ethical or environmental concerns, spiritual well-being, religious belief, or gustatory arguments (Bedford and Barr, 2005; Curtis and Comer, 2006; Fessler *et al.*, 2003; Fox and Ward 2008; Hoek *et al.*, 2004; Hussar and Harris, 2010; Lindeman and Sirelius, 2001; Wilson *et al.*, 2004). These factors are often presented as the ends and purposes of animal product limiting (APL), e.g. flexitarian, vegetarian, and vegan diets, but people's understandings, knowledge of and commitments to APL are also influenced by other factors as well such as significant others, e.g. family, friends, idols, mass media and music (Boyle, 2007; Devine *et al.*, 1998; Haverstock and Forgays, 2012; Jabs *et al.*, 1998; 2000; Larsson *et al.*, 2003; McDonald, 2000; Rozin *et al.*, 1997), food traditions (Devine *et al.*, 1998; Haverstock and Forgays, 2012), significant life transitions in location or social roles (moving to a new area, attending a new educational institution, marriage, getting a divorce, pregnancy, and change of employment) (Beardsworth and Keil, 1992; Devine *et al.*, 1998; Jabs *et al.*, 1998), and reinforced by social support (Barr and Chapman, 2002; Devine *et al.*, 1998; Haverstock and Forgays, 2012).

A considerable amount of social science and psychological research on vegetarianism and veganism has investigated motivational factors in order to understand why people start to follow vegetarian or vegan diets. By and large, these lines of research have interpreted such forms of APL as markers of distinct identities from rather individualistic perspectives. This approach accounts for some stability in people's foodways and discussions of how and why the dietary practices of Animal product limiters (APLs) alter

EurSafe 2018 – DOI 10.3920/978-90-8686-869-8_78, © Wageningen Academic Publishers 2018

and potentially disappear over time have been neglected in much research on the subject. Most people do not follow the same dietary route throughout life and APLs do not seem to be an exception to this. Thus, even though APL is often driven by strongly held ethical viewpoints, health beliefs, and/or motivations to change dietary behaviours, recent research suggests that adopting a vegetarian or vegan diet typically is a dietary phase rather than a lifetime commitment (Boyle, 2007; Humane Research Council, 2014). Furthermore, vegetarians do not seem to follow one specific type of plant-based diet over time but include and exclude food items interchangeably – some vegetarians tend to move in directions in which more animal-derived products are excluded over time (Barr and Chapman, 2002; Haverstock and Forgays, 2012), while others move in directions in which less animal product are excluded over time (Boyle, 2007). However, we do not know much about the social dynamics of APL in an everyday context and how the dietary practices of APLs are challenged or reinforced over time – resulting in more robust or changing foodways. This project aims to fill this gap.

Theoretical approach

The project aims to work with two overall yet interconnected analytical foci that feed into each other: (1) APL in everyday life and (2) The history and institutionalization of APL. APL is understood as a distinct foodway with different manifestations, e.g. flexitarian, vegetarian, and vegan, that is enacted and embedded in a dynamic field of social practices in everyday life and institutionalized eating standards for (in)appropriate eating that regulate how APL is done. This way of approaching and framing APL is mainly inspired by the works of Elizabeth Shove and colleagues. Flexitarians are defined here as people who primarily eat vegetarian diets with the occasional inclusion of (potentially any) meat. Vegetarians are defined as people who do not eat seafood, pork, poultry or other meat but include dairy products and/or eggs in their diets while vegans are defined as people who abstain from eating any animal products (with the possible exception of honey).

Shove and others (2012) argue that practices are organized by a dynamic combination of the following elements: materials (e.g. objects such as food and the animals from which the food is made, tools, technologies, cookbooks, social media, the domestic infrastructure of the home, and the wider infrastructures of e.g. food supply), competences (e.g. knowledge of how to cook without meat and/or other animal derived products, embodied skills, techniques, creativity and improvisations), and meanings (e.g. symbolic and socially shared meanings of protein-dense diets/plant-based diets, cultural conventions of meat-eating/vegetarianism/veganism, ideas, expectations, aspirations) (p. 14). These three elements permeate embodied and discursive processes although not as specific features of the individual human beings but rather as characteristics of the practices that the individuals engage in (Reckwitz, 2002).

A practice theoretical approach to APL challenges ideas of how individual choices foster new behaviour that seem to be specifically influential in contemporary consumption discourses and policymaking as well as in much research on APL. Such ideas take for granted that individuals have control over their surroundings and that they are solely responsible for their 'own' attitudes and actions. Instead, human action and structure are seen as recursively related (Giddens, 1984): social structure is produced and reproduced by everyday behaviour while it at the same time limits and enables everyday behaviour. It is due to the recursively organization of practices that 'the constitution of agents and structures are not two independently given sets of phenomena, a dualism, but represents a duality' (Giddens, 1984, p. 25). These statements represent an important starting point for an integrated analysis of how and why APL is done and change over time where the individual is always viewed in a social context.

APL in everyday life

Stressing the first analytical focus, APL in everyday life, the dietary practices of APLs are framed as 'bundles of practices': 'loose knit patterns based on co-location and co-existence' (Shove *et al.*, 2012: 17,81). Various forms and levels of, for example, planning, cooking, shopping, and eating practices are loosely integrated in these bundles of practices, and shape how APL is done and divided into different manifestations of APL. In order to understand how and why the dietary practices of APLs appear, alter, and potentially disappear in an everyday context the following questions are raised: 'how are flexitarian, vegetarian, and vegan practices organized, e.g. what are the materials, competences, and meanings (Shove *et al.*, 2012) that coordinate these practices?', 'how do the materials, competences, meanings, and the coordination between them differentiate between flexitarian, vegetarian, and vegan practices and how do they change over time?', and 'how are flexitarian, vegetarian, and vegan practices challenged or reinforced by other social practices in everyday life?'. The sources of material for this part of the analysis are interviews with a group of newly practicing (<12 months) APLs and a questionnaire survey (depending on financial support).

According to Shove and others, the 'elements which constitute the practice as entity' are interdependent in distinct configurations and it is through the recurrent performances that these configurations are reproduced and sustained over time (Shove *et al.*, 2012: 7). By implication, practices are then susceptible to change when the connections between the elements are broken, made, and when new combinations occur (Shove *et al.* 2012: 21). Several studies have found that there are different learning phases, processes, or stages that develop after people have commenced vegetarian diets (e.g. Boyle, 2007; Roth, 2005; Rozin *et al.*, 1997). For example, what begins as merely a food preference can turn into more comprehensive vegetarian practices in which health, the environment, animals, humanitarianism, and spirituality become important. Such transformations in the meanings of the vegetarian practices where more pronounced sensitivities to food, meat, and the body typically occur could then stimulate a modification in materials by the exclusion of more animal products which subsequently could lead to the formation of vegan practices. On the other hand, certain levels of 're-routinization' of APL in everyday life without a continual reaffirmation of the underlying rationalizations are also expected to occur (Warde, 2016: 82) which could lead to a continuance of current practices.

In much the same way as elements rely on certain interdependencies for a specific type of practice to exist, practices also interrelate in time and space and thereby influence each other. 'Bundles and complexes of practices' can appear and disappear due to the 'collaboration and/or competition between practices' (Shove *et al.*, 2012: 88). By repeating certain combinations of practices, the terms on which practices are performed, compete or collaborate, are thought to change. Based on these assumptions, it is suggested here that competing and collaborating practices in the everyday lives of APLs, e.g. parental, relationship, cooking, shopping, and work practices challenge or reinforce how APL is done. In this regard, studies have found that most of the opposition to vegetarianism comes from family members (Jabs *et al.*, 1998; Roth, 2005) which could indicate that the elements of family practices are not compatible with the elements of vegetarian practices which subsequently pose a threat to the family's traditions. This further supports the initial suggestion, that attention should not only be paid to individual attitudes and choices in order to understand why dietary change occur among APLs but also to the interrelations between occurring practices in everyday life.

The history and institutionalization of APL

One implication of approaching APL from a practice theoretical perspective is to look specifically to the characteristics of the dietary practices of APLs without neglecting the embodied experiences of the individual APLs and also to connect the everyday practices of APLs to the macro-institutional context.

Some APLs are expectedly more likely to identify themselves as 'flexitarian', 'vegetarian', and 'vegan' than others, some are more explicit and insisting in their knowledge dissemination and attempts to recruit other people while others may act silently and reticently. However, contemporary ways of practicing APL – of cooking, shopping, and eating plant-based foods, are believed to be inextricably linked to the social and national history in which these practices occur (Delormier, 2009; Graça, 2016; Halkier, 2011) and guided by the ways in which APL is institutionalized. The second analytical focus, 'The history and institutionalization of APL', therefore asks 'what are appropriate performances of flexitarian, vegetarian, and vegan practices?' and 'how do the dynamics that power these appropriate performances work against the consumption of animal-based diets and how have they developed throughout history?'. A combination of primary and secondary sources of material will be the basis for this analysis, i.e. interviews with APLs, historical documents, and academic papers presenting the results from other empirical and historical studies.

Although there is no authoritative model for ways of practicing APL, APLs participate in flexitarian, vegetarian, and vegan practices in ways that are socially organized and institutionalized – by certain sets of shared norms, conventions, rules, standards, and justifications (Warde, 2014). This organization and institutionalization is a feature of its history and of the social and continuous process of defining specific food items and practises as (un)healthy, (non)ethical, and (in)appropriate which are influenced by a number of entities and bodies such as interest groups and organizations, cookbooks, how-to videos, the food industry, celebrities, grocery stores, menus at cafes and restaurants, dietitians, doctors, etc. For example, recent changes in the ways in which know-how of plant-based cooking is distributed through the extensive stream of how-to videos on several digital platforms have made learning easier and more accessible and likely contribute to setting standards for the competences of plant-based cooking and aesthetics of plant-based meals. Other developments that likely play a part in the institutionalization of APL are potential changes in the symbolic meanings of meat and plant-based food where the latter seems to evolve into more and more appropriate foods and meat the opposite, changes in the logistics of flexitarian, vegetarian, and vegan practices potentially moving from a more private setting to a public where plant-based meals are more acceptable and accessible, and positive changes in wider infrastructures of the supply of plant-based foods and meat subsidies.

Outro

In order to shed light on how and why changes in the dietary practices of APLs occur, this project aims to study a group of current APLs over a period of four to five years which has not been done before, to my knowledge. Thirty informants (10 flexitarians, 10 vegetarians and 10 vegans) between 15-60 years of age who are active partakers in both shopping and cooking are recruited for the qualitative data collection. They are recruited through a recruitment bureau and social media. The project runs a total of six years (May 2016-May 2022) and five rounds of interviews with six months/one-year intervals are implemented. Based on the insights gained from the interviews (and depending on financial support) a survey will be carried out in the Danish population to quantify the findings from the qualitative part. The social organization and dynamics of APL are explored from a practice theoretical approach which will hopefully provide new insight into how flexitarian, vegetarian, and vegan practices are challenged or reinforced in everyday life and how the eating conventions and standards that are acting out in APLs' social circles and spheres regulate the dietary routes of APLs. The project is also carried out at a time when technical and material innovations such as meat subsidies have become increasingly available and popular.

Acknowledgments

I would like to thank my supervisors professor Peter Sandøe, professor Lotte Holm, and associate professor Thomas Bøker Lund from the Department of Food and Resource Economics, University of Copenhagen, for their knowledgeable comments and helpful suggestions for improvements.

References

Barr, S.I. and Chapman, G.E. (2002). Perceptions and practices of self-defined current vegetarian, former vegetarian, and nonvegetarian women. Journal of the American Dietetic Association102: 354-360.

Beardsworth, A. and Keil, T. (1992). The vegetarian option: Varieties, conversions, motives and careers. The Sociological Review 40: 253-293.

Bedford, J.L. and Barr, S.I. (2005). Diets and selected lifestyle practices of self-defined adult vegetarians from a population-based sample suggest they are more 'health conscious'. International Journal of Behavioral Nutrition and Physical Activity 2: 4.

Boyle, J.E. (2007). Becoming vegetarian: An analysis of the vegetarian career using an integrated model of deviance. Doctoral Thesis. Faculty of the Virginia Polytechnic Institute and State University, Virginia, US, 209 pp.

Curtis, M.J. and Comer, L.K. (2006). Vegetarianism, dietary restraint and feminist identity. Eating Behaviors 7: 91-102.

Delormier, T., Frohlich, K.L. and Potvin, L. (2009). Food and eating as social practice -understanding eating patterns as social phenomena and implications for public health. Sociology of Health & Illness 31: 215-228.

Devine, C.M., Connors, M., Bisogni, C.A. and Sobal, J. (1998). Life-course influences on fruit and vegetable trajectories: Qualitative analysis of food choices. Journal of Nutrition Education 30: 361-370.

Fessler, D.M., Arguello, A.P., Mekdara, J. M. and Macias, R. (2003). Disgust sensitivity and meat consumption: A test of an emotivist account of moral vegetarianism. Appetite 41: 31-41.

Fox, N. and Ward, K. (2008). Health, ethics and environment: A qualitative study of vegetarian motivations. Appetite 50: 422-429.

Giddens, A. (1984). The constitution of society. Outline of the Theory of Structuration. University of California Press, Berkeley and Los Angeles, US, 402 pp.

Graça, J. (2016). Towards an integrated approach to food behaviour: Meat consumption and substitution, from context to consumers. Psychology, Community & Health 5: 152-169.

Halkier, B. and Jensen, I. (2011). Methodological challenges in using practice theory in consumption research. Examples from a study on handling nutritional contestations of food consumption. Journal of Consumer Culture 11: 101-123.

Haverstock, K. and Forgays, D.K. (2012). To eat or not to eat. A comparison of current and former animal product limiters. Appetite 58: 1030-1036.

Hoek, A.C., Luning, P.A., Stafleu, A. and de Graaf, C. (2004). Food-related lifestyle and health attitudes of Dutch vegetarians, non-vegetarian consumers of meat substitutes, and meat consumers. Appetite 42: 265-272.

Humane Research Council (2014). Study of current and former vegetarians and vegans. Initial findings. December 2014. Available at: https://faunalytics.org/wp-content/uploads/2015/06/Faunalytics_Current-Former-Vegetarians_Full-Report.pdf.

Hussar, K.M. and Harris, P.L. (2010). Children who choose not to eat meat: A study of early moral decision-making. Social Development 19: 627-641.

Jabs, J., Devine, C.M. and Sobal, J. (1998). Model of the process of adopting vegetarian diets: Health vegetarians and ethical vegetarians. Journal of Nutrition Education 30: 196-202.

Jabs, J., Sobal, J. and Devine, C.M. (2000). Managing vegetarianism: Identities, norms and interactions. Ecology of Food and Nutrition 39: 375-394.

Larsson, C.L., Rönnlund, U., Johansson, G. and Dahlgren, L. (2003). Veganism as status passage. Appetite 41: 61-67.

Lindeman, M. and Sirelius, M. (2001). Food choice ideologies: The modern manifestations of normative and humanist views of the world. Appetite 37: 175-184.

McDonald, B. (2000). 'Once you know something, you can't not know it' an empirical look at becoming vegan. Society & Animals 8: 1-23.

Reckwitz, A. (2002): Toward a theory of social practices. A development in culturalist theorizing. European Journal of Social Theory 2: 243-263.
Roth, L.K. (2005). 'Beef. It's what's for dinner': Vegetarians, meat-eaters and the negotiation of familial relationships. Food, Culture and Society: An International Journal of Multidisciplinary Research 8: 181-200.
Rozin, P., Markwith, M. and Stoess, C. (1997). Moralization and becoming a vegetarian: The transformation of preferences into values and the recruitment of disgust. Psychological Science 8: 67-73.
Shove, E., Pantzar, M. and Watson, M. (2012). The dynamics of social practice: Everyday life and how it changes. Sage Publications, London, UK, 208 pp.
Warde, A. (2014). After taste: Culture, consumption and theories of practice. Journal of Consumer Culture 14: 279-303.
Warde, A. (2016). The practice of eating. Polity Press, Cambridge, UK, 220 pp.
Wilson, M.S., Weatherall, A. and Butler, C. (2004). A rhetorical approach to discussions about health and vegetarianism. Journal of Health Psychology 9: 567-581.

79. Moral individualism in veterinary practice – a preliminary investigation into normative foundations in small animal clinics

S. Böhm, S. Springer and H. Grimm*
Unit of Ethics and Human-Animal Studies, Messerli Research Institute, University of Veterinary Medicine, Vienna; Medical University of Vienna; University of Vienna, Veterinaerplatz 1 1210 Vienna, Austria; sophia_boehm2003@yahoo.de

Abstract

This contribution aims to empirically investigate how veterinarians morally justify their treatment of animals in small animal practice. Are they exclusively committed to animal centered reasoning like the welfare of the animal, or do other reasons, such as the financial situation of the animal owner or the relationship of the pet owner to the animal play an important role as well? To answer this question, Todd May's (2014) account on two types of reasons for the moral consideration of animals is used: capacity-based reasons (CBRs) and relation-based reasons (RBRs). Against the backdrop of his argument, it was investigated to what extent veterinarians base the justification of their actions on CBRs, RBRs or hybrids of CBRs and RBRs. The guiding hypothesis is: CBRs are the dominant normative source in veterinary practice. Semi-structured interviews were held with six veterinarians from Lower Austria. After a quantitative analysis of the results as well as a qualitative content analysis according to Philipp Mayring (2008), the guiding hypothesis can be supported that CBRs are the main normative source for justifying veterinary treatment. When working with the transcripts, it became clear that a significant number of coded statements could not be grouped under CBRs or RBRs. It was therefore necessary to implement a third 'Hybrid' category to mirror the complexity of data in the analysis. Out of 117 coded justifications for veterinary treatment, 66% could be assigned to CBRs, while 19% corresponded to the 'Hybrid' category and 15% to RBRs.

Keywords: capacity-based reasons, relation-based reasons, veterinary ethics, interview study

Introduction and theoretical background

The distinction of capacity-based reasons (CBRs) and relation-based reasons (RBRs) was originally formulated by Todd May (2014) to group all approaches in animal ethics according to the kind of reasons they prioritize to argue for animals' moral status. CBRs are in line with moral individualism. Within this framework authors argue that animals possess a moral status because of their capacities and properties which make a moral difference. To give a prominent example: Peter Singer argues that the principle of equal consideration of equals requires us to take all (not only human) suffering into account and treat interests in a morally relevant way, irrespectively whether they are the interests of animals or humans. The ability to suffer serves as the key to enter the moral community.

Todd May (2014) categorized such positions and approaches under the group CBRs. Authors like Peter Singer, Tom Regan, Jeff McMahan are gathered under this category. In contrast to moral individualists, who argue animal centred, other authors hold that the source for the moral consideration of animals lies in the relations of humans and animals. Reasons that find their source of normative power in particular relations were categorized as RBRs by May (2014). He categorizes authors like Cora Diamond, Alice Crary, Mary Midgley, Elisabeth Anderson, Clare Palmer in this group. To give an example for relational based reasoning: My moral responsibility to care for children strongly differs if the children are mine or someone else's. E.g. under normal circumstances, I would not pay for the education of some else's children. The reason is not that they lack a particular capacity (CBR), but my specific relation to them.

EurSafe 2018 – DOI 10.3920/978-90-8686-869-8_79, © Wageningen Academic Publishers 2018

In contrast, my refusal to pay for the education of my own children can be questioned on the basis of my relation to them of being a parent. May (2014) argues that neither RBRs nor CBRs alone can adequately comprise the complexity of our morality. In consequence, his account suggests that both kinds of reasoning play a role in practice. Whether the two categories are comprehensive has been questioned in Grimm and Aigner (2016).

Against this background, a study was designed to investigate to what extent veterinary professionals refer to RBRs and CBRs when challenged with morally difficult situations. In particular, the study focused on situations, which were described as morally challenging in veterinary ethics e.g. being torn between the commitment to the animals' welfare and wishes of the pet owners. Hence, the theoretical distinction between CBRs and RBRs was operationalized in order to contribute to clarifying the normative foundations of the veterinary profession in a particular context.

Material and methods

In total, six semi-structured interviews with veterinarians from Lower Austria have been conducted in order to answer the question whether veterinarians are more or less inclined to base their normative accounts on CBRs or RBRs or hybrids. Since it is very likely that the specific human-animal relation makes a difference with regards to RBRs, a group of veterinarians working in the field of small animal medicine was invited for interviews. By means of an interview guide a fixed sequence of thirteen questions were used to ensure that relevant topics and challenging questions of veterinary practice would be thematised. The veterinarians answered to questions on the following topics: Euthanasia, obesity, oncological treatment, difficulty of the owner to pay for treatment. These were the core issues the semi-structured interviews comprehended. Furthermore, participants were asked to answer to a teaching goal in the curriculum of the Vetmeduni Vienna, namely: ethically flawless treatment of animals, humans and nature.

Demographic data such as age, duration of work experience, gender, work in rural or urban areas and the proportion of working hours spent with small animals were collected in order to enable a descriptive statistic of conducted interviews despite the small number of interviews. From March to May 2017 six interviews took place in the veterinarians' practices and recorded. The interviews were transcribed and transferred to the software MAXQDA Analytics Pro 12 (portable license) for qualitative data analysis.

The qualitative analysis was carried out along the methods described by Philipp Mayring (2008). The method is known as subject-oriented qualitative content analysis. The interviews were coded, analysed and interpreted with respect to three deductively formed categories: CBR, RBR and Hybrid CBR-RBR. CBRs and RBRs categories were defined as follows:

- CBRs: Reasons for/against veterinary action that refer to capacities or properties of the animal. E.g. '(...) the dog has been having diarrhoea for three months, the owners wants to put the dog to sleep, five-year-old dog, where we say: No, we won't do it.' (CBRs can't be reduced to RBRs.)
- RBRs: Reasons for/against veterinary action that refer to the relationship between the veterinarian, pet owner and animal. E.g. '(...) there was an old woman, she was almost ninety, with a dachshund, whom we have managed to keep alive for a year for her, because we knew that when the dog dies she might not live much longer than him.' (RBRs can't be reduced to CBRs.)
- The 'Hybrid CBR-RBR' category was introduced during the evaluation process, because not every statement allowed a clear distinction between CBRs and RBRs. This goes along with the idea of Todd May (2014), who holds that CBRs are supplemented by RBRs in particular situations. E.g.: 'it [oncological treatment] is a prolongation, really a prolongation of life for a few months. For me the problem somehow is, but this is really just my personal opinion, how much do I help the animal and in how far do I do it for the owner.'

Descriptive analysis of data

Through the use of semi-structured interviews, variables of participated veterinarians were collected as shown in Table 1.

Although the same number of female and male veterinarians was interviewed, 58% of the coded statements originated from the three interviews with female veterinarians and 42% from the three interviews with male veterinarians. This observation becomes increasingly important, as the total survey time of female veterinarians has been lower on average. In absolute numbers, the female veterinarians achieved 68 codings within a total interview time of approximately 65 minutes. Male veterinarians achieved 49 codings within a total interview time of 112 minutes.

The results of the semi-structured interviews indicate that 66% of all coded responses have a clear CBR tendency, understood as animal centred reasoning. Furthermore, 15% of all coded responses could be assigned to RBRs and 19% to the hybrid category. Based on the analysed data, one can preliminary conclude that veterinarians are primarily referring to CBRs to justify their clinical practice in this context. Thus, the initially formulated hypothesis that CBRs is the dominant normative source in veterinary practice can be supported by the descriptive analysis. However, given the limitation of six narrative interviews it would of course be far-fetched to generalize this result.

Qualitative content analysis

Owners having problems to pay

The previously determined CBR tendency varies widely with regard to the topics addressed in the interviews. By means of an in depth qualitative content analysis it could be shown that in general financial difficulties of the pet owners do not serve as a reason not to treat a patient. The following animal centred statements support that veterinary treatment is strongly based on capacity based reasoning:

> We somehow find a solution that is financially viable. We have a cooperation with a private dogs and cats association. They gain their money through private donations and use it to nurse animals. So we have the possibility to finance a treatment even if the owner has no money. (CBR)

> If the pet has a serious illness and there is a treatment with witch it can be cured, I try to find a solution. (CBR)

Table 1. Overview of different variables of all six participants.

Interview (length in minutes)	Number of codings	Gender	Age	Years of experience	Working with small animals in %	Working-place
Interview one (29:54)	11	male	42	18	100	urban
Interview two (30:58)	14	male	60	30	50	rural
Interview three (25:49)	21	female	50	10	100	urban
Interview four (18:51)	27	female	31	5	100	urban
Interview five (51:28)	24	male	60	32	100	urban
Interview six (20:13)	20	female	39	10	100	urban

> If the pet owners are honest about their financial situation and ask if they can pay fifteen euros a month or something, I'm the last one to turn them down. Or if you really know, it's an emergency and the pet owner would do anything for the pet, but just can't, I would probably do that for free with the agreement of my supervisor. (CBR)

Obesity

Furthermore, veterinarians were asked if they were often confronted with obese animals and if they could depict a specific case. Moreover, it was inquired what they advised the owners to do. Thereby, a strong CBR tendency can be observed, when it comes to the topic of animal's obesity. As with the previous core topic, veterinarians focused on the animal's wellbeing when faced with the problem of obesity, which can be supported by the following statements:

> Yes, I inform them about the medical disadvantages of obesity like cardiovascular, fatty liver or orthopaedic problems (...). Yes, I recommend a weight loss program. There are also special dietary supplements and tables with actual weight and target weight, how many grams per day are recommended for the dog or the cat and so on. And I suggest more exercise, yes, as with humans, it's more exercise and fewer calories. Yes, they often do not know what that means when the cat is three pounds overweight. If you put that in relation: For a human that would mean having twenty, thirty kilos too much. Yes, that does not sound like much now, but if a cat has four kilos, five kilos, that's when a man with 80 kg has one hundred instead, that's 20%, say, in that range. (CBR)

> Or, other cases are cats that develop bladder infections due to, um... obesity. A normal fat cat can't stop urine, they often have crystals or bacterial infections, and because of that most of them have diabetes or they have inflammation, just these interstitial inflammations, which is also sometimes the case with women. It is not really possible to find a reasonable cause. This happens quite often and we always suggest that the cats have to lose weight on the post-treatment sheet. And additionally we try to write a nutrition plan or we transfer the cats to a colleague, then. (CBR)

The core topic 'ethical flawless treatment' also shows a clear tendency for CBR-statements. This focus refers to statements in relation to questions about veterinarians' understanding of ethically flawless and whether it is even possible to act ethically flawless. Furthermore, they were asked if they consider the competence – ethically flawless treatment of animals, humans and nature – from the curriculum of the Vedmeduni, Vienna – as useful and possible.

> Yes, to euthanize animals that are not ill, where there is no clear medical indication for euthanasia – yes, that would be ethically unreasonable. (CBR)

> That I am the advocate of the animals. That I speak for those who can´t speak. (...) But I have to think about natural alternatives; critical use of antibiotics for example, critical use of all drugs – everything has to be done in harmony with nature. I do not have to fight everything with chemical substances. I can also fall back on nature. For non-critical cases, I can wait and see what action is really needed. (...) Likewise, for example, with remedies for fleas and ticks. If you do not necessarily need that, then you should perhaps limit the period of treatment and yes, that's my attitude. (...) For me the animal is in the first place. I am responsible for the welfare of the animal and so I act. (CBR)

Euthanasia

Hugh variation could be seen in answers to the question regarding euthanasia. Participants were asked if they could report a challenging case of euthanasia. Almost half of all hybrid encodings (9 out of 22) were identified in context of this topic.

> For me, when it comes to euthanasia the most important thing is to do make it as clam and stress-free as possible for the animal and also for the pet owner. (RBR-CBR)
>
> Of course sometimes you realize that the bond between the owner and the animal is particularly tight and you often realize at the same time that the life of the animal is going to end because not everything is curable. Sometimes the pet owner is quite old as well, so all family members are dead and the cat or the dog is really the only caregiver left; Yes that is always very, very challenging, I would say. To treat the animal and then make it clear, that the pet owner has to let the animal go. That can be very stressful [for the veterinarian]. (RBR-CBR)
>
> But of course if children are with the pet and when owners are also very connected, for example: If the father of the pet owner may have died and the animal is the last thing he really cares for – it is a very big challenge to explain to the owner that it really is a relief for this animal to die. (RBR-CBR)

Impact of different variables

As mentioned above, the interviews of the female veterinarians contained 16% more coded statements than those of their male colleagues and that, although as many female veterinarians as male veterinarians were interviewed and the interview time of the male veterinarians was 42% higher than those of the female veterinarians. How can this imbalance be explained?

Explanation 1: Influence of gender

One possible explanation can be seen in the response behaviour of male veterinarians who had a much more technical and straightforward approach to the CBR-RBR dilemma. In most situations, male veterinarians focused on the medical challenges in case of euthanasia and did not refer to capacities of the animal nor to relations. Therefore, statements like the following could not be coded:

> (...) because I realized during my studies, that there are many more medical challenges with small animals. The small animal sector is increasingly approaching human medicine, while in large animal practice; the cost pressure is very strong, from the farmer side, not from the vet side. (male)
>
> (...)You face lots of therapeutically and diagnostically challenges every week, but not so much on an interpersonal level with the pet owner. (...). (male)
>
> So the medical-technical challenge was to get the rod out of the body [the cat] again (...). (male)

In contrast, the surveyed female veterinarians immediately referred to CBRs and RBRs. Following CBR coded statements illustrate this point:

> And then it is my job to speak for the animal. I have to be there for the animal (...). (female; CBR)
>
> (...) the first priority is to make sure that I make the best for the animal. (female; CBR)
>
> That I'm actually the advocate of the animals. That I speak for those who can't speak for themselves. I am first obliged to the animals (...). (female; CBR)
>
> For me, the animal is in the first place (...) I am responsible for the welfare of the animal and so I act. (female; CBR)

Explanation 2: Influence of years of experience

Within the survey, 76% of all occupational years are attributable to male veterinarians and only 24% to their female colleagues. A possible explanation is that the male veterinarians who have worked longer have become more acquainted to moral challenges and deal with the complexity of CBRs and RBRs in a habituated way. However, another explanation is that male respondents answer differently than female respondents because of their different views of morality. (Carol Gilligan (2016): In a different voice. Psychological Theory and Women's Development.)

Explanation 3: Influence of education

The imbalance in the number of years of employment allows for another interpretation. The study or training in veterinary medicine has changed during the last years and decades. The topic of animal ethics has gained in importance in the curricula and therefore it can be assumed that veterinarians with significantly fewer years of professional experience have been made much more aware of this topic during their training. The temporal proximity or distance to the obtained training can also explain the difference in observed response behaviour.

Conclusions

Each of these explanations is acceptable based on the existing data, but the sample size is too small to verify the true significance of these factors. However, the preliminary observations formulated here and the resulting hypotheses may be the starting point for a more comprehensive quantitative analysis. Hence, this study provides a number of helpful insights and leads to further questions for the research field of veterinary ethics and can serve as a basis for more extensive quantitative analyses.

References

Bohnsack, R. (2007). Rekonstruktive Sozialforschung. Opladen: Barbara Budrich publisher, p. 235.

Gläser, J. and Laudel, G. (2009). Experteninterviews und qualitative Inhaltsanalyse. Wiesbaden: VS Verlag für Sozialwissenschaften, pp. 197-198.

Grimm, H. and Aigner, A. (2016): Der moralische Individualismus in der Tierethik. Maxime, Konsequenzen und Kritik. In: Köchy, K; Wunsch, M; Böhnert, M [Hrsg.]: Philosophie der Tierforschung Band 2: Maximen und Konsequenzen. München, Verlag Karl Alber, pp. 25-63.

Mayring, P. (2008): Qualitative Inhaltsanalyse. Grundlagen und Techniken. Weinheim und Basel: Beltz publisher, pp. 42-84.

May, T. (2014). 'Moral Individualism, Moral Relationalism, and Obligations to Non-human Animals'. Journal of Applied Philosophy, 31:2, pp. 155-168.

80. The influence of dietary preferences of veterinary medicine students on opinions forming in various ethical dilemmas

L. Mesarčová[*]*, L. Skurková, J. Kottferová, J. Kachnič and A. Demeová*
University of Veterinary Medicine and Pharmacy, Komenského 73, 041 81 Košice, Slovakia; lydia.mesarcova@uvlf.sk

Abstract

The aim of this study is to investigate the hypothesis that people with restricted meat consumption are more sensitive to and strict about animal rights, animal protection and animal welfare in comparison to consumers of meat products. Therefore, we compared attitudes and opinions about the most substantial ethical questions of veterinary ethics in a group of students of veterinary medicine according to their dietary preferences. The research was conducted using a questionnaire, in which students were divided into groups according to whether they consume or do not consume meat. Students then answered questions about basic ethical issues and had the opportunity to express their opinion about topics such as animal euthanasia, animal experiments, animal welfare and welfare of livestock and pets, the impact of meat production on the environment and similar topics.

Keywords: vegetarianism, ethical opinions, animal welfare, animal rights, veterinary students

Introduction

In the human society, choosing a diet is associated with the formation of views on various life situations and circumstances. Unlike most animals, who instinctively know which foods to choose and which to avoid, humans must learn this distinction, which is often closely associated with the tradition and culture of a society (Rozin, 1990).

People who avoid or refuse the consumption of meat and all meat products are commonly called vegetarians. People with a diet that excludes all animal products (milk, milk products, eggs, honey) are called vegans. Vegetarians reject a widespread social norm and that is what makes them different from the others (Back and Glasgow, 1981). There are basically two main motivating reasons for refusing the consumption of meat and other meat and animal products, ethical concerns and health considerations (Fox and Ward, 2008; Janda and Trocchia, 2001). Ethical motivations have been found to be more effective than health reasons for sticking to the decision to eat this way (Ogden *et al.*, 2007) and ethical vegetarians show greater dietary restriction compared to those motivated by health considerations (Rozin *et al.*, 1997). Vegetarianism and veganism are not only influencing food choices, but are part of a growing social movement regarding animal welfare and animal rights (Pfeiler and Egloff, 2018). The popularity of vegetarian diets is on the rise in many countries, although vegetarians are still minority in most cultures (Cultivate Research, 2008; Datamonitor, 2009). Regarding the study of Nielsen Atmosphere agency about diet preferences in Slovakia, which questioned 500 respondents (minimum age of 15 years), 1% of the population in Slovakia are vegans, 2% are vegetarians and nearly 11% of the population are trying to reduce their meat consumption. The main reasons for reduction of meat consumption regarding the study are health benefits and critical impact on the environment, but also ethical issues associated with animal use in the meat industry.

The aim of this study was to investigate the relation between dietary groups of veterinary students and their attitudes and opinions about different ethical issues related to animal welfare, animal rights, using

EurSafe 2018 – DOI 10.3920/978-90-8686-869-8_80, © Wageningen Academic Publishers 2018

animals in scientific research and few more additional questions important for veterinary students on a daily basis.

The other objective of the study was to compare differences of opinion between Slovak students who are vegetarians (including vegans) and those who are meat eaters.

Material and methods

Participants

Participants in the survey consisted of students in their first or third year of General Veterinary Medicine study at the University of Veterinary Medicine and Pharmacy (UVMP) in Košice, Slovakia. A total of 174 students completed the survey. 106 participants (60.9%) were first-year students and 68 (39.1%) were third-year students. The exploratory study was started in 2017. The choice of survey participants was built on the assumption that students of the first year of the UVMP have not yet had knowledge about animal consciousness, welfare and use in research, whereas students of the third year have already had a class in Professional Ethics, in which they discuss all the topics covered by the questionnaire. The results of the study could be affected by this.

Questionnaire / measures

A questionnaire was compiled by a team of workers at the Institute of Applied Ethology and Professional Ethics (IAEPE) at The University of Veterinary Medicine and Pharmacy in Košice, Slovakia. It contained 25 questions including demographic data, data about food preferences of the respondents and their opinions on animal welfare, handling with animals, procedures performed on animals and animal experimentation. For interpretation of the results, the participants were divided into two groups: omnivores and vegetarians (including vegans). All measures were self-reported questionnaires completed by the participants.

Statistical analysis

Results have been evaluated using Chi-Square test. Results were significant at $P<0.05$.

Results

Demographic data and food preferences

Of the total number of students, 39 were men (22.4%) and 135 were women (77.6%). Of all participants, 83.9% were self-reported omnivores (without diet preference), 13.2% were self-reported vegetarians and 2.9% were self-reported vegans. Of 28 vegetarian participants (including 5 vegans) 20 gave ethical reasons for non-consumption of meat (71.4%), two participants gave health reasons and 6 participants marked both as reasons for their nutritional preference. Participants completed further questions about their origin (town, village) and omnivore students registered their weekly frequency of meat and meat products consumption (every day, 4 times a week or 1-2 times a week). Table 1 and 2 show the demographic characteristics and the dietary patterns of the respondents.

Prevalence of vegetarians and vegans

Of the total number of students of first and third year of study of UVMP in Košice, n=174, the number of individuals reported being vegetarians was n=23 (13.2%), the number of vegans was n=5 (2.9%),

Table 1. Demographic characteristics of the survey respondents (n=174).

	Year of study		Sex		Origin	
	First	Third	Male	Female	Urban	Rural
n	106	68	39	135	97	77
%	60.9	39.1	22.4	77.6	55.7	44.3

Table 2. Dietary preferences of the survey respondents (n=174).

	Dietary preference			Frequency of meat consumption			Motivation for vegetarianism (vegan) n=28		
	Vegetarian	Vegan	Omnivore	Every day	4 times/week	1-2 times/week	Ethical	Health	Both
n	23	5	146	22	72	52	20	2	6
%	13.2	2.9	83.9	15.1	49.3	35.6	71.4	7.1	21.5

while n=146 individuals (83.9%) indicated that they consumed meat (omnivores). Ethical reasons for rejection of meat consumption was reported by n=20 vegetarians (71.4%) including vegans, health motivation was reported by n=2 (7.1%) and n=6 respondents identified both reasons for their diet preference. Of the total number of respondents who reported eating meat n=146 (83.9%), 22 students (71.4%) reported every day meat consumption, 72 students (49.3%) eat meat four times a week and the rest of respondents n=52 (35.6%) reported that they eat meat one to two times a week.

A statistically significant difference was shown in gender between both groups (Figure 1). Of the total number of vegetarians (including vegans) n=28 (16.1%), n=27 (96.4%) were women and there was 1 man (7.6%) (The *P*-value is 0.009051. The result is significant at $P<0.05$.)

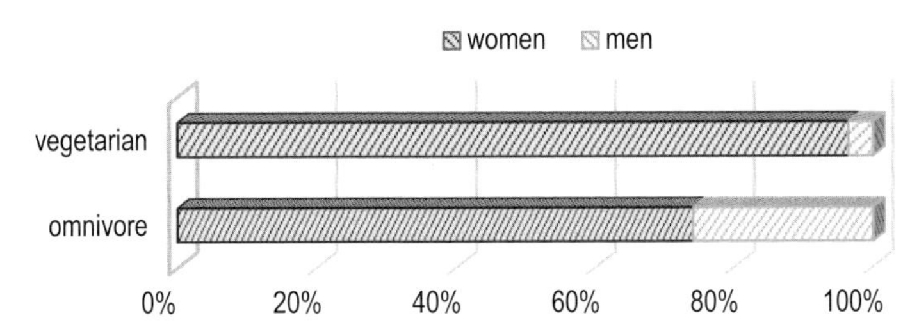

Figure 1. Gender differences. The chi-square statistic is 6.8128. The P*-value is 0.009051. The result is significant at* P*<0.05.*

Opinions about the use of animals in research

In answer to the question whether it is morally permissible to use animals in scientific research, only 35.6% of omnivores agreed and 74.4% of them disagreed. But between vegetarians only 10.7% agreed, and 89.3% of vegetarians (including vegans) did not agree (Figure 2). We found statistically significant differences between omnivores and vegetarians (statistically significant at $P<0.05$.)

Acceptable areas of experiments with animals were research and development of medicine (40.4% meat-eaters, 32.1% vegetarians), education (24% meat-eaters, 21.4% vegetarians), animals diseases research (60.3% meat-eaters, 53.6% vegetarians), human medical research (22% meat-eaters, 21.4% vegetarians), other fields of research (4.1% meat-eaters, 7.1% vegetarians) and research of cosmetic products (0.68% meat-eaters, 0% vegetarians).

Opinions about the impact on the environment

From the group of meat-eaters (n=146), 41.8% agreed that the meat industry has an considerable influence to environment. 50.7% of them believed that the influence is quite small, and the rest (7.6%) believed that the meat industry has no effect on the environment.

In contrast to the omnivores, 82.1% of vegetarians (including vegans) believed that the meat industry and production of animal products has an extreme impact on the environment and only 17.9% of them believed that the impact is not essential.

Opinions about procedures performed on animals

The practice of cropping dogs' ears is currently prohibited in Slovakia. In our questionnaire, we asked participants about was regarding their opinions on ear cropping for cosmetic purposes 72.6% of omnivores disagreed with the practicae and 27.4% of them agreed, but on the other hand, 71.4% of vegetarians (including vegans) disagreed.

Regarding the next question about castration and sterilization, 50% omnivores agreed with this intervention but only in specific reasonable cases and 2.1% omnivore students disagreed. In contrast, 53.6% of vegetarians and vegans agreed with sterilization and castration and 46.4% agreed only in specific cases.

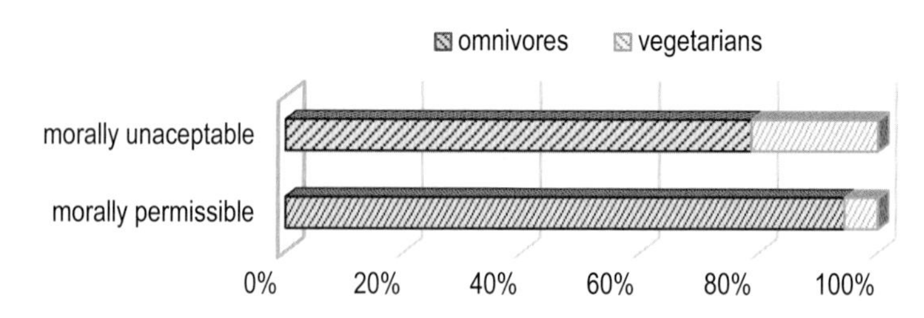

Figure 2. The admissibility of animal use in research. The chi-square statistic is 6.5358. The P-value is 0.010572. The result is significant at P<0.05.

Opinions about the breeding, use and killing of animals

Euthanasia of animals was completely inadmissible for 4.8% omnivores and for 7.2% vegetarians, acceptable for 10.3% of omnivores and 10.7% vegetarian. 84.9% of omnivores and 82.1% of vegetarians agreed that it might be acceptable in specific cases.

Regarding euthanasia of unwanted dogs in dog shelters, 56.2% of omnivores and 53.6% of vegetarians disagreed, 4.1% of meat-eaters and 7.1% of vegetarians agreed and 39.7% of meat-eaters and 39.3% of vegetarians agreed only in specific cases.

Regarding the regulation of bear population, 52.1% of omnivores and 60.7% of vegetarians disagreed, 9.6% of meat-eaters and 3.6% of vegetarians agreed and 38.3% of omnivores and 35.7% of vegetarians agreed only in daily justified cases.

For a question regarding the hunting of wild animals, 45.2% of meat-eaters and only 7.1% of vegetarians expressed a positive attitude, and 54.8% of omnivores and 92.9% of vegetarians expressed a negative attitude.

In response to a question about the rearing of fur animals in cages, 91.8% of meat-eaters and all vegetarians disagreed, 8.2% of omnivores agreed.

On a question regarding keeping animals in circuses and their performances, 93.2% of meat-eaters and all vegetarians disagree, and 6.8% of meat-eaters expressed a positive attitude.

With raising of animals in ZOOs, 69.2% of meat-eaters and only 50% of vegetarians expressed a positive attitude and 30.8% of omnivores and 50% of vegetarians disagreed and differences between both groups were statistically significant (at $P<0.05$.).

In the next question, 65.8% meat-eaters and only 10.7% vegetarians agreed with using animals for slaughter purposes and 34% omnivores and 89.3% vegetarians (including vegans) were against.

Opinions about food production

From all the meat-eaters, 50.7% of them preferred bio products when buying groceries, while among all vegetarians it was 60.7% of the respondents.

When buying eggs, 64.4% of meat eaters and 53.6% of vegetarians showed interest in the origin.

Discussion

The study was focused on the identification and determination of differences in views of veterinary students. Students were divided into two groups, meat eating students and vegetarians/vegans, and their preferences in eating habits were elected using a questionnaire, with the aim of revealing whether their dietary preferences had an impact on their sensitivity to ethical issues related to the handling of animals. Vegetarians are a minority in most cultures, but this group is gradually expanding due to various reasons such as health concerns associated with eating meat (Beardsworth and Keil, 1991), and concerns about animal welfare, and environmental sustainability (Rozin *et al.*, 1997). Studies of vegetarianism have revealed that moral concern regarding the raising and slaughter of animals is the principal motivation for eliminating meat consumption (Amato and Partridge, 1989; Ruby, 2012).

In this study, gender differences were significant in the group of vegetarians There was only one man in the group of vegetarians. This finding was in agreement with study of Ruby (2012), in which it was confirmed that women have more positive attitudes toward vegetarianism than men. The main reason for vegetarianism or veganism is the ethical motivation present in the 71.4% of vegetarians (including vegans). The motivation to refuse eating meat dates back to Greek philosophers (Spencer, 1993), and together with health reasons, is the most common reason for avoiding meat consumption (Rozin *et al.*, 1997).

There was a significant difference in opinions about experiments on animals between the group of 'meat eaters' and the group of vegetarians (or vegans). Only 10.7% of vegetarians confirmed that it is morally admissible to use animals for experimentation and scientific purposes compared to meat eaters.

Regarding the impact of meat production on the environment, the group of vegetarians was stricter. 82.1% respondents from this group believed that meat production has a significant impact on the environment. Only 50.7% of students from the group of omnivores had the same opinion.. This finding is in line with conclusion of the study of Mullee *et al.* (2017). In this study, more Vegetarians than non-vegetarians believed that meat production is bad for the environment and meat consumption is unhealthy.

Views on the breeding, using and killing of animals did not differ significantly between the group of non-vegetarians and vegetarians. Significant differences were found in opinions on the breeding of fur animals in cages and in opinions on circus animals. All of 28 vegetarians expressed disagreement with such usage of animals. Vegetarians also expressed more negative opinions about the of regulation of predatory animals, the regulation of the number of bears, wildlife hunting and slaughtering.

Question about preferences in food quality showed that consumption of organic food (wholefood or food from ecological farming) had importance for 50.7% of omnivores and 60.7% of vegetarians, quality and origin of eggs was important for 50.7% of non-vegetarians and 53.6% of vegetarians, and 42.8% of vegetarians do not eat eggs at all.

This study shows that omnivores and vegetarians think about meat in very different terms. Whereas omnivores have positive attitudes towards meat, associating it primarily with luxury, status, taste, and good health, vegetarians tend to link meat with cruelty, killing, disgust, and poor animal health (Ruby, 2012). Findings of our study verified higher ethical sensibility and perception in the students with vegetarian or vegan preferences towards questions about breeding, treatment and using animals for various purposes. Limitations of the relevance of our results might be influenced by low number of respondents and relatively small sample of students′ not consuming meat. Therefore studies with more students should be performed in the future.

References

Amato, P.R. and Partridge, S.A. (1989). The New Vegetarians. Promoting Health and Protecting Life. Springer Publisher, New York, 282 pp.

Back, K.W. and Glasgow, M. (1982). Social Networks and Psychological Conditions in Diet Preferences: Gourmets and Vegetarians. Basic and Applied Social Psychology 2: 1-9.

Beardsworth, A.D. and Keil, E.T. (1991). Vegetarianism, veganism and meat avoidance. Recent trends and findings. British Food Journal 93: 19-24.

Cultivate Research. (2008). Vegetarian consumer trends report, 2008. Olympia: Author.

Datamonitor. Trends in protein intake. Attitudes and behaviors, 2009, New York: Author.

Fox, N. and Ward, K. (2008). Health, ethics and environment. A qualitative study of vegetarian motivations, Appetite 50: 422-429.
Janda, S. and Trocchia, P.J. (2001). Vegetarianism. Toward a greater understanding. Psychology and Marketing 18: 1205-1240.
Mullee, A., Vermeire, L., Vanaelst, B., Mullie, P., Deriemaeker, P., Leenaert, T., De Henauw, S., Dunne, A., Gunter, M.J., Clarys, P. and Huybrechts, I. (2017). Vegetarianism and meat consumption: A comparison of attitudes and beliefs between vegetarian, semi-vegetarian, and omnivorous subjects in Belgium. Appetite 114: 299-305.
Nielsen Admosphere / SNaP, internetová populácia 15+, n=505, august 2016 DOI: http://www.nielsen-admosphere.sk/press/ts-slovaci-na-jedle-radi-usetria-ale-su-aj-ochotni-si-za-kvalitu-priplatit-a-maju-zaujem-o-zdravy-zivotny-styl.
Ogden, J., Karim, L., Choudry, A. and Brown, K. (2007). Understanding successful behaviour change. The role of intentions, attitudes to the target and motivations and the example of diet Health Education Research 22: 397-405.
Pfeiler, T.M. and Egloff, B. (2018). Examining the 'Veggie' personality: Results from a representative German sample. Appetite 120: 246-255.
Rozin, P., Markwith, M. and Stoess, C. (1997). Moralization and becoming a vegetarian. The transformation of preferences into values and the recruitment of disgust. Psychological Science 8: 67-73.
Ruby, M.B. (2012). Vegetarianism. A blossoming field of study. Research Review. Appetite 58: 141-150.
Spencer, C. (1993). The Heretic's feast. A history of vegetarianism. Published by London: Fourth Estate.

81. Teaching of ethics, animal welfare and concerning legislation in the Slovak veterinary education

J. Kottferová[*]*, D. Takáčová, L. Mesarčová, L. Skurková, J. Kachnič, B. Peťková and D. Vajányi*
University of Veterinary Medicine and Pharmacy, Komenského 73, 041 81 Košice, Slovakia; jana.kottferova@uvlf.sk

Abstract

The study of ethics and animal welfare as a part of veterinary education is becoming more important. Public awareness of animal welfare is increasing. There are increasing demands not for professional skills and knowledge only, but also for the emotional sensitivity of veterinarians to their patients. This trend was adapted by the University of Veterinary Medicine and Pharmacy in Košice, which incorporated into its curriculum the study of ethics, animal welfare and concerning legislation for students of all study programs. Content of courses, their stage, study programs and teaching methods will be discussed in this article. Using survey it will be identified how the opinions of veterinary students changed after the passing of the subject Professional Ethics.

Keywords: professional ethics, education, survey

Introduction

The study of professional ethics, animal welfare and legislation as a part of veterinary education is growing in importance. Public awareness of animal welfare is increasing. There are increasing demands not for professional skills and knowledge only, but also for the emotional sensitivity of veterinarians to their patients. This trend was adapted by the University of Veterinary Medicine and Pharmacy in Košice, which incorporated into its curriculum the study of ethics, animal welfare and concerning legislation for students of all study programs.

The University of Veterinary Medicine in Košice, as a public higher educational institution, is the only institution of this kind providing undergraduate and postgraduate veterinary education in the Slovak Republic. The study programme General Veterinary Medicine is provided only in full-time form, the study lasts 6 years (12 semesters). It was stated to include the study of veterinary ethics and ethology in study syllabus on the basis of evaluation and approximation of education results in the field of veterinary medicine between the Slovak Republic and EU, carried out by Commission EU in 2002.

Ethics, animal welfare and legislation is an integral part of veterinary undergraduate curriculum. Teaching of ethics, animal welfare and concerning legislation are included in the course Professional Ethics mainly, however the topic of animal welfare is included in course of Animal Ethology, Animal Husbandry, Veterinary Legislation and clinical courses, as well.

Our aims were to promote ethics, animal welfare and concerning legislation in the Slovak veterinary education, focusing on professional ethics and identifying what knowledge brings to veterinary students passing the subject Professional Ethics.

EurSafe 2018 – DOI 10.3920/978-90-8686-869-8_81, © Wageningen Academic Publishers 2018

What do they learn?

Professional Ethics

The subject of Professional Ethics consists of formal teaching (lectures), 1 hour per week and practical exercises, 2 hours per week, using the teaching through informal discussions. The semester lasts 13 weeks. The subject is included in the 2nd year of study, within preclinical years of study. Objectives of the course Professional Ethics is to give knowledge about bioethics view on animals, their physiological and psychological needs and ethical interpretation of welfare. Students are informed about the fundamental ethical principles of the profession of veterinary surgeon, too.

Ethics is the branch of philosophy that investigates morality. Ethics is concerned with what is good for individuals and society and it is also described as moral philosophy. First part of theoretical teaching within the course explains the varieties of thinking by which human conduct is guided and appraises the rightness or wrongness of actions, virtue or vice, the motives that prompt them, and the goodness or badness of the consequences to which they might give rise. All frameworks for applying ethics are explained with reference to animal use or veterinary work. Another part of lectures gives information about the development of thinking and opinions on animal rights and animal welfare. Theoretical base of animal welfare of farm animals, companion animals, veterinary ethics, oath, veterinary imperative are the other lecture topics. The role of professional ethics, which concerns the relationships among veterinarians, their clients, their patients, and other members of their profession professionalism in veterinary practice is also an important part of Professional Ethics courses. The base of lecture teaching covers the basic information, however more interactive is small group teaching during seminars. This is an effective method for transferring information, developing communication skills of students. Students learn to analyse different positions and to argue them. The topics are experiments on animals, euthanasia, slaughter, stray animals and communication in veterinary practice, ethical issues of fees and others. Very popular are debates and role-playing sessions, using cases from practice. The advantage is that students see for themselves the ethical problems faced by others involved with animal care and use (Main *et al.*, 2005).

Animal ethology

An understanding of animal behaviour is important in animal welfare science. The teaching of Animal Ethology consists of formal teaching (lectures), 2 hours per week and practical exercises, 2 hours per week, by using the practical trips at farms, zoo and shelters. The subject is included in the 3rd year of study. Students will gain knowledge of the principles of animal behaviour, will be familiar with the behaviour of domesticated animals, they will have information on the prevention and treatment of primary and secondary abnormal behavioural forms of animals at the subject Animal ethology. The practical assessment of behaviour in different systems can also provide information about preferences of animals, about motivation of an animal. Students will learn to estimate the welfare state of an animal and factors that may affect it. They will have to train the measurements, methods of assessing well-being, indicators of good and poor welfare.

Animal hygiene and welfare

The teaching of Animal Hygiene and Welfare consists of formal teaching (lectures), 2 hours per week and practical exercises, 3 hours per week, by using the practical trips at farms, zoo and shelters. The subject is included in 3rd year of study. During the course, the role of veterinarians is explained in the process of siting and designing animal farms and individual animal houses, hygienic requirements on building materials and constructions intended for animal production, veterinary-hygienic protection at animal farms, hygiene and technology of housing of farm animals (cattle, pig, sheep, poultry, horses), welfare problems in different housing systems and animal welfare, too. EU-Legislation as well as National laws are also key topics, laying the focus on transport and slaughter of animals, hygiene and technology of laboratory and companion animals.

Legislation

Veterinary Legislation is taught both in the 2nd and 5th year of study. Public Veterinary Medicine taught in the 2nd year (3rd semester; 1 hour of lecture and 2 hours of lessons) gives to students´ knowledge about the mechanism of both state and public administration, principal function of legislation, creation of the legal rules in the EU. The stress is laid on veterinary care, protection of animals, welfare of animals, and food hygiene. Veterinary legislation that is taught in the 5th year of study (9th semester; 2 hours lectures and 2 hours lessons) focuses mainly at a package of measures to strengthen the enforcement of health and safety standards for the whole agri-food chain.

Survey

We were interested in the influence of the Professional Ethics' passing on the formation of the students' opinions. To explore how their priority has changed a survey for students was developed. Questions were focused on different ethical issues. Students answered yes/no.

The last two questions had to be answered in written form. The topics were as follows:

Training of the subject Professional Ethics has changed my opinion on:
- perception of animal welfare issues;
- problems related to experiments on animals;
- livestock breeding;
- various areas of animal breeding;
- pets breeding;
- reasons for animal euthanasia;
- the role of a vet in the society;
- meat and meat products consumption;
- which topic from the whole semester you were most interested in;
- which topic did you meet in Professional Ethics classes for the first time.

All students enrolled in Professional Ethics course at the Veterinary University in Košice participated in the survey. 87 students filled the survey the last week of winter semester in 2017. There were only few male students participating, therefore gender difference in answering the survey has not been evaluated.

Survey responses

Regarding the first group of questions – 'perception of animal welfare issues' (castration, ear cropping, tail docking) – the majority of students answered 'no' (64, 67, 61 and 56%): training of the subject Professional Ethics did not change their opinion on the topic.

Also 'problems related to experiments on animals' 56% of the students answered 'no'. However, some students indicated this topic as the most interesting one. Some said that they had met this topic for the first time.

The topic 'livestock breeding' significantly changed students' opinion: the majority marked 'yes'. They changed their opinion on ethical issues about poultry farming (62%), pig farming (59%), and cattle breeding (40%).

Regarding topic 'various areas of animal breeding', most students did not change their opinion after Professional Ethics classes (breeding of fur animals 60%, using of animals in circuses 55%, breeding of animals in zoos '50%). The 'pets breeding' ethical issues (dressing, somatic changes) did not change their opinion either. 50.5% of students changed their opinions on the 'reasons for animal euthanasia'.

'The role of a vet in the society' changed the students' opinions: 51% of students signed the importance of communication with the client, the importance of guiding the client's decision on euthanasia 56%, the importance of ethics within the work of a vet 62%. All those students changed their opinion after passing the subject Professional Ethics. On the questions 'meat and meat products consumption', only 17% confirmed changing opinions.

Written answers to a question are summarized as follows:

- 'Which topic from the whole semester you were most interested in?' Answers were especially: stray animals, euthanasia, ethical issues of breeding, experiments, communications.
- 'Which topic did you meet in Professional Ethics classes for the first time?' Students specified: welfare issues, genetic engineering, euthanasia, slaughter, the role of veterinarians.

The students were confronted with different ethical issues during the whole semester. The results of our study show that especially the topic 'animal welfare' is new for students. Teaching about this topic can lead to the attitudinal changes toward animals. Few studies have investigated attitudes toward animals in veterinary students (Hazel *et al.*, 2011; Freire *et al.*, 2016) and demonstrated changes to the human-animal attitude and also human oriented empathy after animal-welfare courses.

The findings of some other studies (Freire *et al.,* 2017, Degeling *et al.,* 2017) suggest that veterinary students consider more important to have the necessary practical skills and knowledge to function as a veterinarian, but attitudes toward animals are important in influencing how animals are treated, too.

It is no doubt, that in 2nd year of study it is very early to give them all knowledge for clinical practice. This confirms the work of Paul and Podberscek (2000). But teaching veterinary students about ethical issues related to the animals can positively influence their attitude towards animals and provide helpful guidance to the students faced with ethical dilemmas or for decision making in everyday clinical practice.

References

Degeling, C., Fawcett, A., Collins, T., Hazel, S., Johnson, J., Lloyd, J., Philips, C.J.C., Stafford, K., Tzioumis, V. and McGreevy, P. (2017). Students' opinions on welfare and ethics issues for companion animals in Australian and New Zealand veterinary schools. Australian Veterinary Journal 95, 6:189-193.

Freire, R., Phillips, P., Verrinder, J. *et al.* (2016). The importance of animal welfare science and ethics to Australian and New Zealand veterinary students. Journal of Veterinary Medical Education 44, 2:208-216.

Hazel, S.J., Signal, T.D. and Taylor, N. (2011). Can teaching veterinary and animal-science students about animal welfare affect their attitude toward animals and human-related empathy? Journal of Veterinary Medical Education.38, 1:74-83.

Main, D.C.J., Thornton, P. and Kerr, K. (2005). Teaching animal welfare science, ethics, and law to veterinary students in the United Kingdom. Journal of Veterinary Medical Education 32,4: 505-508.

Paul, E.S. and Podberscek, A.L. (2000). Veterinary education and students' attitudes towards animal welfare. The Veterinary Record 10: 269-72.

82. Food safety and responsible education: a dedicated concept at the Veterinary Nurse School at the University of Veterinary Medicine, Vienna

M.H. Scheib[*], L.N. Buxbaum and Y. Moens
Veterinary Nurse School, University of Veterinary Medicine, Vienna, Veterinaerplatz 1, 1210 Vienna, Austria; scheibm@staff.vetmeduni.ac.at

Abstract

The relationship between humans and animals has changed in a distinctive way over the last decade. Therefore, today's veterinary technicians, nurses and animal care takers have to face completely new duties and challenges. Through their professional handling they share the responsibility of keeping the animals healthy and adequately treated. It evidences that this profession is irremissible for the food safety and further the whole supply chain itself. Students of the Veterinary Nurse School at the University of Veterinary Medicine, Vienna are educated to be able to meet these requirements through detailed, specific and modern schooling and training. The educational concept of the School puts on top of the acquisition of technical competencies special emphasis on ecological awareness, ethical responsibility, animal protection and welfare as well as on the special 'Human-Animal Bond'. This is realized by cross-referencing all teaching contents – either for general education or specific information – with the relevant principles of 'Ethics for Animals' and the 'Human-Animal Bond'. This concept aims to enhance the role of veterinary technicians, nurses and animal care takers with regards to food safety and animal welfare. Aware of developments in politics and food production they can critically identify chances and threats and serve as experts who give animals a voice.

Keywords: animal care takers, vet nurses, animal welfare education, agricultural animal husbandry, holistic education

High requirements for candidates in the area of food safety

To prepare young people adequately for the increasing demands of today's profession of nurses and animal care takers as well as veterinary technicians, it needs more than professional and technical qualifications. The Veterinary Nurse School at the University of Veterinary Medicine, Vienna offers young people who are interested in occupations in the food safety sector an optimal job preparation.

Systematic quality control and risk management in the food chain rely on the Hazard Analysis Critical Control Point (HACCP)-system. This system in turn relies on guaranteed application of 'Good Practices' like Good Veterinary Practice (GVP), Good Hygiene Practice (GHP) and especially in the section of animal production Good Farming Practice (GFP) (Sofos, 2002).

Well educated animal care takers in the latter sector can help veterinarians and other specialists as well as the farmer with the implementation of GFP in the pre-harvest animal production sector and realization of an HACCP-like system.

This means e.g. that these persons must have adequate knowledge about ante mortem transmissibility of pathogens and all relevant aspects of animal welfare in animal husbandry to enhance the farm-to-table food safety assurance safety strategy of our food supply (Noordhuizen and Collins, 2002).

Svenja Springer and Herwig Grimm (eds.) ***Professionals in food chains***
EurSafe 2018 – DOI 10.3920/978-90-8686-869-8_82, © Wageningen Academic Publishers 2018

Students should be able to successfully face the challenges and responsibilities of a profession in the food safety sector with well-grounded technical qualifications and a reflective personality.

The Veterinary Nurse School at the University of Veterinary Medicine, Vienna

Main goals of the Veterinary Nurse School are to enhance the job description of vet nurses but also to emphasize her/his role in the sectors 'Animal Protection' and 'Welfare'.

One major part of the job description is that the professional educated vet nurses are able to realize and judge current socio-political developments and to give animals a voice in the sector 'food safety'.

In their capacity as animal care takers they link science and practice. The professionals have to implement and to question critically new findings in the field of animal care based on a functional and ethical level.

With a profound education in both, natural science and human science, the future animal care taker is also a supervisory body especially within the scope of food safety. Therefore the new curriculum includes the acquisition of technical competencies, special emphasis on ecological awareness, ethical responsibility, animal protection and welfare as well as raising awareness for the special 'Human-Animal Bond'.

All the topics above-mentioned are part of the new curriculum. The graduates will be conscious of all their responsibilities in their future jobs, because of their specified education.

Animal care takers have different kinds of options after finishing school. They have the opportunity to work in zoos, animal shelters and laboratories, they can train animals or be a part of animal-assisted therapies, but also are able to work in the agriculture department.

A new curriculum

The Veterinary Nurse School is optimally set up for these occupations in a part of a new implemented curriculum, which started in 2017 (Figure 1). It is a three-year, general-education private school with public law for adolescents aged 15-25.

The Veterinary Nurse School offers students a diversified combination of general education and high quality professional education. The full-time school is based on a modular curriculum, which enables students to specialize in one specific sector as well. In the first year of training general educational subjects are taught and basic knowledge of the subjects *Animal Welfare Law*, *Ethics and Human-Animal Relations*, *Human Health Education*, *Ecology* and *Environmental Education* as well as *Fundamentals of Economy* are imparted. Compared to the old curriculum *General Mathematics* or *Social Studies* were replaced by *Applied Mathematics* and *Political Education*. Central theme of the new curriculum is *Ethics* in every teaching subject. The focus on the human-animal-interaction is particularly emphasized in the context of subjects like *Animal Welfare*, *Animal Welfare Law* and *Ethics of Human-Animal Interaction*. Moreover the new curriculum provides the opportunity to gather practical skills and knowledge by visiting dairy farms and pig farms. In the modular third year opportunities for specific workshops led by recognized experts which include food safety in correlation with tasks of animal care takers are offered.

The purpose during the following second year of training is to acquire subject-specific skills. In addition to compulsory subjects like *Biochemistry*, *Pharmacology and Animal Testing Technique*, selected content in the areas of *Zoology*, *Anatomy and Physiology*, *Behavioural Science*, *Animal Husbandry and Animal* Care, *Animal Nutrition*, *General* and *Specific Illness* and *Animal Healthcare* or *Animal Protection*, *Hygiene*

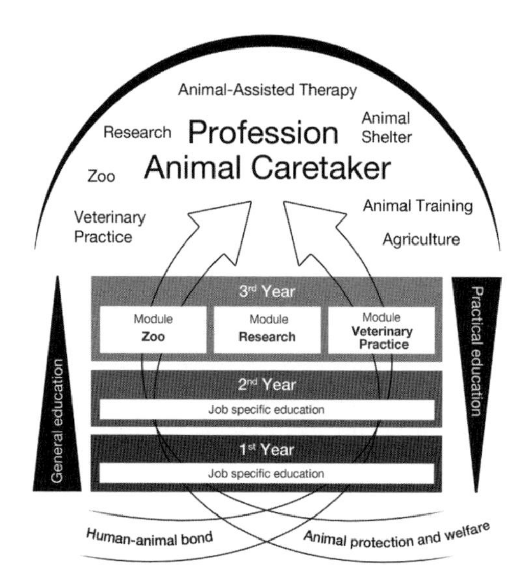

Figure 1. Curriculum of the Veterinary Nurse School Vienna 2017.

and *Hereditary Science* (*Genetics*) and *Animal Breeding* are being taught. Last but not least, the students get the possibility to compete at an international level, thanks to the class *Subject-Specific English*.

Starting with the third and last year of training the professionals to be can choose one out of three specialization modules to be perfectly prepared for their preferred professional field.

The three specializations are:
1. Qualified Veterinary Employee;
2. Animal Care of Experimental Animals/Animal Care in Research;
3. Animal Care in Zoos/ Animal Care of Wildlife;

Especially the modules (1) and (3) offer fundamental knowledge in agricultural animal husbandry, in the food sector and related occupational fields in food safety and serve as preparation for further education in the sectors mentioned.

Case discussions for the specific simulations linked to the working areas zoo, veterinary practice and animal research are also provided.

The structure/principles of learning of the new curriculum

Particularly in the domain of Human-Animal Bond animal care is an important contact point. He/she delivers a major contribution to increase the animals' wellbeing and is responsible for animal-friendly husbandry. Therefore, topics of the animal protection and welfare as well as of the Human-Animal Bond draw a red line through the disciplines taught in the new schedule of the Veterinary Nurse School.

In the General education and the practical the content is context-oriented. That means teachers are working out the roll of animals and that of the human to animal relationship and convey responsible husbandry and Human-Animal handling. The relevant thematic of the cooperation between animal and human in the current association is very important for the personality development of the students.

They should be qualified to think and act responsible and holistic. By reason that the global rotation, diverse ecosystems and sustainable animal husbandries are strong cooperating components, our pertinent training school is exerted to convey an extensive knowledge of these subject areas.

Sustainable benefit of a holistic education

Misguided developments, whose negative consequences showed up at the Fipronil-scandal 2017 in Europe, can be counteracted by high quality education. The holistic education of the Veterinary Nurse School should strengthen the individual responsibility and moral courage of the students and also make them aware of ecological correlations, ethical behaviour relating to the human-animal-bond, animal-friendly husbandry, animal rights and animal welfare in particular in relation to livestock farming. With the contemporary education future vet nurses are able to rote the gregariousness of humans and animals on an ethic and on a professional level. They are also able to scrutinize the state of things, upcoming trends and academic perceptions critically. They are going to implement decisions responsible.

Therefore, one of the final goals of the veterinary nurse school is to teach students to raise their voice for animals and consumers, what in general. Means to, in animal welfare, analyse specific subjects and to act corresponding.

In this case, the winners are the animals, the environment and the humans.

References

Curriculum of the Veterinray Nurse School Vienna (2017). Available at: http://www.vetmeduni.ac.at/de/tierpflegeschule/lehrplan/. Accessed 26 January 2018.

Noordhuizen, J. and Collins, J. (2002). Pre-harvest health and quality monitoring, risk assessment and their relevance to the food chain. In: Smulders, F., Collins, J. (ed.) Food safety assurance and veterinary public health. Vol. 1, Wagenigen Academic Publishers, pp. 115-124.

Sofos, J. (2002). Approaches to pre-harvest food safety assurance. In: Smulders, F., Collins, J. (ed.) Food safety assurance and veterinary public health. Vol. 1, Wageningen Academic Publishers, pp. 23-48.

83. Edible insects in food and feed – far from being well characterized – step 1: a look at allergenicity and ethical aspects

I. Pali-Schöll[1], S. Monsó[2], P. Meinlschmidt[3], B. Purschke[3], G. Hofstetter[1], L. Einhorn[1,4], N. Mothes-Luksch[5] E. Jensen-Jarolim[1,4,5] and H. Jäger[3]*
[1]Comparative Medicine, The Interuniversity Messerli Research Institute of the University of Veterinary Medicine Vienna, Medical University Vienna and University Vienna, Austria; [2]Institute of Philosophy, Karl-Franzens-Universität Graz, Graz, Austria; [3]Department of Food Science and Technology, Institute of Food Technology, University of Natural Resources and Life Sciences (BOKU) Vienna, Vienna, Austria; [4]AllergyCare, Allergy Diagnosis and Study Centre Vienna, Vienna, Austria; [5]Institute of Pathophysiology and Allergy Research, Centre for Pathophysiology, Infectiology and Immunology, Medical University Vienna, Vienna, Austria; isabella.pali@vetmeduni.ac.at

Abstract

Insects become interesting as alternative nutrient source for humans and animals. Safety aspects like allergenicity as well as ethical aspects need to be addressed. We evaluated the cross-recognition of IgE from patients allergic to crustaceans, house dust mite or flies of house cricket *Acheta domesticus*, desert locust *Schistocerca gregaria* and mealworm. We further elucidated changes of immunoreactivity in terms of IgE-binding in processed insect extracts. For migratory locust *Locusta migratoria* different extraction methods, enzymatic hydrolysis or thermal processing were used. IgE from patients with crustacean allergy shows cross-recognition of *A. domesticus*, *S. gregaria* and stable flies, and house dust mite allergics' IgE binds to *A. domesticus* and *S. gregaria*. The known cross-reactivity to *L. migratoria* can be completely deleted by hydrolyzation with different enzymes and heat treatment (cooking, autoclaving). Our results show that crustacean-, HDM- and stable flies-allergic patients cross-recognize desert locust and house cricket proteins, and crustacean-allergic patients also flies proteins. Furthermore, we confirm that the appropriate food processing method of insect proteins can reduce the risk of cross-reactivity for crustaceans- and house dust mite-allergic patients. Although evaluation of allergenicity is an important first step, ecological as well as ethical aspects need to be considered in the future.

Keywords: allergic patients, ethical considerations, immunoreactivity, insect proteins

Introduction

Entomophagy (eating insects) in our industrialized countries has become a trend of modern society during the recent years. Many products can already be found on the market and in restaurants, from whole grasshoppers to insect snacks. However, several concerns have not been addressed. The optimal rearing and production conditions, biological needs, and – much less so – ethical considerations have not been taken into account. As with January 1st, 2018 edible insects have to be processed and assessed according to the Novel food regulation of the EU (Regulation (EU) 2015/2283; http://tinyurl.com/zfzxd5w). Among the aspects to be addressed according to this directive is their hygienic status (toxic or microbiological contaminants) as well as their allergenic potential. Insects have come into our homes also as feed for our novel pet animals, mainly reptiles, which are fed for instance live grasshoppers. Importantly, we could show that this raises the risk for respiratory sensitization of the pet owner to the fed insects (Jensen-Jarolim *et al.*, 2015). As insects have only recently begun to be eaten in the Western part of the world, there is only limited information on the allergenic potential of these species when consumed by the population. Some recent work has shown that there is a high risk for shrimp- and house dust mite-allergic patients to cross-react with mealworm and grasshoppers (Broekman *et al.*, 2015, 2016,

Svenja Springer and Herwig Grimm (eds.) ***Professionals in food chains***
EurSafe 2018 – DOI 10.3920/978-90-8686-869-8_83,

2017a,b; Verhoeckx *et al.*, 2014, 2016; Van Broekhoven *et al.*, 2016). Due to this limited knowledge, it is important to further investigate the cross-reactivity potential and, furthermore, how production and processing technologies, e.g. thermal treatment and enzymatic hydrolysis, might influence protein integrity, the immunoreactivity and/or allergenicity of processed insect species. In the present work we describe the binding capacity of IgE from sera of shrimp-allergic patients to insect proteins from migratory locust and Yellow mealworm after treatment by different pH, centrifugation, or enzymatical hydrolysis. Following that, we briefly discuss some ethical aspects connected to edible insects.

Materials and methods

Extract preparation and hydrolysis of migratory locust

Wings and legs from frozen adult migratory locusts (*Locusta migratoria*; LM) were removed (purchased from NGN BV, Helvoirt, the Netherlands). The remaining bodies were mixed with Millipore water (1:1 w/w) and ground in a kitchen blender (C-Series 5200, Vitamix, Cleveland, USA). The pH value was set to 3.8 with citric acid (0.36 g/100 g LM). Thereafter, filtration was performed through a flexible filter with a pore size of 1 mm to separate chitin, and the filter rinsed once with Millipore water (10% w/w of the initial slurry mass). The filtrate was centrifuged (3,130×*g* for 20 min at ambient temperature) for fat separation and the fat layer removed. The residual slurry was freeze-dried, ground using a mortar grinder and stored at 3 °C until further analysis. For hydrolysis, endo- and exoprotease from *Aspergillus oryzae* (Flavourzyme® 1000 LAPU/g) were used (kindly provided by Novozymes A/S, Bagsvaerd, Denmark). In addition, papain, a cysteine-protease from papaya latex, was used (≥10 units/mg, E.C. 3.4.22.2, Sigma no P4762, St. Louis, MN, USA) for a two-step process: LM-dispersion was pre-heated (50 °C) and 0.05% (w/w) papain was added for 1 h of pre-digestion, thereafter, 1% (w/v) Flavourzyme was added, and the hydrolysis was continued for another 1 h. Enzymatic hydrolysis was stopped by heating (90 °C, 10 min) and samples were stored at -20 °C.

Production of different Tenebrio molitor *larvae protein concentrates*

Raw Yellow mealworms in final larvae stage (*Tenebrio molitor*, TM) were frozen at -30 °C (obtained from Dragon Terraristik, Duisburg, Germany). Frozen larvae were crushed using a blender (Braun 4162 MQ300, Braun, Poland) and mixed with deionized water (1:1.7 w/v). The pH of the TM slurry was adjusted to 7.0, 8.75, 10.5 or 11.32 (Table 1) using 1 M NaOH or 1 M HCl. The slurry was stirred for 2 h to solubilize the proteins and centrifuged (Sigma Laborzentrifugen, Type 4-15, Osterode, Germany)

*Table 1. Parameters of processing for migratory locust (*Locusta migratoria*) and Yellow mealworm (*Tenebrio molitor*).*

No.	Species	Processing
1	*L. migratoria*	Acid extraction
2	*L. migratoria*	Two-step hydrolysis with papain followed by endo-and exoproteases from *Aspergillus* (Flavourzyme) (E/S: 1.0 and 0.05%)
3	*T. molitor*	-
4	*T. molitor*	Centrifugation (pH 11.32, 4,800×*g*, 20 min)
5	*T. molitor*	Centrifugation (pH 8.75, 11,611×*g*, 20 min)
6	*T. molitor*	Centrifugation (pH 8.75, 4,800×*g*, 20 min)
7	*T. molitor*	Centrifugation (pH 10.50, 9,190×*g*, 30 min)
8	*T. molitor*	Centrifugation (pH 7.00, 9,190×*g*, 30 min)

under defined conditions (Table 1, Sample 3-8) for separation of protein extract. The insoluble fractions (solid chitin pellet, fat layer) were removed after centrifugation. The liquid protein extract was finally frozen at -30 °C, freeze-dried, ground using a mortar grinder and stored at 3 °C until further use.

Silver staining

Protein extracts with equally adjusted concentration (1.5 µg/slot) were separated via SDS-PAGE (15%) under reducing conditions, and visualized by silver staining (modified from (Merril *et al.*, 1981)).

Patients' sera

Sera were collected from patients (n=3, with written informed consent) during routine allergy testing in an allergy-outpatient clinic (Vienna) with diagnosed crustacean allergy.

Immunoblotting

Processed extracts of migratory locust and mealworm were blotted individually onto a nitrocellulose membrane (140 µg/blot with accompanying protein weight marker) and blocked with TBST/3% BSA for 30 min, thereafter washed 4× with TBST. Sera from patients with allergy to crustacean (n=3, pooled) or non-hypersensitivity serum as negative control (n=1; data not shown) were applied at a final dilution of 1:20 in TBST/0.1% BSA. Blots were incubated with sera o.n. at 4 °C, and washed 4× with TBST. Goat anti-human IgE-HRP (Invitrogen, REF A18793) was diluted 1:3,000 and incubated for 2 h at RT for detection of IgE-binding. Visualization of reaction was performed with ECL (Bio-Rad, Clarity Western ECL, 1705061) according to manufacturer's instructions. All experiments were performed twice.

Results and discussion

Extracts from edible migratory locust and Yellow mealworm larvae were used for processing, as they are among the most frequently consumed insect species in Western Countries.

Influence of processing on IgE-binding to LM

For LM, IgE-binding of sera from shrimp-allergic patients, i.e. cross-recognition, was found to proteins at a molecular weight around 30 and 38 kDa in acidic protein extract (Figure 1, lane 1). Importantly, this IgE-binding was completely lost when enzymatic hydrolysis was applied to the extract (Figure 1, lane 2).

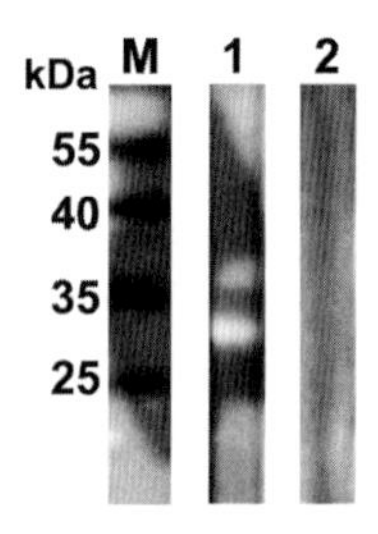

Figure 1. Immunoblot of migratory locust extracts. IgE from shrimp-allergic patients binds to proteins at around 30 and 37 kDA in acidic extract (lane 1), but diminishes completely when enzymatic hydrolysis is applied (lane 2). M: protein marker.

Influence of different pH and centrifugation on protein composition and IgE binding to TM

When Yellow mealworm (Figure 2) was extracted and subjected to different pH and centrifugation speed (conditions see Table 1, samples 3-8), several proteins remained in the extract and could be visualized by silver staining (Figure 3A, lanes 3-8).

In addition, also IgE-binding remained to these mealworm proteins (Figure 3B, lanes 3-8). The most prominent bands detected by IgE in immunoblotting had a molecular weight around 37-38 up to 40 kDa. These results clearly show that not all food processing and treatment options have the capacity to influence the IgE-binding from allergic patients. The appropriate food-processing method needs to be chosen if IgE-binding and therefore most likely also cross-reactivity and allergenicity shall be reduced in edible insects.

Ethical aspects

In addition to allergenicity, the ethical aspects of eating insects have to be considered (reviewed in Pali-Schöll *et al.*, submitted). Until recently, insects were thought to make no moral claims on us because of their presumed lack of sentience. However, recent evidence suggests that insects may actually possess some level of consciousness (Yarali *et al.*, 2008; Bateson *et al.*, 2011; Barron and Klein, 2016; Klein and

Figure 2. Yellow mealworm larvae (AdobeStock: Picture Partner, #37970600).

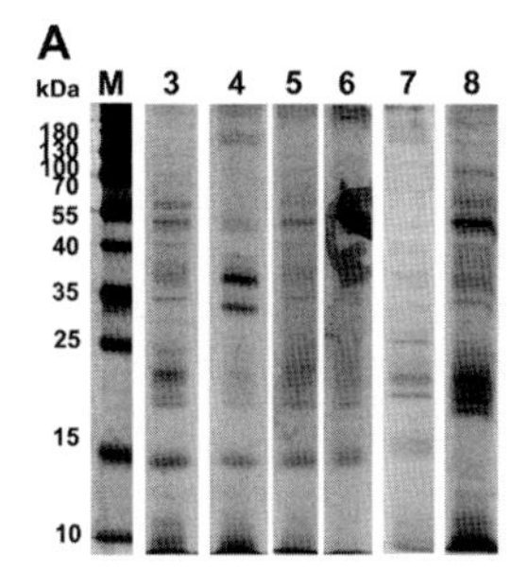

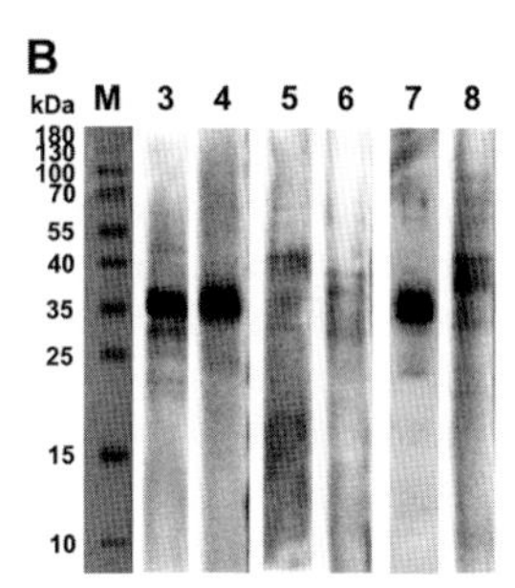

Figure 3. Silver staining and immunblot of Yellow mealworm. A variety of proteins from mealworm extract subjected to different pH and centrifugation conditions remains intact, as revealed by silver staining (A). IgE from shrimp-allergic patients recognizes mainly proteins at 37-38 and around 40 kDa in immunblot (B). Conditions for lanes 3-8 can be found in Table 1. M: protein marker.

Barron, 2016). If insects are conscious, this might mean that they possess sentience, or the ability to feel pain and pleasure. The possession of sentience is usually considered the threshold condition for being a member of the moral community. This is because sentience can be plausibly viewed as both a necessary and a sufficient condition for possessing interests (Singer, 2009). Having the capacity to feel pain and pleasure means feeling certain things or treatments as aversive or attractive, which in turn implies having an interest in stopping or continuing these treatments. Once a being has interests, she can make moral claims on us because whatever happens to her matters to her.

When we speak about the allergenicity of insects, we are discussing our own (human) interests. As consumers of insects we have an interest, for instance, in not experiencing an allergic reaction to them, but this has nothing to do with the insects' own experience. The insects themselves are not (directly) affected by their own allergenicity. If insects are sentient, this means that they should not be seen as mere inanimate objects that lack a perspective of their own. If they indeed are sentient, their interests need to be taken into account. This means that, in addition to consider how eating insects might affect us, we need to consider how rearing insects for food and feed would affect them. At the very least, this means that we need to consider whether all the basic needs of the insects are being met. In order to ensure that we are taking the insects' needs into account, and thus treating them ethically, we need to first find out what these needs are. Thus, we need to determine, with respect to the different species reared for food and feed, whether there are reasons to suppose that they not only possess nociception but can also experience pain subjectively, and what conditions should be met by the husbandry systems for their five freedoms (Brambell, 1965) to be guaranteed. Currently there is simply not enough information to ensure that edible insects will be reared under acceptable welfare conditions. In some cases, such as the case of migratory locusts, reaching such welfare conditions may not be possible, since they might have basic needs, such as the need to migrate over great distances, that cannot be catered for in captivity. With our existing knowledge, we cannot confidently assert that we are not inflicting any suffering on these animals. Thus, a significant amount of research effort needs to go into filling this knowledge gap. It is especially important to make this research effort at the present time, since the consumption of insects in Europe is still in its early stages and is likely to grow.

Conclusion

Our results show that crustacean-allergic patients cross-recognize migratory locust protein in acidic extract, and also different proteins in Yellow mealworm when subjected to different pH and centrifugation conditions. The IgE-binding to migratory locust proteins diminishes completely when enzymatic hydrolysis is applied, but not to mealworm extract subjected to different pH and centrifugation conditions. This implies that the appropriate food processing method of insect proteins can reduce the risk of cross-reactivity for crustacean-allergic patients; however, more allergic patients have to be tested individually, and especially also *in vivo* allergy testing needs to be performed to assess the allergenicity (or reduction thereof) of edible insect proteins. Regarding the ethical aspects of using insects as a novel food source, there is a great need for research in the area of insect welfare, especially regarding species-specific needs, health, farming systems and humane methods of killing. From an animal protection point of view these issues should be investigated and satisfyingly solved before propagating and establishing intensive husbandry systems for insects as a new type of mini-livestock factory farming.

References

Barron, A.B. and Klein, C. (2016). What insects can tell us about the origins of consciousness. Proceedings of the National Academy of Sciences, 113(18):4900-4908.

Bateson, M., Desire, S., Gartside, S.E. and Wright, G.A. (2011). Agitated Honeybees Exhibit Pessimistic Cognitive Biases. Current Biology, 21(12):1070-1073.

Brambell, R. (1965). Report of the Technical Committee to Enquire Into the Welfare of Animals Kept Under Intensive Livestock Husbandry Systems, Great Britain Parliament, 1-84.

Broekman, H., Knulst, A., den Hartog Jager, S., Monteleone, F., Gaspari, M., de Jong, G., Houben, G. and Verhoeckx, K. (2015). Effect of thermal processing on mealworm allergenicity. Molecular Nutrition and Food Research, 59 (9):1855-64. 10.1002/mnfr.201500138.

Broekman, H., Verhoeckx, K.C., den Hartog Jager, C.F., Kruizinga, A.G., Pronk-Kleinjan, M., Remington, B.C., Bruijnzeel-Koomen, C.A., Houben, G.F. and Knulst AC. (2016). Majority of shrimp-allergic patients are allergic to mealworm. Journal of Allergy and Clinical Immunolology, 137 (4):1261-3. 10.1016/j.jaci.2016.01.005.

Broekman, H., Knulst, A.C., de Jong, G., Gaspari, M., den Hartog Jager, C.F., Houben, G.F. and Verhoeckx, K.C.M. (2017a). Is mealworm or shrimp allergy indicative for food allergy to insects? Molecular Nutrition and Food Research, 61 (9):10.1002/mnfr.201601061.

Broekman, H., Knulst, A.C., den Hartog Jager, C.F., van Bilsen, J.H.M., Raymakers, F.M.L., Kruizinga, A.G., Gaspari, M., Gabriele, C., Bruijnzeel-Koomen, C., Houben, G.F. and Verhoeckx K.C.M. (2017b). Primary respiratory and food allergy to mealworm. Journal of Allergy and Clinical Immunology, 140 (2):600-03 e7. 10.1016/j.jaci.2017.01.035.

Jensen-Jarolim, E., Pali-Scholl, I., Jensen, S.A., Robibaro, B. and Kinaciyan, T. (2015). Caution: Reptile pets shuttle grasshopper allergy and asthma into homes. World Allergy Organization Journal, 8 (1):24. 10.1186/s40413-015-0072-1.

Klein, C. and Barron, A.B. (2016). Insects have the capacity for subjective experience. Animal Sentience: An Interdisciplinary Journal on Animal Feeling, 1(9). Available at: http://animalstudiesrepository.org/animsent/vol1/iss9/1. Accessed 30 March 2018.

Merril, C.R., Goldman, D., Sedman, S.A., Ebert, M.H. (1981). Ultrasensitive stain for proteins in polyacrylamide gels shows regional variation in cerebrospinal fluid proteins. Science, 211 (4489):1437-8.

Singer, P. (2009). Animal liberation: The definitive classic of the animal movement (Reissue edition). Harper Perennial Modern Classics, New York, USA.

Van Broekhoven, S., Bastiaan-Net, S., de Jong, N.W., Wichers, H.J. (2016). Influence of processing and *in vitro* digestion on the allergic cross-reactivity of three mealworm species. Food Chemistry, 196 1075-83. 10.1016/j.foodchem.2015.10.033.

Verhoeckx, K.C., van Broekhoven, S., den Hartog-Jager, C.F., Gaspari, M., de Jong, G.A., Wichers, H.J., van Hoffen, E., Houben, G.F. and Knulst AC. (2014). House dust mite (Der p 10) and crustacean allergic patients may react to food containing Yellow mealworm proteins. Food Chemistry and Toxicology, 65 364-73. 10.1016/j.fct.2013.12.049.

Verhoeckx, K., Broekman, H., Knulst, A. and Houben, G. (2016). Allergenicity assessment strategy for novel food proteins and protein sources. Regulatory Toxicololgy and Pharmacology, 79 118-24. 10.1016/j.yrtph.2016.03.016.

Yarali, A., Niewalda, T., Chen, Y., Tanimoto, H., Duerrnagel, S. and Gerber, B. (2008). 'Pain relief' learning in fruit flies. Animal Behaviour, 76(4):1173-1185.

84. Kant on food, physical satisfaction or humanity choice?

Y. Guo
Philosophy Department, Munster University, Paulus Str. 14, 33428 Harsewinkel, Germany; 121684092@qq.com

Abstract

Kant's topic on food was enormous but has been traditionally overlooked. It was generally accepted that Kant's view of food is linked with Bodily Pleasures, as hinted at in the passage on good living, one of the primary purposes of food is to supply physical satisfaction. Recent interpreters try to defend Kant's view on food in several disciplines (psychology, environment, biology), these interpretations I will argue do fully grasp the essence of Kant's humanity. Kant in fact identifies the ability to choose between different kinds of foods as one of the specific features of humanity. This work will provide the most sophisticated account of food in Kant's ethics and introduces the mechanism of the 'the mean' to determine ethical conduct in Kant's humanity. I will argue there are two points to dispute the physical satisfaction charge: first, there are no universal rules that guide its realization. Humanity choice requires tastes in food and drink vary amongst people. The second point, more precisely, the excessive ingestion of food or drink, which leads to a significant weakening or a loss of the capacity to use one's powers, goes against the Kant's explanation of duties to the self.

Keywords: food, Kant, humanity choice

Introduction

In western philosophy history, Kant's topic on food was not enormous and indeed not a central issue in Kant scholarship, but it has its attention within the 'duties to oneself'. However, this has been traditionally overlooked and this lack of comprehension have been redressed, but again and again the disdain for food that had accompanied western thought from the outset won out. It was generally accepted that Kant's view of food is linked with physical satisfaction, as hinted at in the passage on good living, one of the primary purposes of food is to supply physical satisfaction. But there is recently arise becomes a Kantian ambivalence in interpretation of concept of food whether it belongs to reason or nature. The reason interpretation is given that consumption choices fall of individual autonomy, consumers should determine their own food in Kant's 'Anthropology' (1798). For instance, to enjoy art and food socially means to transcend nature – to judge something that is given and structured according to the standards of judgment that are shared among rational beings. Food plays a special role and it is the extraordinary ability to stimulate solidarity through enjoyment.

In this article, my aim is to make good on all that, to discuss some of the instances in which food surfaces in Kant's philosophical discourse, and to predict a brighter future for philosophy and food. The aim of this paper is not only to explain Kant's account of the ideal proportions of good in food issues, but also, and more importantly, to argue that humanity suggests us forbid inappropriate ingestion of food or drink. Recent Kantian defenders have explored the relations between Kant's duty to oneself or autonomy and food, Wood explains perfect duties to oneself fundamentally involves treating humanity in one's own person with contempt, the violation of duties of respect to others involves treating someone else with contempt (Wood, 2009). Campbell argues obedience to the noumenal will was for Kant the only basis for moral acts, and he did not hesitate to claim the importance of duties to oneself, such as avoiding servile behaviour, or not stupefying oneself with food and drink (Campbell, 1996). But I will argue these interpretations do not yet fully grasp the essence of Kant's humanity. Kant strongly advocates for an undeniable sovereignty in food choice and makes it clear that this choice is intrinsically connected

EurSafe 2018 – DOI 10.3920/978-90-8686-869-8_84, © Wageningen Academic Publishers 2018

with individual autonomy. Kant in fact identifies the ability to choose between different kinds of foods as one of the specific features of humanity.

This work will provide one sophisticated account of food in Kant's ethics. Due to the space limited, I will not extensively show how my account differs from other accounts. At least, I will argue there are two points to dispute the physical satisfaction charge: first, there are no universal rules that guide its realization. Humanity choice requires tastes in food and drink vary amongst people. The second point, more precisely, the excessive ingestion of food or drink, which leads to a significant weakening or a loss of the capacity to use one's powers, goes against the Kant's explanation of duties to oneself.

Kant's topic on food was not enormous and indeed not a central issue in Kant scholarship, but it has its attention within the 'duties to oneself' in his 'Metaphysics of morals' (1797). One interesting contributions of Kant's 'Anthropology' (1798) is his conception of food. There is also one trend that Kant emphases duty to oneself in his moral philosophy, for example, both in 'Groundwork' and the Kant's 'Metaphysics of morals', Kant confines those duties that are generated by applying the principle of morality to human nature in general. According to Kant the duties to oneself are distinguished in perfect and imperfect duties. Each one can be further distinguished into duties to one's self as moral and animal being, which makes a total of four duties:

1. Suicide. Kant asks us to imagine someone 'weary of life because of a series of ills that has grown to the point of hopelessness' who asks himself if whether it is 'contrary to duty' to take his own life (Gr.4:422)[1], his maxim of suicide is formulated as follows: 'from self-love I make it my principle to shorten my life if, when protracted any longer, it threatens more ill than it promises agreeableness' (Gr.4:422).
2. False Promises. His maxim reads: 'when I believe myself in need of money, I shall borrow money, and promise to repay it, even though I know it will never happen' (Gr.4:422). Kant argues that the maxim would contradict itself if universalized since 'it would make the promise and the end one may pursue with it itself impossible, as no one would believe he was being promised anything, but would laugh about such an utterance, as a vain pretense' (Gr.4:422).
3. Developing one's Talents. A man harbours a certain talent by which he could make himself 'a useful human being in all sorts of respects', but prefers to give himself up to gratification and idle amusement rather than cultivate his talent (Gr.4:423). He asks himself whether this 'agrees with what one calls duty' (Gr.4:423).
4. Helping Others. In Kant's fourth example, a person who is prospering while others struggle, thinks to himself, 'What's it to me?'(Gr.4:423). His maxim reads: 'May everyone be as happy as heaven wills, or as he can make himself, I shall take nothing from him, not even envy him; I just [will not contribute]...anything to his well-being, or his assistance in need!' (Gr.4:423).

Especially in the third and fourth duty, Kant may have illustrated his attitude to the food. For example, Kant understands gluttony and drunkenness affect/violate the duty to himself as an animal being directly. Insofar as these duties require, as their condition of possibility, consciousness and self-control, anything that impedes these capacities becomes a vice and a violation of the duty to oneself: 'the vices that are here opposed to his duty to himself are such excessive consumption of food and drink as weakens his capacity for making purposive use of his powers.' (Ms.6:427) However, these relations (food, physical

[1] The abbreviations for the English translations are as follows:
Gr. Groundwork of the 'Metaphysics of morals', translated by M.J. Gregor in Practical Philosophy (Cambridge: Cambridge University Press, 1996).
Ms. 'Metaphysics of morals', translated by M.J. Gregor in Practical Philosophy (Cambridge: Cambridge University Press, 1996).
An. Zöller, G., Louden, R. (2007): Anthropology, History, and Education, Cambridge: Cambridge University Press.

satisfaction, duty to oneself as moral being) in Kant's writings is not straightforward and Kant's concept of food has been traditionally overlooked. In this paper I will try to show some potential association between the analysis of Kant's doctrine physical satisfaction, and the comprehension of Kant's doctrine of humanity. I will layout Kant's concept of food in two different uses. First, a biological use can be found as Kant describes food as physical satisfaction. Second, a practical use, refers essentially to human choice and the ability to act voluntarily. These two characters become a Kantian ambivalence of food as nature and reason as I will analyse. At the end of paper I briefly introduce a central characteristic of the Kant's morality and humanity.

Kant's concept of food

Physical satisfaction. In many remarkable passages of 'Anthropology', the perspective which Kant takes on can be neither easily nor completely subjected to the duty to oneself that Kant gave in his other writings. Kant writes eating to excess is vice under the heading of a lack of respect for duty to oneself in Part 2 of the 'Metaphysics of morals', the 'Doctrine of virtue'. In his 'Critique of judgment', Kant also rejects food as objects of contemplative critical appreciation. However, in 'Anthropology', Kant holds an instrument of nature is already presupposed, thus a daily choice (like food) is to be a product of nature. The choice must be withdrawn from nature and makes that choice not the least comprehensible. The relation between Kant's notion of food and duty to oneself is uncertain here: how does Kant make distinctions between the physical satisfaction and the duty to oneself? Why Kant in 'Anthropology' says good living, one of the primary purposes of food enjoyment is to supply physical satisfaction?

Practical use, refers essentially to human choice and the ability to act voluntarily. Insofar as choice for food are an essential part of the anthropological experience. As Kant interprets that physical satisfaction are not merely a source of constraints, they provide invaluable helps to the mind's wonderings. Insofar as the physical satisfaction is the condition of possibility of the continuation of the experience as a whole, the 'first appetite' has to be satisfied at the outset. But this is not the only bodily contribution. Once physical satisfaction has been addressed and turns to the mind's pleasure through conversation, 'A dispute arises which continues to whet the appetite for food and drink.' (An.7:278) here, we find once again the continual interaction between mind and body, but also the beneficial effects produced by the body on the mind. In practical use, Kant does not specify the actual benefits of the food, but we can suppose that it has to do with control oneself in decision as a rational power. Physical satisfaction through food and drink remind the guests that the experience is about enjoyment, but the practical use of concept of food that disagrees such physical satisfaction should not be taken too seriously. As a result, the attainment of a harmony between our physical body and our rational powers seem to hardly consist in finding the right proportions for their and our rational powers, and it further becomes tension in finding the right proportions for their mixture, rather than ignoring one for the benefit of the other.

Food between nature and reason – the Kantian ambivalence

The above passages reveal an interesting tension in Kant's concept of food, which, although remains indistinct in his work, ought to be carefully distinguished. One could expect, Kant believes that there is a tension, and even a conflict, between our humanity and its ethical counterpart human nature. The good living which still seems to harmonize best with virtue is a good meal in good company (and if possible with alternating companions) must not only try to supply physical satisfaction? Which everyone can find for himself?

What I want to distinguish here, is the humanity use of the concept of food in the Kant's 'Anthropology', where related to the practical doctrine of duty to oneself. In light of this canonical or practical definition of life, it is clear we could not endow living beings in general with such a specific capacity-for physical satisfaction, which is characteristic in a narrower sense. At most, this concept of food could be applied were one to grant them the capacity of desire; yet it is evident not thinking about this possibility. In fact, his notion of organism includes plants, animals, and human beings. His claim is rather characteristic feature of organized beings can be considered to in some way to life in a practical sense. What is the similarity that allows this analogy between organisms life in a practical sense? There is one feature I would like to emphasize: relation of living or organized beings to a special kind of morality, to the causality exercised by the humanity, that is, final with an important constraint that nature and reason arose.

The humanity interpretation

The two sections in this work have reinterpreted the most sophisticated account of food in Kant's ethics. I will argue there are two points to dispute the physical satisfaction charge: first, there are no universal rules that guide its realization. The second point, the excessive ingestion of food or drink goes against the Kant's explanation of Duty to the self as animal being (Ms.6: 421ff).

The diversity of food and drink for us has to be restricted by certain ethical rules. For their ingestion has an impact not only on our bodily functions, but more importantly on our understanding humanity as self-mastery. And since, we have the ability to choose what we consume, it is crucial to eat well and appropriately to our bodily needs as well as to the circumstances and the demands they make on us. The consumption of alcohol should not lead to extreme drunkenness, for 'All stultifying drunkenness, such as comes from opium or brandy, that is, drunkenness which does not encourage sociability or the exchange of thought, has something shameful about it' (An.7:209). In particular, we know from experience that certain drinks are intoxicating and hinder our ability to think and control ourselves. These sentences from the 'Anthropology' suggest that although excessive drinking ought to be avoided, moderate drinking can be morally beneficial: 'Drink loosens the tongue (*in vino disertus*). But it also opens the heart wide, and it is a vehicle instrumental to a moral quality, that is, openheartedness. ... Good-naturedness is presupposed when this license is granted to a man to cross the boundary line of sobriety for a short time, for the sake of sociability.' (An.7:201).

So perhaps unexpectedly, Kant is not an advocate of complete sobriety in the context of dinner parties. Sociability together with morality permits and even encourages slight inebriation as long as the drinker is good-natured and so long as it allows sincerity and sociability.

Another aspect to understand humanity on Kant's topic of food would be the excessive ingestion of food or drink, as a significant weakening or a loss of the capacity to use one's powers, goes against the duties to oneself. When stuffed with food he is in a condition in which he is incapacitated, for a time, for actions that would require him to use his powers with skills and deliberation? It is obvious that putting oneself in such a state violates a duty to oneself. Moreover, these excesses lead to a loss of humanity that entails that we are bereaved of our right for respect: 'A human being who is drunk is like a mere animal, not to be treated as a human being.'(Ms.6:551). This is because this man loses what makes a human being properly human, and thus worthy of respect, is precisely his capacity for autonomy. He automatically gives up the right these capacities namely respect by relinquishing them through drunkenness however. By contrast, moral quality is a good grasp of humanity. In this sense, there is a classification of drinks and drugs according to their effect, sociability and virtue. For 'what matters is only the kind of relationship whereby the inclination to good living is limited by the law of virtue.' (An.7:278) this is because the 'behaviour [of not revealing oneself completely] betrays the tendency of our species to be evil-minded towards one another ... [it] does not fail to deteriorate gradually from presence to intentional deception, and finally to lying.' (An.7:320).

Finally, the ultimate resolution of the conflict between our physical body and our moral powers, Kant takes into account that the physical satisfaction has not only an impact but a constraining effect on the operations of the mind: 'because reasoning is always a kind of work and exertion of energy, this finally becomes difficult after eating rather copiously during the dinner.' (An.7:285) but far from being presented as a difficulty, this suggests that bodily constraints have to be integrated into our understanding of humanity.

In summary

This paper is to explain Kant's account of the ideal proportions of good in food issues. There are two points to dispute the physical satisfaction charge: first, there are no universal rules that guide its realization. Humanity choice requires tastes in food and drink vary amongst people. The second point, the excessive ingestion of food or drink goes against the Kant's explanation of duties to oneself as animal being. Kant's humanity suggests us forbid inappropriate ingestion of food or drink, our ability to choose between different kinds of foods as one of the specific features of humanity.

References

Campbell, R. and Christopher, J. (1996): Moral development theory: A critique of its Kantian presuppositions. Developmental review16: 1-47.

Gregor, M., (1996). Practical Philosophy, Cambridge: Cambridge University Press.

Wood, A. (2009): Duties to oneself, duties of respect to others. The Blackwell guide to Kant's ethics: 229-251.

Zöller, G. and Louden, R. (2007): Anthropology, History, and Education, Cambridge: Cambridge University Press.

Author index

U

V

W